W0050027

The Vestibular System: Function and Morphology

The Vestibular System: Function and Morphology

Edited by
Torquato Gualtierotti
Head, Institute of Human Physiology
University of Milan Medical School
Milan, Italy

SPRINGER-VERLAG
New York Heidelberg Berlin

Torquato Gualtierotti
Institute of Human Physiology
University of Milan Medical School, Italy

Publication of this book was made possible,
in part, by a contribution from the Institute
of Human Physiology, Milan, Italy

Library of Congress Cataloging in Publication Data

The Vestibular System: Function and Morphology
"From the Society for Neuroscience, 8th annual meeting
satellite symposium, cosponsored by NASA and the
University of Pittsburgh School of Medicine."

Includes index
1. Vestibular apparatus—Congresses.
I. Gualtierotti, Torquato. II. Society for Neuroscience.
DNLM: 1. Vestibular apparatus—Anatomy and histology—
Congresses. 2. Vestibular apparatus—Physiology—Congresses.

WV 255 S679v 1981
QP471.V46 559.01'825 80-28212

Softcover reprint of the hardcover 1st edition 1981

9 8 7 6 5 4 3 2 1

ISBN-13: 978-1-4612-5904-6 e-ISBN:13: 978-1-4612-5902-2
DOI: 10.1007/978-1-4612-5902-2

Contents

PART 3 VESTIBULAR FUNCTION

PART 4 VISUAL-VESTIBULAR INTERACTIONS

Preface

Vestibular physiology has acquired new interest in the past few years. The development of fast aircraft, beginning during the Second World War and continuing through modern supersonic jets and space travel, has increased dramatically the acceleration range to which man is subjected, from prolonged weightlessness to several *g*s. The labyrinth measures all kinds of acceleration, including gravity, and therefore has taken on particular importance in the study of human health in the Space Age. It is understandable, then, that NASA sponsored the Symposium on Vestibular Function and Morphology, a satellite of the Eighth Annual Meeting of the Society of Neuroscience, held October 30–November 1, 1978, at the University of Pittsburgh, which also supported the symposium. The present volume was developed from papers originally presented there.

My strategy in editing this book has been to focus upon the vestibular end organ and its direct connections, both peripheral and central, in order to assure a coherent discussion of the basics of vestibular function, structure, and ultrastructure. Fortunately, it was possible to secure the contributions of a number of outstanding investigators from both the United States and abroad, including functionally oriented morphologists, ultrastructuralists, biophysicists, and vestibular physiologists. The result is a comprehensive discussion of the main mechanisms of the labyrinth: the transduction process at the receptor level, in the semicircular canals, and in the utricular and saccular maculae; the information transfer through neuronal pathways to the nuclei and the cerebellum; and the general organization of the system.

Special attention has been given to efferent control, still poorly understood in its function but clearly described here as far as its structure and connections with the central nervous system are concerned. In this respect some provocative thoughts were presented, even challenging the existence of centrally originated terminals on the hair cells. The lively discussions reported here indicate the special interest of this subject.

A number of contributions have been included dealing with the vestibulo-ocular reflexes. This subject is particularly important, not only because of the scientific problems involved, but also because it makes available important techniques for studying vestibular function in man.

Many problems remain unsolved. An acceptable general model of the labyrinth has yet to come. The characteristics, function, and specific activity of the vestibular efferent system are still unclear. In the future, we can look forward to new techniques that will allow further exploration of vestibular structure and function, such as recording the activity of single vestibular units for periods of days or weeks. Such a development would add a new dimension to our understanding of very slow changes, such as the long-term effects of weightlessness and the circadian rhythm.

This volume reviews and updates current understanding of the vestibular system. Unfortunately, debate on how to proceed with basic research programs, particularly the most important ones sponsored by NASA, has been so divisive that crucial momentum has been lost. While it may not be possible to establish a consensus in the near future, if this book sustains and reinvigorates study in this field, its purpose will have been served.

Torquato Gualtierotti

Contributors

P.J. Abbas, Department of Speech Pathology and Audiology, University of Iowa, Iowa City, Iowa

Joanne Ballarino, Section of Neurobiology and Behavior, Cornell University, Ithaca, New York

Robert W. Baloh, Department of Neurology, UCLA School of Medicine, Los Angeles, California

Neal H. Barmack, Neurological Sciences Institute, Portland, Oregon

Michael V.L. Bennett, Department of Neuroscience, Albert Einstein College of Medicine, New York, New York

F. Bracchi, Institute of Human Physiology, University of Milan Medical School, Milan, Italy

Humbert Bracho, Doheny Eye Foundation, Los Angeles, California

Ruben Budelli, Instituto Nacional de Neurologia y Neurocirugia, Mexico City, Mexico

U.W. Buettner, Department of Neurology, University of Zurich, Zurich, Switzerland

J.A. Büttner-Ennever, Brain Research Institute, University of Zurich, Switzerland

U. Büttner, Department of Neurology, University of Zurich, Zurich, Switzerland

Carol A. Christensen, Department of Psychology, Vassar College, Poughkeepsie, New York

Geoffrey M. Clark, Department of Psychology, University of Washington, Seattle, Washington

Manning J. Correia, Department of Otolaryngology, University of Texas Medical Branch, Galveston, Texas

Ian S. Curthoys, Department of Neurology, School of Medicine, University of Sydney, Sydney, Australia

Nancy G. Daunton, NASA-Ames Research Center, Moffett Field, California

Robert J. Douglas, Department of Psychology, University of Washington, Seattle, Washington

A.R. Eden, Department of Otolaryngology, University of Texas Medical Branch, Galveston, Texas

Hans Engström, Department of Otolaryngology, Academy Hospital, University of Uppsala, Uppsala, Sweden

B. Engström, Department of Otolaryngology, Academy Hospital, University of Uppsala, Uppsala, Sweden

Lawrence C. Erway, The University of Cincinnati, Cincinnati, Ohio

Åke Flock, Department of Psychology II, Karolinska Institute, Stockholm, Sweden

Peter J. Fraser, Zoology Department, Aberdeen University, Aberdeen, Scotland

Richard R. Gacek, Department of Otolaryngology, Upstate Medical Center, Syracuse, New York

Frank Galey, Doheny Eye Foundation, Los Angeles, California

Jay M. Goldberg, Department of Pharmacology and Physiological Sciences, University of Chicago, Chicago, Illinois

Torquato Gualtierotti, Institute of Human Physiology, University of Milan Medical School, Milan, Italy

V. Henn, Department of Neurology, University of Zurich, Zurich, Switzerland

Craig K. Henkel, Department of Anatomy, Bowman Gray School of Medicine, Winston-Salem, North Carolina

Vincente Honrubia, Division of Head and Neck Surgery, UCLA School of Medicine, Los Angeles, California

Howard C. Howland, Section of Neurobiology and Behavior, Cornell University, Ithaca, New York

David G. Hubbard, Aberrant Behavior Center, Dallas, Texas

Herman A. Jenkins, Division of Head

and Neck Surgery (Otolaryngology), UCLA School of Medicine, Los Angeles, California

Shyam M. Khanna, Department of Otolaryngology, Columbia University, New York, New York

Paul Kileny, Department of Audiology, Glenrose Hospital, and Faculty of Rehabilitation Medicine, University of Alberta, Edmonton, Alberta, Canada

Young S. Kim, Division of Head and Neck Surgery (Otolaryngology), UCLA School of Medicine, Los Angeles, California

J.P. Landolt, Defense and Civil Institute of Environmental Medicine, Downsview, Ontario, Canada

W. Lang, Brain Research Institute, University of Zurich, Zurich, Switzerland

Clifford Lau, Department of Surgery, UCLA School of Medicine, Los Angeles, California

Otto Lowenstein, Neurocommunications Research Unit, The Medical School, Birmingham University, Birmingham, England

Omar Macadar, Department of Anatomy, UCLA School of Medicine, Los Angeles, California

B.F. McCabe, Department of Otolaryngology and Maxillofacial Surgery, University of Iowa, Iowa City, Iowa

A. Morabito, Institute of Human Physiology, University of Milan Medical School, Milan, Italy

M.-D. Ni, Texas Instruments, Houston, Texas

Dennis P. O'Leary, Department of Otolaryngology, School of Medicine, University of Pittsburgh, Pittsburgh, Pennsylvania

Charles M. Oman, Research Laboratory of Electronics and Man Vehicle Laboratory, Massachusetts Institute of Technology, Cambridge, Massachusetts

Adrian A. Perachio, Departments of Otolaryngology, Physiology, and Biophysics, The University of Texas Medical Branch, Galveston, Texas

Roberta Peterka, Department of Otolaryngology, School of Medicine, University of Pittsburgh, Pittsburgh, Pennsylvania

Vito E. Pettorossi, Institute of Human Physiology, Catholic University of the Sacred Heart, Rome, Italy

Christopher Platt, Department of Biological Science, University of Southern California, Los Angeles, California

Arthur N. Popper, Department of Anatomy, Georgetown University, School of Medicine and Dentistry, Washington, D.C.

Wolfgang Precht, Max Planck Institute for Brain Research, Frankfurt, West Germany

J.L. Rae, Department of Physiology, Rush College of Health Sciences, Chicago, Illinois

Muriel D. Ross, Department of Anatomy, School of Medicine, University of Michigan, Ann Arbor, Michigan

J.H. Ryu, Department of Otolaryngology and Maxillofacial Surgery, University of Iowa School of Medicine, Iowa City, Iowa

John Schairer, UCLA School of Medicine, Los Angeles, California

Carl Sherrick, Princeton University, Princeton, New Jersey

D.D. Thomsen, NASA-Ames Research Center, Moffett Field, California

David L. Tomko, Departments of Otolaryngology and Physiology, School of Medicine, University of Pittsburgh, Pittsburgh, Pennsylvania

Charles G. Wright, University of Texas at Dallas, Dallas, Texas

Laurence R. Young, Department of Aeronautics and Astronautics, Massachusetts Institute of Technology, Cambridge, Massachusetts

Part 1

Morphology

INVITED LECTURES

1

The Structure of the Vestibular Sensory Epithelia

H. ENGSTRÖM AND B. ENGSTRÖM

Our knowledge of the structure of the vestibular sensory epithelia and their function is of old date and excellent descriptions have been given by many authors (32). The publications by Wersäll (41) and Smith (35) gave the first detailed descriptions of the ultrastructure of the vestibular sensory cells. Many later publications have verified and extended these early reports (5, 8, 15–20,26). An extensive report on the vestibular epithelia was given in 1969 by Lindeman (25). Werner (40) should be mentioned, as he was one of the first to give a good report on a subdivision of the macular epithelia and also described the striolae. In this relation Lorente de Nó's descriptions (27) of the innervation of vestibular sensory regions is of interest. An excellent description of the vestibular sensory regions was also given by Watanuki and collaborators (38,39). In recent years several extensive reports on frogs and lizards and other animals have increased our understanding of the morphology and function of vestibular sensory cells (2,3,23,24).

During the last several decades we have seen an augmented interest respecting vestibular structure and function. Several factors are responsible. It has become evident that vestibular dysfunction is an important factor in el-

This study was supported by Grant No. B79-12X-03156 from the Swedish Medical Research Council.

derly people. Also, the problems encountered in space flight with space-sickness of both astronauts and cosmonauts are of importance. The vestibular sensory cells are integrated into a functional system for orientation in space and the perception of linear and angular acceleration. A thorough knowledge of their structure and function in normal and pathologic situations is obviously of great interest.

The vestibular sensory cells are arranged in special regions in the vestibular portion of the inner ear. In each ear there are three semicircular canals, which have their sensory cells in the crista ampullaris. Sensory cells are further found in the maculae of the utricle and of the saccule, one in each ear. The sensory cells are very similar in these different regions. There is, however, a great difference in the structure of the covering material. Each crista is covered by a cupula, a gelatinous, fine fibrillar structure with a specific gravity resembling that of the endolymph. On the maculae there is also a gelatinous covering layer (23,24) which, however, in mammals contains

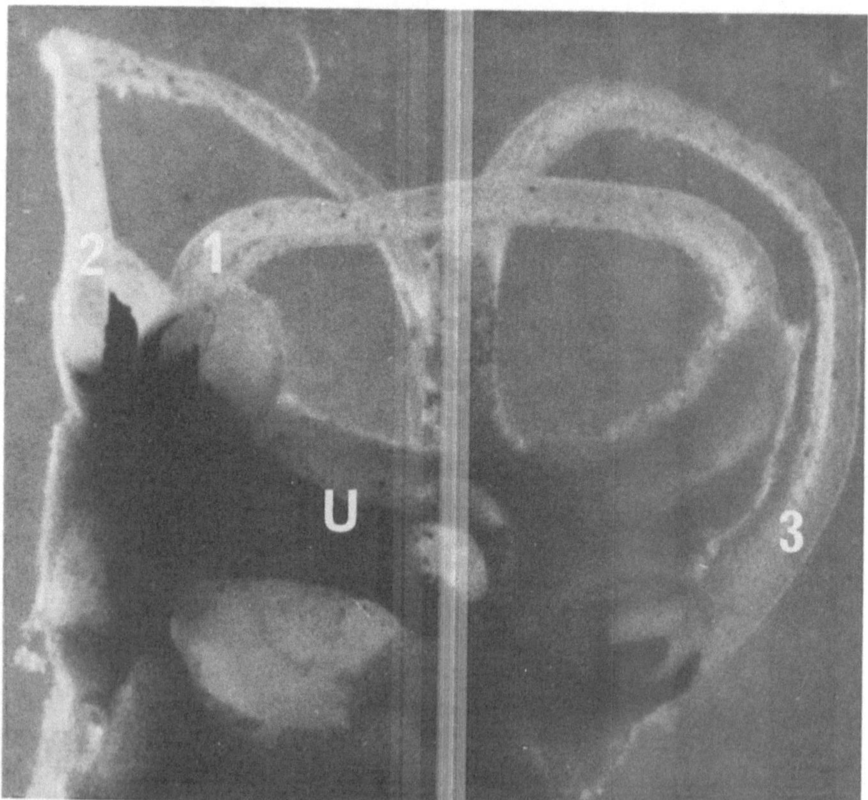

Fig. 1-1. Semicircular canals (*1,2,3*) and utricle (*U*) from the guinea pig labyrinth. The bone has been dissected away.

crystals of calcium carbonate in the form of calcite. These crystals, called *statoconia*, have a specific gravity of about 2.74 (4).

The form of the cristae ampullares as well as of the maculae can vary considerably between different species. Their general form in higher mammals is shown in Figs. 1-1–1-4. The three cristae, one in each of the ampullae of the lateral, the anterior, and the posterior semicircular ducts, are arranged as ridges (Fig. 1-2), covered by sensory epithelium, and at the side of this epithelium there are secretory cells forming a semilunar surface, called *planum semilunatum*. Over the epithelial cells the cupula forms a gelatinous, fibrillar structure. There have been many discussions concerning the height of the cupula and its form.

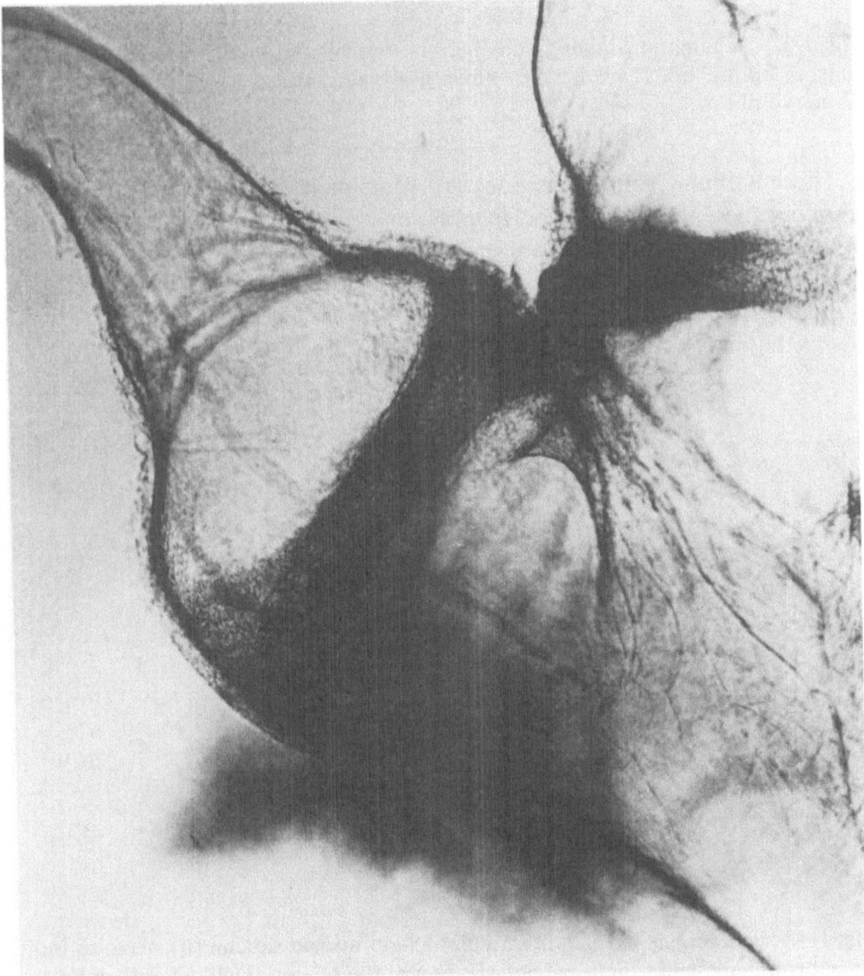

Fig. 1-2. Semicircular canal with its ampulla and crista. Guinea pig.

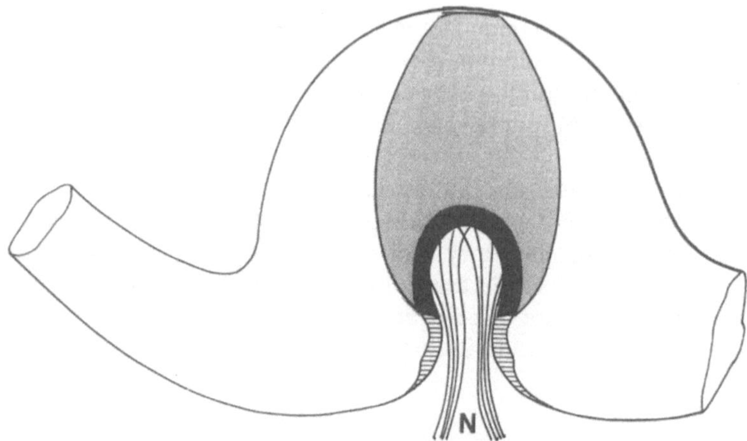

Fig. 1-3. Schematic drawing showing one ampulla with its crista ampullaris. The sensory epithelium is black, the cupula gray, and the secretory epithelium striped. *N*, nerve fibers.

The vestibular sensory regions are fundamental for our sense of equilibrium but they are still only peripheral sense organs, and severe disturbances in equilibrium may arise from damage to, or disturbance of, central neural projections from the peripheral sensory regions.

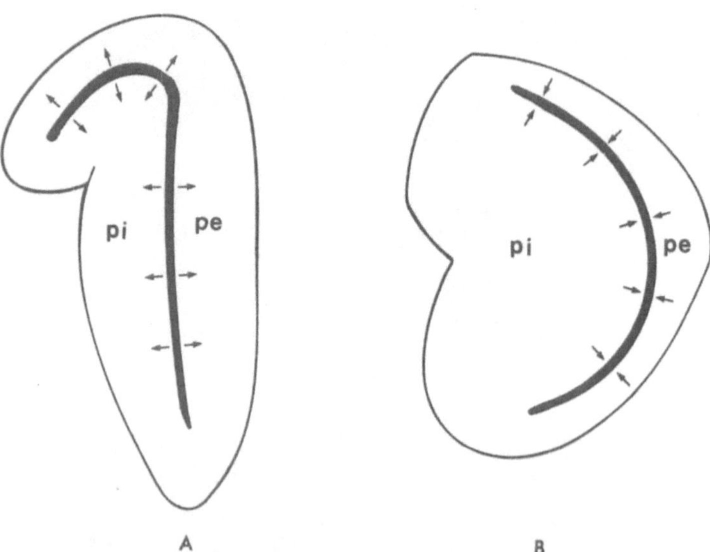

Fig. 1-4. Schematics of macula utriculi (**A**) and macula sacculi (**B**). Arrows indicate the morphologic and functional polarization of the sensory cells. *pi*, pars interna; *pe*, pars externa.

Normally only small gravitational forces act upon the vestibular end organ, but with the development of modern aircraft and space transportation dramatic deviations from normal may occur. As a result of high *g*-force action upon our bodies not only vestibular sense organs are stimulated but complicated interactions between the vestibular receptors and, for instance, blood circulation lead to unexpected responses such as blackouts. In the same manner astronauts are exposed to *g*-force modifications which lead to unforeseen reactions. Thus, "spacesickness" has become a new reality of great physiologic and clinical interest. For our understanding of vestibular dysfunction a thorough knowledge of the morphology of the inner ear is of importance. In a series of earlier papers we have given detailed morphologic descriptions and we will refer to some of these publications (8,11,14). In this paper we shall especially describe the structure of the sensory epithelia, based upon a series of studies conducted over a long period by our research group.

Many different animals have been used in our study (guinea pigs, squirrel monkeys, rhesus monkeys, rabbits, chinchillas, and frogs). They have been studied by light microscopy and scanning (SEM) and transmission (TEM) electron microscopy. Many different techniques have been used. The majority of the TEM material has been "fixed" in osmium tetroxide or glutaraldehyde, embedded in Epon 812, sectioned with diamond knives, and studied in various types of transmission electron microscopes. For the SEM material we often today follow a technique described by Malik and Wilson (28) and Hunter-Duvar (21).

There are many excellent descriptions of the general form and structure of the vestibular labyrinth (25,32). The most important step forward in recent years came when Wersäll (41) gave a thorough description of the structure of the epithelium on the semicircular cristae. Shortly thereafter Smith (35) gave a corresponding description of the macula utriculi. Many later papers have verified Wersäll's observations and added further information. These new observations have stimulated not only morphologic research in this field but also physiologic studies concerning the vestibular labyrinth. In our group, Lindeman (25) made an extensive study of the morphology of the vestibular apparatus.

GENERAL STRUCTURE OF THE VESTIBULAR SENSORY EPITHELIA

As was pointed out by Wersäll and by Smith, the general structural principle is the same in the maculae and the cristae. The great difference is found in the form and arrangement of the cells and in the superstructure over the hairs on the sensory cells. There is also a close resemblance in the structure of the supporting cells found between and below the sensory cells in all the sensory regions.

Wersäll described two different kinds of sensory elements, type I and type II cells. The form and structure of these two kinds of cells can best be observed in Figs. 1-5–1-9. The type I cell is flask shaped with a rounded lower end. At the surface there is a cuticle provided with one kinocilium and many stereocilia. Below the cuticle the cells have a rather narrow neck. The cells widen around the nucleus. The cells are surrounded to a great extent by a nerve chalice which continues as an afferent nerve fiber through the base-

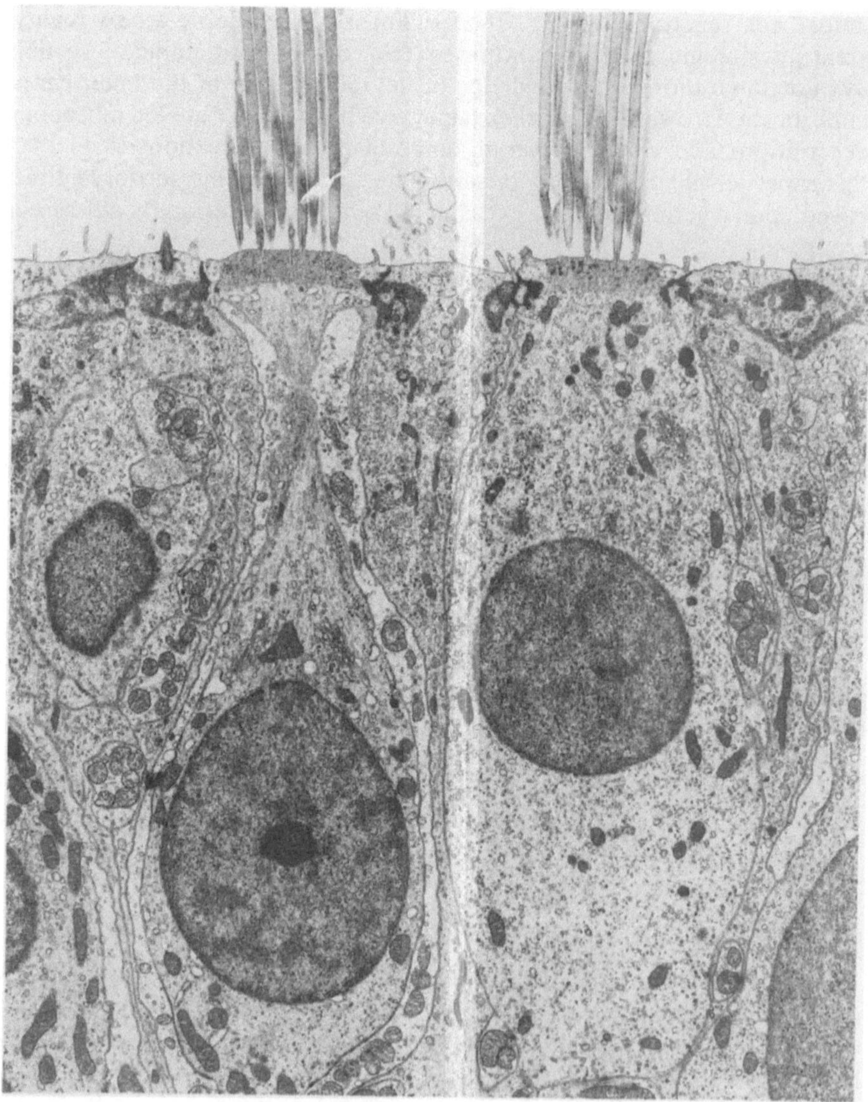

Fig. 1-5. Low-power electron micrograph showing type I (left) and type II (right) cells. The type I cell is surrounded by a nerve calyx; the type II cells have nerve endings in several places along the plasma membrane of the sensory cell.

ment membrane. Below this membrane the fibers become myelinated. One or sometimes several type I cells may be found inside a single chalice. In the same way several separate chalices may send their central nerve fiber to the same myelinated nerve fiber. Between the nerve chalices supporting cells, often with their nuclei close to the basement membrane, separate individual sensory cells.

The type II cell is more cylindric but it has also a rounded lower end. At the surface there is a cuticular plate provided with one kinocilium and many stereocilia. The type II cell has contact with many nerve endings which form synapses along the plasma membrane at different levels both below and above the nucleus.

Thus, the principal difference between type I and type II cells is that the former is in direct contact only with one termination or calyx from an afferent, inward-leading nerve fiber, whereas the type II cell forms direct synaptic contacts with both afferent and efferent fibers. This difference is evident in Figs. 1-6 and 1-7.

ULTRASTRUCTURE OF THE SENSORY CELLS

Type I Hair Cells

The hair cells of type I (Fig. 1-6) have a cuticular plate provided with sensory hairs of two kinds, stereocilia and one kinocilium. The kinocilium has a basal body located in a cuticle-free region. From this basal body the hair proper protrudes, with a length varying in different vestibular regions. Kinocilia are usually of a uniform cylindric shape (Figs. 1-8, 1-10, and 1-11), but in several animals kinocilia with a bulbous tip (Fig. 1-12B) can be found.

The kinocilia are usually the longest of all the hairs on a cell and in the ampullar cristae they may have a considerable length (40 μm) (Fig. 1-13). They are built according to the principle of vibratile cilia with nine peripheral double fibrils and one central double fibril. We have the impression that the central fibrils are not always present throughout the whole length of the cilium. On the other hand we have on many occasions, and especially in some old rhesus monkeys, seen more than 9 + 2 fibrils. The cilia have a short, 0.2 to 0.25 μm neck region immediately above the cell surface. In this area there is a helical arrangement (Figs. 1-15 and 1-17B). The basal body (Figs. 1-11 and 1-17B) is located in a region with no cuticle. From the lower portion of the basal body thin, irregular, cross-striated fibrils reach into the deeper parts of the cell. These cross-striated fibrils are often seen to be present at the basal bodies of different kinocilia, such as those described by Afzelius and Franzén (1).

From the cuticular region many microtubules (Figs. 1-14 and 1-15) originate (22) and we are of the opinion that a large portion of these have their origin close to the basal body. They extend into the neck portion of the cell and further down to the lower part of the cell. In the neck region they form almost straight lines in the electron micrographs, and in between long rows of tiny vesicles are often seen (Fig. 1-14). The stereocilia (Figs. 1-16, 1-17) are arranged in a regular pattern with an outer portion, the hair proper, and a tapering rootlet. This root is fixed in the cuticular plate and there has been much discussion as to whether they also can penetrate into the infracuticular region. The stereocilia have a varying or graded length and the longest stereocilia are always found close to the kinocilium (Fig. 1-18). The arrange-

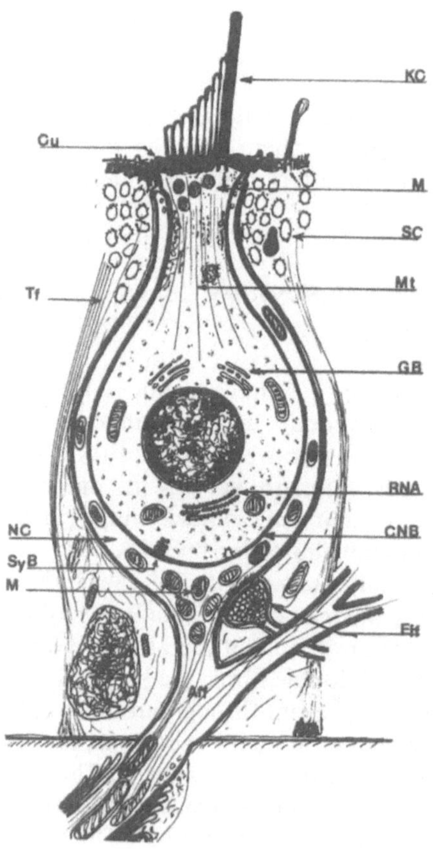

Fig. 1-6. Schematic of a type I cell. The cell is surrounded by a nerve calyx (*NC*). At the surface one kinocilium (*KC*) and several stereocilia are inserted in the cuticle. Below the cuticle are rounded mitochondria (*M*). From this region microtubules (*Mt*) reach far down into the cells. Above the nucleus are Golgi complexes (*GB*). Below the nucleus a rough endoplasmic reticulum (*RNA*) is often present. Between the sensory cell and the nerve calyx (*NC*) there is a double membrane (*CNB*). There are many mitochondria in the lower portion of the nerve calyx. On the border between the calyx and the sensory cell synaptic structures (*SyB*) can be found. On the outside of the nerve calyx efferent endings (*Eff*) are found. In the supporting cell (*SC*) are many granules and many tonofilaments (Tf).

ment, length, and form of both kinds of cilia are beautifully depicted with SEM (Figs. 1-12 and 1-13).

The stereocilia are outwardly bordered by a double plasma membrane (13). From the other layer a number of tiny macromolecular strands radiate, forming a "fuzz" (Fig. 1-19) as described by several authors (see Chapter 15). Inside the plasma membrane each stereocilium contains a large number of microfilaments, which have been shown by Flock and Cheung (18) to contain actin.

The exact number of stereocilia on each hair cell seems to vary in different sensory areas and also in different species. In the squirrel monkey and in the guinea pig we have found between 70 and 80 cilia per cell. We have in several normal animals seen double cilia in the cochlea and in the vestibular regions. In animals exposed to noxious influences such as ototoxic antibiot-

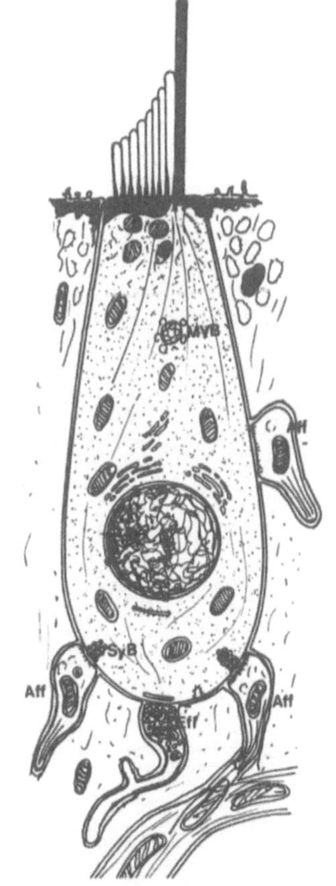

Fig. 1-7. Sensory cell type II with surrounding afferent (*Aff*) and efferent (*Eff*) endings. At the surface are one kinocilium and many stereocilia. At the afferent endings synaptic bars (*SyB*) are often found. *MVB*, multivesiculated body often seen at this level both in type I and II cells.

ics or noise, but also in congenitally deficient animals, pronounced fusion of sensory hairs may occur.

We often distinguish in the sensory cells one infracuticular, one supranuclear, and one infranuclear region. In the infracuticular portion numerous rounded mitochondria with densely packed cristae mitochondriales are seen. In this region there is a rather rich endoplasmic reticulum and the earlier mentioned microtubules (7) and also many small vesicles. The cells have direct contact with the surrounding supporting cells and the reticular framework, which also surrounds the cuticular plate. Occasional osmiophilic granules and lysosomes appear in this region, but more commonly in the supranuclear region. This latter region contains several round, but often oblong mitochondria, rich Golgi complexes, and a few multivesiculated

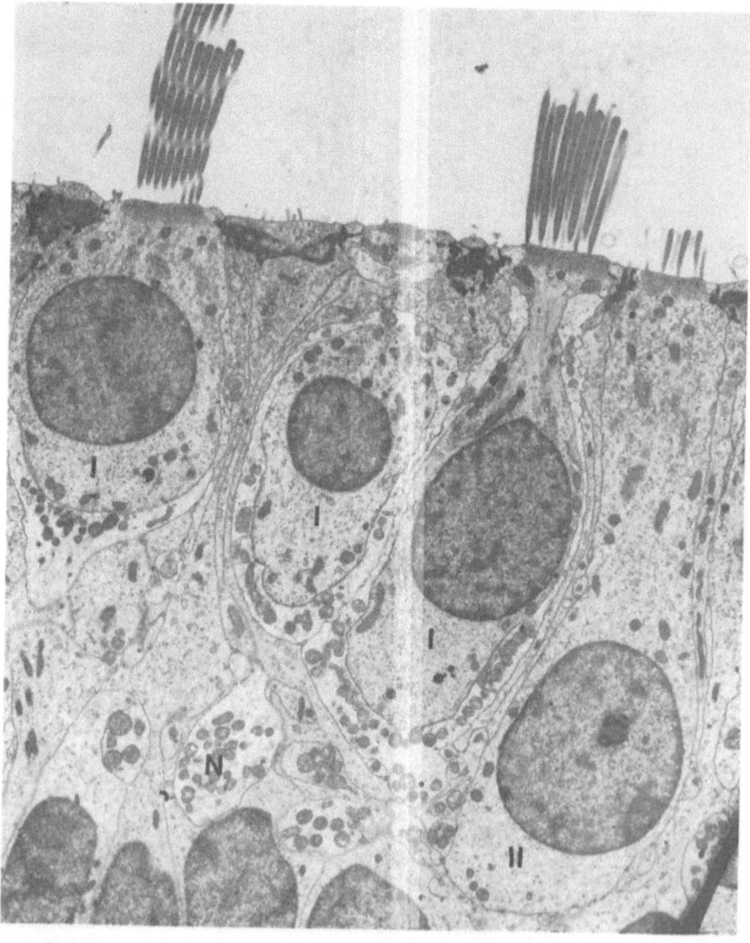

Fig. 1-8. Low-power electron micrograph showing three type I and one type II cells. Macula utriculi; monkey.

bodies. The Golgi complexes form an almost circular or ring-shaped formation above the nucleus (Fig. 1-20).

In the infranuclear region the cytoplasm contains a very fine or sparse endoplasmic reticulum and a few mitochondria. The region rather often contains a few or more endoplasmic cisternae or membranes with ribosomes (Fig. 1-6). Small clusters of ribosomes are found throughout the infranuclear region. In this portion occasional osmiophilic granules are seen (Figs. 1-21 and 1-23). The nucleus of the type I cell is ovoid (Figs. 1-8 and 1-10), often with a slightly pointed upper end.

The type I cell is bordered by a plasma membrane which covers the sensory hairs and continues along the infracuticular and supranuclear region, where the cell is bordered outside by the upper part of the nerve calyx.

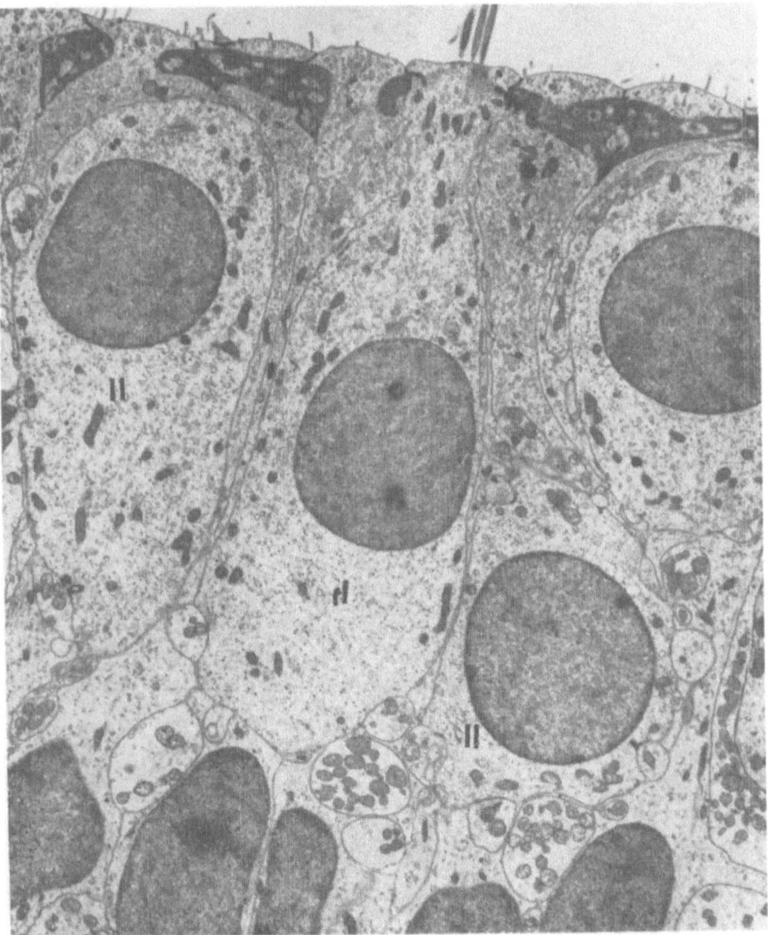

Fig. 1-9. Low-power electron micrograph showing several type II cells. Macula utriculi; monkey.

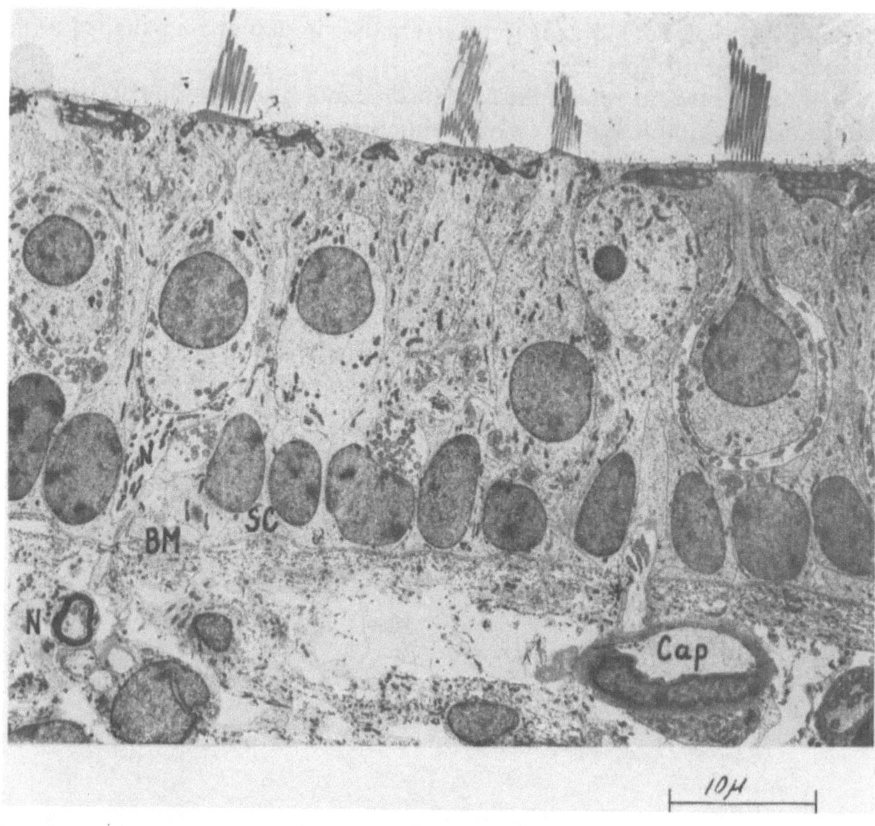

Fig. 1-10. Low-power electron micrograph from macula utriculi showing the general arrangement of sensory and supporting cells (*SC*). Several type I and type II cells of varying shape are shown. Below the basement membrane (*BM*) are capillaries (*Cap*) and nerve fibers (*N*). Macula utriculi; monkey.

About the level of the nucleus the plasma membrane loses its normal structure and a thicker specialized membrane is formed as border to the nerve calyx. This membrane, formed by the calyx and the sensory cell, is very distinct (Figs. 1-25 and 1-26). It has a thickness of about 250 Å, but it varies slightly along its course. It will be further discussed below together with a description of its relation to the synaptic structures, the synaptic bars or bodies which are found along the membranes.

Type II Hair Cells

The hair cells of type II (Fig. 1-22) are more cylindric than are the type I cells. They vary considerably in length, as do the type I cells. Similarly, they also have a cuticle with hairs. Often the cuticle of the type II cell seems to be less distinct than in the type I cells, but this varies also in the different vestibular regions. Thus, we have seen very well-developed cuticles in some cristae ampullares of the squirrel monkey. The type II cells have a richer endoplasmic reticulum than do the type I cells. They often appear denser in the electron micrographs. The mitochondria are often dispersed over a larger portion of the upper part of the cells and the Golgi complexes are usually situated higher up in the supranuclear portion. Otherwise, there is a good resemblance between the two types of cells.

The major difference is their relation to the nerve endings. As was just described, the type I cell has relation to one afferent ending only, the nerve calyx. The efferent endings do not reach the plasma membrane of the sensory cells. In the type II cell there is no specialized thick membrane in the lower part of the cell. Here both afferent and efferent nerve endings form direct synaptic contacts with the sensory cell in different locations along the cell membrane (Figs. 1-5 and 1-7). In the type I cell rough endoplasmic reticulum membranes are often seen in the infranuclear portion. Such cisternae or membranes are very infrequent in the type II cell.

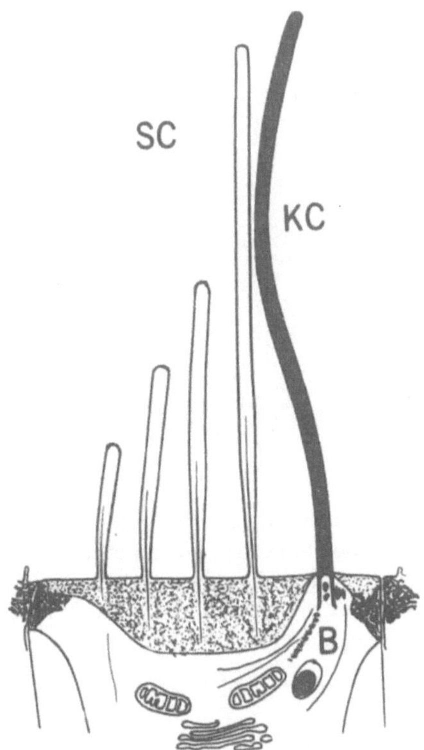

Fig. 1-11. Schematic of the surface of a vestibular sensory cell. There is one long kinocilium (*KC*) with a basal body (*B*) and several stereocilia. The length of the stereocilia increases toward the kinocilium.

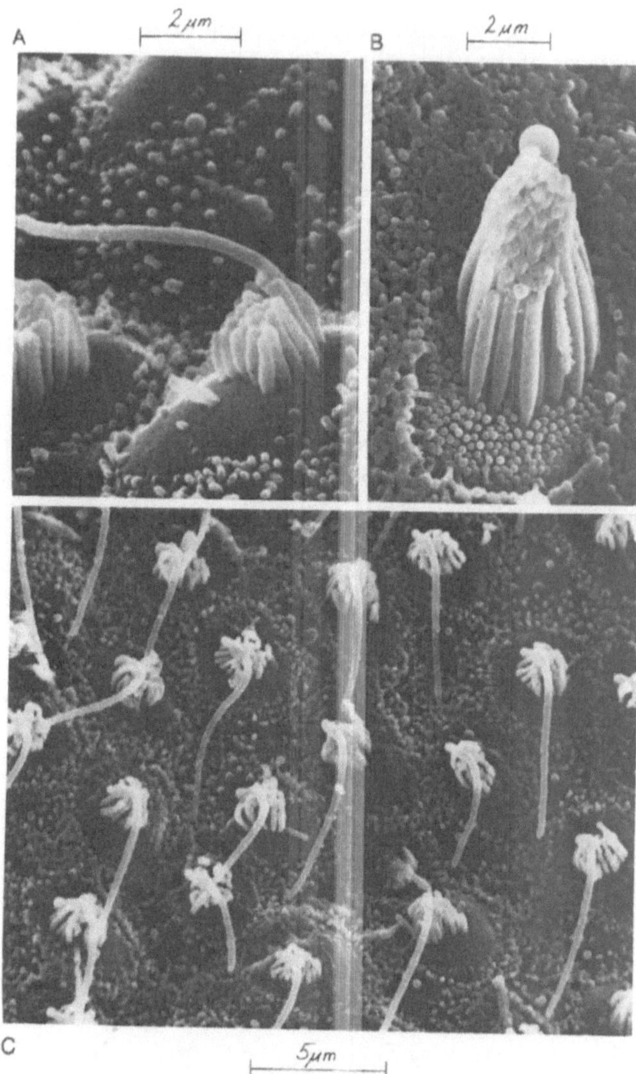

Fig. 1-12. Scanning electron micrographs from the labyrinth of a bullfrog (*Rana catesbeiana*). **A** One long kinocilium and many rather short stereocilia. **B** The stereocilia are almost equal in length to the kinocilium, which has a bulbous tip. Observe the large number of microvilli at the cuticular surface. **C** Several cells demonstrate the morphologic polarization of the cilia.

Innervation of Sensory Regions

Innervation of the sensory regions (Figs. 1-22, 1-23, 1-24, 1-25, 1-26, 1-27) is provided by fibers of the upper and lower division of the vestibular nerve and the ganglion of Scarpa, which contains rather large bipolar ganglion cells

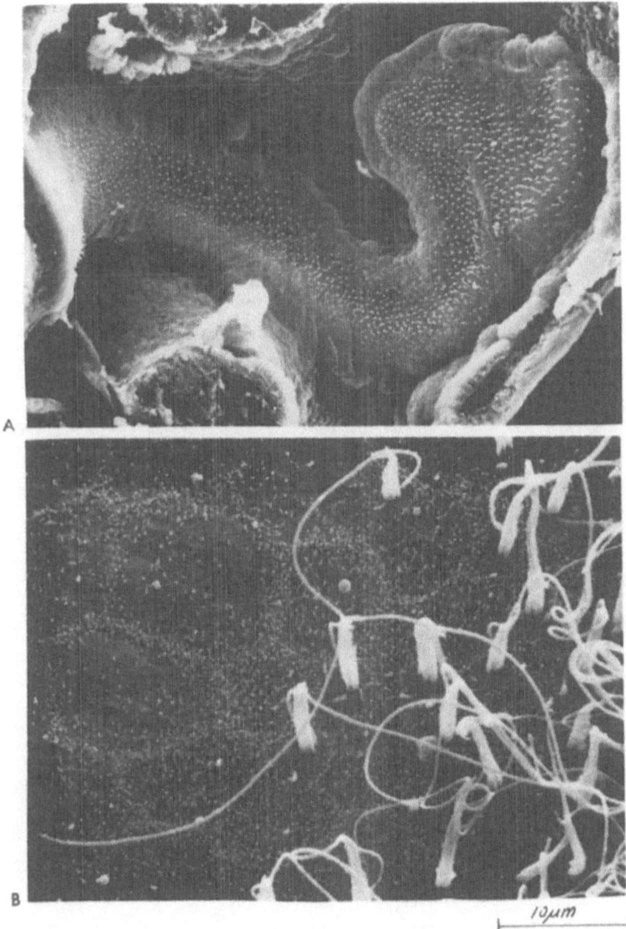

Fig. 1-13. Scanning electron micrographs. **A** Macula neglecta from a bullfrog. Both form and sensory cell population of inner ear sensory regions are demonstrated. **B** A marginal portion of the sensory epithelium on a crista ampullaris shows very long kinocilia and rather short stereocilia.

covered by a myelin layer, similar to that of the cochlear ganglion cells. The sensory cells also have an efferent innervation.

Innervation of Type I Cells

Innervation of type I cells presumably takes place mainly at the lower portion of the sensory cells. In this region there is a double membrane with a thickness of between 200 and 300 Å. This membrane is very distinct, and in certain areas has varying modifications. In certain areas more or less pronounced invaginations are found. In such invaginations the membrane

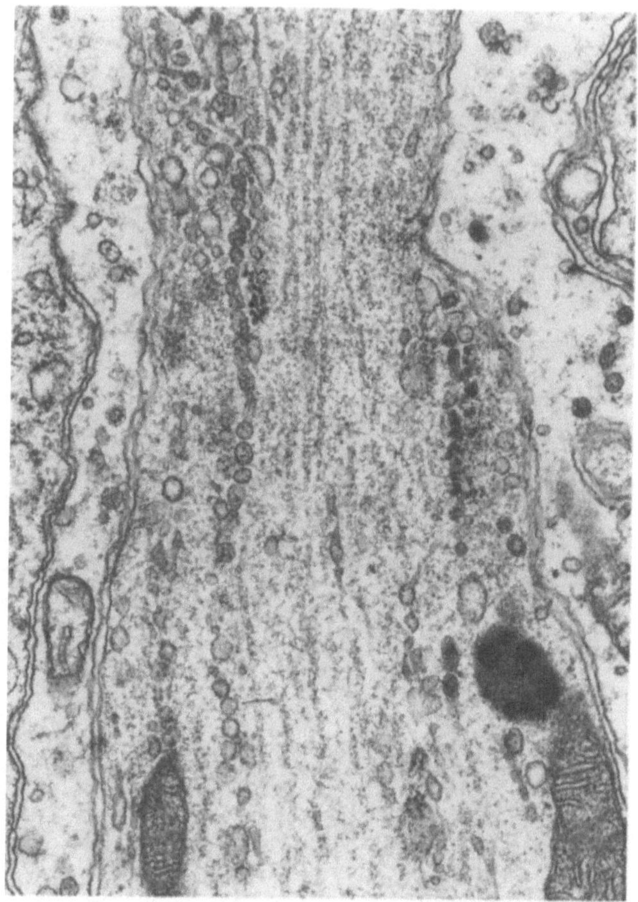

Fig. 1-14. Neck portion of type I cell with many microtubules. Squirrel monkey.

thickness is considerably reduced, occasionally almost to an extent resembling fusion. In some regions the sensory cells contain special differentiations called synaptic bars or bodies. They were observed by Smith and Sjöstrand (35a) in the cochlear hair cells and have been seen by several authors in the vestibular end organ (10). In some animals they form short bars or rounded bodies, perhaps less than 0.2 to 1 μm in length or diameter. In other animals, e.g., squirrel monkeys, very long bars are present (Fig. 1-28). We have seen bars more than 5 μm in length, as well as double and triple bars. In mammals we have almost always seen these synaptic structures in direct contact with the synaptic membrane. In frogs several authors have seen round synaptic balls or bodies moving freely in the cytoplasm of the infranuclear portion.

The synaptic bars or bodies (Fig. 1-28) have a dark center in which several membranous lines can occasionally be seen. Synaptic vesicles are seen

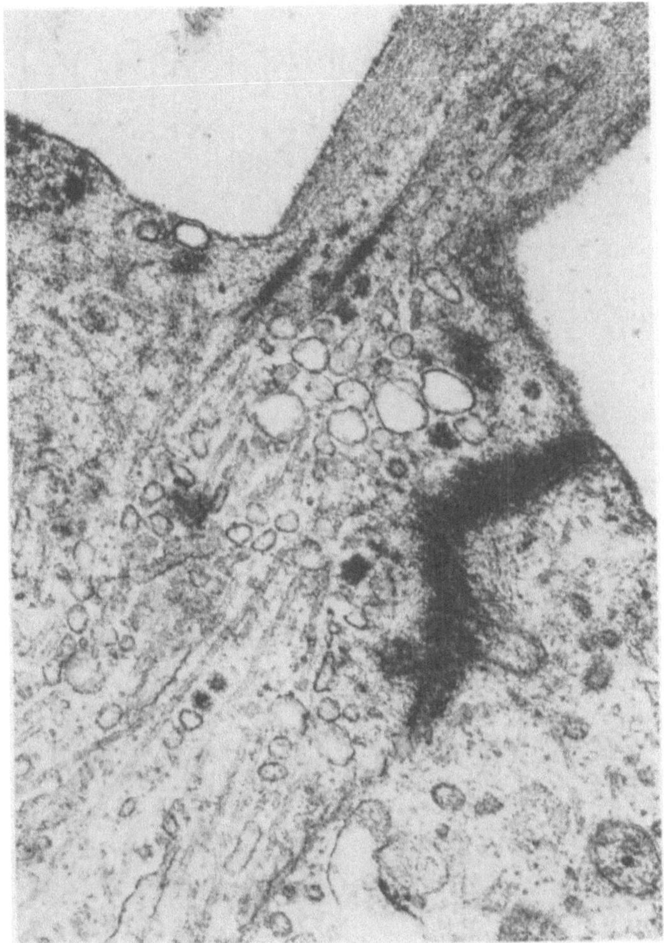

Fig. 1-15. Region below a kinocilium with many microtubules, which here seem to have direct relation to the kinocilium. Guinea pig macula utriculi.

outside these bodies. In the region of contact with the plasma membrane there is a presynaptic thickening, the membrane thickness is reduced, and there is often found a postsynaptic network, which has been described in other synaptic regions.

In the neighborhood of the synaptic areas several infoldings are frequently observed. These have a coating of small spines, as can be seen, for example, in Fig. 1-26. Whether these are indications of reabsorption or recycling can not be determined from the electron micrographs.

Direct contact between efferent endings and the type I cells does not exist, but efferent endings are found outside the nerve chalice, usually at its lower portion (Figs. 1-8 and 1-9). In specimens stained according to Maillet it has been shown that efferent endings form loops or rings around the lower

Fig. 1-16. Stereocilia at the surface of a macula utriculi from a rhesus monkey. The graded length is evident, as is the method of insertion of the rootlets of the hairs into the cuticle of the type I cell (HCI). The arrows indicate the ring-shaped reinforcement or support for the cuticle. The uppermost portion of a nerve calyx can also be seen.

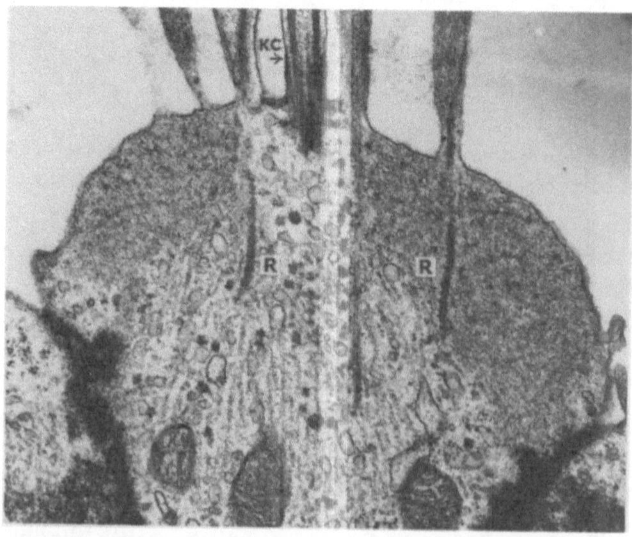

Fig. 1-17. Surface of a type I cell with a kinocilium (*KC*) inserted in a cuticle-free zone. Several rootlets (*R*) from stereocilia can be seen. A few small cross-striated rootlets are seen in the center below the kinocilium. Many microtubules are seen in the infracuticular area.

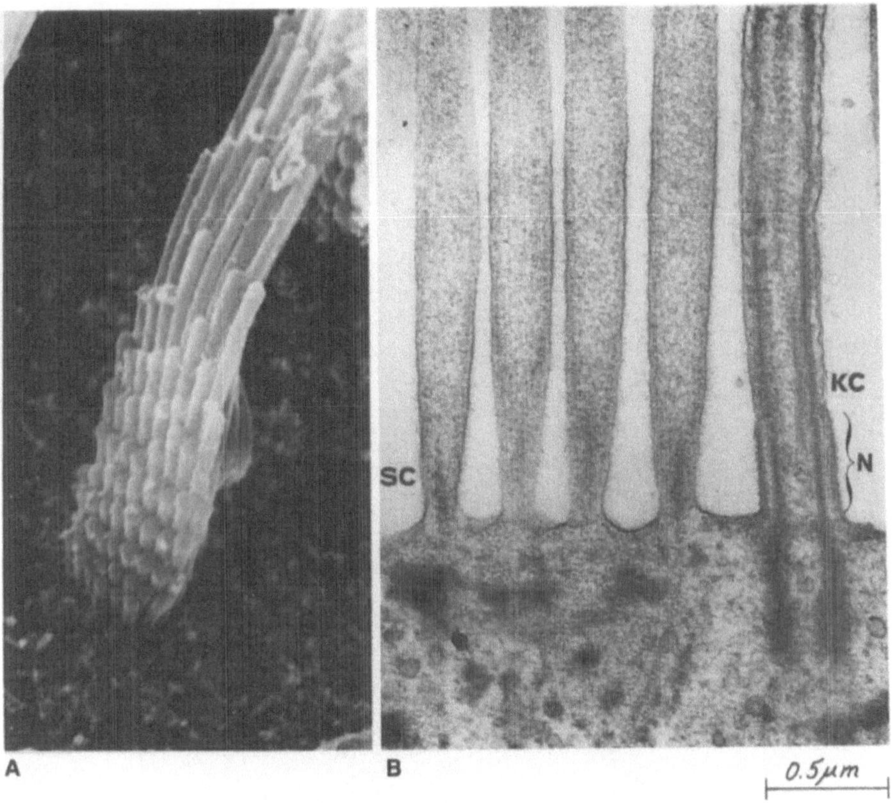

Fig. 1-18. **A** SEM showing the graded length of stereocilia. **B** One kinocilium (*KC*) and several stereocilia (*SC*). The inner tubular structure of the kinocilium, its neck-portion (*N*), and its basal body can be seen.

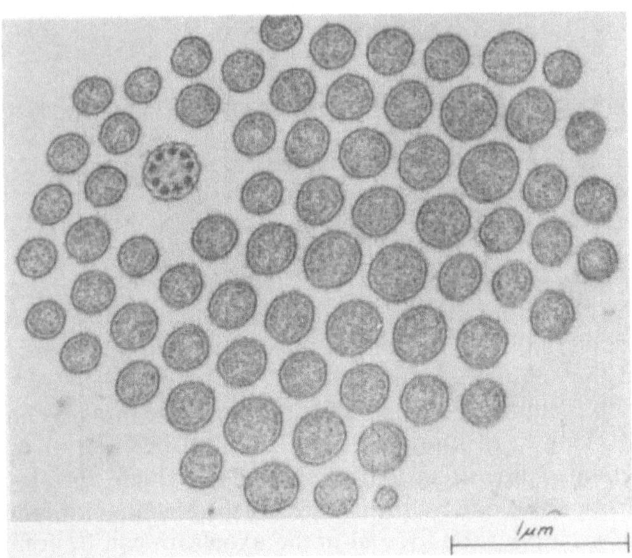

Fig. 1-19. One kinocilium and many stereocilia. Observe the "fuzz" around the stereocilia.

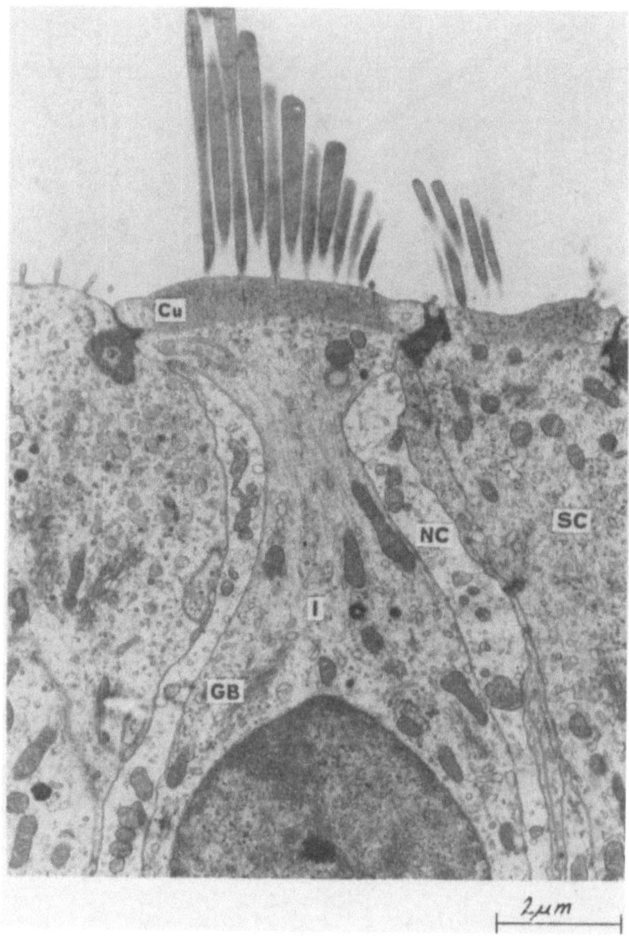

Fig. 1-20. Upper part of a type I cell (*I*) surrounded by its nerve calyx (*NC*). Below the cuticle (*Cu*) are some round mitochondria. The arrows indicate supporting reinforcements. In the supranuclear region a symmetric Golgi complex (*GB*) is seen. In the supporting cell (*SC*) many granules are observed.

part of the calyx or its afferent fiber. A detailed picture of the different nerves is difficult to obtain because of the rich network of fibers found in the lower half of the sensory epithelia.

Inside the epithelium all the nerve fibers are devoid of a myelin sheath. (It has, however, long been known that in certain fishes and also some other animals myelinated fibers can be followed rather high up into the epithelium.) Myelin begins close to, or just below, the basement membrane (Figs. 1-29 and 1-30). Nerve tubuli found in the axoplasm can be followed rather high up into the nerve chalices, as observed by Wersäll (41).

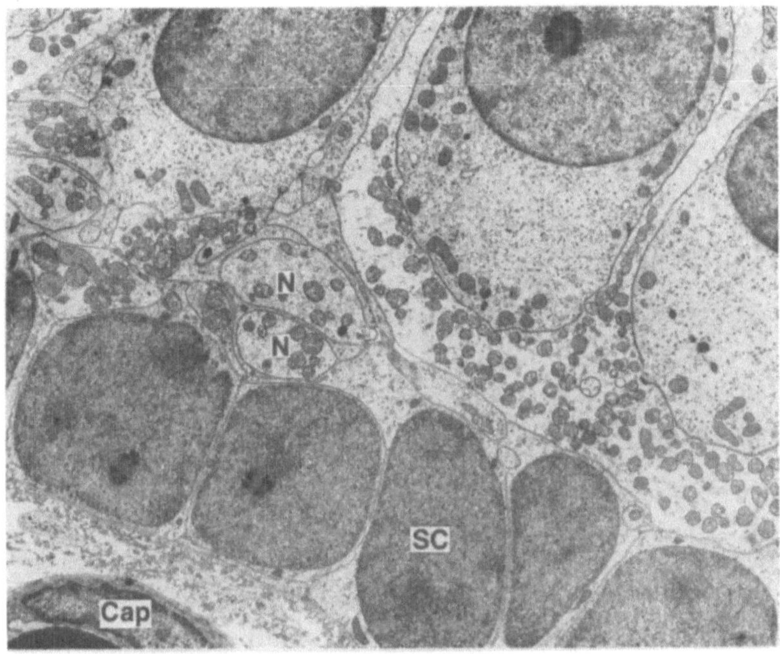

Fig. 1-21. Two type I cells in a double calyx and one type II cell, showing the lower portion of these cells and their relation to nerve fibers (*N*) and supporting cells (*SC*). A capillary (*Cap*) is seen below the basement membrane. Monkey macula utriculi.

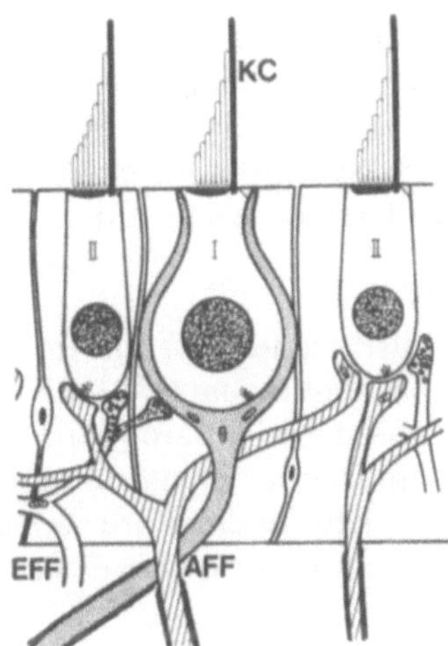

Fig. 1-22. Schematic of some sensory cells and their relations to nerve endings and nerve fibers. *AFF*, afferents; *EFF*, efferents. *KC*, kinocilium.

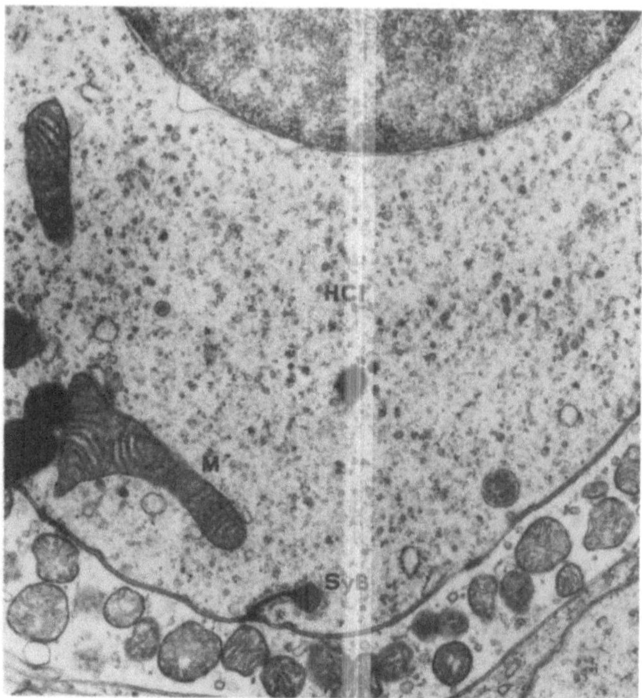

Fig. 1-23. Lower portion of a type I cell (*HCI*) with some mitochondria (*M*). On the calyx–sensory cell border a well-developed synaptic body or bar (*SyB*) is seen.

The efferent endings are filled with synaptic vesicles, which we have observed to be mainly of a round shape. The mitochondria in the efferent endings are distinctly smaller than those in the nerve chalices. On the border of the nerve calyx and the efferent ending presynaptic clusterings of vesicles are often very distinct. Dense-cored vesicles, which are very numerous in certain parts of the efferent system in the cochlea, are few in the vestibular labyrinth. They are of the large type with a diameter of about 600 to 800 Å.

Innervation of Type II Cells

Innervation of type II cells is quite different from that of type I cells. The former cells form direct synaptic contacts with both afferent and efferent endings. The majority of these endings are found in the infranuclear portion, but it is quite common to find nerve endings forming synaptic contacts above the nuclear region. Occasionally, endings are seen even close to the reticular membrane. The difference in structure between afferent and efferent synaptic areas is very distinct.

Afferent endings have their presynaptic structure inside the sensory cell and postsynaptic differentiation inside the nerve ending. The presynaptic structure is often a short bar or rounded body; the long bars seen in type I

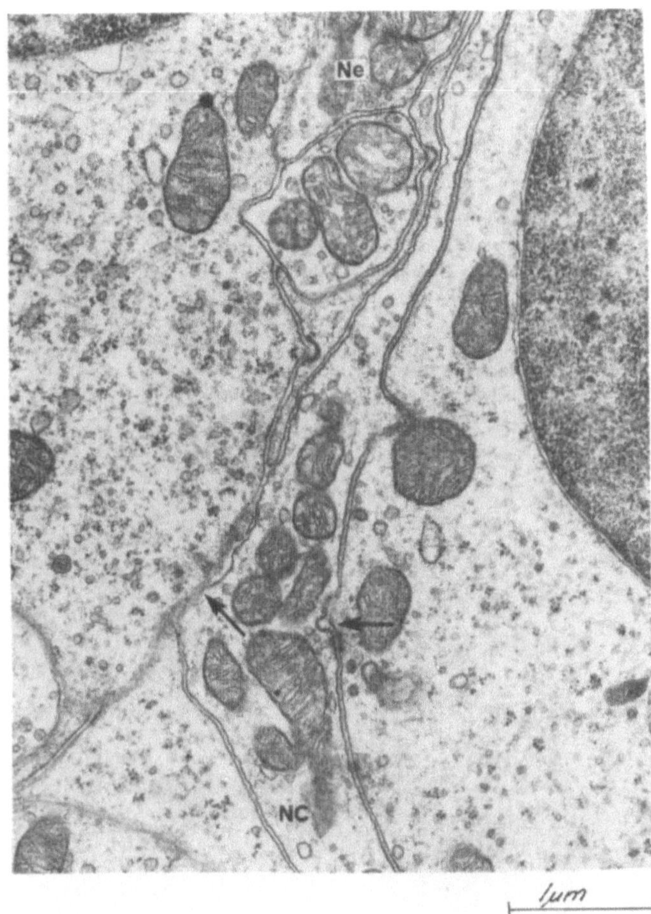

Fig. 1-24. Parts of one type I and one type II sensory cell. The nerve calyx (*NC*) is in contact not only with the type I cell but also with the type II cell (*arrows*). The type I cell has different modifications of the double membrane. One smaller nerve ending (*Ne*) is also in contact with the type II cell. Macula utriculi; monkey.

cells have been rarely seen by us. There are distinct presynaptic thickenings. The intrasynaptic cleft can vary in structure, depending upon fixation (Figs. 1-26, 1-31). This is well known from many other regions (7).

Efferent endings may be the same size as the afferent ones, but they are often smaller and have considerably smaller mitochondria (Fig. 1-32). The presynaptic thickening or grid is found inside the nerve ending and the postsynaptic structure is often a thin subsynaptic cistern (Fig. 1-29). Such cisterns are often well developed at the large efferent endings seen at outer hair cells in the cochlea. The efferent fibers run a complicated, irregular course inside the epithelium. Below the basement membrane they run as thin myelinated fibers (31).

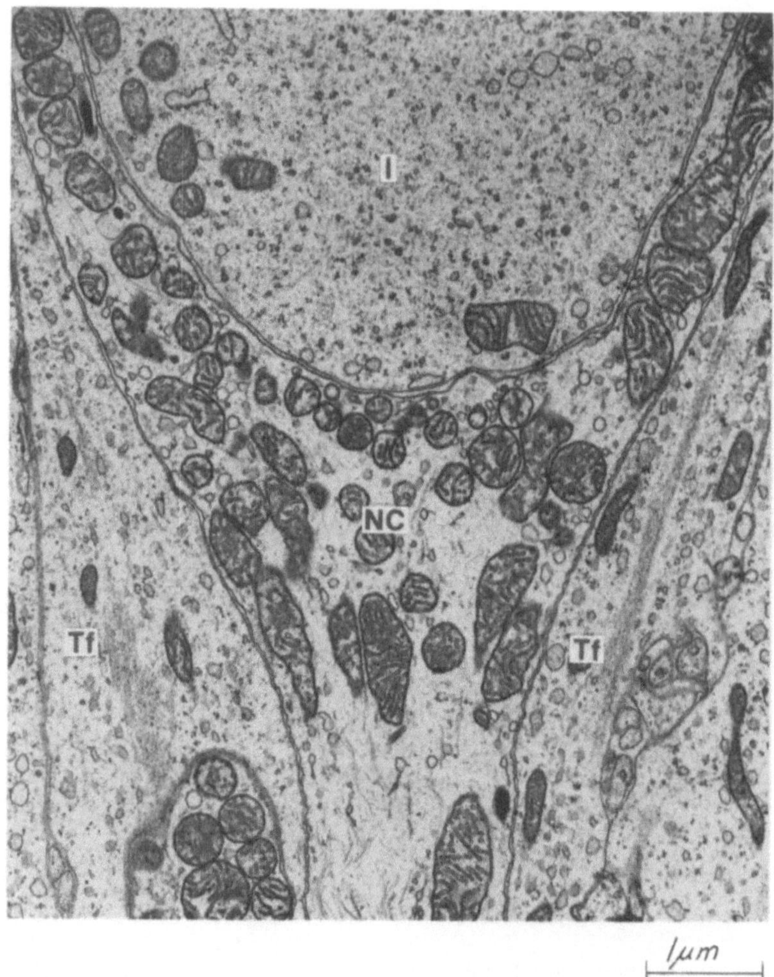

Fig. 1-25. Lower portion of a type I cell and its nerve calyx (*NC*). The large number of mitochondria found in the calyx is demonstrated, indicating a high enzymatic activity and energy turnover. In the supporting cells tonofilaments (*Tf*) can be seen. Macula utriculi; monkey.

We have observed elsewhere that presynaptic bars or bodies can be seen close to efferent endings (ref. 11, Fig. 29E; ref. 12, Fig. 8). This has been noted in many specimens and the question is whether or not these are reciprocal synapses. They have been seen by us in many cases in the American bullfrog. Reciprocal synapses have also been discussed by Dunn (6, 7) who has also seen them in the bullfrog. Although there are two main types of hair cell in the vestibular sensory regions, intermediate types can also be seen. (Fig. 1-22) Thus, it is not uncommon to observe that a nerve calyx forms synaptic contacts with both a type I and a type II cell (Fig. 1-24). Such in-

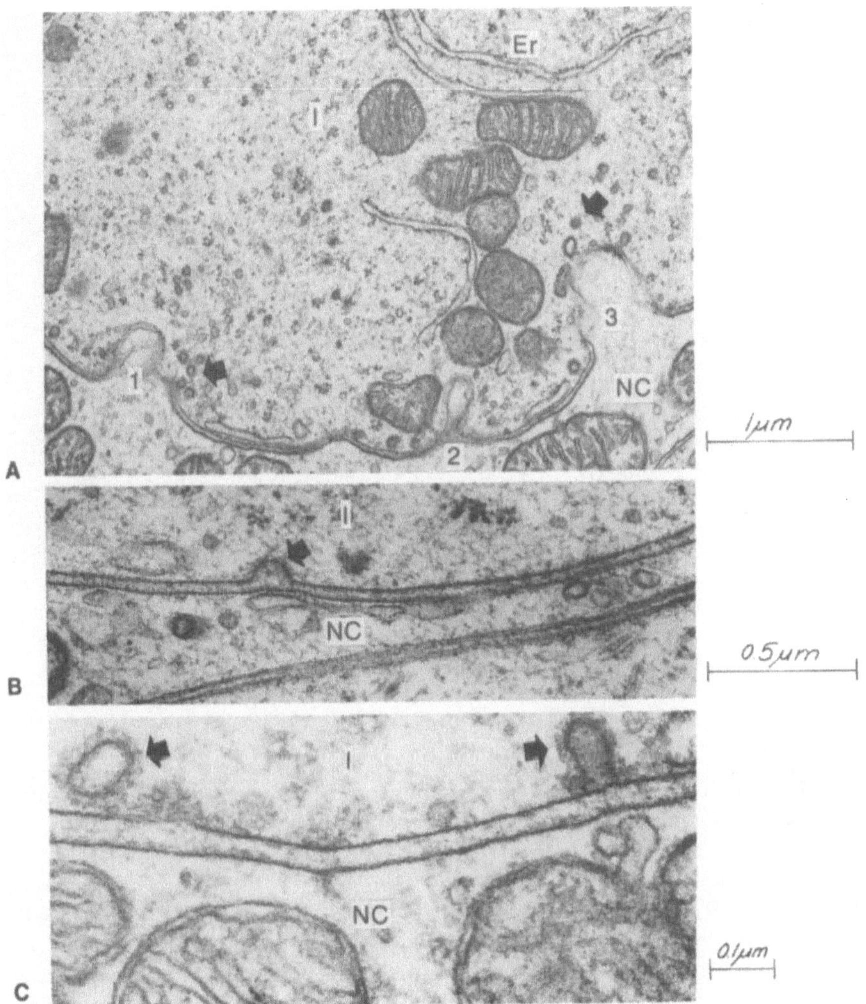

Fig. 1-26. **A–C** Variations in the structure of the membrane between nerve calyx and sensory cell.

terrelations between the two types of cell and their nerve endings are presumably evidence of development of the type I cell from the type II cell. It has been repeatedly noted that the type II cell is the phylogenetically primary cell and is found in more primitive animals as the only type (41).

AGE VARIATIONS

Age variations in human vestibular regions are much more pronounced than was previously believed. Studies by Bergström (3a) have shown that there is a considerable reduction in the number of vestibular nerve fibers in the aged.

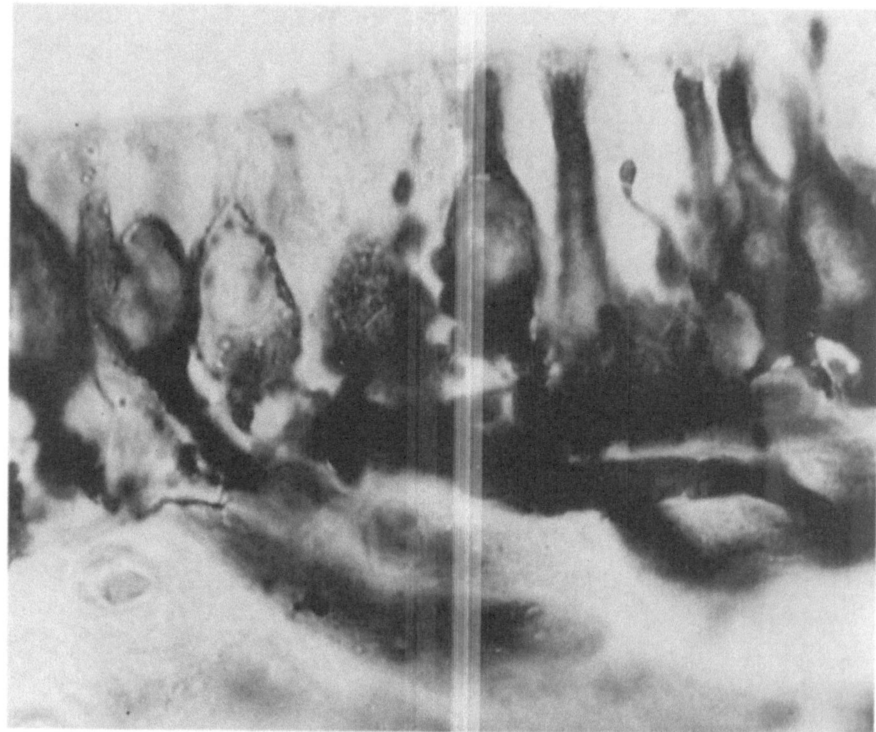

Fig. 1-27. Maillet-stained nerve calyces showing the general form and structure of nerve endings. Guinea pig. Produced in cooperation with H. Lindeman, Jr.

In the ampullar nerves this reduction may reach 40% and in the macular nerves 20%. The number of sensory cells is also considerably reduced in a corresponding manner (33,34). It is of interest to know that remaining cells also demonstrate pronounced structural changes (9,12). These modifications are at present being studied by our group with rhesus monkeys as test animals.

SUPPORTING CELLS IN THE VESTIBULAR LABYRINTH

The sensory areas in the vestibular labyrinth contain sensory cells and surrounding supporting elements (Figs. 1-10 and 1-20). The supporting cells and their arrangement have been much discussed by Lindeman (25) and Engström et al. (11). These cells have a form which can vary considerably. Many of them reach from the basement membrane to the surface of the epithelium. The basement membrane is usually a flat, smooth layer, but in old animals it can be thickened and irregular. The border between this membrane and the supporting cells is very distinct. The supporting cells are considerably more numerous than the sensory cells. They are grouped

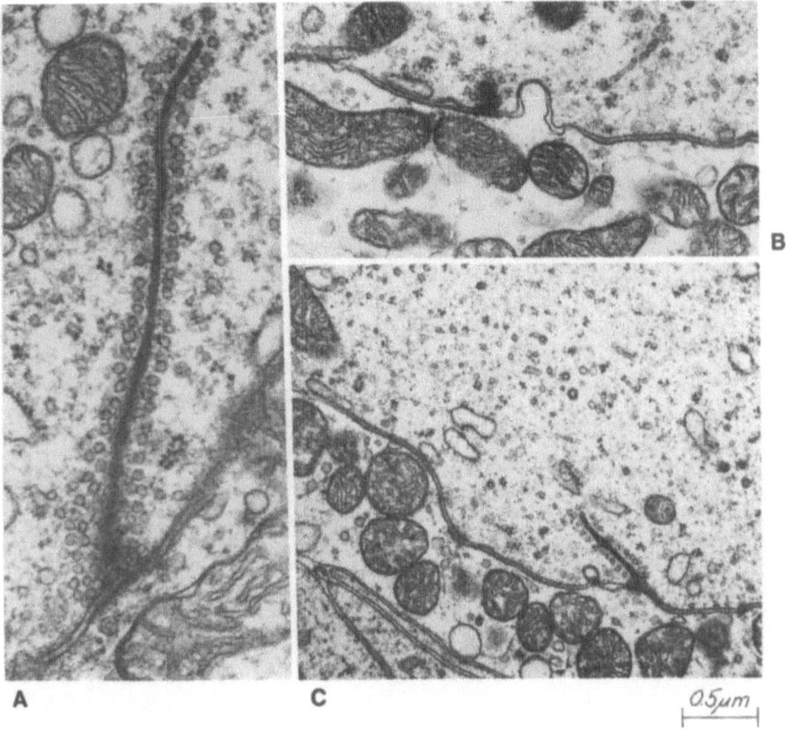

Fig. 1-28. Three different forms of synaptic bars in afferent endings from the macula utriculi of the squirrel monkey (**A**) and rhesus monkey (**B** and **C**). Thick sections.

around the type II cells and their nerve endings and around the nerve chalices of type I cells. At the surface the cells are provided with a small number of rather short microvilli. It is very common to find a rudimentary kinocilium on these cells.

The nucleus of a supporting cell lies close to the basement membrane (Fig. 1-10). From this region minute tonofilaments (Fig. 1-33) reach up toward the reticular framework, which forms an important support for the sensory cells. In this reticular lamina the hair cells are inserted in a similar manner to that of the cochlear hair cells. Lindeman (25) and Watanuki (38) have given good descriptions of the surface structure of the vestibular sensory regions. The inner structure of the supporting cells can be seen in several of the electron micrographs (Figs. 1-5, 1-8, 1-9, 1-10). Of interest are the large number of granules or vesicles which fill to a great extent the upper portion of the supporting cells. They often have ribosomes along their surfaces and thus have a certain resemblance to endoplasmic reticulum, but their round shape is more of a granular type. Neither their chemical composition nor their function is known.

Supporting cells as well as sensory cells undergo pronounced modifica-

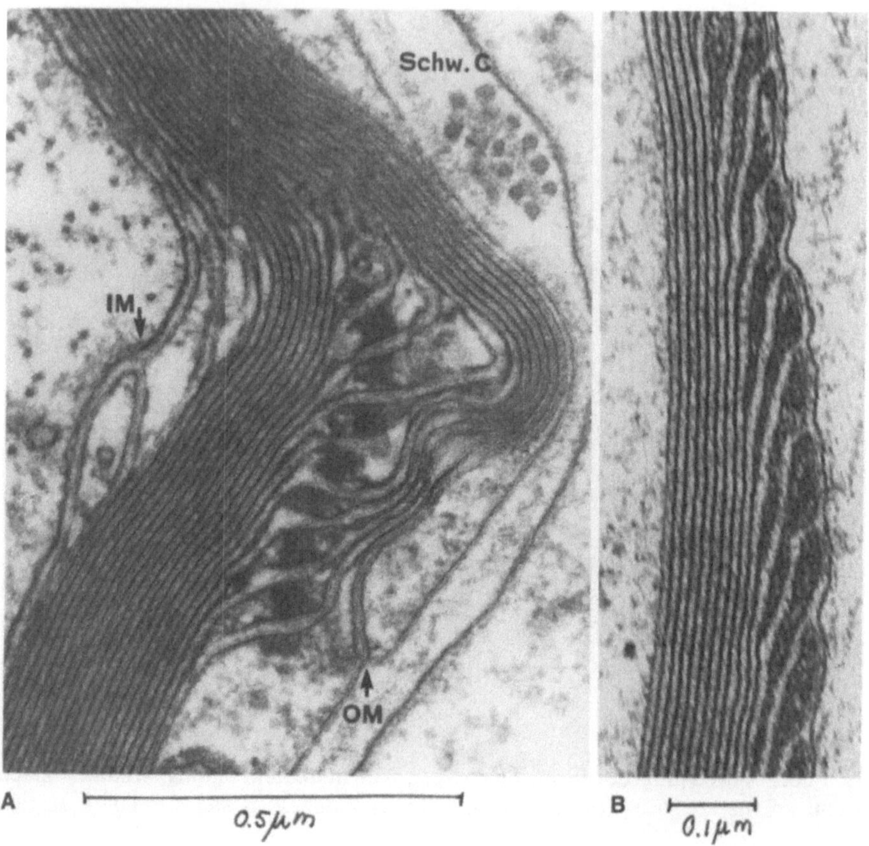

Fig. 1-29. Myelin sheath arrangements in regions of demyelinization below the macula utriculi of a squirrel monkey. **A** The outer (*OM*) and inner (*IM*) mesaxons are recognized. The myelin sheath Schwann cell cytoplasm (*Schw.C*). **B** Demyelinization. Stepwise reduction of the myelin layers.

tions with increasing age. These changes include agglomerations of lysosomes, cystlike formations, and the occasional appearance of large numbers of lipofuscin granules. There also appear laminated membranes at the reticular membrane.

MORPHOLOGIC POLARIZATION OF VESTIBULAR SENSORY CELLS

It has already been mentioned that there is a morphologic and functional polarization of vestibular sensory cells. The earliest evidence of this came from studies made by Loewenstein and Wersäll (26), who demonstrated that the sensory cells on the cristae had a characteristic orientation of functional significance. It was also known that the maculae have a subdivision with a more or less central striola and one portion on each side. These regions have

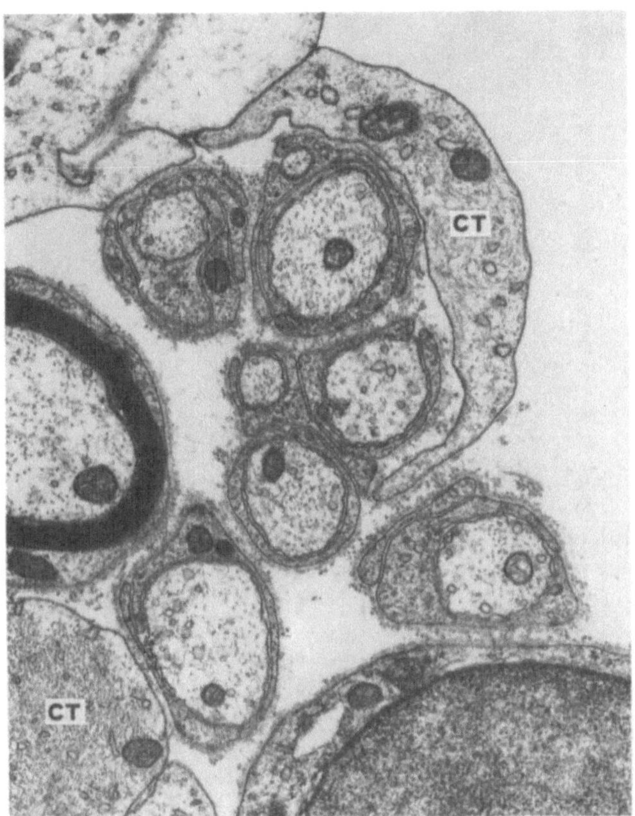

Fig. 1-30. Several unmyelinated nerve fibers and one myelinated nerve fiber in the connective tissue (*CT*) below a macula utriculi of the squirrel monkey.

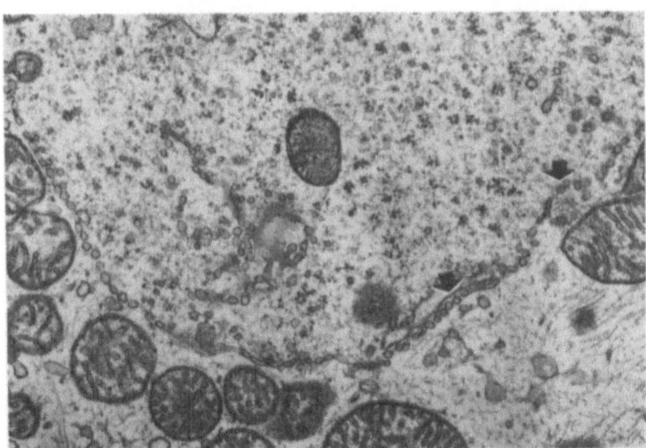

Fig. 1-31. Region where the synaptic membrane has become interrupted and granular (*arrows*). This can occasionally be observed in squirrel monkeys (macula utriculi). We have not been able to determine whether this is a functional stage or an artifact.

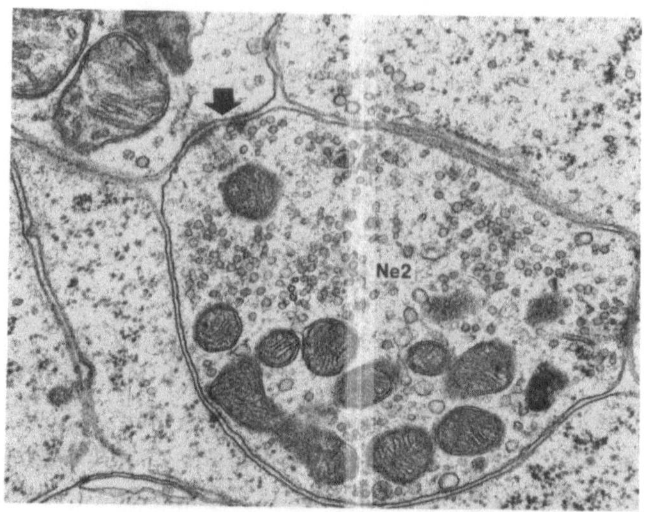

Fig. 1-32. Efferent ending (*Ne 2*) at type II cell. The nerve ending contains many vesicles. There is one dense-cored vesicle and also an indication of a neuroneuronal synapse (*arrow*).

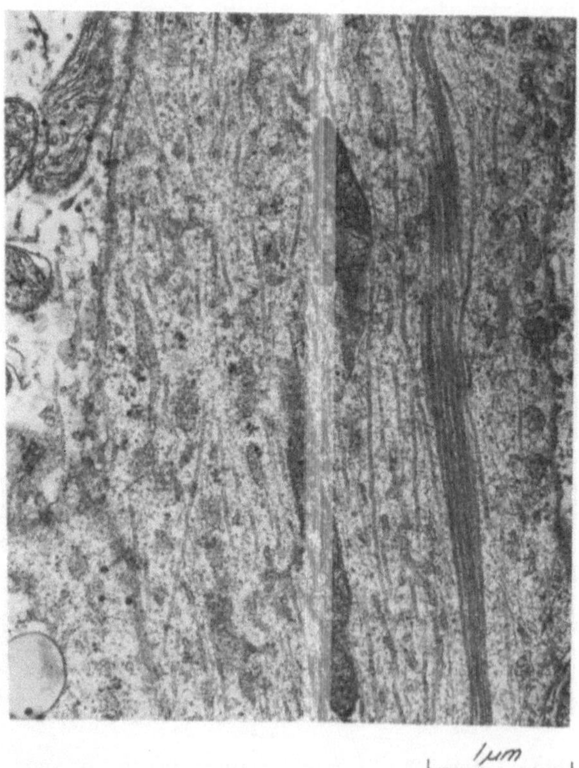

Fig. 1-33. Tonofibrils in a supporting cell of a macula sacculi of a squirrel monkey. Glutaraldehyde fixation demonstrates the fibrils very well.

received different names. We are using the nomenclature recommended by Lindeman, who has named them pars interna and pars externa (Fig. 1-4). Also, use of the electron microscope has elucidated a pattern of structural polarization in the maculae. Engström et al. (10) demonstrated that the orientation of the sensory hairs on the maculae was rather uniform over large areas. At a certain region this pattern became reversed and the two portions of the maculae thus had an opposite polarization. In 1964 Flock provided a thorough description of this polarization (15) and shortly thereafter Spoendlin (36) verified his observations. Subsequently many publications have appeared describing structural and functional properties of vestibular maculae in many animals (15,17,29,30,37).

It has long been known that the cells of the acousticovestibular system and the lateral line systems showed many functional similarities. It was often pointed out that the hair cells could be regarded as directionally sensitive displacement detectors. When the morphologic and functional studies clearly showed that the structural polarization was of great significance for the neurophysiologic response, series of studies were carried out to clarify the polarization concept and many important discoveries were made concerning the complicated structural patterns in different sensory regions.

The arrangement of the sensory cells in the mammalian vestibular end organ is described by Lindeman (25) and an excellent survey of the related functional properties is provided by Flock (17). Schematic drawings (Fig. 1-4) show the major features of the structural polarization. Recent SEM studies have contributed to a detailed investigation of the polarization and many interesting publications have been forthcoming. These studies have also become of interest from a clinical point of view.

References

1. Afzelius, B. and Franzen, Å.: The spermatozoon of the jelly-fish nausithoë. J. Ultrastruct. Res. 37:186, 1971.
2. Bagger-Sjöbäck, D. and Flock, Å: Freeze-fracturing of the auditory basilar papilla in the lizard calotes versicolor. Cell Tissue Res. 177:431, 1977.
3. Bagger-Sjöbäck, D. and Wersäll, J.: The sensory hairs and tectorial membrane of the basilar papilla in the lizard calotes versicolor. J. Neurocytol. 2:329, 1973.

3a. Bergström, B.: Morphological studies of the vestibular nerve. Acta Universitatis Upsaliensis. Suppl. 159:1, 1973.

4. Carlström, D. and Engström, H.: The ultrastructure of statoconia. Acta Otolaryngol. (Stockh.) 45:14, 1955.
5. Dijkgraf, S.: The functioning and significance of the lateral-line organs. Biol. Rev. 38:51, 1962.
6. Dunn, R.F.: Cytoplasmic organization in the receptor cells of the crista ampullaris. Proc. Electronmicros. Soc. Am. vol. 60, 1972.
7. Dunn, R.F.: Reciprocal synapses in the crista ampullaris: a possible mechanism

for hair cell interaction. Trans. Am. Acad. Ophthalmol. Otolaryngol. 82:188, 1976.

8. Engström, H.: On the double innervation of the sensory epithelia of the inner ear. Acta Otolaryngol. (Stockh.) 49:109, 1958.
9. Engström, H., Ades, H.W., Engström, B., Gilchrist, D., and Bourne, G.: Structural changes in the vestibular epithelia in elderly monkeys and humans. Adv. Otorhinolaryngol. 22:93, 1977.
10. Engström, H., Ades, H.W., and Hawkins, J.E., Jr.: Structure and functions of the sensory hairs of the inner ear. J. Acoust. Soc. Am. 34:1356, 1962.
11. Engström, H., Bergström, B., and Ades, H.W.: Macula utriculi and sacculi in the squirrel monkey. Acta Otolaryngol. (Suppl.) (Stockh.) 301:75, 1972
12. Engström, H., Bergström, B., and Rosenhall, U.: Vestibular sensory epithelia. Arch. Otolaryngol. 100:411, 1974.
13. Engström, H. and Engström, B.: Structure of the hairs on cochlear sensory cells. Hearing Res. 1:49, 1978.
14. Engström, H. and Wersäll, J.: The ultrastructural organization of the organ of Corti and of the vestibular sensory epithelia. Exp. Cell Res. 5:460, 1958.
15. Flock, Å.: Structure and function of the macula utriculi with special reference to directional interplay of sensory responses as revealed by morphological polarization. J. Cell Biol. 22:413, 1964.
16. Flock, Å.: Electron microscopic and electrophysiological studies on the lateral line canal organ. Acta Otolaryngol. (Suppl.) (Stockh.) 199:1, 1965.
17. Flock, Å.: Sensory transduction in hair cells. In Loewenstein, W.R. (ed.): Handbook of Sensory Physiology. Principles of Receptor Physiology. Berlin, Springer-Verlag, 1971, pp. 396–441.
18. Flock, Å. and Cheung, H.C.: Actin filaments in sensory hairs of inner ear receptor cells. J. Cell Biol. 75:339, 1977.
19. Flock, Å., Flock, B., and Murray, E.: Studies on the sensory hairs of receptor cells in the inner ear. Acta Otolaryngol. (Stockh.) 83:85, 1977.
20. Flock, Å. and Wersäll, J.: A study of the orientation of the receptor cells of the lateral line organ of fish. J. Cell Biol. 15:19, 1962.
21. Hunter-Duvar, J.: Electron microscopic assessment of the cochlea. Acta Otolaryngol. (Suppl.) (Stockh.) 351:1, 1978.
22. v. Ilberg, C.: Tubuläre Strukturen in äusseren Haarzellen. Arch. Klin. Exp. Ohr-Nas-Kehlk Heilk. 194:408, 1969.
23. Lim, D.J.: Ultrastructure of the otolithic membrane and the cupula. Adv. Otorhinolaryngol. 19:35, 1973.
24. Lim, D.J.: Formation and fate of the otokonia. Ann. Otol. Rhinol. Laryngol. 82:23, 1973.
25. Lindeman, H., Jr.: Studies on the morphology of the sensory regions of the vestibular apparatus. Adv. Anat. Embryol. Cell Biol. 42:1, 1969.
26. Loewenstein, O. and Wersäll, J.: A functional interpretation of the electron-microscopic structure of the sensory hairs in the cristae of the Elasmobranch Raja clavata in terms of directional sensitivity. Nature 184:1807, 1959.
27. Lorente de Nó, R.: Etudes sur l'anatomie et la physiologie du labyrinth de l'oreille et du VIIIe nerf. II. Trab. Inst. Cajal Invest. Biol. Madrid 24:53, 1926.

28. Malik, L.E. and Wilson, R.B.: Evaluation of modified technique for SEM-examination of vertebrate specimens without evaporated metal layers. II-TRI/SEM.: 259, 1975.

29. Popper, A.N.: A scanning electron microscopic study of the sacculus and lagena in the ears of fifteen species of teleost fishes. J. Morphol. 153:397, 1977.

30. Popper, A.N.: Scanning electron microscopic study of the otolithic organs in the bichir (*Polypterus bichir*) and shovel-nose sturgeon (*Scaphirhyncus platorynchus*). J. Comp. Neurol. 181:117, 1978.

31. Pratt, L.L.: A histochemical study of the course and distribution of the efferent vestibular fibers to the maculae of the saccule and utricle. Laryngoscope 9:1515, 1969.

32. Retzius, G.: Das Gehörorgan der Wirbelthiere. II. Stockholm, Centraltryckeriet, 1884.

33. Rosenhall, U.: Vestibular mapping in man. Ann. Otol. Rhinol. Laryngol. 81:339, 1972.

34. Rosenhall, U.: The vestibular sensory regions in man. Acta Univ. Upsaliensis (Suppl.) 191:1, 1974.

35. Smith, C.: Microscopic structure of the utricle. Ann. Otol. Rhinol. Laryngol. 65:450, 1956.

35a. Smith, C. and Sjöstrand. F.S.: Synaptic structure in the hair cells of the guinea pig cochlea. J. Ultrastruct. Res. 5:184, 1961.

36. Spoendlin, H.: Organization of the sensory hairs in the gravity receptors in utricle and saccule of the squirrel monkey. Z. Zellforsch. 62:701, 1964.

37. Spoendlin, H.: Ultrastructural studies of the labyrinth in squirrel monkeys. In: The Role of the Vestibular Organs in the Exploration of Space. NASA, SP-77, 1965, pp. 7–22,

38. Watanuki, K.: Distribution pattern of the sensory cells in the vestibular otolithic organs. Otol. Fukuoka 16:252, 1970.

39. Watanuki, K. Meyer, Z., and Gottesberge, A.: Toxic effects of streptomycin and kanamycin upon the sensory epithelium of the crista ampullaris. Acta Otolaryngol, (Stockh.) 72:59, 1971.

40. Werner, C.F.: Das Labyrinth. Leipzig, Thieme Verlag, 1940.

41. Wersäll, J.: Studies on the structure and innervation of the sensory epithelium of the cristae ampullares in the guinea pig. Acta Otolaryngol. (Suppl.) (Stockh.) 126:1, 1956.

Discussion

GOLDBERG: In what animals are reciprocal synapses found? Also, could you estimate the frequency of occurrence of synaptic bars in mammalian type I and type II hair cells?

ENGSTRÖM: Reciprocal synapses: We have studied different animals. The largest number of animals we have are squirrel monkeys and rhesus monkeys. We have seen synaptic arrangements which could be reciprocal synapses in monkeys: we have seen the same thing in guinea pigs; we have seen the same thing in the frog.

But they are not very numerous. But, as I said before, there is something present which very much resembles reciprocal synapses in all those animals we have studied. The second question, frequency of synaptic bars—that is, in type I and type II hair cells. I think there is a difference in the number of synaptic bars or synaptic bodies in different portions of the vestibular labyrinth. As to the frequency and number of them, I can't give the correct answer. But very often I have seen the nicest arrangement of synaptic bodies in the central portion of both macula sacculi and macula utriculi. My personal guess is that there are more in the central portion than in the periphery. But I can't give you any figures on this and I don't believe anyone can do that yet. We have to find a way to study this. These are basic questions which have to come from neurophysiologists: how many are there: where are they located?

PLATT: I was wondering if it was possible that the synaptic body is in fact a disk, so that when sectioned in certain planes it looks round, and in other planes it looks like a synaptic bar. Is there any evidence one way or the other?

ENGSTRÖM: This is a question I can definitely answer. I think that the synaptic arrangement can vary, but mainly it is a kind of bar which sticks out as shown in our figures. Some synaptic bodies are definitely round. But they can also have a rectangular form. These must have some kind of longer arrangement in one direction than in the other. And there is an inner structure in the synaptic bar also, which has been described by Dunn (in the eye) and by myself (in the ear).

KHANNA: The kinocilium you said is present in the vestibular system?

ENGSTRÖM: Yes.

KHANNA: But not in the auditory system. Do you have any thoughts about the role of the kinocilium in the two organs? That would be very welcome.

ENGSTRÖM: I cannot answer that question. I have written about it, but I think that the person who can answer more precisely is Dr. Åke Flock, who has been studying this very carefully. And I think it is better that Åke should speak about the function. I have discussed the morphology and Åke Flock can perhaps talk about physiology.

PRECHT: When you look at the receptor cell and the synapse between receptor and afferents in very young animals or newborn animals, do you find them fully developed, or is there, in the first months or weeks, a gradual increase in the number of stereocilia or in the development of the synapse between receptor and afferents?

ENGSTRÖM: This is a very interesting question, and there is just now an application for funds from the National Research Council to study this structure. I think this has to be developed. It will take a very long time, in reality, to study the development of the synaptic regions and I have rather little knowledge about what results can be expected. I know that it's very different in different animals. As to the cilia, there is a clear difference; some animals have, as adults, remnants of the kinocilium. In the rhesus monkey, for instance, about 30% of the cells in the middle portion of the cochlea still have remnants of the kinocilium sticking up from the surface. In the lower portion, they are very seldom present; in the top they are also seldom present. So there is a reduction there; how long that goes on I cannot answer. As to the number of synaptic structures in very young animals and older animals, I don't know that anyone can answer that question yet.

WIEDERHOLD: I guess I'll follow up on Jay Goldberg's question, at the risk of getting too interdisciplinary. Could you comment on synaptic bars or accumulation of vesicles on cochlear outer hair cells?

ENGSTRÖM: You mean about the accumulation of them under certain different circumstances?

GOLDBERG: Where do you see them? It has been reported that synaptic bars are not present in the outer hair cells of the cat.

ENGSTRÖM: Oh, they are very common in outer hair cells. But they are not very evident. There is nothing compared to the very long ones I saw in squirrel monkeys. But there is a small synaptic bar in very many outer hair cells. In those animals I have been studying, it's quite common to see a synaptic bar. Sometimes it lies very close to the synaptic membrane.

WIEDERHOLD: The cat is one of the animals we have used infrequently. But we've used many monkeys. It's very common in squirrel monkeys, in rhesus monkeys, and in guinea pigs.

HOWLAND: I'd like to return for just a second to the otoconia and ask you about them in relation to the work of Lim. Have you any comments on the organic matrix of otoconia? Have you seen any shrunken otoconia in any of your preparations?

ENGSTRÖM: This is a rather difficult question. It seems that for many years the structure of the statoconia has been discussed. Igarashi and Muriel Ross have been studying it also. There is some kind of arrangement with a central core, a kind of nucleus described since, I think, 50 years ago. We have also found a center in the statoconia. And there is some kind of matrix also. I think that the papers by Lim, Ross, Igarashi, and some others can answer your question.

2

The Afferent and Efferent Vestibular Pathways: Morphologic Aspects

RICHARD R. GACEK

Our knowledge of the course and connections of the neurons serving vestibular reflexes has been greatly expanded over the last ten years by the use of modern neuroanatomic techniques. Anterograde methods utilizing the phenomenon of wallerian degeneration have been employed to demonstrate the termination of first order vestibular neurons in the vestibular nuclei and the projection of second order neurons in the vestibular nuclei serving vestibuloocular and vestibulospinal reflexes (3,18). These anterograde techniques employed selective silver impregnation methods (the Nauta method) to trace fine axon terminals in the central nervous system and a myelin stain (Sudan black) to overcome technical difficulties involved in the study of neurons surrounded by the temporal bone. Although these methods successfully demonstrated the course and termination of first and second order vestibular neurons, the development of a reliable retrograde technique was necessary to identify the precise location of cells within the vestibular nuclei that project to the effector organs. The retrograde axoplasmic tracer method utilizing horseradish peroxidase has fulfilled this need and has been utilized widely as a standard neuroanatomic technique (14). It has been a great aid in the elucidation of the precise location of central nervous system neurons. This discussion will review the neuroanatomic features of the vestibular system in the cat that have been revealed by these techniques.

FIRST ORDER AFFERENT NEURONS

In the cat there are approximately 12,000 myelinated first order vestibular neurons. The myelinated axons of these first order afferents measure from 1 to 10 μm in diameter with the vast majority (88%) measuring 1 to 5 μm. The posterior and superior canal ampullary nerves each contain approximately 2500 fibers, the lateral canal ampullary nerve about 2200 fibers, the utricular nerve about 2700, and the saccular nerve 1800 fibers (10). Since these true bipolar neurons resist retrograde degeneration, lesions must be placed in the ganglion in order to produce wallerian degeneration of both the peripheral and central processes. Such degeneration techniques have been performed to map out the peripheral course and the central termination of the first order afferents supplying the vestibular sense organs (Fig. 2-1). These experiments (8) show that the afferents from all the canal sense organs converge in the rostral half of the vestibular nerve before entering the brainstem. The afferents from the posterior canal crista are caudal to those serving the cristae

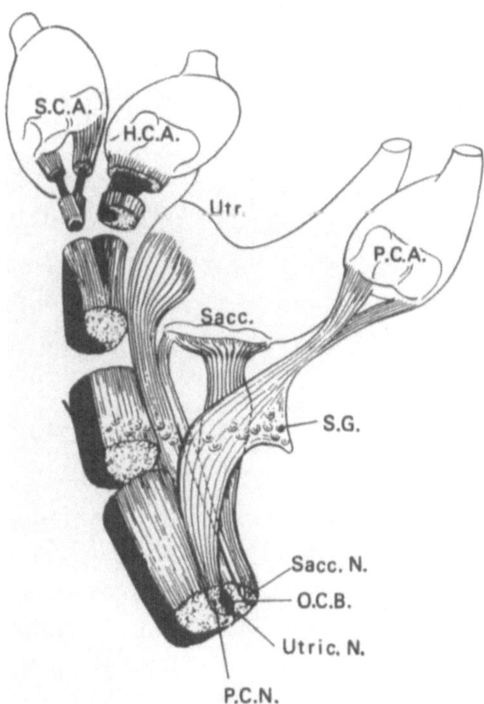

Fig. 2-1. The location of first order neurons supplying the vestibular sense organs in the cat. Dark areas represent location of thick fibers in the superior vestibular division. *P.C.N.*, posterior canal nerve; *O.C.B.*, olivocochlear and vestibular efferent bundle; *S.C.A.*, superior canal crista; *H.C.A.*, horizontal canal crista; *P.C.A.*, posterior canal crista; *Utr.*, utricle; *Sacc.*, saccule; *S.G.*, Scarpa's ganglion; *Sacc. N.*, saccular nerve; *Utric. N.*, utricular nerve.

of the lateral and superior canals. Afferents from the utricular and saccular maculae course in the caudal one-half or one-third of the vestibular nerve prior to penetrating the brainstem.

A closer analysis of the wide spectrum of fiber diameters in the first order vestibular neurons reveals that there are two populations of canal afferents: a small population of large-caliber fibers and a larger population of the small-size myelinated fibers. There is a consistent pattern to the distribution of the large- and small-caliber fibers in the cristae. The large fibers occupy the center portion of the ampullary nerve branch innervating the cristae, whereas small fibers are concentrated at the periphery of the nerve (Fig. 2-2). The large fibers penetrate the top of the crista where they terminate as large calyciform endings on type I hair cells (17,19), whereas the small afferent fibers terminate on the type II hair cells by means of bouton-type endings (Fig. 2-3). Large and small fibers also exist in the macular nerves but their arrangement is more diffuse. However, large fibers there also terminate on type I hair cells with the smaller ones innervating type II hair cells.

Since the ganglion cells of the large afferents to the cristae supplied by the superior division are located in the rostral portion of the vestibular ganglion and the small afferent ganglion cells are located caudally, it is possible to selectively lesion these two areas and to make observations on the central

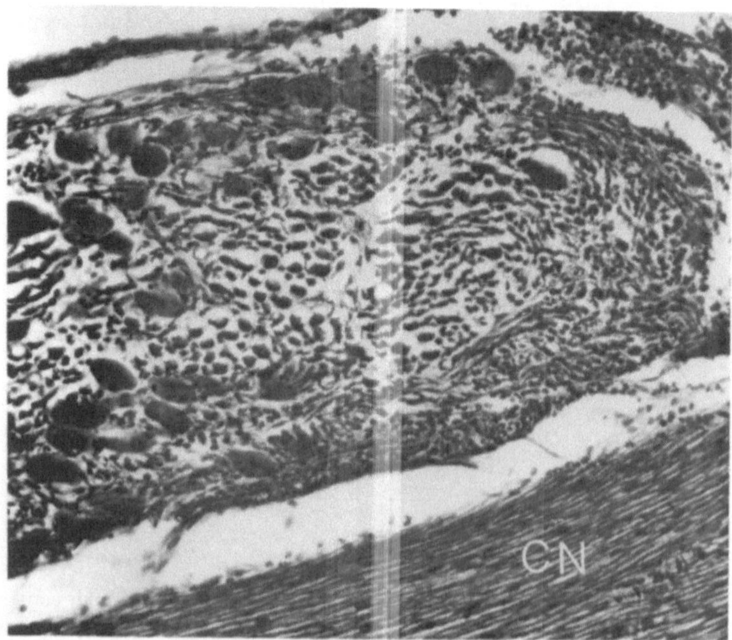

Fig. 2-2. Photomicrograph of the posterior ampullary nerve as it crosses the cochlear nerve (*CN*), demonstrating the central location of thick axons with smaller axons forming the periphery of the nerve.

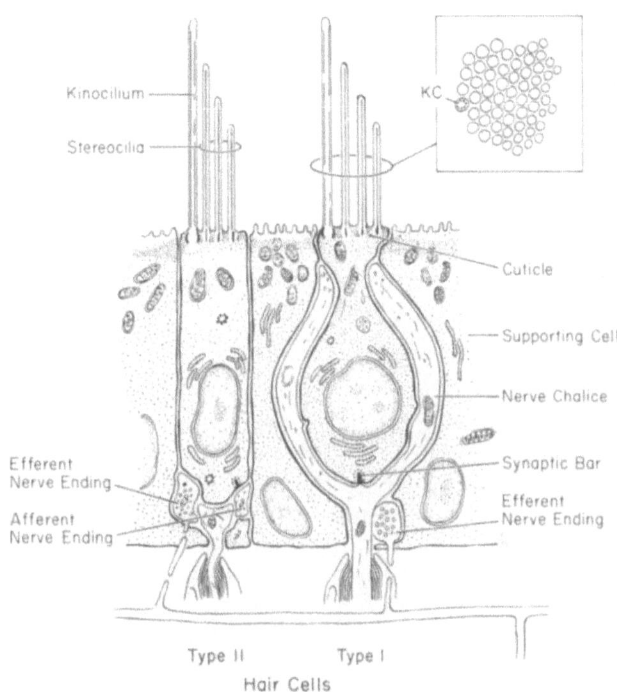

Fig. 2-3. Type I and II vestibular hair cells, with their afferent and efferent innervation.

termination and course of these two fiber populations to the superior division cristae (8). It is not possible to similarly study large and small fibers to the maculae.

CENTRAL TERMINATIONS OF FIRST ORDER VESTIBULAR AFFERENTS

The central terminations of first order vestibular afferents are concerned with the four major vestibular nuclei (superior, medial, lateral, and descending) as well as two minor vestibular nuclei (the interstitial nucleus of the vestibular nerve and the group Y nucleus). A review of some salient anatomic features of these nuclei will be helpful in understanding the termination of first order afferents as well as the location of second order vestibular neurons.

Major Vestibular Nuclei

Superior Vestibular Nucleus

The superior vestibular nucleus is well delineated and the most rostral of the major vestibular nuclei. It is characterized by the presence of heavily

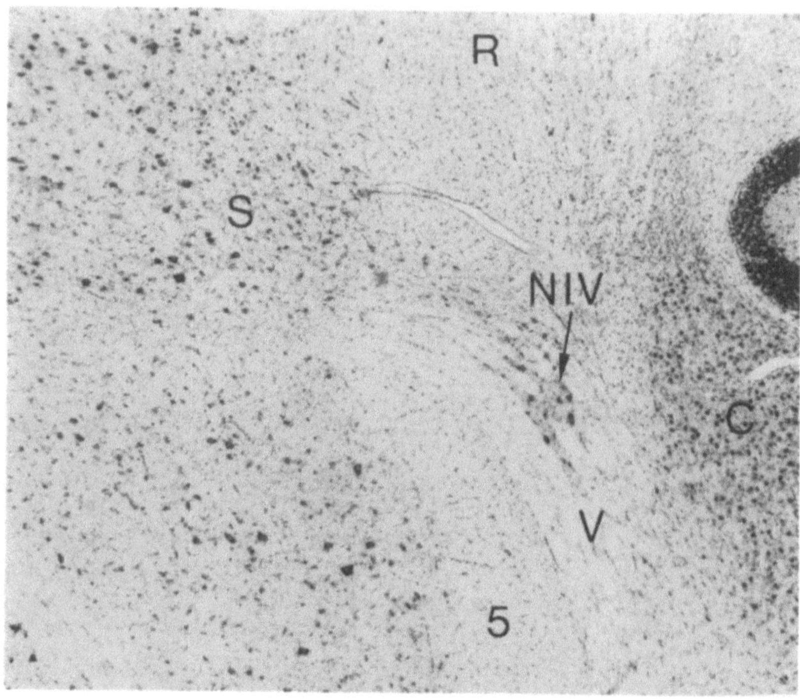

Fig. 2-4. Vestibular nerve root (*V*) as it enters the superior vestibular nucleus (*S*). Note the larger neurons in the central region of the nucleus with smaller neurons located laterally. *NIV*, interstitial nucleus of the vestibular nerve; *R*, restiform body; *C*, cochlear bundle; *5*, descending trigeminal root.

myelinated fiber bundles (of primary afferents) coursing from ventrolaterally to dorsomedially through the center of the nucleus. Large and medium-size neurons with heavy Nissl substance surround these fiber bundles in the central portions of this nucleus. The peripheral regions, particularly laterally and ventromedially, are densely packed with small neurons (Fig. 2-4).

Lateral Vestibular Nucleus

The lateral vestibular nucleus is easily recognized by the large multipolar neurons with heavy Nissl substance that populate this largest of the vestibular nuclei. It extends from the superior vestibular nucleus to the beginning of the medial and descending vestibular nuclei. Medially, it is bounded by the dorsal acoustic stria which separates it from the rostral portion of the medial nucleus. Laterally, it is limited by the restiform body and penetrated by the incoming vestibular nerve root. The large and medium-size neurons which comprise this nucleus easily distinguish it from the bordering vestibular nuclei. It can be subdivided into a larger dorsal and a smaller ventral division (Fig. 2-5). The dorsal division has a massive cerebellar input but does not receive any first order vestibular afferents, whereas the smaller ventral division receives the primary input from the periphery.

Medial Vestibular Nucleus

The medial vestibular nucleus is probably the most complex of the vestibular nuclei and has a large triangular portion which extends from the caudal margin of the lateral nucleus to a tapered end in the caudal brainstem. It also has a rostral extension which forms a narrow strip of neurons between the fourth ventricle and dorsal acoustic stria and reaches rostrally into the superior vestibular nucleus (Fig. 2-5). This portion of the medial nucleus is populated by scattered medium-size neurons and packed densely with small neurons whose processes are aligned in a rostrocaudal direction. The main body of the medial nucleus is densely populated by both medium and small neurons.

Descending Vestibular Nucleus

The descending vestibular nucleus is also triangular shaped and parallels the body of the medial nucleus, but lies lateral to it. It is characterized by

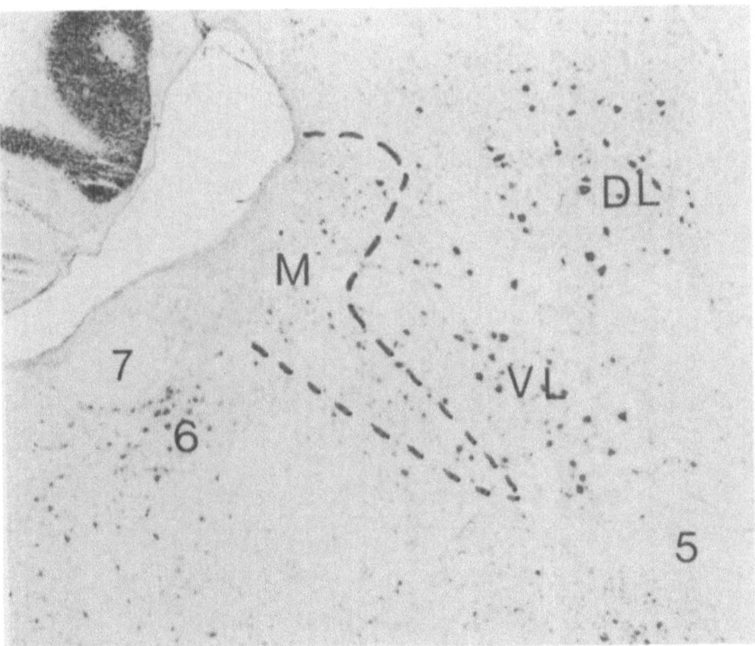

Fig. 2-5. Transverse section through the medial (*M*) and lateral (*DL* and *VL*) vestibular nuclei. The portion of the medial nucleus which undercuts the ventral division of the lateral nucleus (*VL*) contributes to the ascending tract of Deiters. *7*, facial nerve genu; *6*, abducens nucleus; *5*, descending trigeminal root.

medium- and large-size neurons separated by longitudinally running fiber bundles of the descending vestibular root.

Minor Vestibular Nuclei

The Interstitial Nucleus of the Vestibular Nerve

The interstitial nucleus is a small, compact nucleus which has two divisions and which is embedded in the fiber bundles of the incoming vestibular root as it crosses over the descending trigeminal nerve and nucleus (Fig. 2-4). It is made up almost uniformly of medium- and large-size neurons.

The Group Y Nucleus

The group Y nucleus (2) is a compact triangular nucleus of small spindle-shaped neurons oriented in a medial-lateral direction, located between the lateral vestibular nucleus and the restiform body (Fig. 2-6). The dorsal acoustic stria forms a cap over this nucleus. Immediately above this nucleus are loosely arranged medium-size neurons which form an ill-defined nucleus

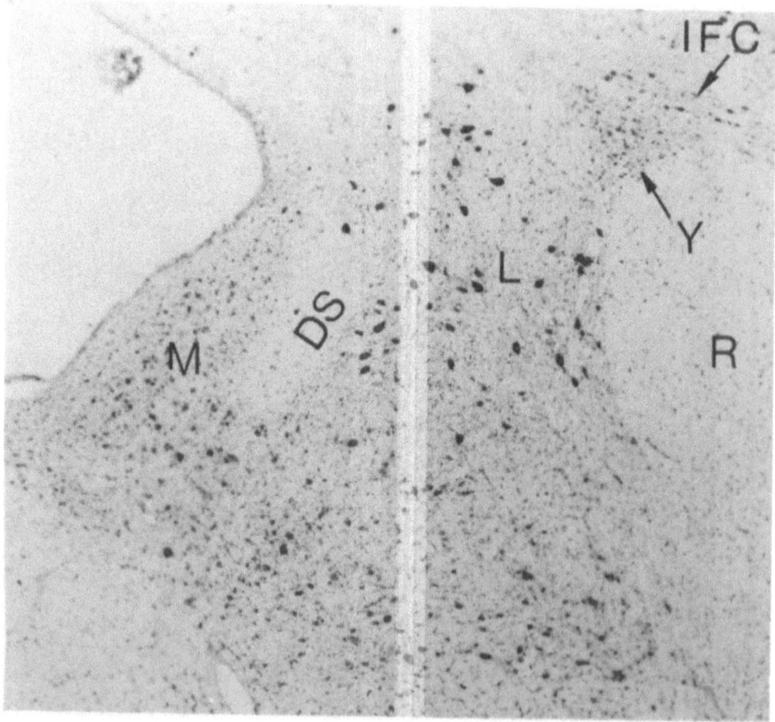

Fig. 2-6. This transverse section through the lateral (*L*) and medial (*M*) vestibular nuclei also shows the location of the infracerebellar (*IFC*) and group Y (*Y*) nuclei. *R*, restiform body; *DS*, dorsal acoustic stria.

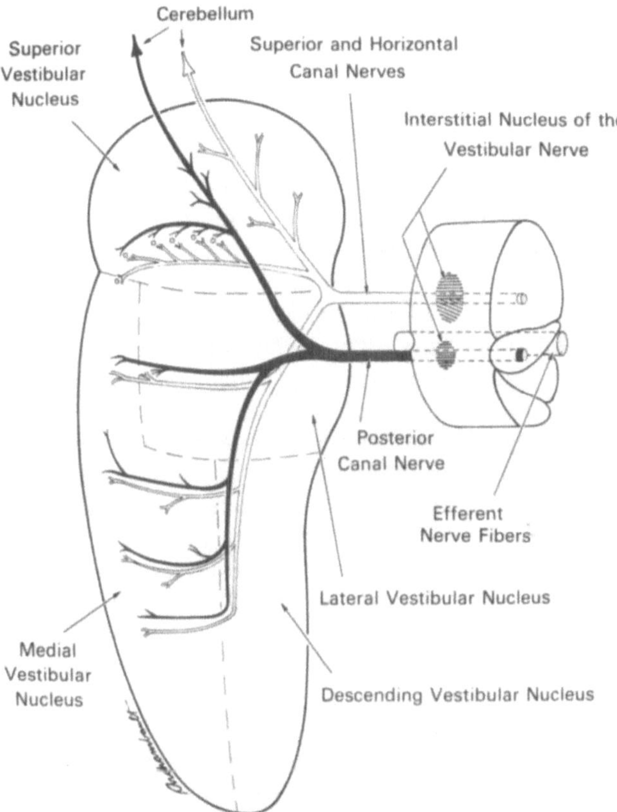

Fig. 2-7. Schematic of the central termination of canal afferents.

with a tenuous connection to the dentate nucleus. This aggregate of cells is called the infracerebellar nucleus and is embedded within the fiber bundles of the fasciculus angularis.

As the primary canal afferents enter the brainstem in the rostral one-half of the vestibular nerve, they give off rich collateral terminations in the interstitial nucleus of the vestibular nerve (8). The fibers belonging to the superior division cristae and those from the posterior canal crista give off collaterals to two separate portions of the interstitial nucleus (Fig. 2-7). The posterior canal crista afferents travel caudal to the afferents from the other cristae and penetrate the lateral vestibular nucleus more medially than the other canal afferents. At this point, all canal afferents bifurcate with an ascending branch coursing through the superior vestibular nucleus in a dorsomedial direction. The posterior canal fibers terminate around large and medium cells in the caudal and central portion of the superior nucleus, whereas those from the superior division terminate more rostrally. However, all ascending branches from the canal afferents give off long collaterals in the caudal portion of the superior nucleus which reach to cells in the ventromedial region of the superior nucleus. Collaterals from all canal afferents converge on large neurons which are located here. The ascending rami of canal afferents then continue on through the superior nucleus and penetrate the brachium conjunctivum in two or three groups of fibers to reach vestibular portions of the cerebellum. The smaller primary afferents from the canal sense organs terminate more laterally and ventrally in the superior nucleus where small cells are concentrated. They also give off fine collaterals which are directed toward the medial portion of the superior nucleus. The continuations of these small afferents also have a cerebellar destination.

The descending branches of primary canal afferents turn in a sharply caudal direction in the descending vestibular root (Fig. 2-7). They first give off collaterals which penetrate the lateral vestibular nucleus in a medial direction to terminate on cells in the rostral extension of the medial nucleus. As these descending rami reach the main body of the medial nucleus, they give off numerous collaterals which terminate in all portions of the medial vestibular nucleus. Within the descending vestibular root, the fibers of the posterior canal afferents are located dorsal to those from the superior division canal afferents. The collateral branches from these descending rami also travel across the medial nucleus at dorsal and ventral levels respectively. However, they appear to converge on the same cells or groups of cells within this nucleus.

The primary afferents from the utricular macula bifurcate within the lateral vestibular nucleus (Fig. 2-8). The ascending branch traverses the ventral division of the lateral nucleus making rich contact with the large neurons located in this region and the medium-size neurons in the rostral extension of the medial nucleus. These branches do not continue into the cerebellum. The descending branch traveling in the lateral part of the descending vestibular root makes contact with cells in the medial nucleus by way of collateral branches. The utricular neurons do not give off any terminals to the interstitial nucleus of the vestibular nerve.

Finally, the primary saccular afferents reach the brainstem in the caudal edge of the vestibular nerve before bifurcating; the ascending branch then rises directly to terminate in the group Y nucleus, whereas the descending branch arches in a medial and caudal direction to terminate in the medial vestibular nucleus as well as some portions of the lateral nucleus (Fig. 2-8). Saccular afferents also do not form any connections with the interstitial nucleus in the vestibular nerve. The medial vestibular nucleus is the one area where collaterals of descending branches from all canal and otolith organ afferents converge on second order vestibular neurons.

VESTIBULOOCULAR PROJECTIONS

The vestibuloocular (VO) projections have been studied in considerable detail by both anterograde (3,18) and retrograde (5,6,11,15) methods. Several

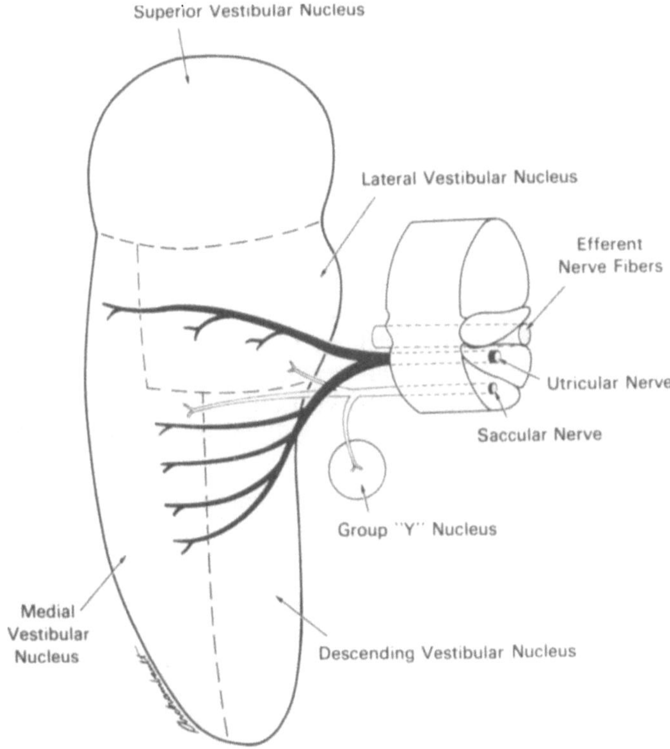

Fig. 2-8. Schematic of the central termination of otolith afferents.

new features regarding these projections have been revealed and are to be emphasized in this discussion.

1) Although the vast majority of the VO neurons are located in the major vestibular nuclei (superior and medial vestibular nuclei, lateral vestibular nucleus), several smaller neuron groups have been shown to project to the extraocular nuclei (6,11,15). These are the nucleus praepositus hypoglossi, the infracerebellar nucleus, and the interneuron in the abducens nucleus.

2) Vestibuloocular neurons serving horizontal eye movements by supplying the medial and lateral recti muscles are located caudally in the brainstem, whereas those VO neurons serving vertical and rotatory eye movements are located in the superior and medial vestibular nuclei (5).

3) The location of VO neurons in the superior nucleus shows some degree of organization.

Horizontal Eye Movements

Horizontal eye movements are served by the medial and lateral recti muscles which are innervated by the abducens nucleus and the medial rectus subgroup of the oculomotor nucleus. It is now quite clear that the abducens nucleus receives bilateral afferent input from both medial vestibular nuclei (5,15). The ipsilateral input is somewhat greater than the contralateral (Fig. 2-9). The neurons serving this projection are medium-size neurons located in those portions of the medial nucleus which are adjacent to the dorsal acoustic stria. These include the rostral extension of the medial nucleus and that part of the medial nucleus which projects laterally from the dorsal acoustic stria undercutting the ventral division of the lateral nucleus. Some abducens VO neurons from the rostral extension of the medial nucleus are located in the medial region of the superior nucleus where they can be contacted by long collaterals from incoming primary canal afferents entering the superior nucleus. It is possible that a small number of first order vestibular ganglion cells in the superior division of the vestibular nerve project directly to the abducens nucleus (5). Most of these appear located in that part of the ganglion which serves the utricular macula.

The medial rectus subgroup in the oculomotor nucleus is supplied by ascending afferents located (1) in the abducens nucleus (6,11,13) and (2) in the region of the medial nucleus (3,6) which undercuts the ventral division of the lateral vestibular nucleus (Figs. 2-5 and 2-9). The interneuron located within

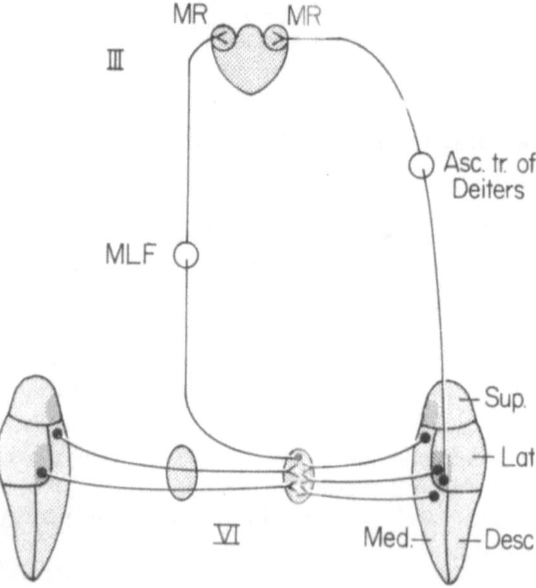

Fig. 2-9. Schematic of vestibuloocular neurons serving the lateral and medial recti muscles. Heavy stippling designates extension of the medial nucleus into the region of adjacent superior and lateral nuclei.

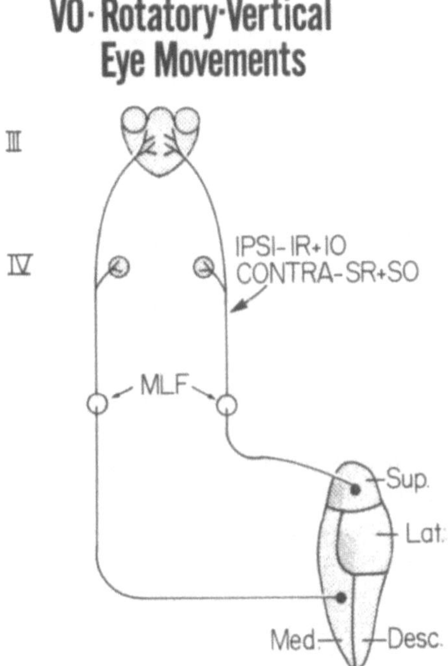

Fig. 2-10. Schematic of vestibuloocular neurons serving the vertical and oblique extraocular muscles.

the abducens nucleus projects contralaterally in the medial longitudinal fasciculus (MLF) to terminate in the medial rectus subgroup. The pathway from the medial vestibular nucleus ascends ipsilaterally along with a small contribution from the ventral portion of the lateral nucleus as the "ascending tract of Deiters" to supply the medial rectus subgroup (3,6).

Other afferents to the abducens nucleus are the nucleus praepositus hypoglossi and the contralateral reticular formation (5,15). The role of these latter neurons is unclear, whereas the bilateral contributions from the medial vestibular nuclei are likely inhibitory or excitatory in function (1).

Rotatory and Vertical Movements

Neurons serving rotatory and vertical movements are those which project to the trochlear nucleus and the subgroups of the oculomotor nucleus which serve the superior and inferior recti and the inferior oblique muscles. The VO neurons to these motor neuron groups are located in the superior and medial vestibular nuclei (3,12,18) and project through the MLF (Fig. 2-10). The pathway arising in the superior nucleus is ipsilateral, whereas the medial nucleus projects contralaterally to these extraocular nuclei. Evidence from retrograde tracer studies suggests that the VO neurons to these eye nuclei are organized differentially in the superior nucleus (SN). VO neurons to the subgroups of the oculomotor nucleus are located bilaterally in the SN, whereas those to the trochlear nucleus are located only ipsilaterally in the SN.

1) The VO neurons to the superior rectus subgroup are ipsilaterally dispersed throughout the central and rostral regions of the superior nucleus. However, a layer of neurons in the contralateral superior nucleus just ventral to the brachium conjunctivum (21) also projects to the superior rectus motoneuron pool (Fig. 2-11). The latter group probably projects by way of the brachium conjunctivum (21).

2) Trochlear VO neurons are located only ipsilaterally in the superior nucleus. They also occupy the stratum of neurons ventral to the brachium conjunctivum, but in addition they are diffusely arranged in the central region of the rostral half of the superior nucleus (Figs. 2-12 and 2-13).

3) VO neurons supplying the inferior rectus and inferior oblique are compactly arranged in the caudal and medial portions of the superior nucleus (Fig. 2-14A and B) at the dorsal levels, but are diffusely arranged in central portions of the nucleus at ventral levels (Fig. 2-15). The intermixing of these VO neurons in the central zone of the superior nucleus allows for contact by primary afferents from more than one crista. The contralateral VO supply to

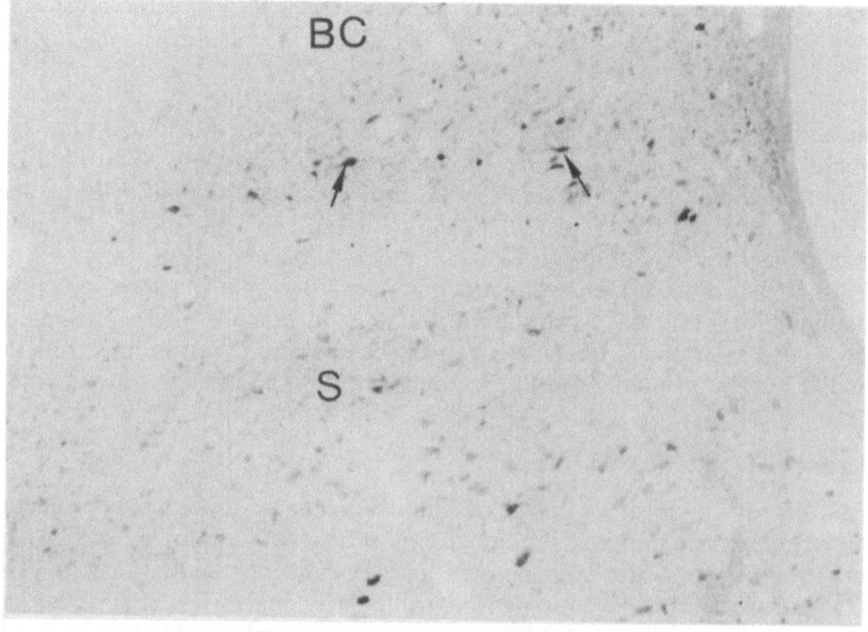

Fig. 2-11. Photomicrograph of VO neurons (*arrows*) in the contralateral superior nucleus (*S*) labeled after injection of horseradish peroxidase into the superior rectus subgroup of the oculomotor nucleus. *BC*, brachium conjunctivum.

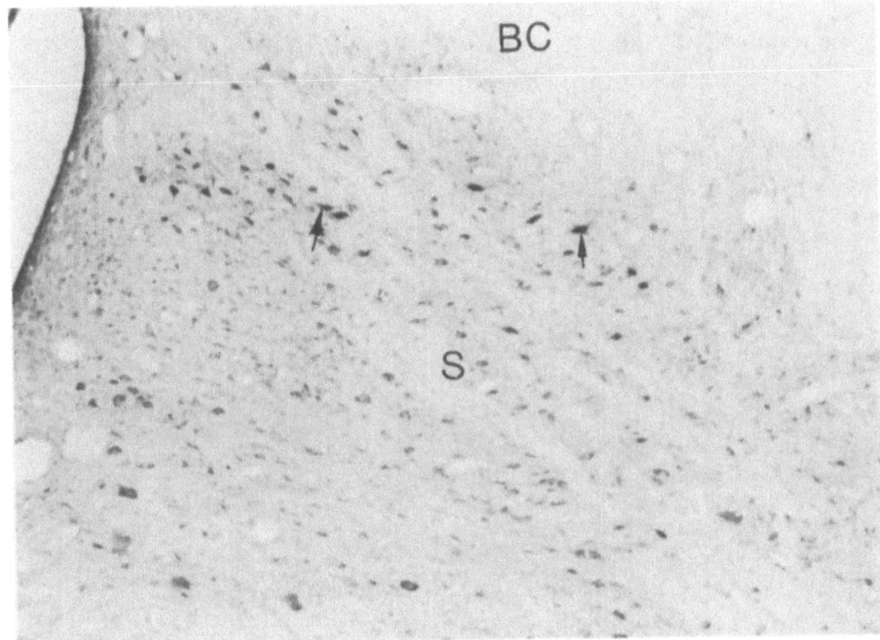

Fig. 2-12. VO neurons (*arrows*) in the caudal part of the ipsilateral superior nucleus (*S*) labeled after injection of horseradish peroxidase into the trochlear nucleus.

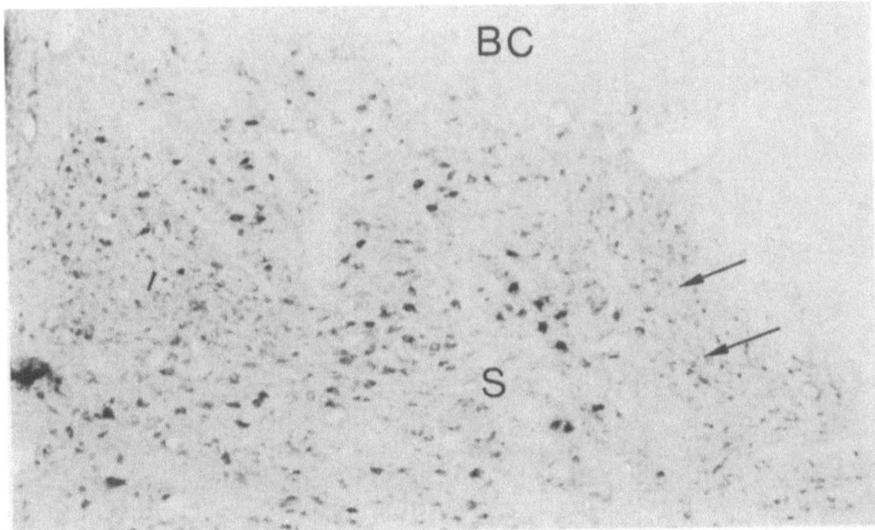

Fig. 2-13. Labeled VO neurons in more rostral sections of the superior nucleus (*S*) are located centrally. The small neurons (*arrows*) in peripheral zones of the nucleus are unlabeled because they are not vestibuloocular.

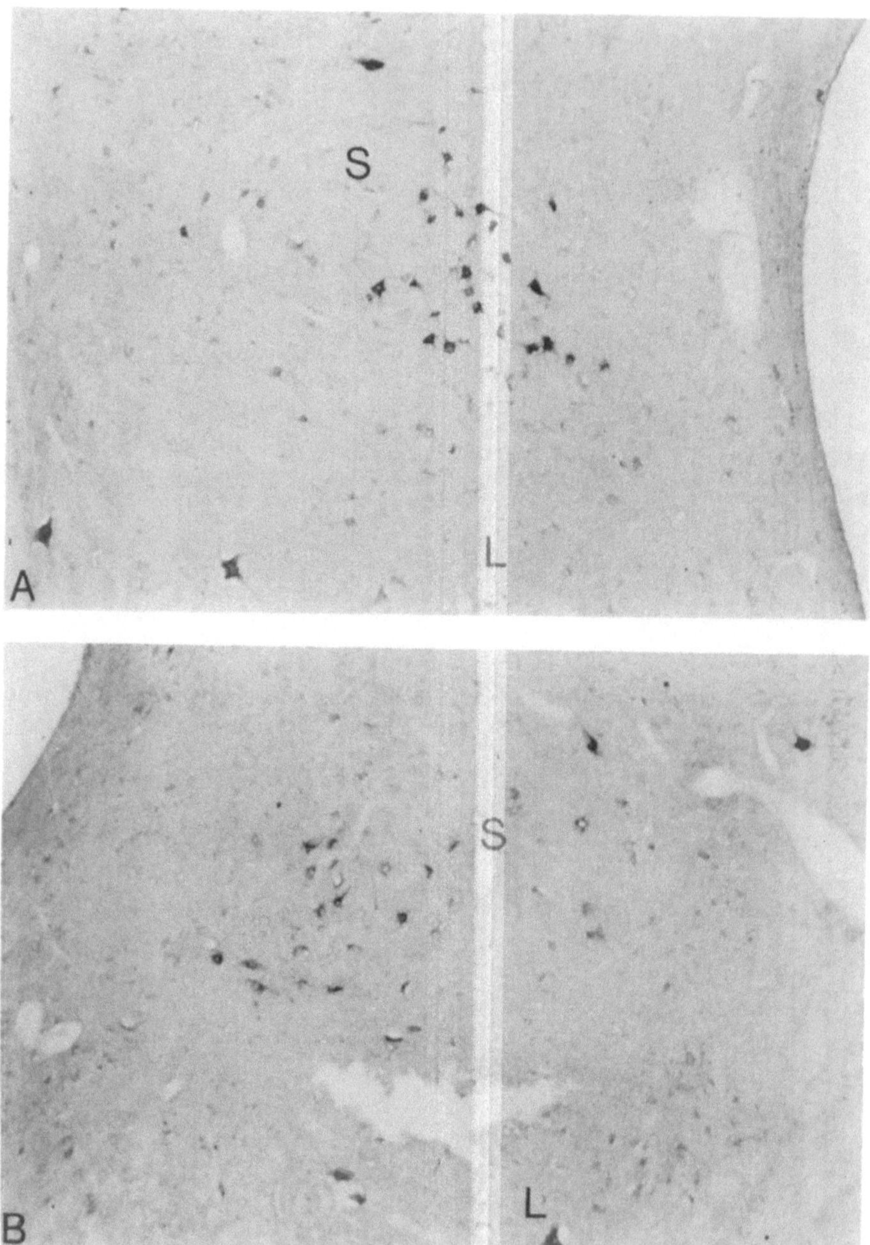

Fig. 2-14. **A** Horizontal section shows VO neurons labeled in the caudomedial part of the superior nucleus (*S*) after injection of tracer into the inferior rectus and inferior oblique subgroups of the oculomotor nucleus. These ipsilateral cells are more intensely labeled than are the contralateral cells in **B**. *L*, lateral vestibular nucleus. **B** Similar location of labeled VO neurons in the contralateral superior nucleus in same animal as in **A**. Note lesser intensity of labeling.

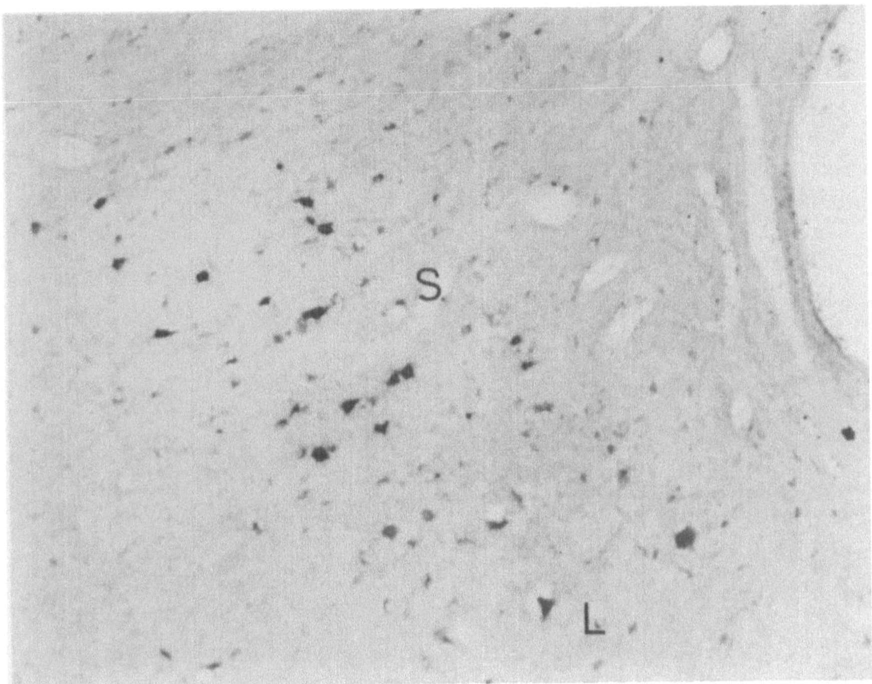

Fig. 2-15. Horizontal section at a ventral level shows that labeled VO neurons are diffusely arranged in the superior nucleus (*S*) following placement of tracer in the inferior rectus and inferior oblique subgroups.

the fourth and third nuclei arises mainly from that part of the medial vestibular nucleus which is caudal to the dorsal acoustic stria. Most functional studies have indicated that the contralateral arc to these nuclei is an excitatory one whereas the ipsilateral one is inhibitory (12).

The ipsilateral nucleus praepositus hypoglossi and the infracerebellar nucleus (Fig. 2-16) also project to the third nucleus by way of the MLF and the fasciculus angularis respectively (6). Since these two minor nuclei may be contacted by primary afferents from the canals and the saccule, they play some role in the VO response. In addition to these inputs, the oculomotor nucleus receives a direct input from the nucleus interpositus of the cerebellum (6). This input travels to the contralateral superior rectus subgroup by way of the brachium conjunctivum (3).

VESTIBULOSPINAL PROJECTIONS

Vestibulospinal projections are provided by the lateral, medial, and descending vestibular nuclei (Fig. 2-17). The lateral vestibular nucleus is the source of the most prominent of these descending projections (2,3,20). The dorsal

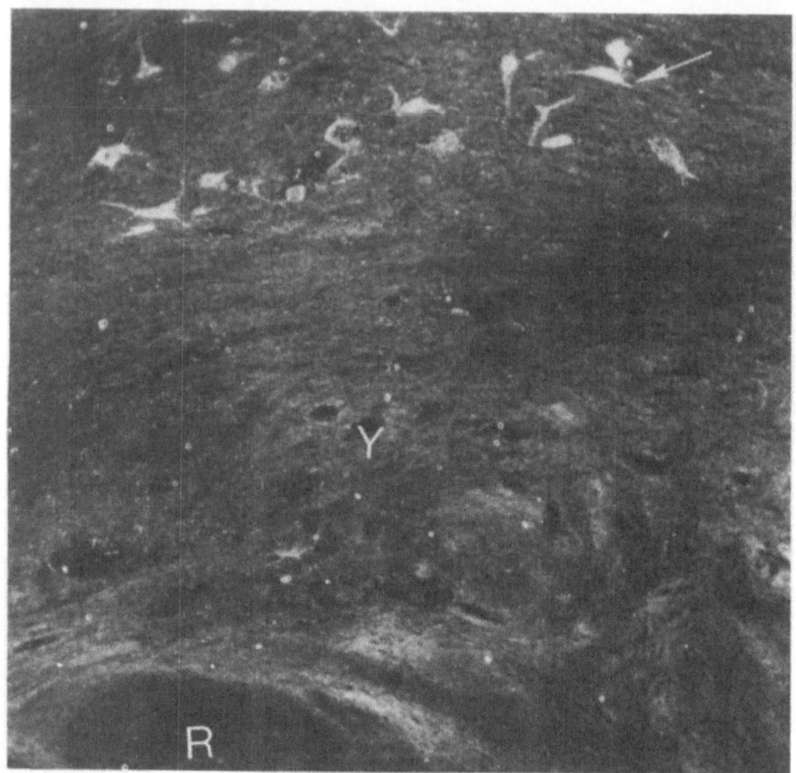

Fig. 2-16. Dark field photomicrograph of a transverse section showing labeled infracerebellar nucleus neurons (*arrow*) following injection of tracer in the ipsilateral oculomotor nucleus. *Y*, group Y nucleus; *R*, restiform body.

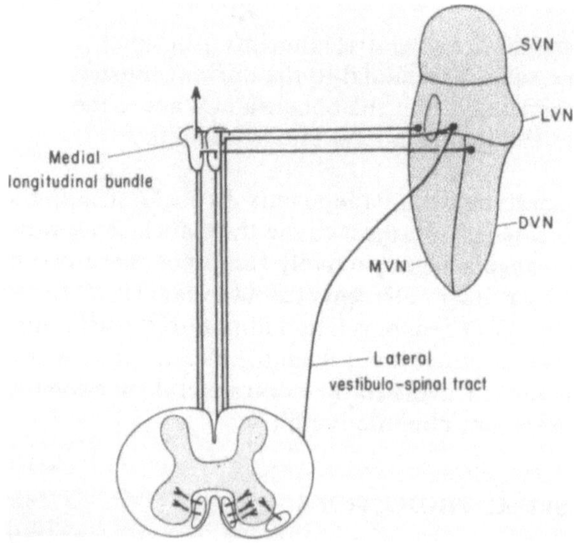

Fig. 2-17. Schematic summarizing the contributions of the vestibular nuclei to the vestibulospinal tracts.

division of the lateral vestibular nucleus is responsible for that part of the lateral vestibulospinal tract which descends to all levels of the spinal cord. The smaller ventral division of the lateral vestibular nucleus projects to the cervical and upper thoracic levels of the spinal cord by way of the lateral vestibulospinal tract, but also projects into the medial longitudinal fasciculus as the medial vestibulospinal tract which descends to cervical cord levels. Contributions to the medial vestibulospinal tract (in the medial longitudinal fasciculus) are also provided by the medial and descending vestibular nuclei (3,20).

COMMISSURAL PROJECTIONS

Retrograde tracer studies (7) have shown that commissural vestibular neurons are located primarily in the superior and medial vestibular nuclei and the group Y nucleus (Figs. 2-18 and 2-19). Small neurons in the lateral and ventral portions of the superior nucleus, the rostral extension of the medial nucleus, and most of the caudal portion of the medial nucleus provide the commissural projections. No commissural neurons are located in the lateral vestibular nucleus and very few, if any, are found in the descending vestibular nucleus (7). The caudal commissural neurons located in the medial vestibular nucleus directly cross the midline a few millimeters below the floor of the fourth ventricle. The commissural units in the rostral extension of the medial nucleus, the superior vestibular nucleus, and the group Y nucleus traverse the midline in the commissure of the superior vestibular nuclei to reach the contralateral vestibular complex. Their location in the superior and medial vestibular nuclei permits these commissural neurons to receive their input from the primary canal afferents. Long collaterals from the descending rami of canal afferents contact commissural neurons in the ros-

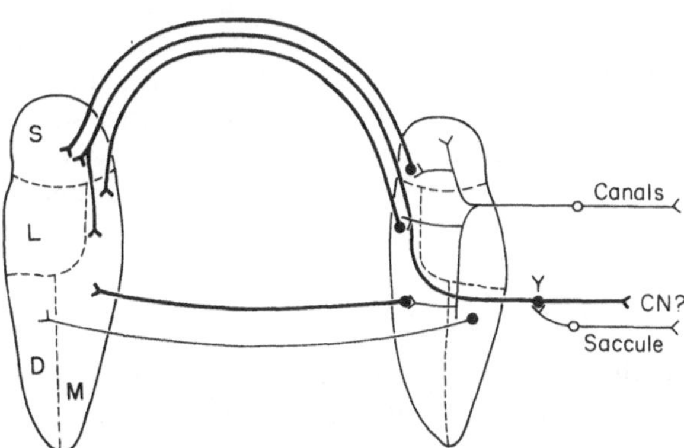

Fig. 2-18. Schematic of the location and projection of commissural neurons in the vestibular nuclei. Probable peripheral inputs to these units are represented at the right. *CN*, cochlear nucleus.

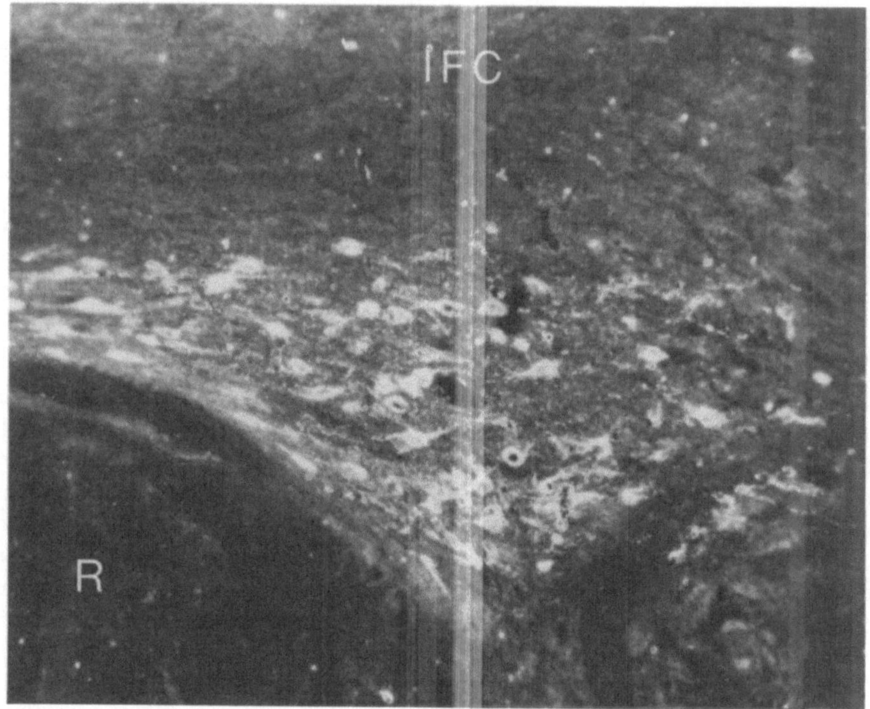

Fig. 2-19. Dark field photomicrograph of transverse section showing labeled commissural nerurons in group Y nucleus. *IFC*, infracerebellar nucleus; *R*, restiform body.

tral medial nucleus extension, whereas the commissural neurons in the periphery of the superior nucleus are probably contacted by canal afferents projecting to this region. Primary canal afferents projecting to the periphery of the superior nucleus are the small-caliber afferents supplying the type II hair cells in the cristae. It is thus possible that primary vestibular afferents supplying type I and type II hair cells in the cristae may project to functionally different populations of the neurons in the superior vestibular nucleus. These are, respectively, the large vestibuloocular neurons in the central portion of the nucleus and the small commissural neurons in the peripheral zone of the nucleus.

EFFERENT VESTIBULAR NEURONS

The efferent vestibular pathway has been firmly established by numerous anterograde (4) and retrograde (9) studies. These approaches have been complementary in describing the pathway from origin to termination. The efferent vestibular pathway has a bilateral origin from small neurons located lateral to the abducens nucleus and ventral to the medial vestibular nucleus (Fig. 2-20). These small cells give off approximately 300 small-caliber (1 to 3 μm) myelinated axons to each labyrinth. The fibers merge with the

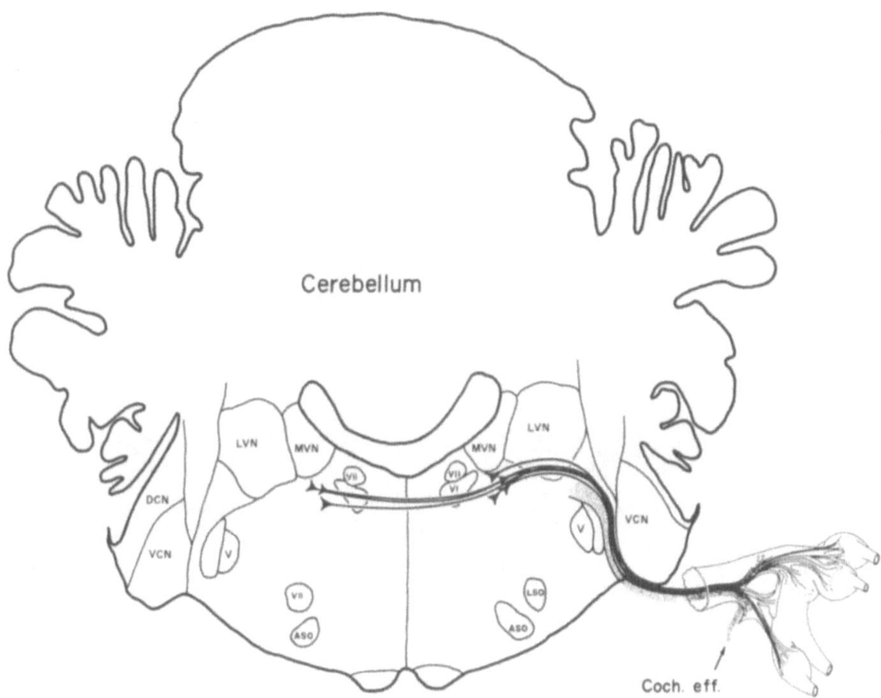

Fig. 2-20. Drawing of the origin and course of the vestibular efferent pathway.

olivocochlear efferent fibers before joining the vestibular nerve root in the brainstem. As a compact bundle of labyrinthine efferent fibers, they emerge from the brainstem in the central and ventral portion of the vestibular nerve (Fig. 2-21). When they reach the level of the vestibular ganglion, the olivocochlear nerve fibers leave in the vestibulocochlear anastomosis in order to enter the cochlea. Vestibular efferent fibers at this point disperse as scattered fibers or groups of fibers into the superior and inferior divisions and finally into the individual nerve branches supplying vestibular sense organs (Fig. 2-22). These efferent fibers become scattered among the afferent fibers as they proceed peripherally to a sense organ and may form branching patterns before approaching the neuroepithelium. Upon penetrating the basement membrane, they branch more abundantly and provide contacts in the form of vesiculated bouton-type terminals at the base of type II hair cells and on the afferent calyciform endings of type I hair cells in all vestibular sense organs (Fig. 2-3).

DESCENDING NEURONS TO THE VESTIBULAR NUCLEI

A number of descending neuronal projections from the diencephalon have been demonstrated following the placement of the horseradish peroxidase tracer in various areas of the vestibular (7) and the abducens (5,16) nuclei.

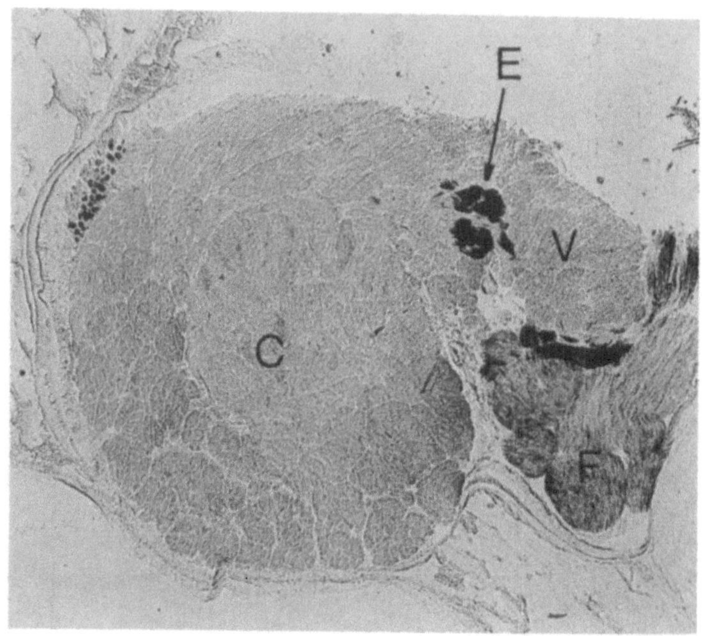

Fig. 2-21. Cross-section of the internal auditory canal showing acetylcholinesterase localization in the efferent cochlear and vestibular bundle (*E*) within the vestibular nerve (*V*). *C*, cochlear nerve; *F*, facial nerve.

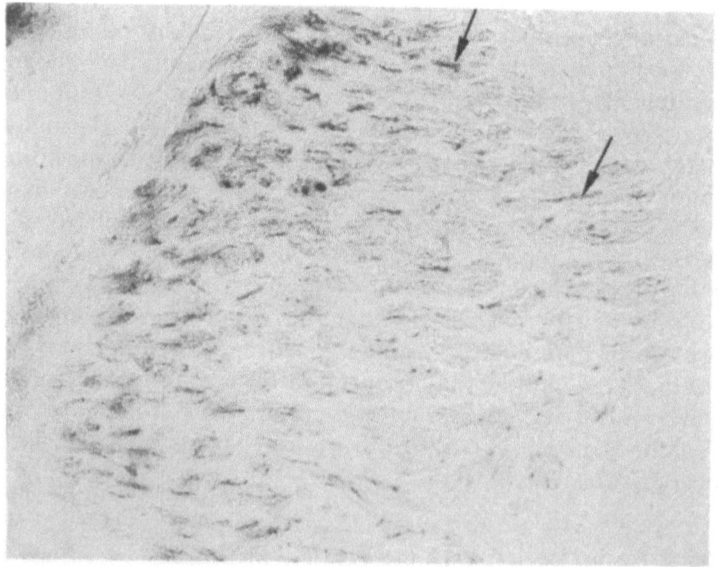

Fig. 2-22. Scattered efferent fibers (*arrows*) in a peripheral vestibular branch are demonstrated by the acetylcholinesterase method.

1) Following placement of the tracer into the abducens nucleus, a small number of labeled neurons have been consistently identified within the oculomotor nucleus, particularly near the medial rectus subgroup. This is interpreted as representing a complementary interneuron from the medial rectus subgroup to the abducens nucleus, forming a reciprocal to the interneuron located in the abducens nucleus. The same experiment revealed numerous small neurons labeled dorsal to the periphery of the oculomotor nucleus and scattered into the central gray substance around the aqueduct of Sylvius.

2) Following injections of peroxidase material into the vestibular nuclei, a substantial number of labeled neurons have been regularly found in the Edinger-Westphal nucleus of the oculomotor nucleus (Fig. 2-23), in the interstitial nucleus of Cajal (Fig. 2-24), and in the nucleus of Darkshevich. These descending neuron links provide evidence of the interaction of visual centers with the vestibular complex.

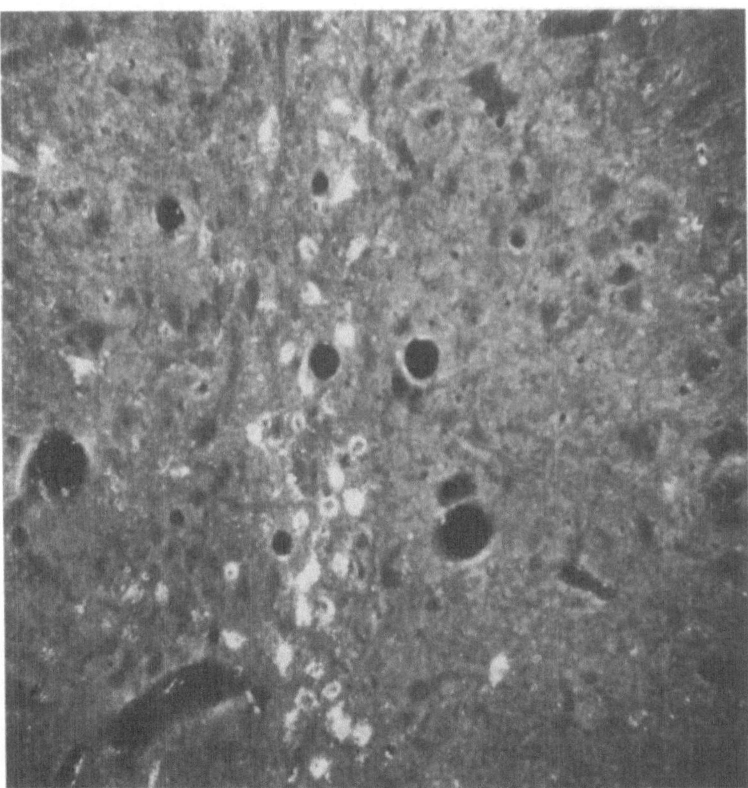

Fig. 2-23. Horizontal section through the oculomotor nucleus showing labeled neurons in the Edinger-Westphal nucleus following injection of tracer into the vestibular nuclei.

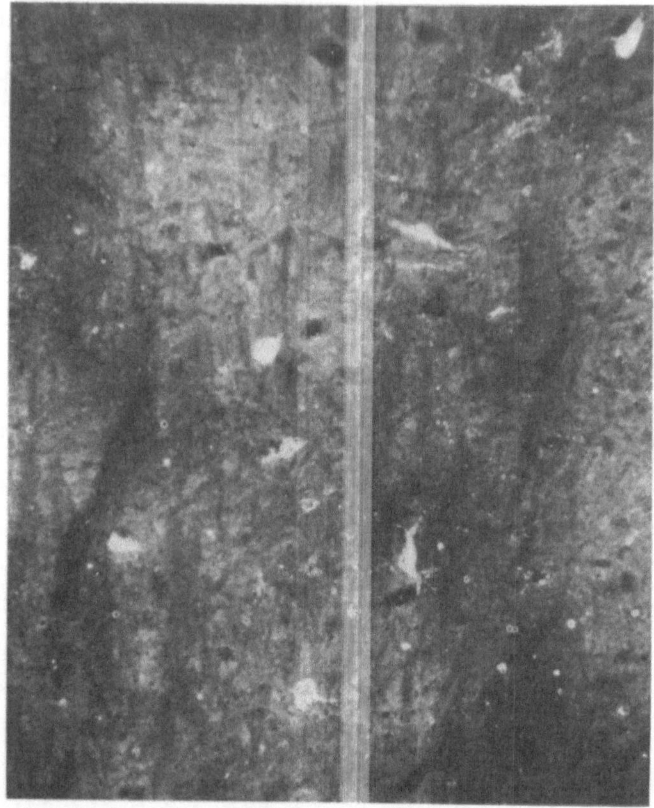

Fig. 2-24. Horizontal section of labeled cells among fiber bundles in the interstitial nucleus of Cajal after placement of tracer in the vestibular nuclei.

References

1. Baker, R., Mano, N., and Shimazu, H.: Postsynaptic potentials in abducens motoneurons induced by vestibular stimulation. Brain Res. 15:577, 1969.
2. Brodal, A. and Pompeiano, O.: The vestibular nuclei in the cat. J. Anat. 91:438, 1957.
3. Gacek, R.R.: Anatomical demonstration of the vestibulo-ocular projections in the cat. Acta Otolaryngol. (Stockh.) (Suppl.) 293:1, 1971.
4. Gacek, R.R.: Efferent component of the vestibular nerve. In Rasmussen, G.L. and Windle, W.F. (eds.): Neural Mechanisms of the Auditory and Vestibular Systems. Springfield, Ill., Thomas, 1960.
5. Gacek, R.R.: Location of abducens afferent neurons in the cat. Exp. Neurol. 64:342, 1979.
6. Gacek, R.R.: Location of brainstem neurons projecting to the oculomotor nucleus in the cat. Exp. Neurol. 57:725, 1977.
7. Gacek, R.R.: Location of commissural neurons in the vestibular nuclei of the cat. Exp. Neurol. 59:479, 1978.

8. Gacek, R.R.: The course and central termination of first order neurons supplying vestibular end organs in the cat. Acta Otolaryngol. (Stockh.) (Suppl.) 254: 1, 1969.

9. Gacek, R.R. and Lyon, M.J.: The localization of vestibular efferent neurons in the kitten using horseradish peroxidase. Acta Otolaryngol. 77:92, 1974.

10. Gacek, R.R. and Rasmussen, G.L.: Fiber analysis of the statoacoustic nerve of guinea pig, cat, and monkey. Anat. Rec. 139:455, 1961.

11. Graybiel, A.M. and Hartwieg, E.A.: Some afferent connections of the oculomotor complex in the cat: an experimental study with tracer techniques. Brain Res. 81:543, 1974.

12. Highstein, S.M.: Organization of the inhibitory and excitatory vestibuloocular reflex pathways to the third and fourth nuclei in the rabbit. Brain Res. 32:218, 1971.

13. Highstein, S.M. and Baker, R.: Termination of internuclear neurons of the abducens nuclei on medial rectus motoneurons. Neurosci. Abs. 2:398, 1976.

14. Kristensson, K. and Olsson, Y.: Uptake and retrograde axonal transport of peroxidase in hypoglossal neurons. Acta Neuropathol. 19:1, 1971.

15. Maciewicz, R.J., Eagen, K., Kaneko, C.R.S., and Highstein, S.M.: Vestibular and medullary brainstem afferents to the abducens nucleus in the cat. Brain Res. 123:229, 1977.

16. Maciewicz, R.J., Kaneko, C.R.S., Highstein, S.M., and Baker, R.: Morphological identification of interneurons in the oculomotor nucleus that project to the abducens nucleus in the cat. Brain Res. 96:60, 1975.

17. Spoendlin, H.: Some morphofunctional and pathological aspects of the vestibular sensory epithelia. Second Symposium on the Role of the Vestibular Organs in Space Exploration. NASA, 1966, pp. 99–115.

18. Tarlov, E.: Organization of vestibulo-oculomotor projections in the cat. Brain Res. 20:159, 1970.

19. Wersäll, J.: Electron micrographic studies of vestibular hair cell innervation. In Rasmussen, G.L. and Windle, W.F. (eds.): Neural Mechanisms of the Auditory and Vestibular Systems. Springfield, Ill., Thomas, 1960, pp. 247–257.

20. Wilson, V.J.: Physiological pathways through the vestibular nuclei. Int. Rev. Neurobiol. 15:27, 1972.

21. Yamamoto, M., Shimoyama, I., and Highstein, S.M.: Vestibular nucleus neurons relaying excitation from the anterior canal to the oculomotor nucleus. Brain Res. 148:31, 1978.

Discussion

BARMACK: You showed no afferent projections from the utricle to the superior vestibular nucleus. What kind of information do you think the afferents from the utricle are carrying and what kind of contribution do you think the secondary cells of the medial vestibular nucleus, which receive utricular afferents in your studies, make to vestibular reflexes?

GACEK: The main type of information would be the kind of information that is necessary for vestibulospinal reflexes, the vestibulospinal reflexes being mediated through the lateral and medial vestibular nuclei. These nuclei contribute quite substantially to vestibulospinal reflexes and the utricular terminations in these nuclei are very important for this type of reflex. The opportunity for reflexes through the extraocular nuclei, especially those concerned with horizontal eye movements, is also made possible by the termination of the utricular afferents in the medial vestibular nucleus.

CLARK: The reasons for my question will be clear this afternoon. Do you have any evidence for direct commissural fibers from one Y nucleus to the other specifically?

GACEK: No, I don't, but I think that is possible. The reason I say I don't know is that when these injections of tracer are made, one can say nothing about the precise termination of these commissural units. One can only say something about their origin. However, the precise termination would be of considerable interest.

GOLDBERG: The Y group is the main termination of saccular afferents. From the data you presented today, it would appear that the group is mainly concerned in commissural connections. How, then, might you imagine that saccular impulses could reach vestibuloocular pathways? Also, you mentioned a group of neurons dorsal to group Y. Can you tell us anything about its connections and whether it should be considered a part of the vestibular nuclei or of the deep cerebellar nuclei?

GACEK: To answer the last question first, yes, I think it [infracerebellar nucleus] is a derivative of the lateral [dentate] nucleus in the cat. There is no reliable evidence of its afferent supply, but it is an important connection. There are a number of possible pathways to the extraocular nuclei from the saccule. Its known termination in group Y may project to the opposite vestibular nuclei, perhaps the superior, and then be extended to the extraocular reflex pathway in that way. It is also possible that some saccular afferents may terminate in the infracerebellar nucleus which is located just dorsal to group Y. Some of the cells in the infracerebellar nucleus have very long dendritic afferents. As I indicated in my presentation, the IFC labels intensely when the trochlear and oculomotor nuclei are injected.

PRECHT: I was particularly intrigued by your demonstration of direct afferents to the abducent nucleus, and if I can remember correctly from your slides, it is only to the ipsilateral abducent. That would in some way correspond to findings in some fish and reptiles; there it has been shown that there are direct primary fibers to the nuclei of the third and sixth nerves. When you record from abducent motoneurons, after eighth nerve stimulation as we did, you find in almost any ipsilateral motoneuron both IPSPs, which represent the horizontal canal projection, the inhibitory VOR, and EPSPs, which often have quite short latencies. These EPSPs, we know, come specifically from the utricle. My question now is, could it be that these ganglion cells that you labeled by retrograde means are located in the area in the ganglion where the utricle is represented?

GACEK: Thank you for the information on those connections in lower forms. You are quite correct in suggesting direct connections from the utricle to the sixth nucleus. The labeled first order neurons in our experiment were located in the ventral part of the superior division ganglion. This is where the utricular ganglion cells are located.

BÜTTNER: Did you find evidence that vestibular afferents from the nerve ended in the brainstem outside the vestibular nucleus, as was described by Hangle-Hanssen in Lergeb. Anat. Entwichl.-Gersh. 40:1, 1968?

GACEK: You are asking about vestibular afferents entering the brainstem and terminating outside the boundaries of the vestibular nuclei. The experiment in which the first order neuron projection was traced required exposure of the vestibular ganglion in the internal auditory after retracting the cerebellum. This retraction produces a substantial amount of anterograde degeneration which terminates in the vestibular nuclei and surrounding areas of the brainstem. So it is very difficult to say whether some of the degeneration that was present outside the vestibular nuclei was from cerebellar degeneration or from the first order afferents. I really can't answer that question.

ATKINSON: Lorente de Nó described five groups of fiber bundles in the vestibular nerve, groups I and II arising from the cristae and groups IV and V from the maculae. The origin of group III was rather nebulous. Could this group correspond to the efferent bundle that you have described?

GACEK: I think all those groups of Lorente de Nó really refer to afferent fiber projections coming into these vestibular nuclei. Some of the five groups correspond to different end organ projections and some of them may correspond to different fiber size projections (large and small) from an individual sense organ. I do not think that the efferents are to be confused with group III.

FREE PAPERS

3

Otolith Organ Receptor Morphology in Herring-like Fishes

CHRISTOPHER PLATT AND ARTHUR N. POPPER

Fishes of the order Clupeiformes, which includes the herrings, sardines, and anchovies, have unusual structures associated with the inner ear that may function to enhance acoustic sensitivity (27,30,35). A pair of slender, hollow ducts extend forward from the gas bladder and connect to closed gas-filled portions of special otic bullae. Both the pro-otic and pterotic bullae have surfaces contacting the wall of the utricle within the membranous labyrinth.

Although the utricle in fishes is usually considered to be a gravistatic organ (15,32), the connection from the gas bladder to the utricular region in herring has led to speculation that their utricle may be an important auditory organ (1,3,20,35), rather than just the saccule and lagena, as is suggested in other teleosts (25,30,31). Structural features of the ducts and bullae in clupeiforms support models suggesting enhanced sensitivity to certain auditory frequencies (3), and the physiological evidence now also implies some auditory role for the utricle (4,5).

In the present study we have explored another facet of the otolithic organs in clupeiforms, that of the ultrastructure of the sensory surfaces. In clupeids, as in other vertebrates, the sensory epithelia contain a large number of sensory hair cells, each cell having an apical ciliary bundle projecting into the lumen of the otolith organ. Each bundle contains a single kinocilium arising at one end of an array of stereocilia; intracellular recordings in vestibular hair cells have confirmed that this structural asymmetry is a correlate of physiological directional sensitivity (9,18,34).

Comparative studies utilizing the scanning electron microscope (SEM) show some recurring ultrastructural properties in the form and organization of the hair cell ciliary bundles within each otolith organ macula in teleost fishes. We have now done a similar study on herring (26) and other clupeiform species and found some consistent and remarkable differences in macular organization from that in other teleosts, as well as in tetrapods.

Tissue preparation was done as in previously reported studies (21,23). Briefly, after anesthesia and decapitation, the otolith organs were exposed, fixed in situ with glutaraldehyde in a buffered solution, then rinsed and post-fixed with buffered osmium tetroxide. After opening the pouches to remove the otoliths, tissues were dehydrated through either alcohol or acetone, dried by the critical-point method using CO_2, and mounted and gold-coated for SEM observation.

Two species were examined representing the two major clupeiform families. The Marquesan sardine, *Sardinella marquesensis*, from the central and south Pacific, is in the family Clupeidae, and the Pacific northern anchovy, *Engraulis mordax*, is in the family Engraulidae. Some supporting observations were made on the Pacific herring, *Clupea harengus pallasi*, also in the family Clupeidae. The spatial relations of the inner ear sensory maculae are shown in Fig. 3-1.

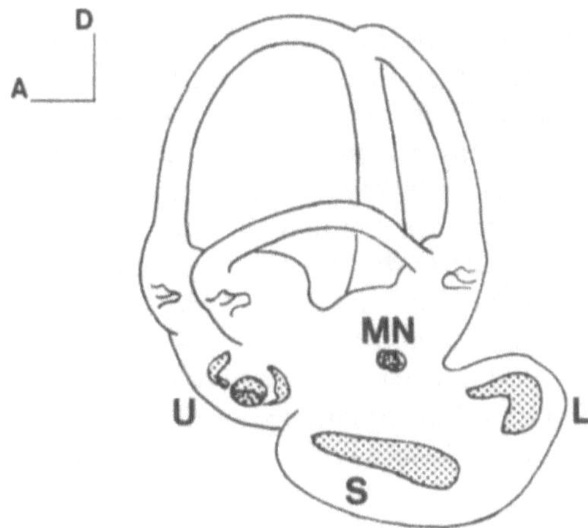

Fig. 3-1. Diagram of the inner ear of a sardine (*Sardinella marquesensis*). Left ear, lateral view, shows the relative sizes and positions of the sensory maculae (stippled) of the otolith organs; canal cristae also are indicated for reference. The utricular macula (*U*) is tripartite; the macula neglecta (*MN*) is a small pair of spots; the saccular (*S*) and lagenar (*L*) maculae are roughly vertical.

UTRICLE

The clupeiform utricle has been described on the basis of gross examination and light microscopy to have a sensory macula that is trilobed or tripartite, forming three laterally directed fingers known as the maculae anterior, media, and posterior (20,35). The SEM observations confirm that the three lobes are in fact three completely separated hair cell populations. There is a gap without hair cells or even rudimentary cilia between each of the maculae on the medial utricular surface. Since the fish we have examined are adults, we do not believe this gap is filled in later by developing hair cells. The slimmer macula anterior and macula posterior curve slightly around the broader macula media (Fig. 3-2). Whereas the middle and posterior maculae lie across the floor of the utricle, the anterior macula lies on the rising curve of the anterior utricular wall. The otolith itself lies directly over the middle and posterior maculae, but the otolith membrane extends over the anterior macula (20,35).

One striking feature of the utricular maculae is the widespread distribution of hair cells with tall ciliary bundles; this form in other fishes has been labelled "F3" (23). The clupeiforms seem to lack any distinct striola, a region in which a band of specialized utricular hair cells is present in other vertebrates. The utricular macula media is densely covered with cells, each having a ciliary bundle with a kinocilium that is 5 to 6 μm long; the tallest stereocilia are roughly 0.5 μm shorter than the kinocilium (Fig. 3-3A). Toward the edges of the macula media most of the ciliary bundles are taller, with proportionally taller stereocilia.

Similar tall ciliary bundles occupy large areas of the maculae anterior and posterior, although these maculae also have fairly wide peripheral regions characterized by another bundle form, "F2" (23), in which the kinocilium of 5 to 7 μm has associated stereocilia reaching only to 1 to 2 μm (Fig. 3-3B). We also find this latter form of ciliary bundle in the macula neglecta of clupeiforms, a small sensory patch posterior to the utricle near the utriculosaccular duct (27). However, not enough data were available to establish orientation patterns for the macula neglecta.

A second striking feature of the clupeiform utricular maculae is the organization of the morphological polarization of hair cell populations. Each hair cell can be assigned an orientation vector pointing toward the side of the bundle where the kinocilium arises. Instead of the utricular cells forming the usual two regional populations having opposed orientations, in these fish the utricle has six such populations, two for each of the three macular parts (Fig. 3-4).

In the utricular macula media, hair cells of the anterior half have vectors "facing" posteriorly, and those of the posterior half face anteriorly. The opposing populations meet facing each other along an imaginary boundary line that runs medial to lateral, nearly bisecting the macula media. A similar bisection into two opposing populations occurs in both the maculae anterior

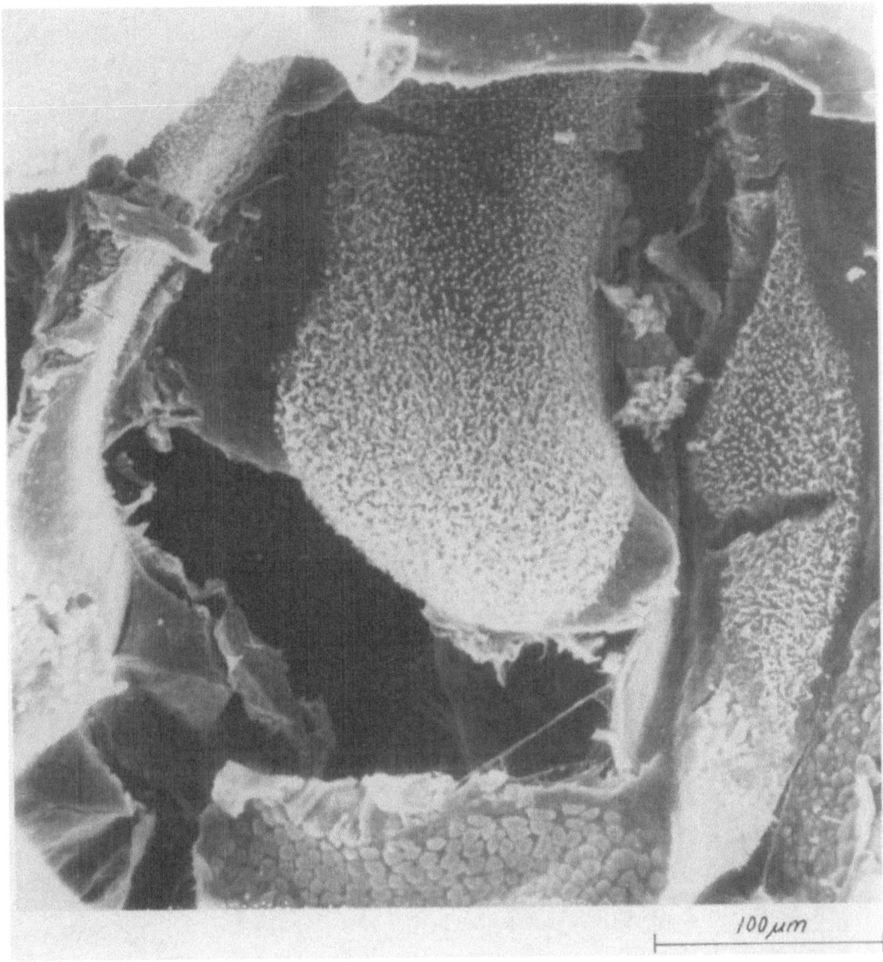

Fig. 3-2. Maculae utriculi of an anchovy (*Engraulis mordax*). Dorsal view by SEM, anterior is to the left, lateral is downward, scale bar is 100 μm. The macula anterior and macula posterior curve around the macula media. The large tear in tissue anterior to the macula media is in the area of the pro-otic fenestra.

and posterior; in these maculae, however, the line is more curved and follows the curve of the maculae. Again, the anterior regions of each macula contain cells facing posteriorly, and the posterior regions have cells facing anteriorly. Since the macula anterior in fact lies well up on the anterior utricular wall, the hair cell populations in this macula in situ must be considered to face almost dorsally and ventrally.

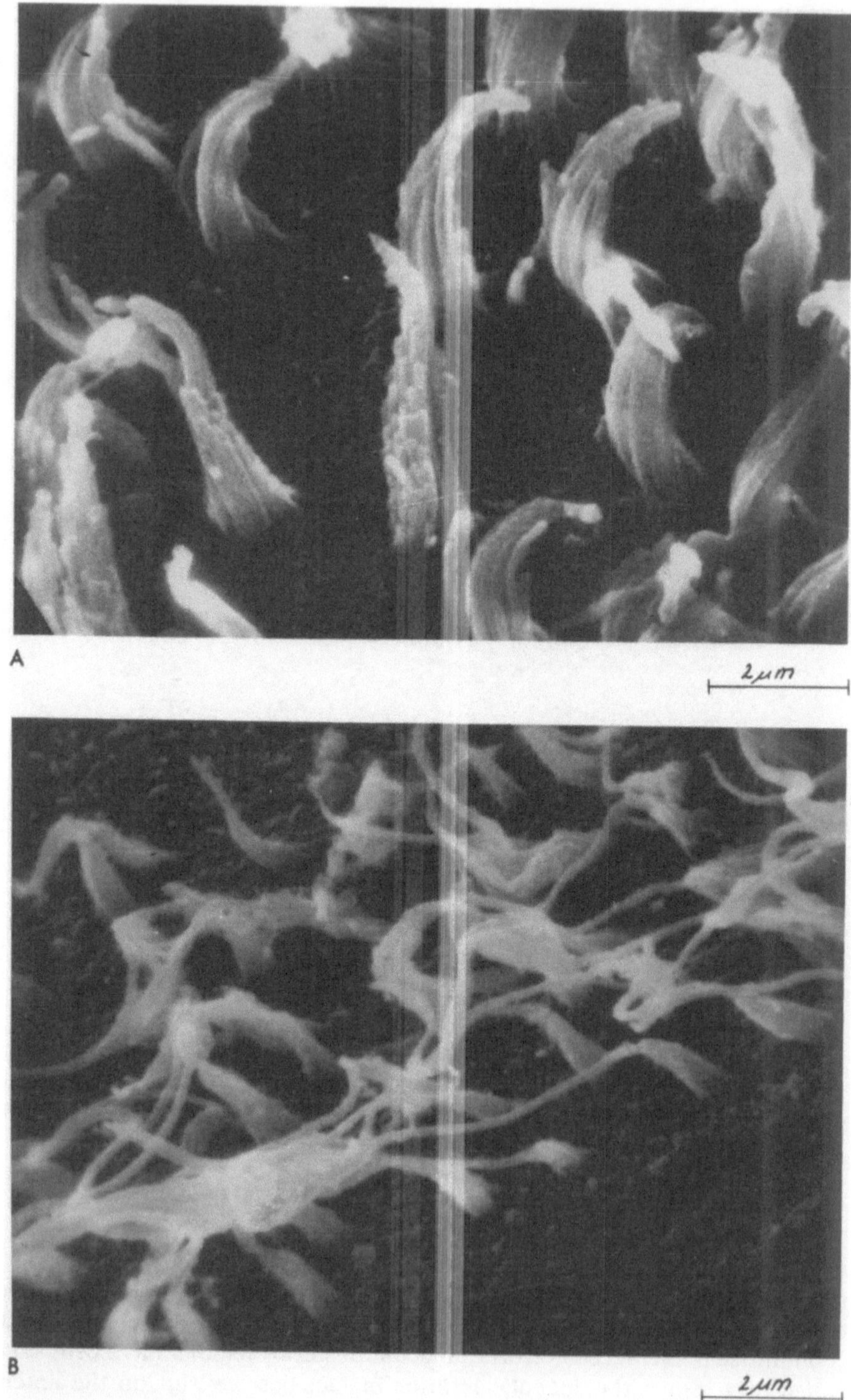

Fig. 3-3. Hair cell ciliary bundles in the utricle of the anchovy. SEM; scale bar is 2 µm. **A** Tall bundles (F3 form) in the macula media. **B** Bundles with long kinocilia and short stereocilia (F2 form) in the periphery of the macula posterior.

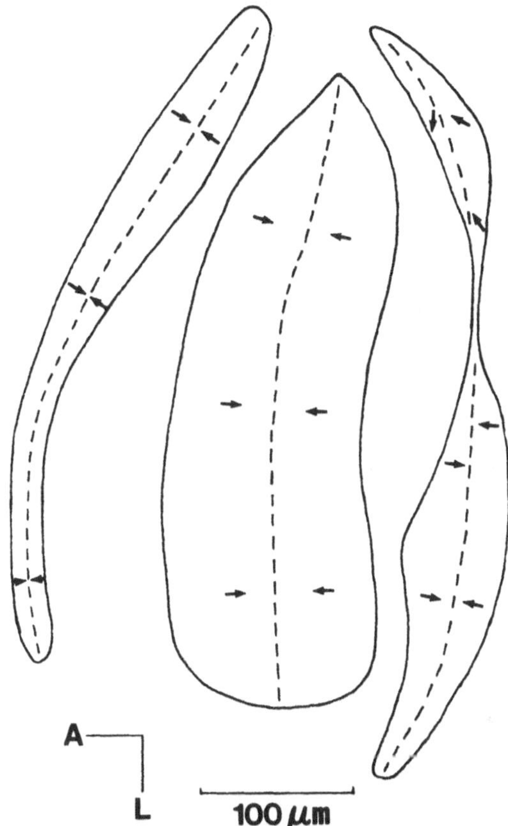

Fig. 3-4. Utricular pattern of the hair cell orientations in the left macula of the anchovy, *Engraulis mordax*. Two opposed populations occur in each of the three maculae.

SACCULE

The saccule in the clupeiforms we have studied is similar to that in many other non-ostariophysine teleost fishes (23). The saccular sensory macula is much smaller than the extent of the saccular otolith, which is the largest of the three otoliths in these fishes. Lying near the vertical plane on the medial wall of the saccule, the macula is elongated in a rostral-caudal direction.

Many hair cell ciliary bundles in the saccular macula are of the tall F3 form with a kinocilium of 6 to 8 μm and the tallest stereocilia nearly that length (Fig. 3-5). Some shorter, but otherwise similar ciliary bundles of "F1" form (22) are found centrally on the macula (Fig. 3-5B), and there is a narrow region of F2 cells all along the very edges of the macula. These F2 cells have a slightly longer kinocilium and much shorter stereocilia than those on F3 cells. There is no obvious striolar-like region.

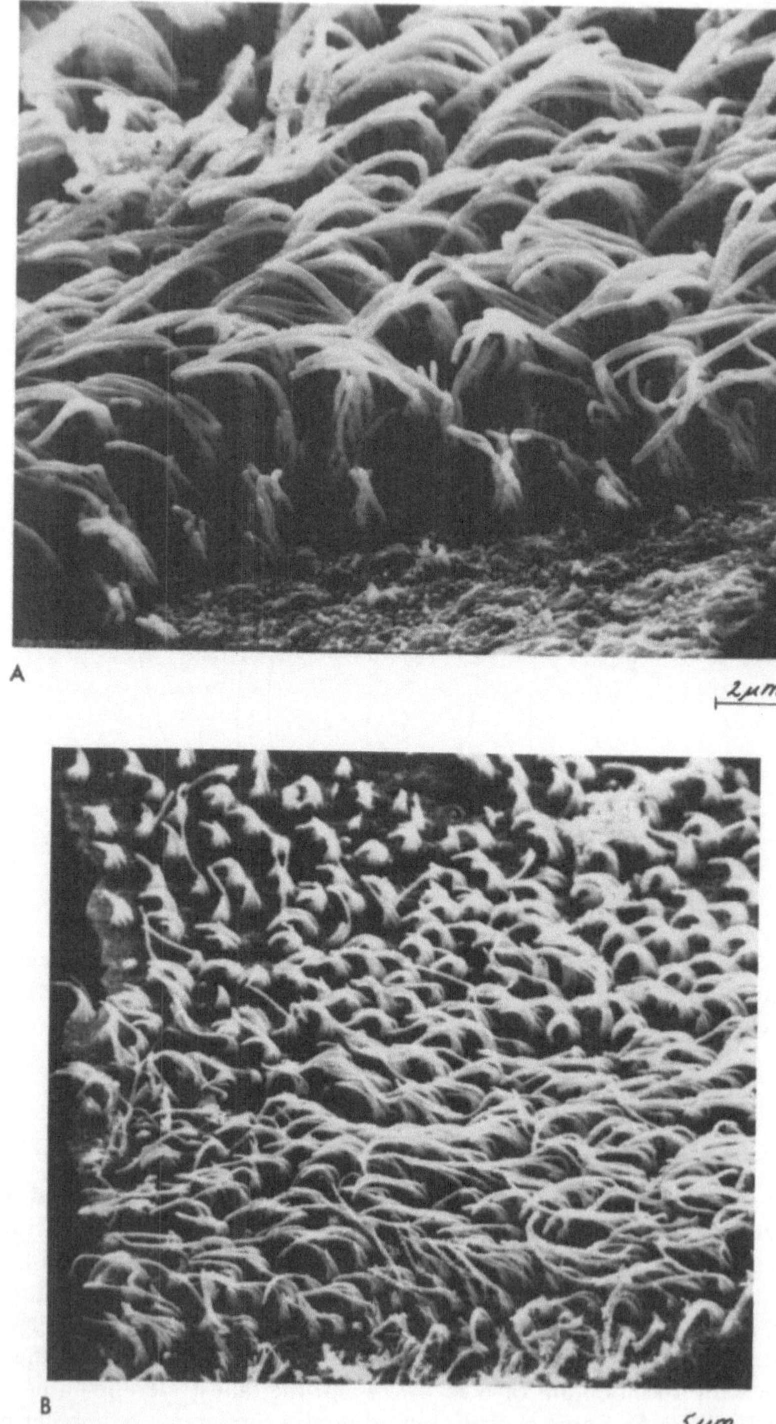

Fig. 3-5. Hair cell ciliary bundles in the saccule of *Sardinella*. **A** Tall (F3) bundles oriented rostrally (toward *right*). Some ciliary bundles at the lower edge were sheared off in tissue preparation. **B** Tall (F3) bundles on the edge of the saccular macula (*bottom*) along with shorter (F1) bundles located more medially on the macula.

Hair cell morphological polarization on the saccular macula is similar to that in non-ostariophysine teleosts, forming a quadrant orientation pattern (Fig. 3-6). The anterior half of the macula has a dorsal population of cells facing posteriorly and a ventral population facing anteriorly. The posterior half of the macula has a dorsal population that faces dorsally, away from the ventral population that faces ventrally. The boundary lines that can be drawn between the four populations divide the macula nearly in half rostrocaudally, and nearly in half dorsoventrally.

LAGENA

The lagenar macula in these clupeiforms is similar to that of several other teleosts (23), and is covered by an otolith that extends only slightly beyond the macular edge. The lagena is an outpouching off the medial posterior saccular wall; its curved macula lies on the medial wall of this rounded pouch. The lagenar macula is shaped something like the outline of a chemist's retort flask, having a long anterior limb extending from a large, rounded, ventrally curving posterior area.

Hair cell ciliary bundles of the lagenar macula are similar to those of the saccular macula in form and distribution, most cells having a tall kinocilium, with tall stereocilia centrally and short stereocilia peripherally. Hair cell orientations in the lagenar macula form two separable opposing populations that curve (Fig. 3-7). Dorsal cells in the anterior limb form a population with a polarization facing posteriorly, but showing a gradual change to a ventrally facing population in the posterior region. Conversely, in the ventral part of the anterior limb, hair cells face anteriorly, with a gradual change to dorsally facing cells in the posterior area.

COMPARATIVE ASPECTS

Three structural features of the clupeiform utricular macula appear to be unusual among the vertebrates. These features include the tripartite structure

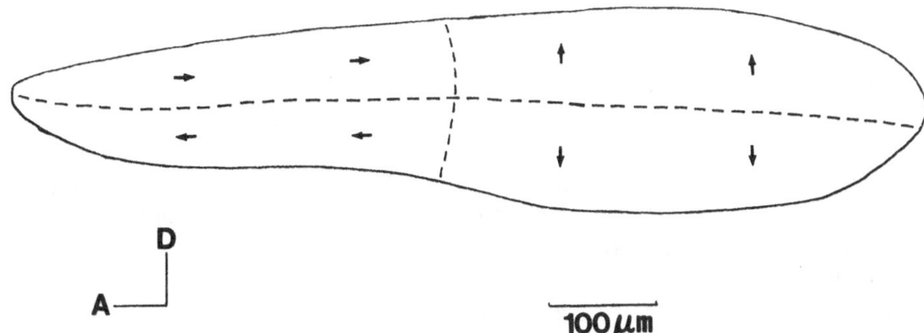

Fig. 3-6. Saccular pattern of hair cell polarization in the sardine (left macula), showing quadrant division of oriented populations.

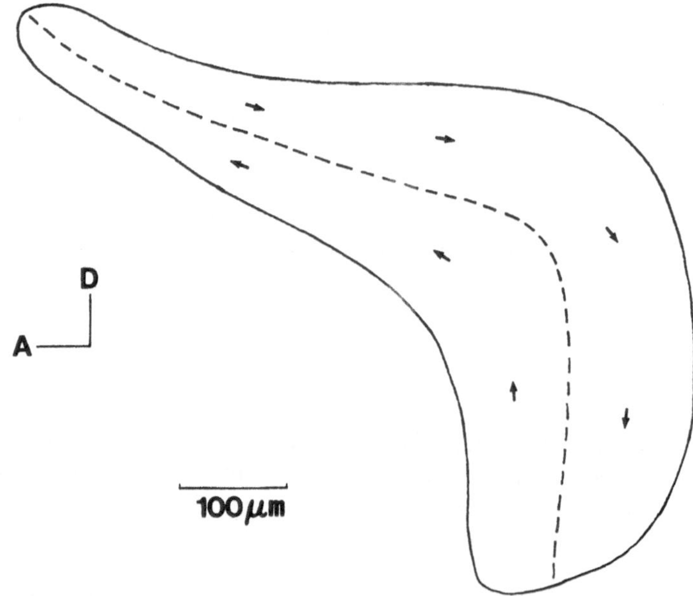

Fig. 3-7. Lagenar pattern of hair cell polarizations in the sardine (left macula), showing two opposing populations.

of the macula, the three separate lines of opposition between oriented hair cell populations, and finally the presence of hair cell ciliary bundles that are more similar in form and distribution to those found in the saccule and lagena than is the case in other teleosts.

Although we must be cautious in determining what is "unusual" among 20,000 or more fish species when only a few dozen yet have been examined, the literature does seem to include samples from a wide enough range of groups to see some trends and set the clupeiforms in perspective. Similarly, we should be cautious in making comparisons of fish to other vertebrates, but again there is now data on the inner ear from at least some representatives of every vertebrate class to enable at least preliminary comparisons to be made.

In most vertebrates, the utricular macula is a single dish-like or crescent-shaped area, which also may have a posterolateral extension, the lacinia (16). There is a single line that can be drawn between a pars interna containing hair cells oriented anteriorly and laterally and a pars externa containing cells facing posteriorly and medially inward (7,13). This dividing line between cells facing each other usually parallels the general curve of the anterior-lateral macular edge, then swings to bisect the lacinia (when present) along its long axis. There are no reported exceptions to this general form within the bony fishes (2,21,24) and tetrapods (12,13). However, an elasmobranch, the ray *Raja clavata*, has a utricular macula with a patchy distribution of opposing orientations (17), although this macula is not structurally split as in the clupeiforms.

The orientation pattern of the tripartite clupeiform utricular macula appears to be at least derivable from the more common vertebrate form. The pattern of polarized cell populations facing one another is a feature shared with other vertebrate utricles, but not with saccules where the cells face away from a dividing line. The origins of the maculae anterior and posterior of the clupeiform utricle are not clear, but at least the macula posterior shares its gross location and polarized pattern with the lacinia of other vertebrates. Comparative studies on larval clupeiforms might clarify these origins, whether arising independently or as offshoots from the macula media.

Orientation patterns in the clupeiform saccular and lagenar maculae are quite similar to those seen in many other teleost species (2,6,11,22–24). The ostariophysine fishes (e.g., goldfish, catfish) and tetrapods differ from this pattern in having a saccule with only two separable hair cell populations, with dorsal cells facing dorsally and ventral cells facing ventrally (8,10,13,21,28).

Since the morphological polarization of a hair cell ciliary bundle lies along the axis of maximum response sensitivity to shearing, the orientation patterns have functional significance. In the clupeiform utricle, the macula anterior, by having two opposed populations on the anterior utricular wall, could provide a bidirectional detector for dorsoventral shearing displacements, whereas the macula media could provide the major bidirectional detection of rostrocaudal displacements. In this context, it is notable that the pro-otic bulla has a pro-otic fenestra forming a fluid opening into the utricle right between the macula anterior and macula media (3,20).

The clupeiform saccule and lagena have orientations that could maximize sensitivity for either dorsoventral or rostrocaudal displacements. This property may be utilized by other non-ostariophysine fishes (29) that do not have a structural link between gas bladder and saccule. In the Ostariophysi, though, saccular auditory sensitivity has been shown to depend on the presence of weberian ossicles directly coupling the gas bladder to the inner ear; and it may be that only dorsoventral orientations are needed by their saccule and that rostrocaudal information might not be passed through the mechanical linkage (25).

Hair cell ciliary bundles show a variety of forms in a variety of vertebrates, and it is not yet clear what correlation exists between structure and function. In amphibians and reptiles, the auditory organs, but not the equilibrium organs, have hair cells with a distinctive bulbed process at the apical tip of the kinocilium (12,19). Such a feature has been looked for in many fishes but never found (21,23,24). Another suggested correlation is that taller cilia may have higher frequency sensitivity (14,33). This issue is made more complex by tall kinocilia occurring with either tall or short stereocilia. However, in many teleosts where the utricle is considered gravistatically sensitive, it typically contains many bundles with the tallest stereocilia roughly half as tall as the kinocilium, of the F1 form, as well as a distinct striola where the bundles are significantly larger than elsewhere in the macula (21). In contrast, the saccule in many teleosts is considered acoustically sensitive and typically contains many bundles with the tallest stereocilia nearly as tall as the kinocilium, of the F3 form, in addition to numerous F1 bundles, and usually has no striola (23).

The clupeiform inner ear is unusual in that the utricle shares ciliary bundle structural features with the saccule, having many F3 bundles and no clear striola. If these features in the saccule are functionally related to audition in fishes, the clupeiform ear indeed might be utilizing the similar features of its utricle for acoustic reception, as suggested by the structure and mechanics of the gas-filled bullae near the utricle. The specialized role of the otolith and otolith membrane in such reception remains unknown.

CONCLUSIONS

On the basis of the ultrastructure of the inner ear sensory surfaces in relatives of the herring, we believe that the hair cell orientations and ciliary bundles in the utricle, but not in the saccule and lagena, are distinctly different from the corresponding macular organization in other fishes. We suggest that unusual similarity of ciliary bundles between utricle, saccule, and lagena supports the hypothesis that the utricle may share functional acoustic sensitivity with the saccule and lagena.

ACKNOWLEDGMENTS

This work was supported by grants from NINCDS (1 RO1 NS13946) to C.P.; from NINCDS (7 RO1 NS15090) and NSF (BNS 78-22411) to A.N.P.; and a Research Career Development Award from NIH (1 KO4 NS-00312-01) to A.N.P.

References

1. Allen, J., Blaxter, J.H.S., and Denton, E.J.: The functional anatomy and development of the swimbladder–inner ear–lateral line system in herring and sprat. J. Marine Biol. Assoc. U.K. 56:471, 1976.
2. Dale, T.: The labyrinthine mechanoreceptor organs of the cod *Gadus morhua L.* (Teleostei: Gadidae). Nord. J. Zool. 24:85, 1976.
3. Denton, E.J. and Blaxter, J.H.S.: The mechanical relationships between the clupeid swimbladder, inner ear, and lateral line. J. Marine Biol. Assoc. U.K. 56:787, 1976.
4. Denton, E.J. and Gray, J.A.B.: The analysis of sound by the sprat ear. Nature 282:406, 1979.
5. Enger, P.S.: Hearing in herring. Comp. Biochem. Physiol. 22:527, 1967.
6. Enger, P.S.: On the orientation of haircells in the labyrinth of perch (*Perca*

fluviatilis). In: Schuijf, A., and Hawkins, A.D. (eds.): Sound Reception in Fish. Amsterdam, Elsevier, 1976, pp. 49–62.

7. Flock, Å.: Structure of the macula utriculi with special reference to directional interplay of sensory responses as revealed by morphological polarization. J. Cell Biol. 22:413, 1964.
8. Hama, K.: A study on the fine structure of the saccular macula of the goldfish. Z. Zellforsch. Mikrosk. Anat. 94:155, 1969.
9. Hudspeth, A.J. and Corey, D.P.: Sensitivity, polarity, and conductance change in the response of vertebrate hair cells to controlled mechanical stimuli. Proc. Natl. Acad. Sci. U.S.A. 74:2407, 1977.
10. Jenkins, D.: A transmission and scanning electron microscopic study of the saccule in five species of catfishes. Am. J. Anat. 154:81, 1979.
11. Jørgensen, J.M.: Hair cell polarization in the flatfish inner ear. Acta Zool. 57:37, 1976.
12. Lewis, E.R. and Li, C.W.: Hair cell types and distributions in the otolithic and auditory organs of the bullfrog. Brain Res. 83:35, 1975.
13. Lindeman, H.H.: Studies on the morphology of the sensory regions of the vestibular apparatus. Ergeb. Anat. Entwickl.-Gesch. 42:1, 1969.
14. Lowenstein, O.: The electrophysiological study of the response of the isolated labyrinth of the lamprey (*Lampetra fluviatilis*) to angular acceleration, tilting and mechanical vibration. Proc. R. Soc. Lond. [Biol.] 174:419, 1970.
15. Lowenstein, O.: The labyrinth. In Hoar, W.S. and Randall, D.J. (eds.): Fish Physiology, Vol. V. New York, Academic Press, 1971, pp. 207–240.
16. Lowenstein, O.: Comparative morphology and physiology. In Kornhuber, H.H. (ed.): Handbook of Sensory Physiology, Vol. VI, Vestibular System, Pt. 1. New York, Springer, 1974, pp. 75–120.
17. Lowenstein, O., Osborne, M.P., and Wersäll, J.: Structure and innervation of the sensory epithelia of the labyrinth in the thornback ray (*Raja clavata*). Proc. R. Soc. Lond. [Biol.] 160:1,1964.
18. Lowenstein, O. and Wersäll, J.: A functional interpretation of the electron-microscopic structure of the sensory hairs in the cristae of the elasmobranch *Raja clavata*, in terms of directional sensitivity. Nature 184:1807, 1959.
19. Miller, M.R.: Scanning electron microscope studies of some lizard basilar papillae. Am. J. Anat. 138:301, 1973
20. O'Connell, C.P.: The gas bladder and its relation to the inner ear in *Sardinops caerulea* and *Engraulis mordax*. U.S. Fish Wildlife Service Fishery Bull. 56:505, 1955.
21. Platt, C.: Hair cell distribution and orientation in goldfish otolith organs. J. Comp. Neurol. 172:283, 1977.
22. Popper, A.N.: Ultrastructure of the auditory regions in the inner ear of the lake whitefish. Science 192:1020, 1976.
23. Popper, A.N.: A scanning electron microscopic study of the sacculus and lagena in the ears of fifteen species of teleost fishes. J. Morphol. 153:397, 1977.
24. Popper, A.N.: Scanning electron microscopic study of the otolithic organs in the bichir (*Polypterus bichir*) and shovel-nose sturgeon (*Scaphirhynchus platorynchus*). J. Comp. Neurol. 181:117, 1978.

25. Popper, A.N. and Fay, R.R.: Sound detection and processing by fish: A critical review. J. Acoust. Soc. Am. 53:1515, 1973.

26. Popper, A.N. and Platt, C.J.: The herring ear has a unique receptor pattern. Nature 280:832, 1979.

27. Retzius, G.: Das Gehörorgan der Wirbelthiere: morphologisch–histologische Studien. I. Das Gehörorgan der Fische und Amphibien. Stockholm, Samson and Wallin, 1881.

28. Saito, K.: Fine structure of macula of lagena in the teleost inner ear. Kaibogaku Zasshi Acta Anat. 48:1, 1973.

29. Sand, O.: Directional sensitivity of microphonic potentials from the perch ear. J. Exp. Biol. 60:881, 1974.

30. Tavolga, W.N.: Sound production and detection. In: Hoar, W.S. and Randall, D.J. (eds.): Fish Physiology, Vol. V. New York, Academic Press, 1971, pp. 135–205.

31. von Frisch, K.: Über die Bedeutung des Sacculus und der Lagena für den Gehörsinn der Fische. Z. Vergl. Physiol. 25:703, 1938.

32. von Holst, E.: Die Arbeitsweise des Statolithenapparates bei Fischen. Z. Vergl. Physiol. 32:60, 1950.

33. Weiss, T.F., Mulroy, M.J., Turner, R.G., and Pike, C. L.: Tuning of single fibers in the cochlear nerve of the alligator lizard: Relation to receptor morphology. Brain Res. 115:71, 1976.

34. Wersäll, J., Flock, Å., and Lundquist, P.G.: Structural basis for directional sensitivity in cochlear and vestibular sensory receptors. Cold Spring Harbor Symp. Quant. Biol. 30:115, 1965.

35. Wohlfahrt, T.A.: Das Ohrlabyrinth der Sardine (*Clupea pilchardus* Walb.) und seine Beziehungen zur Schwimmblase und Seitenlinie. Z. Morphol. Oekol. Tiere 31:371, 1936.

4

Is the Growth of the Otolith Controlled by Its Weight?

H.C. HOWLAND AND J. BALLARINO

The utricular otolith of the chicken is composed of many otoconial crystals lying on or embedded in a gelatinous layer which in turn rests on the hair cells of the spoon-shaped macular bed.[1] In dissection the otoliths are readily recognized by their very white crystalline appearance. Figure 4-1 shows a freshly dissected utricular otolith with its neighboring ampulla of the horizontal semicircular canal and the adjacent saccular otolith.

Under the scanning electron microscope (SEM) critically point-dried utricular otoliths of the 1-day-old chick look very similar to those of the guinea pig (7), except that the otoconia do not appear to be embedded in the otoconial membrane. The longer diameter of the macula here is about 1 mm (Fig. 4-2).

At higher magnifications under the SEM one can see that the otoconia range in size from 5 to 20 μm (Fig. 4-3), with rounded, rough sides and angular ends as in pigeons and mammals (10,11), and that they are covered on their rounded sides by a rough globular substance which spreads between some of the otoconia like a thin membrane (Fig. 4-4).

Because the otoconia are extracellular material the regulation of their size

[1]Following Carlström (2) we use the term "otolith" in the broad sense to designate "any dense body in the labyrinth."

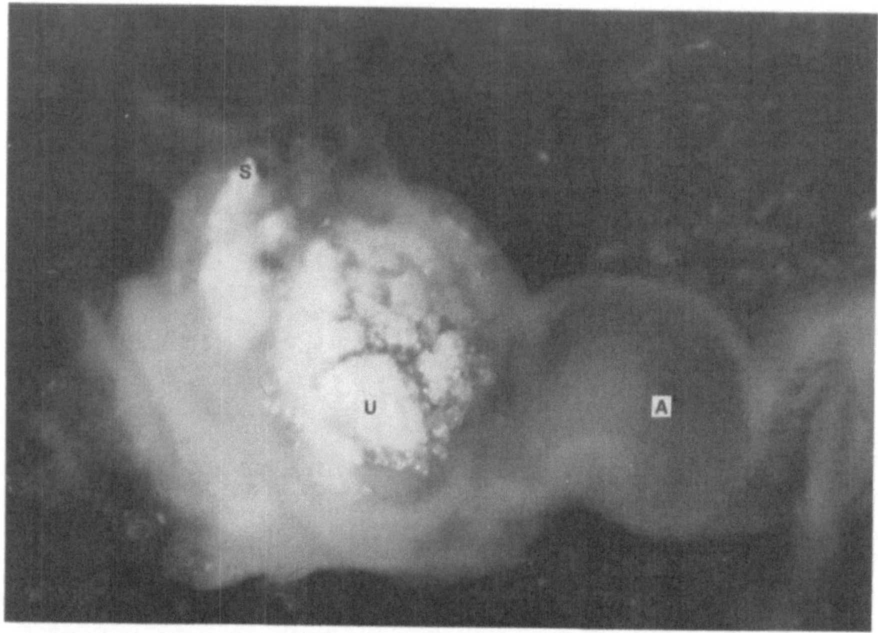

Fig. 4-1. Dissection showing utricular (*U*) and saccular (*S*) otoliths with attached ampulla of the horizontal semicircular canal (*A*).

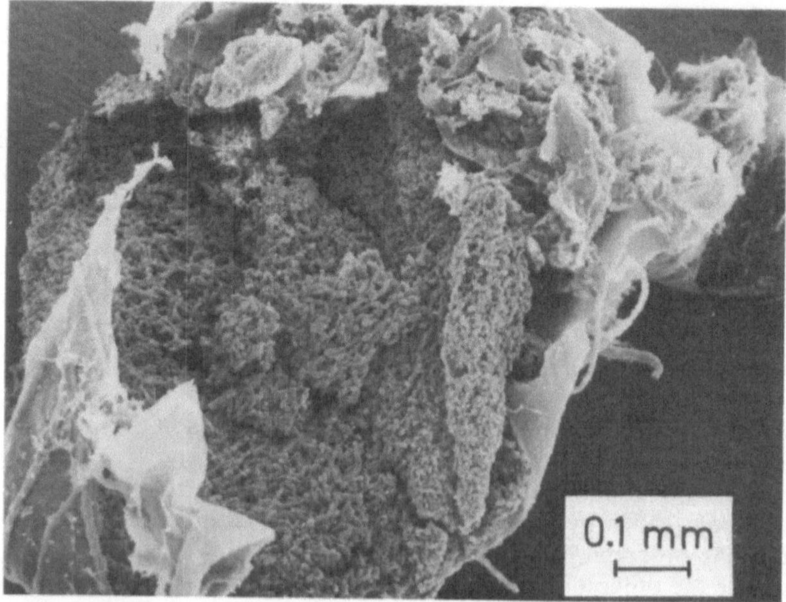

Fig. 4-2. Scanning electron micrograph of the utriculus of a 1-day-old chick. The length of the otoconial membrane is about 1 mm.

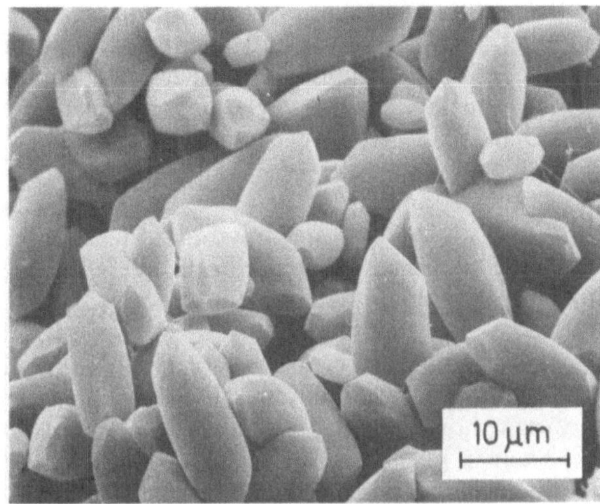

Fig. 4-3. Scanning electron micrograph of a section of the utricular otoconial membrane showing the individual otoconia, ranging in size from 5 to 20 μm, and a variety of crystalline shapes of otoconia, such diversity being also typical of calcite crystals.

must present unusual problems for the organism. Since the total mass of the otoconia apparently constitutes an important parameter in the proper functioning of the otolithic organ, the means by which this extracellular, inert crystalline mass is regulated seemed to us worthy of investigation.

One possible hypothesis is that the size of the otoconial mass might be regulated by its weight. This hypothesis could be tested by allowing animals

Fig. 4-4. A Scanning electron micrograph showing flat and smooth end faces of the otoconia and round and rough side surfaces. **B** Enlarged portion of **A** (*arrow*) showing rough globular membrane investing and connecting two neighboring otoconia.

to develop otoliths under artificial gravity in a centrifuge. For these experiments the developing chick is a particularly appropriate vertebrate in that (1) it grows largely under water—a condition which both shock-mounts it and reduces weight-dependent exercise and the possible associated increase in bone calcification; and (2) it grows exceedingly rapidly, being a precocial bird with fully functional motor and sensory systems at 21 days.

Unfortunately, little is known about the mechanism and dynamics of otoconial growth (6). Lim (10) has suggested that the calcium carbonate of the otoconia may be reabsorbed by the dark cells of the macula, as he found shrunken otoconia associated with these cell surfaces. Balsamo et al. (1), on the basis of tetracycline-labeling studies, proposed that carbonic acid may be generated by carbonic anhydrase in the endolymphatic duct and the resulting crystals of calcium carbonate migrate into the sacculus and utriculus. It is not clear, however, how this theory can be reconciled with that of the organic otoconial matrix proposed by Lim (10), which would seem to demand crystallization at the otoconial membrane.

It should be noted that our nonmorphologic evidence that chick otoconia are indeed composed of calcium carbonate includes (1) observing their effervescence in dilute hydrochloric acid; (2) the detection of a strong calcium line in the energy spectrum of scattered X-ray emission in microprobe analysis; and (3) the failure to find metallic cations other than calcium as well as the failure to find phosphorus in the same test. This, of course, is in agreement with Carlström's classic study of vertebrate otoliths (2).

In any event, there are a number of possible ways in which the deposition or removal of calcium carbonate could be regulated. These are (1) alteration of the concentration of calcium ion in the endolymph; (2) alteration of the concentration of the carbonate ion via carbonic anhydrase; or (3) alteration of the carbonate ion concentration via changes in pH of the endolymph and thus promotion or inhibition of the dissociation of the carbonic acid and the bicarbonate ion.

METHODS

Batches of 45 eggs of white leghorn chickens (*Gallus domesticus*, Cornell strain "s") were placed in the incubators on the arms of a Genisco model C-159 centrifuge. We centrifuged the eggs at 40 rpm at radii of 34 to 39.5 inches generating net accelerations of 1.8 to 2.1 *g*. The eggs were maintained at 37 to 39 °C by a standard incubator heater approximately 10 inches from the bed of the eggs which were placed on a plane inclined at 63° from the true horizontal (Fig. 4-5).

At the same time we maintained a control set of eggs in a stationary incubator again at 37 to 39° C. It was not possible to guarantee that the temperature fluctuations in the incubators were identical. The data we report below are based on a study of three such experiments involving 270 eggs.

The centrifuge ran continuously except for brief periods three times a day

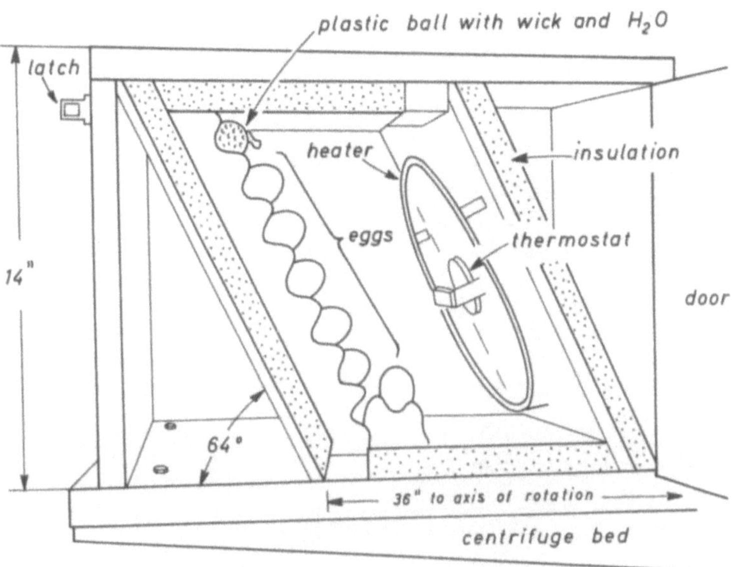

Fig. 4-5. View of incubator mounted on one arm of Genisco centrifuge.

when it was stopped to allow turning and removal of the eggs. [Turning the eggs is necessary to prevent sticking of embryonic membranes to the shell of the egg (4).]

Embryos were removed at various intervals from the centrifuge and control incubators, at which time their behavior was observed and the morphologic age determined by comparison with their appearance to standard pictures of embryos (4). They were then placed in 10% formalin buffered to a pH of approximately 8.9 with sodium borate. Between 1 and 7 days later the embryos were removed from the formalin, weighed, and their otoliths dissected out. Wet mounts were made of the utricular otoliths and the underlying maculae and these were photographed under a polarizing microscope with linear crossed polarizers (Fig. 4-6). The otoliths and attached maculae were then air dried and weighed on a Cahn Model 4100 electrobalance to an accuracy of $\pm$ 0.5 μg. Otoliths from a normal chick one day after hatching were used for the scanning electron micrographs (Figs. 4-2–4-4) taken on an AMR 1000-A scanning electron microscope. These were placed in buffered formalin for approximately three days and then run through an alcohol series, critically point-dried in CO_2, and sputter-coated with gold.

Areas of otoliths from photographs of wet mounts such as that in Fig. 4-6 were measured either with a digital planimeter or computed from weighed tracings of photographs.

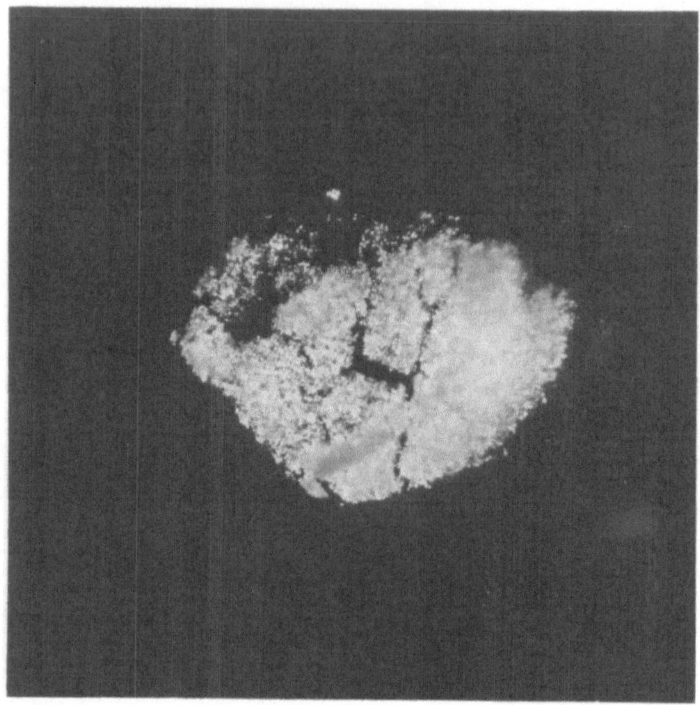

Fig. 4-6. Light micrograph of utricular otoconia under crossed linear polarizers. Maximum diameter of otolith is approximately 1 mm.

RESULTS

We found *no* significant difference in the weights of the centrifuged and control embryos as a function of age (Fig. 4-7) (*F* tests for residual mean squares, slopes, and elevations; ref. 14). Likewise, we found *no* differences in the behavior of the centrifuged and control embryos using the classification of embryonic behaviors according to Kuo (8). Indeed, the two chicks which were allowed to hatch from the centrifuged eggs were indistinguishable in their behavior from the hatched control animals.

The rates of development determined from a comparison of morphologic ages of the embryos with their chronologic ages did not differ much between the two groups. The small differences that did occur could be attributed to differences in temperature between the two incubators.

With regard to gross otolithic weights (i.e., otoliths plus attached maculae) our data show appreciable scatter (Fig. 4-8) due in part, perhaps, to the difficulty of the dissection. There appears to be no difference in the otolithic weights between the control and experimental animals up to approximately 20 g body weight or day 17.

However, for the heavier, older embryos all of the otoliths of the experimental animals (12 otoliths, seven animals) are lighter than those of the con-

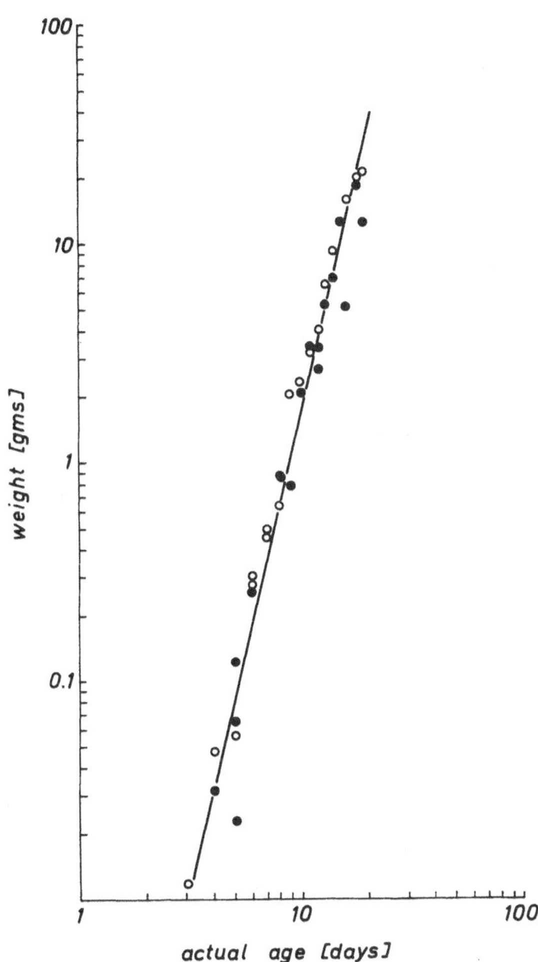

Fig. 4-7. Plot of embryo weights vs. actual age for controls (○) and experimental animals (●). The differences between the two regression lines are not significant. The joint regression line for all data is: Log weight (g) = −4.136 × 4.362 log (age in days).

trols of the nearest comparable weight. This difference is statistically significant ($p \leq 0.01$, one-tailed binomial test).

With regard to the areas of the otoliths, again there was considerable scatter in the data (Fig. 4-9), but there was no difference between the regressions of the otolith areas vs. body weights for the control and experimental animals (F tests for residual mean squares, slopes, and elevations; ref. 14).

DISCUSSION

Our initial hypothesis was that the growth of the otolith was controlled by its weight in that, as the mass of the otolith increased, its increased weight

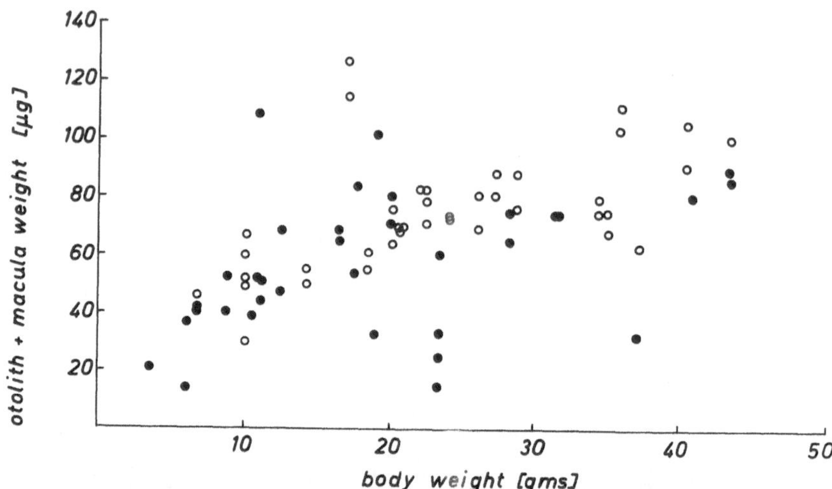

Fig. 4-8. Plot of gross (otoconial membrane plus macula) weights of otoliths vs. body weight for controls (○) and centrifuged (●) animals. The weights of animals above 20 g differ significantly ($p \leq 0.01$) in control and experimental animals, the centrifuged embryos having lighter otoliths.

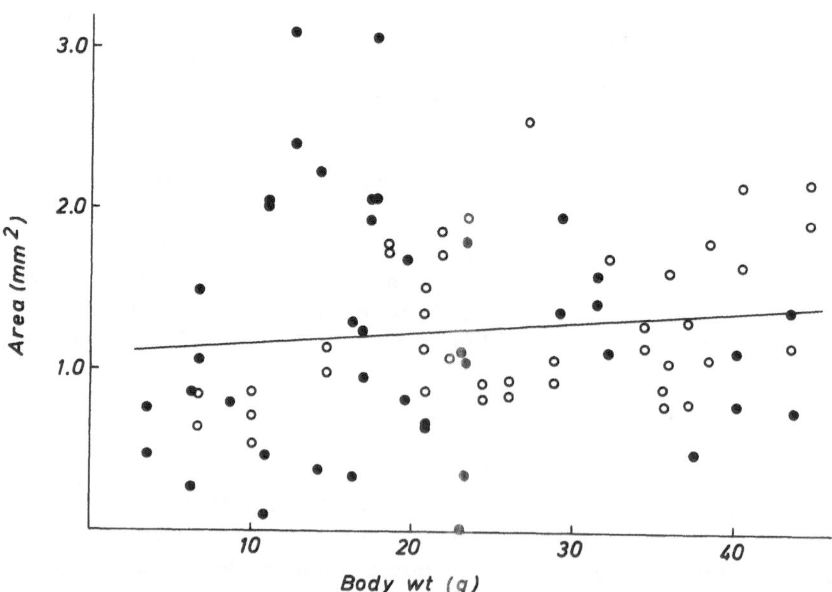

Fig. 4-9. Plot of utricular otolithic areas vs. body weight for control (●) and experimental (○) animals. The differences between the two regressions are not significant. The joint regression line for all data is: Otolithic area in mm^2 = 1.102 + 0.007077 × weight in g.

would in some way cause a feedback reduction in the deposition of calcium carbonate in the otoconia. There are a number of possible mechanisms by which such a feedback loop could work including (1) direct mechanical pressure on membranes which transport calcium, hydrogen ions, or bicarbonate ions, or on cells mediating the production of carbonic anhydrase; or (2) an indirect neurally mediated feedback loop via the normal mechanical stimulation of the hair cells.

Our results support the hypothesis that there is a weight-dependent regulation of otolithic growth in the later stages of chick development. This conclusion, however, must be regarded as preliminary in that we believe that improvements in technique will allow us to obtain weights of the otoliths without the underlying sensory epithelium.

Furthermore, there is an additional control experiment which should be performed before we can be certain that the results we observe are due to the increased centripetal acceleration: that is to measure the otolith weights of embryos which have been continuously rotated at 40 rpm at 1 *g* (i.e., at zero centrifuge radius). This is because there is always the possibility (although we deem it unlikely) that the passive rotation itself, perhaps compounded with head movements of the embryos, could affect the growth of the otolith.

Although the sensory cells of the utricular macula become ciliated as early as day 8.5 (12) and vestibular excitability by way of a postrotational head nystagmus has been reported in chicks of 8 days (ref. 15 cited in ref. 5), our results do not show a difference in rate of growth between experimental and control otoliths until the chicks are approximately 17 days old. Similarly, Decker (3) has shown that extirpation of the otocysts on day 3 or 4 does not affect the cyclic motor activity of the chicks until they have reached day 17, and Kuo (8) observed that chicks do not behaviorally respond to inversion of the egg until the egg is pipped on day 20, although general body movements start very early in development on day 4.

It is thus possible that a neuronal feedback loop controlling the growth of the otolith must wait upon the development of the same centers mediating the integrated control of motor behavior and righting responses.

Our finding that chicks which have been allowed to develop in the egg in a 2-*g* environment show normal behavior in a 1-*g* environment after hatching is not without its parallel in the centrifugation of hatched chicks. Smith and Burton (13) reported that 75% of white leghorn chicks which had been centrifuged at 1.5 *g* for a week showed a normal posture upon return to a 1-*g* environment, whereas the remaining 25% showed repeated somersaulting for 12 h after centrifugation. It is possible that this sort of variance in response to increased acceleration is also reflected in our otolith weight data of Fig. 4-9.

Lastly, we should comment on the scanning electron micrographs of the utricular otoconia of a 1-day-old chick. With the highest magnification (Fig. 4-4B) one may clearly see a membranous structure with a rough globular surface investing two neighboring otoconia, which also appears to form a part of the coating of the rounded sides of the otoconia. We have not seen this membrane in other published otoconial micrographs (7,9–11). It is interesting to speculate that this membrane may function as cement to hold the otherwise loose otoconia together. What role it plays in the deposition or removal of calcium carbonate from the otoconia must remain a subject of future studies.

CONCLUSIONS

1. Chick embryos raised in an incubator centrifuged at a resultant net acceleration of approximately 2 *g* showed no differences in body weight, rate of development, or behavior from uncentrifuged controlled embryos.
2. There also appeared to be no differences in the gross weights of utricular otoliths and underlying maculae in embryos up to the age of 17 days or approximately 20 g.
3. However, embryos of more than 20-g body weight showed a significant ($p \leq 0.01$) difference in utricular otolith weights, those of centrifuged animals being lighter than those of control animals.
4. Although it is believed that this effect may be due to the action of a weight-dependent mechanism regulating the deposition or removal of calcium carbonate from the otoconia, additional control experiments and measurements of otoconial weights without maculae are deemed necessary to confirm this finding.
5. In the scanning electron micrographs of the utricular otoconia of a 1-day-old chick a rough membrane connecting neighboring otoconia and covering their round surfaces was observed. It was speculated that this membrane functions to hold the otoconia together and hence helps to attach them to the otoconial membrane.

ACKNOWLEDGMENTS

We wish to thank Dr. M. Parthasarathy for the use of the scanning microscope, and Dr. C. Pueschel for technical assistance with it. Drs. T. Eisner and G. Hausfater loaned us essential equipment for this project which we acknowledge with thanks. We are also indebted to S. Prensky, J. Krueger, and M. Howland for technical assistance, and to Dr. Edward Brothers for a most helpful critical reading of the manuscript.

References

1. Balsamo, G., de Vincentiis, M., and Marmo, F.: The effect of tetracycline on the processes of calcification of the otoliths in the developing chick embryo. J. Embryol. Exp. Morphol. 22(3):327, 1969.

2. Carlström, D.: A crystallographic study of vertebrate otoliths. Biol. Bull. 125:441, 1963.
3. Decker, J.D.: The influence of early extirpation of the otocysts on the development of the behavior of the chick. J. Exp. Zool. 174:349, 1970.
4. Freeman, B.M. and Vince, M.A.: Development of the Avian Embryo. London, Chapman and Hall, 1974.
5. Gottlieb, G.: Prenatal behavior of birds. Q. Rev. Biol. 43:148, 1968.
6. Harada, Y. and Sugimoto, Y.: Metabolic disorder of otoconia after streptomycin intoxication. Acta Otolaryngol. 84:65, 1977.
7. Kellerhals, B., Marti, E., and Villiger, W.: Surface view of the guinea pig otolithic membrane. A scanning electron microscope observation. Pract. Oto-rhino-laryngol. 32:65, 1970.
8. Kuo, Z.-Y.: Ontogeny of embryonic behavior in aves: Chronology and general nature of behavior in the chick embryo. J. Exp. Zool. 61:395, 1932.
9. Lim, D.J.: Three dimensional observation of the inner ear with the scanning microscope. Acta Otolaryngol. (Suppl.) 255:1, 1969.
10. Lim, D.J.: Formation and fate of the otoconia: Scanning and transmission electron microscopy. Ann. Otol. 82:23, 1973.
11. Lim, D.J.: The statoconia of the non-mammalian species. Brain Behav. Evol. 10:37, 1974.
12. Romanoff, A.L.: The Avian Embryo. New York, Macmillan, 1960, p. 374.
13. Smith, A.H. and Burton, R.R.: Chronic acceleration of animals. In Gordon, S.A. and Cohen, M.J. (eds.): Gravity and the Organism. Chicago, Univ. of Chicago Press, 1971, pp. 371–388.
14. Snedecor, G.W. and Cochran, W.G.: Statistical Methods, 6th ed. Aims, Iowa, Iowa State Univ. Press, 1967, pp. 432–436.
15. Visintini, F. and Levi-Montalcini, R.: Relazione tra differenziazione strutturale e funzionale dei centri e delle vie nervose nell'embrione di pollo. Schweiz. Arch. Neurol. Psychiat. 43:1, 1939.

Discussion

LIM: Just one comment. In the animal—which is a genetic mutant—that Larry Abel, David Clark, and I have been examining, some animals had totally deficient otoconia and some animals had some otoconia left. One thing we learned from that study was until the otoconia are lost, more than 70% of the animals do not show real remarkable behavioral disorders, so to speak. They have problems swimming; and one parameter which we thought might be mostly correlated to the fish otoconia (pardon, Professor Engström) deficiency is that their righting reflex is very sensitive. This correlates with the swimming behavior. So, to get back to the experiment that you are presenting here, two things may have to be kept in mind. One is the hypogravity itself: even though the total number of crystals may be slightly reduced, total gravity of force to the senses may not be. It might be actually accelerated. I don't know exactly what behavioral test you plan to do; but that is a very complex problem and I don't know how it can be tested.

5

Morphologic Observations of Human Otoconial Membranes

CHARLES G. WRIGHT AND DAVID G. HUBBARD

The otoconial membranes which cover the neuroepithelia of the saccule and utricle provide mechanical stimulation for the receptor cells in response to linear acceleration. These structures are composed of sheets of gelatinous material covered by thousands of tiny calcium carbonate crystals known as otoconia (7,9,10,12).

Since revival of the microdissection method and the advent of scanning electron microscopy, the otoconia have received increased attention in human temporal bone studies (5,16,18). The microdissection technique permits examination of intact otoconial membranes by either light or electron microscopy and it avoids lengthy storage of specimens in preservatives or acidic solutions which may alter the otoconial crystals.

This report presents a survey of observations made on human otoconial membranes during an investigation concerning the development and pathology of the static organs in man.

METHODS

Altogether, temporal bones from 125 individuals, ranging in age from 14 weeks of gestation to 62 years, were examined in the course of this study. Specimens were obtained at autopsy and fixed by perilymphatic perfusion with either 1% osmium tetroxide or 2% glutaraldehyde; both fixatives were veronal buffered at pH 7.4. With minor modifications, all materials were prepared for study by the microdissection technique as described by Hawkins and Johnsson (6). The vestibule was opened via the oval window to permit the saccular otoconial membrane to be photographed in situ. The anterior portion of the utricle was then carefully removed from the vestibule and trimmed so as to allow examination of the utricular otoconial membrane.

Specimens studied by scanning electron microscopy were removed from the maculae and either air or critical-point dried. They were sputter coated with gold-palladium before examination with a JEOL Model 35 microscope operated at 15 kV.

RESULTS AND DISCUSSION

The microdissection technique for specimen preparation is especially useful for study of the otoconial membranes since both light and scanning electron microscopic observations may be made on a single tissue sample.

Figure 5-1, for example, shows a light micrograph of a saccular otoconial membrane lying in situ on macula sacculi. With care, the entire membrane, including both the crystalline and gelatinous layers, can be dissected off the macula, dried, and then examined by scanning electron microscopy. Figure 5-2 shows a series of electron micrographs of this specimen at increasing levels of magnification.

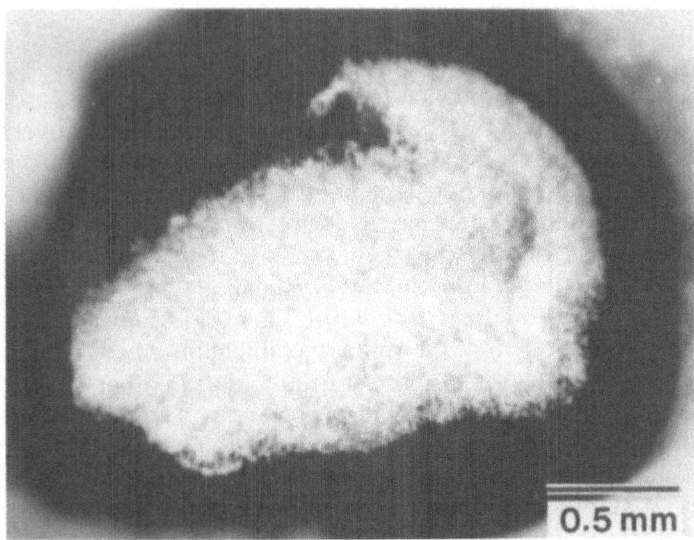

Fig. 5-1. Light micrograph of a saccular otoconial membrane from a 2-year-old infant. Here the upper surface of the otoconial mass is viewed through the enlarged oval window.

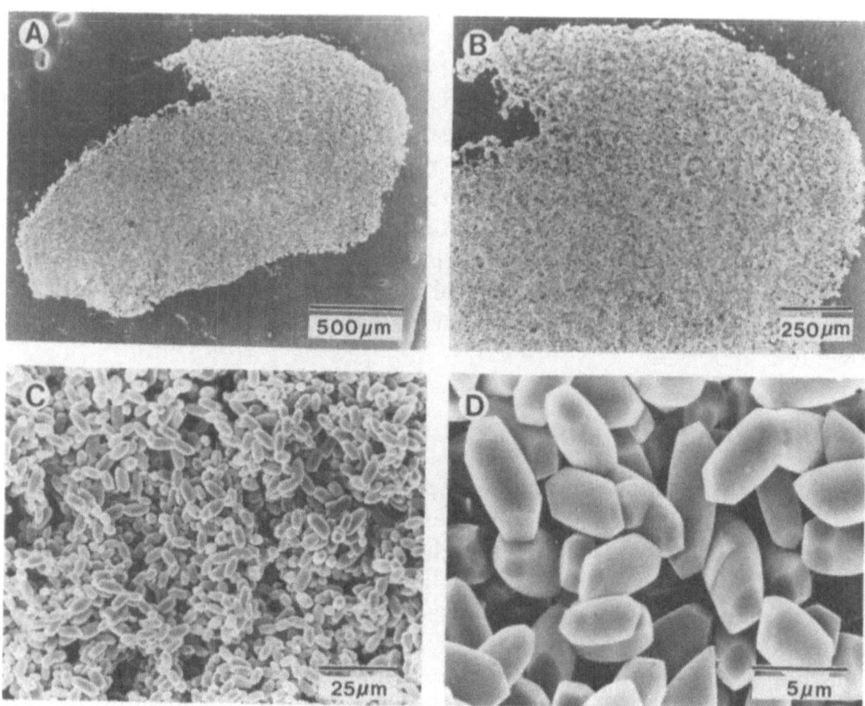

Fig. 5-2. Scanning electron micrographs of the specimen shown in Fig. 5-1. At higher magnifications (**C** and **D**) the morphology of individual otoconia may be conveniently studied.

Light micrographs of otoconial membranes from the saccule and utricle of a young adult are seen in Fig. 5-3. The otoconial membranes cover the entire macular surface and conform closely to the shape of the underlying neuroepithelia. The characteristic "hook" shape of the saccular otoconial membrane is due to an extension at its anterior end called the dorsal lobe; this feature is quite variable in size and shape between different individuals.

In both the saccule and utricle, the crystalline layer is thickened along a curved ridge known as the "snowdrift" line (1). This contour directly overlies the striola region of the sensory epithelium, where type I hair cells are concentrated (12). It is of interest to note that in other species, such as the guinea pig, the "snowdrift" is also present as a thickened band on the saccular otoconial membrane; however, the corresponding area of the utricular membrane is usually quite thin or even lacking in crystals (12,17).

Figure 5-4 shows a utricular otoconial membrane from a human fetus of 14 weeks gestational age compared to a fetal specimen at approximately term. Although the otoconial layer of the smaller membrane is noticeably thinner, it has a well-developed "snowdrift" line. The more mature otoconial membrane is approximately 3.4 times larger in surface area than the

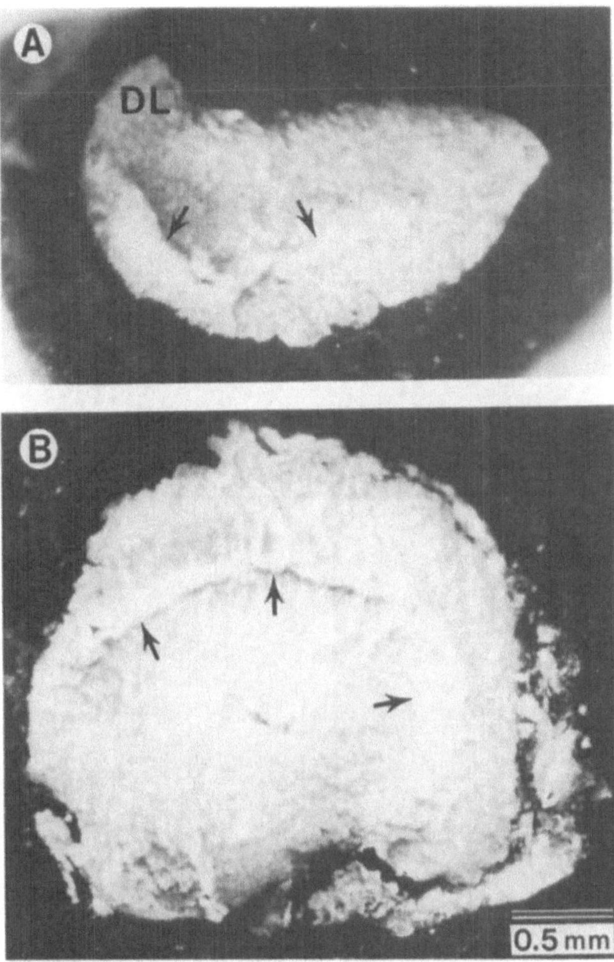

Fig. 5-3. Saccular (**A**) and utricular (**B**) otoconial membranes from a 31-year-old adult. *DL*, dorsal lobe of saccular otoconial membrane; arrows indicate surface contours known as "snowdrift" lines. From Wright, C.G., et al.: Ann. Otol. Rhinol. Laryngol. 88:267, 1979.

14-week specimen. Our observations on otoconial membrane growth during the second and third trimesters have shown that these structures increase some three to four times in surface area in the course of this period of fetal development (19). This growth is illustrated by the saccular otoconial membranes shown in Fig. 5-5 which were obtained from fetuses of 14 to 32 weeks gestational age. Comparison of fetal material with specimens from infants and children has shown that the otoconial membranes reach their full surface area by term. In fact, growth in size appears to be complete by the 26th or 27th week of gestation. This finding is in accord with the observa-

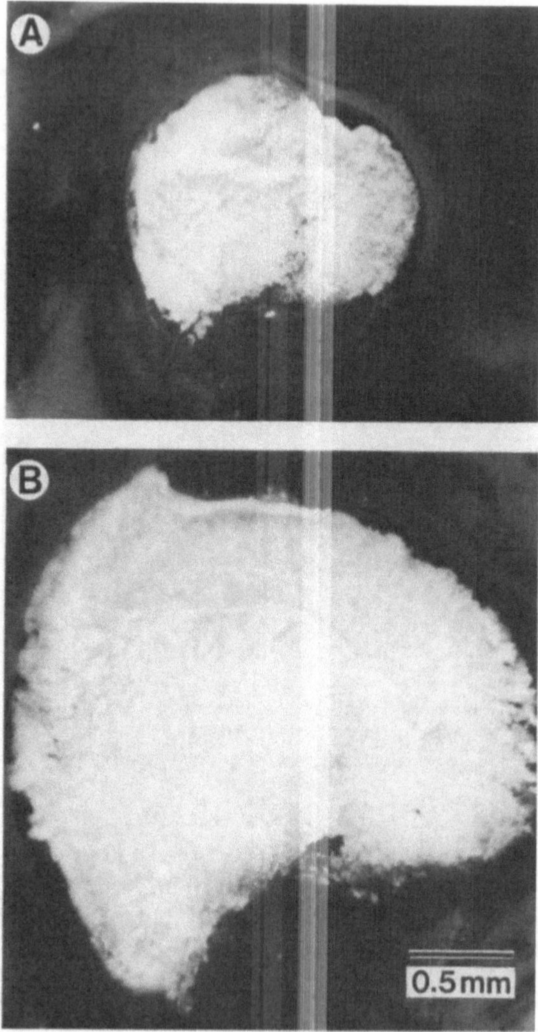

Fig. 5-4. Utricular otoconial membranes from fetuses of 14 weeks (**A**) and 36 weeks (**B**) gestational age. From Wright, C.G., et al.: Ann. Otol. Rhinol. Laryngol. 88:267, 1979.

tions of Rosenhall (15), who found the surface area of the macular neuroepithelia in human infants to be the same as that in adults.

Scanning electron microscopic studies revealed an abundance of very small, apparently immature otoconia on the surfaces of fetal otoconial membranes. In Fig. 5-6, many tiny crystals are seen near the edge, and adjacent to the "snowdrift" of an 18-week utricular membrane. Figure 5-7 shows immature otoconia on the surface of an otoconial membrane from a specimen of 20 weeks gestational age. In this case, one can see that the crys-

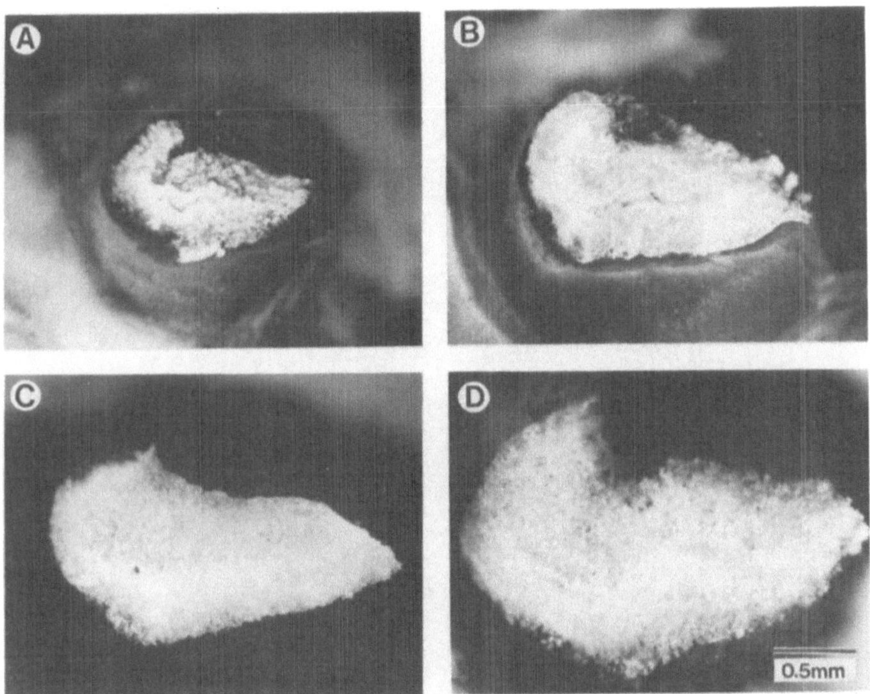

Fig. 5-5. Growth of fetal otoconial membranes illustrated by saccular specimens of 14 weeks (**A**), 16 weeks (**B**), 22 weeks (**C**), and 32 weeks (**D**) gestational age.

tals which are partially embedded in the gelatinous substance of the membrane are considerably larger than the tiny otoconia directly at the surface.

We have found little or no evidence to indicate that otoconia are first formed within the gelatinous portion of the otoconial membrane and emerge from it as they mature. Rather, it appears that the crystals develop at the interface between the endolymph and the crystalline layer of the otoconial membrane.

In more mature specimens from infants and children, saccular otoconia are remarkably homogeneous in size and shape, as illustrated in Fig. 5-8 where saccular and utricular otoconia from the same infant are compared. The saccular crystals range from approximately 8 to 12 μm in length and have a highly uniform cylindric shape. On the other hand, as Ross et al. (16) have also observed, utricular otoconia are consistently smaller and more variable in shape. Generally, utricular crystals vary from less than 1 to about 8 or 9 μm in length.

Previous investigations on the effects of aging in man have shown that saccular and utricular otoconia also differ in regard to their susceptibility to pathologic change. Johnsson and Hawkins (8) and Ross et al. (16) have, for

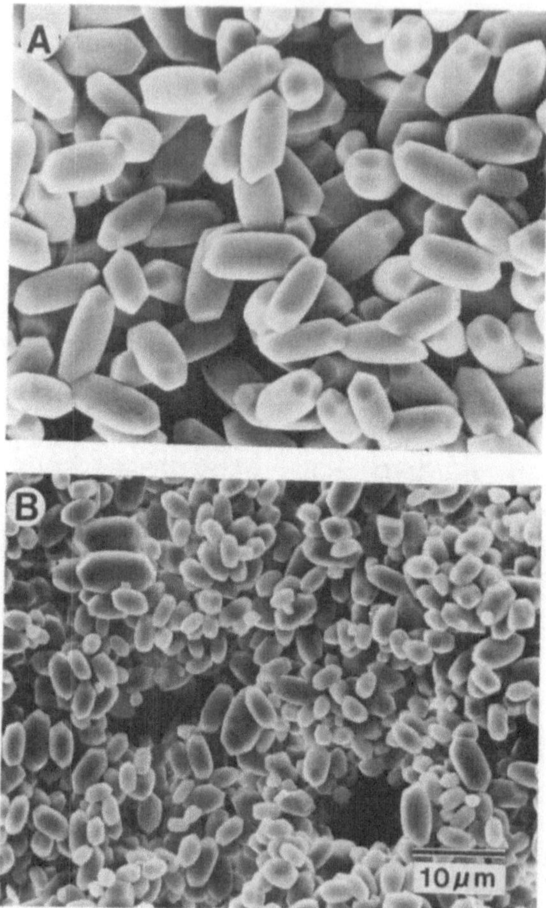

Fig. 5-6. Small, immature otoconia near the edge (**A**) and adjacent to the snowdrift (**B**) of a utricular otoconial membrane from a fetus of 18 weeks gestational age.

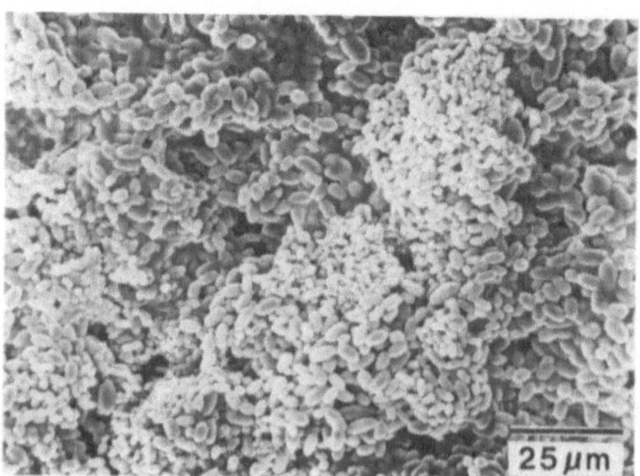

Fig. 5-7. Growing otoconia on the surface of a saccular otoconial membrane of 20 weeks gestational age. Crystals surrounded by gelatinous material (*upper left*) are larger than the very small otoconia directly on the surface (*right center*).

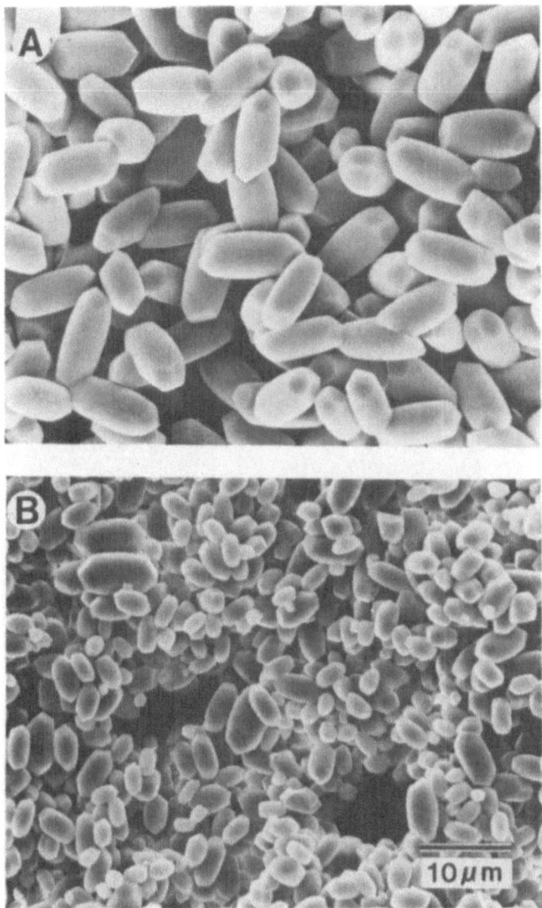

Fig. 5-8. Saccular (**A**) and utricular (**B**) otoconia from a 4-month-old infant. Note smaller size and greater variation in shape of utricular crystals relative to those from the saccule.

example, demonstrated a rather selective loss of saccular otoconia associated with age-related degeneration of the inner ear.

In the present study, abnormal saccular otoconia were observed in several infants with otitis media (18). Giant, malformed crystals from such a case are shown in Fig. 5-9. The peculiar cobblestone-like arrangement of large otoconia illustrated in Fig. 5-9B was visible through the oval window at the time of temporal bone fixation. Thus, it cannot be the result of fixation or storage artifact. From a total of 30 temporal bones with otitis media involving severe inflammation and edema of the middle ear mucosa, four were found to contain abnormal saccular otoconia; in these cases the utricular crystals were normal.

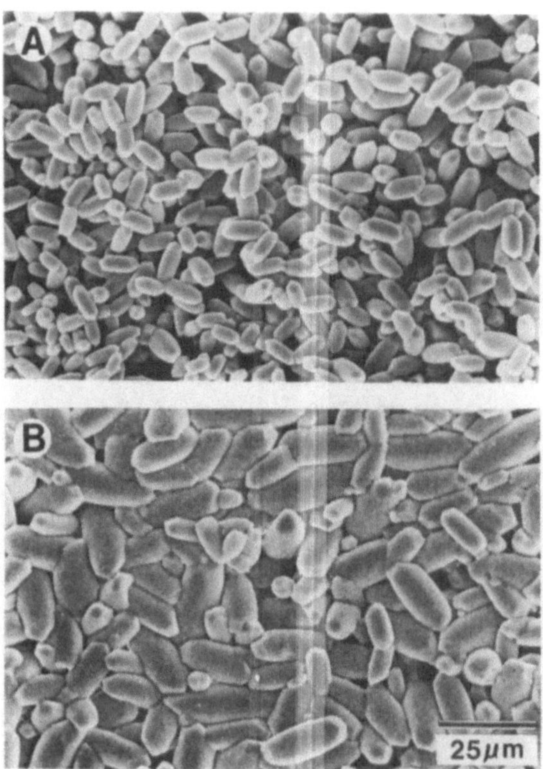

Fig. 5-9. **A** Normal saccular otoconia from a 2-year-old infant. **B** Giant, malformed otoconia from a 6-week-old infant with severe otitis media. Note dense, cobblestone-like packing of crystals. From Wright, C.G. and Hubbard, D.G.: Acta Otolaryngol. 86:185, 1978.

These findings, together with recent reports of sensorineural hearing loss in both acute and chronic otitis media (2,14), provide evidence that toxic substances from middle ear infections may occasionally reach the endolymph and adversely affect the inner ear, including the otoconia.

Figures 5-10 and 5-11 are photomicrographs of material from a case in which otoconia were apparently congenitally absent. This anomaly occurred in a 6-week-old victim of sudden infant death syndrome (20). At the time of temporal bone fixation, it was determined that all four maculae were devoid of otoconia, as illustrated in Fig. 5-10 where the right utricular macula is seen. Although otoconia were absent, the gelatinous otoconial membranes were intact and in place over the neuroepithelia. Figure 5-11 shows scanning electron micrographs of the gelatinous membrane of the right utricle after critical-point drying.

The absence of otoconia in this infant is not attributable to fixation artifact or to long storage in preservatives, nor is it likely to be the result of postmor-

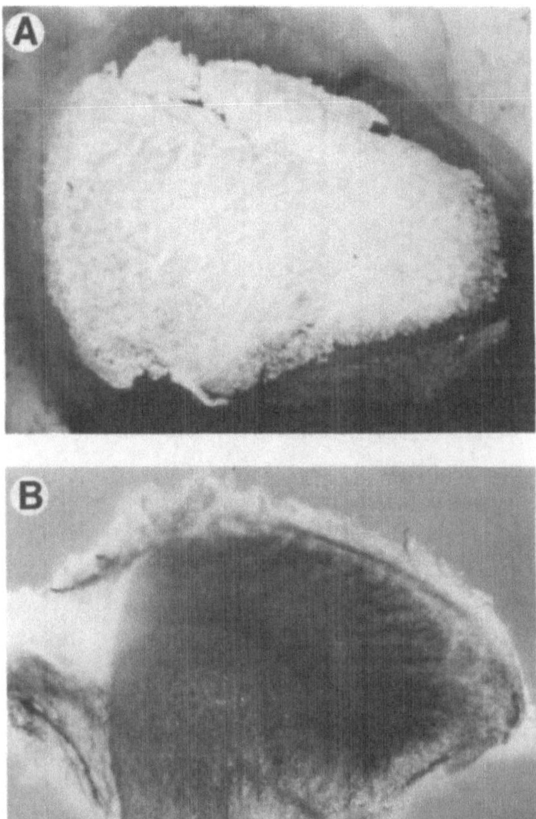

Fig. 5-10. **A** Normal crystalline otoconial membrane covering utricular macula from a 12-week-old infant. **B** Right utricular macula, devoid of otoconia, from a 6-week-old infant in whom otoconia were absent. The darkly stained myelinated innervation is visible beneath the neuroepithelium. From Wright, C.G., et al.: Ann. Otol. Rhinol. Laryngol. 88:779, 1979.

tem autolysis. The postmortem interval before fixation was less than 6 h and this is the first such case we have seen, even though there is wide variation in age at death, premortem pathology, and postmortem time in our material. It might also be noted that the other structures of the membranous labyrinth, including the neuroepithelia and innervation, appeared normal in this infant.

Previous studies on pathology of the otoconia have shown that when otoconia degenerate in vivo or when they are experimentally dissolved in acids, residual organic material or crystalline fragments remain behind as evidence of their former presence on the maculae (10,16). In the infant we studied, no such material was present on the gelatinous otoconial mem-

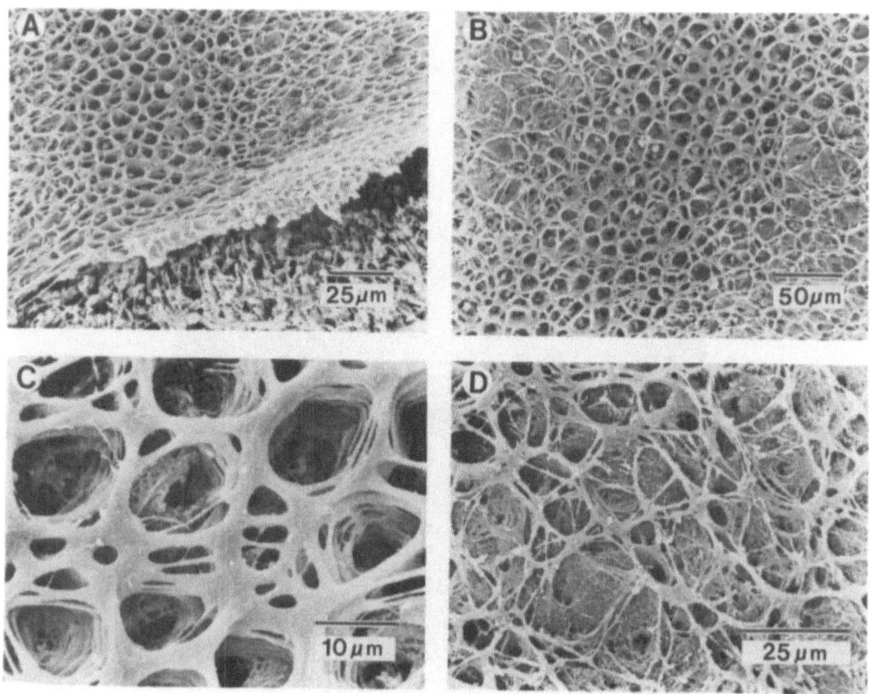

Fig. 5-11. Scanning electron micrographs showing the gelatinous layer of right utricular otoconial membrane from infant with congenital absence of otoconia. **A** Low-power view near anterior end of macula; note surface of neuroepithelium at lower right. **B** Central area of specimen in striola region. **C** Higher-magnification view of openings in the gelatinous layer over the striola. **D** Surface of gelatinous layer in lateral portion of specimen. From Wright, C.G., et al.: Ann. Otol. Rhinol. Laryngol. 88:779, 1979.

branes. It therefore seems probable that the crystals never developed during the fetal period; thus, the lack of otoconia was most likely a congenital, rather than acquired, defect.

Genetic abnormalities involving the otoconia have been investigated in experimental animals such as the pallid mutant mouse (11,13). In severely affected pallid mice, the otoconia never develop; however, the gelatinous otoconial membranes, neuroepithelia, and macular innervation are normal. The highly specific nature of this defect has made the pallid mouse a valuable subject for studies of vestibular anatomy and physiology. In addition to alterations in postural reflexes, these animals show behavioral and neuroanatomic deficits due to impaired vestibular input to the central nervous system (3,4).

The human case described above demonstrates that an inner ear defect strikingly similar to that in the experimental animal model may also occur in man.

ACKNOWLEDGMENT

This work was supported by a grant from the Leland Fikes Foundation to the University of Texas at Dallas subcontracted through the Aberrant Behavior Center. The authors wish to thank La Nece Lomonte for her skillful assistance in preparation of the manuscript.

References

1. Ades, H.W. and Engström, H.: Form and innervation of the vestibular epithelia. In: The Role of the Vestibular Organs in the Exploration of Space. Washington, D.C., NASA SP-77, 1965, p. 23.
2. Arnold, W., Ganzer, U., and Kleinmann, H.: Sensorineural hearing loss in mucous otitis. Arch. Otol. Rhinol. Laryngol. 215:91, 1977.
3. Clark, G.M., Douglas, R.J., Erway, L.C., Wright, C.G., and Hubbard, D.G.: Vestibular nuclei: neuronal loss in mice with otoconial agenesis and evidence of right-left asymmetry. In Gualtierotti, T. (ed.): Vestibular Function and Morphology. New York, Springer-Verlag, 1981, Chap. 6.
4. Douglas, R.J., Clark, G.M., Erway, L.C., Hubbard, D.G., and Wright, C.G.: Effects of genetic vestibular defects upon behavior related to spatial orientation and emotionality. J. Comp. Physiol. Psychol. 93:467, 1979.
5. Harada, Y., Graham, M.D., Pulec, J.L., and House, W.F.: Human otoconia in surgical specimens. Arch. Otolaryngol. 104:371, 1978.
6. Hawkins, J.E., Jr. and Johnsson, L.-G.: Microdissection and surface preparations of the inner ear. In Smith, C.A. and Vernon, J.A. (eds.): Handbook of Auditory and Vestibular Research Methods. Springfield, Ill, Thomas, 1976.
7. Johnsson, L.-G. and Hawkins, J.E., Jr.: Otolithic membranes of the saccule and utricle in man. Science 157:1454, 1967.
8. Johnsson, L.-G. and Hawkins, J.E., Jr.: Sensory and neural degeneration with aging, as seen in microdissections of the human inner ear. Ann. Otol. Rhinol. Laryngol. 81:179, 1972.
9. Lim, D.J.: Formation and fate of the otoconia. Ann. Otol. Rhinol. Laryngol. 82:23, 1973.
10. Lim, D.J.: Ultrastructure of the otolithic membrane and the cupula. Adv. Otorhinolaryngol. 19:35, 1973.
11. Lim, D.J. and Erway, L.C.: Influence of manganese on genetically defective otolith, a behavioral and morphological study. Ann. Otol. Rhinol. Laryngol. 83:565, 1974.
12. Lindeman, H.: Studies on the morphology of the sensory regions of the vestibular apparatus. Ergeb. Anat. Entwicklungsgesch. 42:1, 1969.
13. Lyon, M.F.: Absence of otoliths in the mouse: an effect of the pallid mutant. J. Genet. 51:638, 1953.
14. Paparella, M.M., Oda, M., Hiraide, F., and Brady, D.: Pathology of sensorineural hearing loss in otitis media. Ann. Otol. Rhinol. Laryngol. 81:632, 1972.

15. Rosenhall, U.: Vestibular macular mapping in man. Ann. Otol. Rhinol. Laryngol. 81:339, 1972
16. Ross, M.D., Johnsson, L.-G., Peacor, D., and Allard, L.: Observations on normal and degenerating human otoconia. Ann. Otol. Rhinol. Laryngol. 85:310, 1976.
17. Ross, M.D. and Peacor, D.: The nature and crystal growth of otoconia in the rat. Ann. Otol. Rhinol. Laryngol. 84:22, 1975.
18. Wright, C.G. and Hubbard, D.G.: Observations of otoconial membranes from human infants. Acta Otolaryngol. 86:185, 1978.
19. Wright, C.G., Hubbard, D.G., and Clark, G.M.: Observations of human fetal otoconial membranes. Ann. Otol. Rhinol. Laryngol. 88:267, 1979.
20. Wright, C.G., Hubbard, D.G., and Graham, J.W.: Absence of otoconia in a human infant. Ann. Otol. Rhinol. Laryngol. 88:779, 1979.

Discussion

LIM: I know your work and I had the opportunity to read your manuscript. I think it's a very important area of research. As you know, many of the genetic deafnesses, heredity types of deafness, are often associated with saccular degeneration, known as cochleosaccular degeneration. This is a very clinically important area where a practical clinical test is needed for just the saccular function, since currently it's just not done well. I think this type of animal model just mentioned is very important. Particularly, the finding that total absence of otoconia can occur in certain humans is very important and a strong attempt should be made to define this group of patients. Muriel Ross, as you mentioned, described two types of crystals, a normal type that you described and a type which is presumably early, immature crystal. Have you seen these in your embryonal humans or is it just an abnormality of the rat?

WRIGHT: We haven't observed crystals of that type in material from second and third trimester fetuses. However, otoconia of the immature form described by Dr. Ross might be present earlier in gestation. Unfortunately, it simply has not yet been possible to obtain fetal material suitable for study of the earliest phases of crystal development in man.

6

Vestibular Nuclei: Neuronal Loss in Mice with Otoconial Agenesis and Evidence of Right-Left Asymmetry

GEOFFREY M. CLARK, ROBERT J. DOUGLAS, LAWRENCE C. ERWAY, AND DAVID G. HUBBARD

The importance of sensory systems to brain development has been investigated in terms of transneuronal changes due to sensory deprivation. The bulk of these studies have involved the visual system, in part because of the relative ease with which the animal can be deprived of stimulation. Notably lacking in the literature are reports of the effects of vestibular deprivation. This is most unfortunate since some investigators (32) postulate the vestibular system to play a key role in brain development, whereas others have argued that vestibular deprivation may underlie some mental disorders. A variety of mental disorders, most notably autism and schizophrenia, have been reported by many investigators to be associated with vestibular disorders or abnormalities (2,19,30,40,41,45,50,52,53). Further, increased vestibular stimulation has been reported to significantly ameliorate the symptoms of a number of mental and/or brain disorders (2,7,8,30,37). Thus, there are many reasons why a study of vestibular deprivation ought to be undertaken, but until recently there were no feasible ways of producing this deprivation except through surgical intervention and the attendant stress and trauma.

Since the advent of the space age it has been possible to eliminate the normal functioning of gravity receptors, and a number of investigators have examined the effects of prolonged periods of weightlessness at various stages of development. Vinnikov et al. (55) reported several vestibular defects found in ten rats orbited for 19.5 days aboard the Kosmos 782 satellite. These defects included (1) thinning of otolithic membranes; (2) increased edema vacuolization in the receptor epithelium of the utricle; (3) necrobiosis primarily of type I receptor cells and vacuolization of the cup-shaped nerve endings surrounding them; and (4) a loss of proper crystalline form of the otoconia. Vinnikov and his colleagues (54,56,57) have shown that 6- to 16-day-old fish and amphibian embryos continue to develop apparently normal labyrinths when exposed to 40 h of space flight.

It is only recently that the behavioral effects of weightlessness during early development have been studied (28,29,49,58). Von Baumgarten et al. (58) observed that killifish hatched on the 9.5-day Skylab mission "appeared to maintain their orientation visually immediately upon hatching by swimming with their backs turned toward the light source." These fish unfortunately died before they could be tested at normal gravity and were decomposed beyond useful histologic analysis.

Hoffman et al. (28,29) studied the postflight behavior of killifish carried as fertilized eggs aboard the 9.5-day Apollo-Soyuz space flight. In these studies only the fish which were at the earliest stage of embryonic development (32 h postfertilization) at the time of insertion into orbit were consistently different in their behavior from ground-reared controls. Compared to the latter, the space-reared fish (1) hatched earlier; (2) showed increased negative phototaxis; (3) displayed decreased swimming speed in the clockwise direction while tracking a black and white vertically striped drum; (4) rotated at slower propeller speeds about the longitudinal body axis in prop-wash tests of vertical (x-axis) stability; (5) showed less disorientation to zero gravity in parabolic flight in early test parabolas; and (6) were more sensitive to hypergravity (1.4 g as measured by swimming above the bottom third of the tank). Unfortunately, the results of brain histology on this one group (32-h eggs) have not been published.

There are several problems with the use of space flight as a means of studying vestibular deprivation. The animal is, for example, subjected to the added and confounding effects of excessive g-forces on lift-off and reentry. More importantly, even the static organs are not so much deprived of input as subjected to abnormal input. These problems can be eliminated by the use of animals in which static organ function is eliminated by a genetically caused failure of otoconial development. Such an animal is the pallid mouse (C57BL/6J-*pa*), first studied by Lyon (36) and Erway et al. (15–17). Mice heterozygous for the pallid gene are normal and make excellent littermate controls for those homozygous for the pallid gene. The homozygote for the pallid gene can be completely lacking in otoconia, partially lacking, or normal. Prenatal supplementation with manganese will produce normal otoconial development. We have already investigated the behavior of mice with otoconial agenesis and have found these "pallid" mice to display marked deviations from their normal littermates in terms of both "intellectual" and emotional components of behavior (12). Many of these behavioral deviations are similar to behavior deviations reported in autistic and/or child schizophrenic humans. This behavioral work led to an investigation of possible brain disorders such as transneuronal degeneration in mice with otoconial agenesis.

Cell populations chosen for study were the medial vestibular nucleus because of its utricular input, nucleus Y because of its saccular input (21), and the cerebellar fastigial nucleus because of its possible first order and definite second order static organ input (5,31,48). (See Chap. 2 for an excellent discussion of these neuroanatomic relationships.) In two pilot studies we were able to confirm that two of these regions in the defective animals displayed statistically significant reductions in the number of neurons. These were nucleus Y and the lateral portion of the medial vestibular nucleus (9).

The pilot studies, however, were confounded by the fact that most of the mice with otoconial agenesis had much smaller brains than did the control animals. Further, the controls consisted of mice heterozygous for the pallid gene and thus failed to control for the possibility that cell reduction is an independent effect of homozygosity. In the present study we attempted to replicate and extend the earlier results in a group consisting only of pallid homozygotes with virtually equal-sized brains.

METHODS

Ten mice of the C57BL/6J-*pa* strain were chosen. This sample included three animals with complete bilateral agenesis of otoconia (bilateral defectives) and four with strictly unilateral agenesis (unilateral defectives). Of the latter, two had normal or near normal otoconial development in the left labyrinth and two in the right. A final group consisted of three mice which, although homozygous for the pallid gene, nevertheless had normal otoconial development in both labyrinths. Subjects were coded so that the experimenter was not aware of which subjects were in which group. Neurons were counted in the two vestibular relay nuclei and in the fastigial nucleus of the cerebellum, in a series of 10 μm coronal sections.

This study is concerned with two types of comparisons, i.e., between subject (group) comparisons and within subject comparisons. The unilateral group is both a group within the overall study and a separate study unto itself. Within each of these animals we have both a normal and deafferented group of vestibular nuclei, assuming that primary input is solely ipsilateral (uncrossed). In addition, comparisons between right and left halves of the brainstem (symmetry) were made on all subjects.

Histologic Procedure

The coded brains were extracted, formalin fixed, and forwarded to the University of Washington by Dr. Erway. The temporal bones were dehydrated in alcohol, cleared in oil of wintergreen, and microscopically viewed for otoconia scoring (Fig. 6-1). Upon arrival the mouse brains were dehydrated and embedded in paraplast plus. Serial 10-μm coronal sections were cut through the brainstem in the plane used by Kovac and Denk (35). The paraplast ribbons were mounted on slides, deparaffinized, and stained with cresyl violet.

General Cell Counting Procedure

Neuron nucleoli were counted at 400 × using a Howard disk eyepiece reticule (10 × 10 mm ruled into 36 squares) in a Zeiss or Leitz Dialux 20 microscope. In order to be included within a square, a nucleolus had to be more than half inside the square. Section thickness was measured by focusing on the top and bottom of the cerebellar granule cell layer using the Zeiss plan 100 × oil immersion objective, and determining the thickness from the micrometer on the fine focus knob. Three estimates were averaged for each section.

NUCLEUS Y

Procedure

The Leitz Dialux 20 microscope was used for all cell counts in nucleus Y. In coronal sections nucleus Y tapers off anteriorly and posteriorly and reaches its greatest extent and cell population at a point where the top of the glial cap, which forms the beginning of the dorsal acoustic striae of the dorsal cochlear nucleus, is roughly even with the dorsal surface of the restiform body (the ventral extent of nucleus Y). The center of the nucleus then was the area of its greatest extent and cell population. For nucleus Y a grid measuring 249 μm along each edge was superimposed over the nucleus in the manner shown in Fig. 6-2B, and neuron nucleoli were counted within this grid on consecutive sections extending for 100 μm in a rostrocaudal direction. The central 50 μm, where the presence of nucleus Y was unambiguous, was defined as the "central" portion and a combination of the 25 μm blocks anterior and posterior to this was defined as the "peripheral" portion of the nucleus. The peripheral portion included a larger proportion of cells which were probably not part of nucleus Y, but this nucleus does not have well-defined boundaries.

Results

The results are summarized in the first two rows of Table 6-1. *Y, 100* refers to the total 100-μm sample of nucleus Y and *Y, 50* refers to the central 50-μm portion.

Reduction in cell number was most evident in the central portion of nucleus Y, where the normal mice averaged 538 neurons and the bilaterally

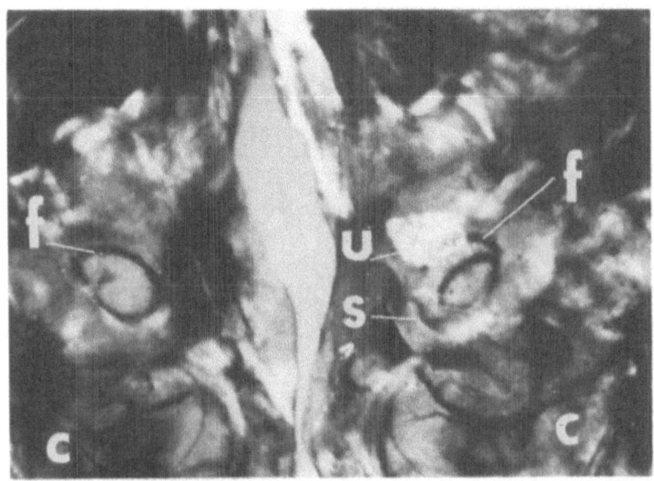

Fig. 6-1. Asymmetric development of otoliths in a pair of labyrinths from pallid mouse. These inner ears from a single pallid mouse were cleared in oil of wintergreen, and the whole mounts were photographed with polarized light, without removal of any of the bony labyrinth. The outline of the cochlea (*c*) and of the fenestra ovalis (*f*) can be seen in each ear. The smaller, inner ring associated with the right fenestra ovalis is the footplate of the displaced stapes; the stapes was removed from the left ear during dissection. The right utricle (*u*) and saccule (*s*) possess a normal, birefringent mass of otoconia. The left ear contains no otoconia which would normally be seen at the edges of the fenestra ovalis (*f*), appearing as a ring in the background.

defective mice 399. This difference, amounting to 26% reduction, in the bilaterally defective mice was statistically reliable ($F = 6.085$; $t = 3.25$; $p < 0.025$). The mice with unilateral otoconial agenesis averaged 422 neurons and were also reliably below the normals in this respect ($F = 6.085$; $t = 3$; $p < 0.01$). As in the pilot studies, the unilaterally defective animals did *not* have a greater cell reduction on the side ipsilateral to otoconial agenesis, but instead had an approximately equal reduction on both sides.The differences between groups were less pronounced in the peripheral portions of nucleus Y and were in no case statistically significant. In the peripheral region, however, a sharp asymmetry was discovered, with the largest number of cells counted on the right side. This was assessed by assigning to each mouse a score consisting of the number of neurons in the right peripheral region divided by the number in the left. Right dominance, as reflected by scores above 1.0, was clearly evident in all groups, although more pronounced in the normals. The overall average of 1.270 was highly significant statistically ($t = 5.6$; $p < 0.001$), as was the mean of 1.163 for the whole 100-μm block ($t = 6.2$; $p < 0.001$). The effect was, however, considerably diluted in the central portions as compared to the peripheral portions.

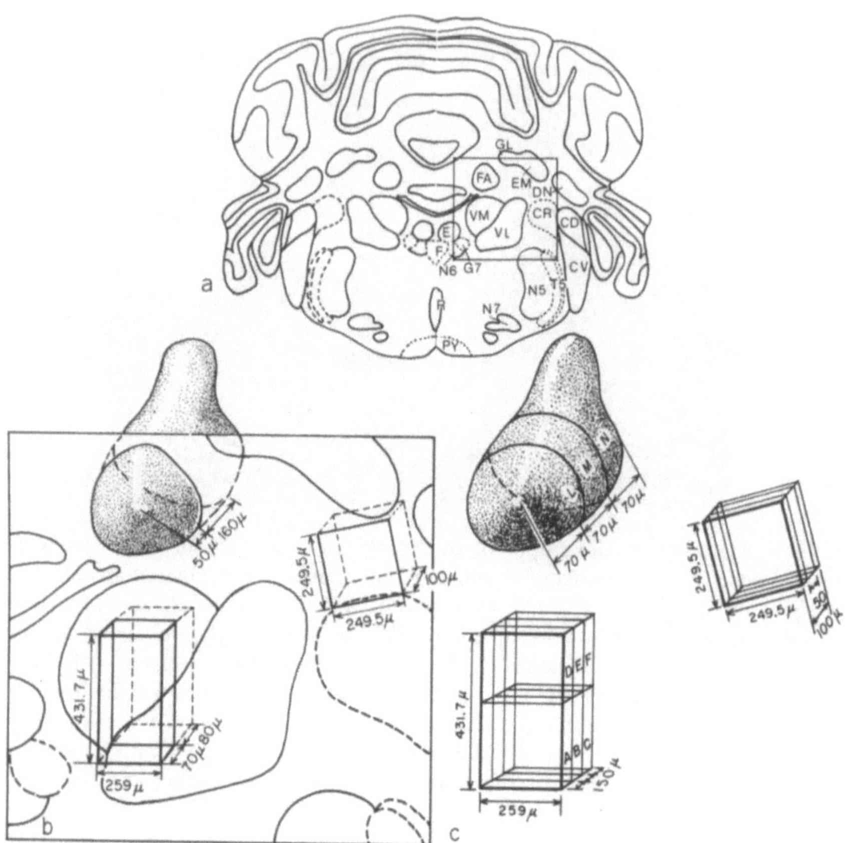

Fig. 6-2. **a** Diagram of a cross section of mouse brainstem seen from a rostral position at the level of abducens nucleus showing the plane of section and rostrocaudal level of sample used in this study. Adapted from Kovac W. and Denk, H.: Der Hirnstamm der Maus. New York, Springer Verlag, 1968, Plate 16. **b** Enlargement of box shown in **a** illustrating the location and *total* sample sizes measured in each nuclear region. **c** Labeling, sizing, and positioning of the subsamples used in data analysis. *CD*, nuc. cochlearis dorsalis; *CR*, corpus restiforme (inferior cerebellar peduncle); *CV*, nuc. cochlearis ventralis; *DN*, nuc. dentatus cerebelli; *E*, nuc. eminentiae medianae; *EM*, nuc. emboliformus cerebelli; *F*, fasciculus longitudinalis medialis; *FA*, nuc. fastigii cerebelli; *GL*, nuc. globosus cerebelli; *G7*, genu n. facialis; *N5*, nuc. tractus spinalis n. trigemini oralis; *N6*, nuc. n. abducentis; *N7*, nuc. facialis; *PY*, tractus pyramide; *R*, nuc. raphe magnus; *T5*, tractus spinalis trigemini; *VL*, nuc. vestibularis lateralis (Dieters nuc.); *VM*, nuc. vestibularis medialis.

Discussion

It was originally hypothesized that the Y nucleus, medial vestibular nucleus, and fastigial nucleus of the cerebellum would show the effects of otoconial agenesis on purely anatomic grounds. It was further hypothesized that the Y

Table 6-1.
MEAN NEURON COUNTS FROM REGIONS IDENTIFIED IN FIG. 6-1.

Region	Sum of Both Sides: Normal	Sum of Both Sides: Unilateral	Sum of Both Sides: Bilateral	Percentage Difference (right [+] vs. left)	Unilateral: Normal Side	Unilateral: Defective Side
Y, 100	1030.0 $n = 3$ $s = 47.3$	849.8[a,h] $n = 4$ $s = 73.9$	807.1[b,g] $n = 3$ $s = 168$	+ 16.4[d,i]	414.9 $n = 4$ $s = 51.2$	434.9 $n = 4$ $s = 63.4$
Y, 50	538.3 $n = 3$ $s = 33.5$	422.3[a,h] $n = 4$ $s = 53.7$	399.0[c,g] $n = 3$ $s = 66.4$	+7.7	205.2 $n = 4$ $s = 34.2$	217.1 $n = 4$ $s = 28.8$
Med. nuc. *A*	444.5 $n = 3$ $s = 33.3$	548.1 $n = 3$ $s = 76.9$	370.4 $n = 2$ $s = 66.7$	−3.0	280.7 $n = 3$ $s = 48.1$	267.4 $n = 3$ $s = 29.5$
B	500.5 $n = 3$ $s = 13.8$	450.5 $n = 4$ $s = 60.4$	455.1 $n = 3$ $s = 70.3$	+8.5	238.9 $n = 4$ $s = 40.3$	211.4 $n = 4$ $s = 28.6$
C	466.6 $n = 3$ $s = 31.0$	442.7 $n = 4$ $s = 65.2$	433.5 $n = 3$ $s = 43.8$	+11.9[d,g]	237.1[e,g] $n = 4$ $s = 29.1$	205.6[e,g] $n = 4$ $s = 38.2$
D	472.2 $n = 3$ $s = 41.5$	467.5 $n = 2$ $s = 108$	463.8 $n = 3$ $s = 84.9$	+0.2	243.9 $n = 3$ $s = 71.3$	223.6 $n = 3$ $s = 37.6$
E	493.0 $n = 3$ $s = 69.2$	383.0 $n = 3$ $s = 72.2$	488.0 $n = 2$ $s = 13.5$	+8.1	201.0 $n = 3$ $s = 57.1$	182.0 $n = 3$ $s = 17.5$
F	448.8 $n = 3$ $s = 29.5$	369.0 $n = 4$ $s = 98.6$	452.1 $n = 3$ $s = 36.6$	+23.6[d,g]	186.7 $n = 4$ $s – 67.2$	182.4 $n = 4$ $s = 34.2$
Fas. nuc. *L*	824.6 $n = 3$ $s = 235$	621.2 $n = 3$ $s = 355$	602.2 $n = 3$ $s = 82.2$	−15.5[f,g]	326.6 $n = 3$ $s = 180$	294.6 $n = 3$ $s = 175$
M	1174.1 $n = 3$ $s = 66.6$	1207.4 $n = 2$ $s = 408$	1214.9 $n = 3$ $s = 408$	−1.0	641.4 $n = 2$ $s = 221$	566.0 $n = 2$ $s = 186$
N	1321.0 $n = 3$ $s = 3\ 4$	1431.1 $n = 3$ $s = 39.3$	1490.7 $n = 3$ $s = 307$	+2.1	744.0 $n = 3$ $s = 44.3$	687.0 $n = 3$ $s = 28.9$
Brain weight (mg)	547.3 $n = 3$ $s = 41.0$	564.0 $n = 4$ $s = 34.5$	563.3 $n = 3$ $s = 11.0$			

[a]Unilaterally defective vs. normal mice: $n = 7$; 1-tail t-test.
[b]Bilaterally defective vs. normal mice: $n = 6$; 1-tail Mann-Whitney "U" test.
[c]Bilaterally, defective vs. normal mice: $n = 6$; 1-tail t-test.
[d]Right vs. left in all 10 mice: 2-tail matched pairs t-test.
[e]Normal vs. defective side within unilaterally defective group: $n = 4$; regions *B* and *C* combined; 1-tail matched pairs t-test.
[f]Right vs. left in bilaterally defective and normal mice: $n = 6$; 2-tail matched pairs t-test.
[g]$p = 0.05$.
[h]$p = 0.01$.
[i]$p = 0.001$.
s, standard deviation.

nucleus would be most affected since it is the "outstanding" site of projection of the saccule (21). Pilot work sampling from the "central" 50 μm of nucleus Y had shown a 33% reduction in cell number in the bilaterally deficient mice as compared to the normals (9). A similar 26% reduction was found for the same comparison in the present subjects and was statistically reliable ($p < 0.025$). The unilaterally defective mice also had significantly fewer cells than the normals (22%; $p < 0.01$). When the unilaterally and bilaterally defective animals are combined and compared to the normals the difference achieves a very high level of statistical significance ($t = 4.8$; $p < 0.001$). When tissue sample size is doubled by adding the anterior and posterior 25-μm blocks to the "central" 50 μm the group differences are still clearly evident but less spectacular. It is worth noting that there was no overlap at all between the normal and defective animals. The lowest scoring normal had more cells than the highest scoring unilateral or bilateral mouse in both the 50-μm and 100-μm cases.

One of the most surprising results is the similarity between the bilateral and unilateral groups, which was found both in the pilot study and in the present investigation. It was hypothesized that the unilaterally defective animals would have normal relay nuclei ipsilateral to normal static organs and cell-deficient nuclei ipsilateral to the defective labyrinths. Averaging the two sides in unilaterally defective mice would then be expected to result in a figure half way between those for the normals and the bilaterally defective mice. Yet, there is clearly no such effect to be found in the present or pilot study data, the unilateral and bilateral animals having very similar scores. Within the unilaterally defective group the nuclei ipsilateral to the normal labyrinth had, in fact, slightly *fewer* neurons than did the nuclei ipsilateral to the defective labyrinth, although the difference did not approach statistical significance.

One possible explanation for bilateral cell loss in nucleus Y in animals only unilaterally lacking otoconia is that each nucleus Y receives input from both saccules. The contralateral input could be direct via primary afferents which cross the midline, indirect via a disynaptic pathway involving direct commissural fibers connecting the two Y nucleus regions, or even polysynaptic. In support of contralateral primary input, Fredrickson et al. (20), and more recently Chan et al. (6), have reported vestibular neurons that responded to contralateral labyrinthine stimulation with very short latencies. Chan et al. (6) found cells in the lateral vestibular nucleus that responded with latencies under 1.4 ms to stimulation of the contralateral saccule. Brodal and Hoivik's (4) small contralateral primary vestibulocerebellar pathway might also provide a rather obscure disynaptic route.

The commissural Y nucleus interchange hypothesis has no confirmation or even direct support in the literature. It seems reasonable, however, since the four major vestibular nuclei have such bilateral interactions. It was pleasing to hear from Dr. Gacek at these meetings that he has determined that nucleus Y is a "commissural nucleus" and he thinks it is likely that the

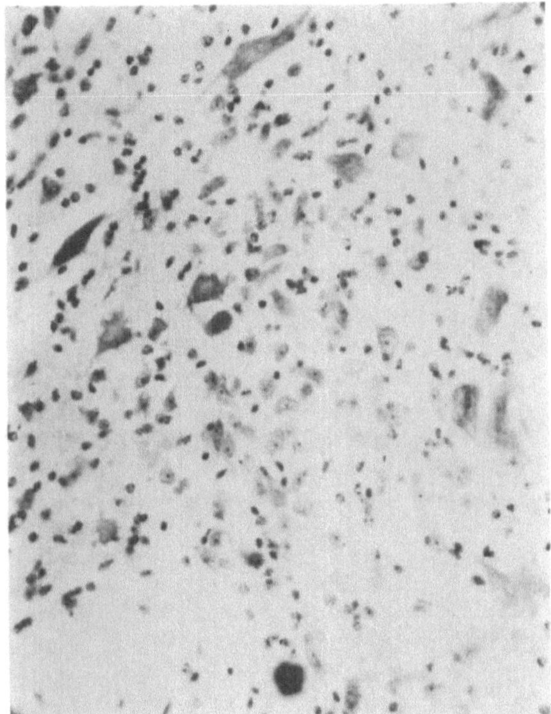

Fig. 6-3. A section through nucleus Y in a pallid mouse (#2) with otoconia in each utricle and saccule. Note that the neurons within the grid are largely medium sized, well distributed, and have a healthy appearance.

fibers reach the contralateral nucleus. Henkel and Martin (27) have located commissural cells in the opossum Y nucleus. While discussing the bilateral input hypothesis it is appropriate to point out that the cell populations within the brain that are most susceptible to atrophy following visual deprivation are those receiving *binocular* input (51).

Figures 6-3–6-5 show respectively fields sampled in nucleus Y in subjects 2 (a normal), 7, and 5 (both totally lacking otoconia). The vacant and dark-staining appearance of Fig. 6-4 when compared to Fig. 6-3 was characteristic of subjects lacking otoconia. Viewing Figs. 6-4 and 6-5, one is left with the impression that neurons are missing and that transneuronal degeneration has taken place. In Fig. 6-5 many neurons appear to have been displaced laterally and compressed into a smaller space. This appearance was particularly typical of the mice with bilateral otoconial agenesis. One of the investigators attempted to blindly (without knowledge of group identification) rank order all ten mice in terms of the appearance noted above. Ranks 1 through 3, indicative of the most severe disorder, were awarded correctly to the three bilateral animals and ranks 4 through 6 to three of the unilaterals.

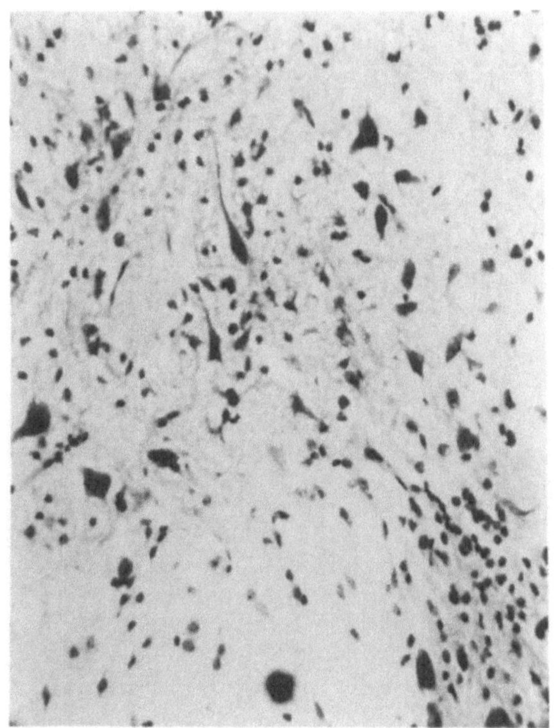

Fig. 6-4. A section through nucleus Y in a pallid mouse (#7) having no otoconia within either labyrinth. Note that the few neurons in evidence are generally small, dark or hyperchromic, shrunken appearing, and widely spaced.

The last unilateral mouse tied with a normal for rank 7.5, while the remaining normals received ranks 9 and 10. The tied rank was due to the unilateral mouse having a nonpathologic appearance. Thus, even though we have employed formal cell counting it is interesting that informal nonquantitative observation can also detect brain changes due to vestibular deprivation.

MEDIAL VESTIBULAR NUCLEUS

Procedure

The Zeiss microscope was used for the medial vestibular nucleus cell counts. The position of the eyepiece grid for the cell counts was determined as follows:

1. The vertical edge of the grid was set parallel to the vertical midline of the brainstem on low (100 ×) power.
2. The dorsoventral center of the lateral edge of the genu of the facial nerve was established at 400 × and the ventromedial corner of the grid was placed in that position.

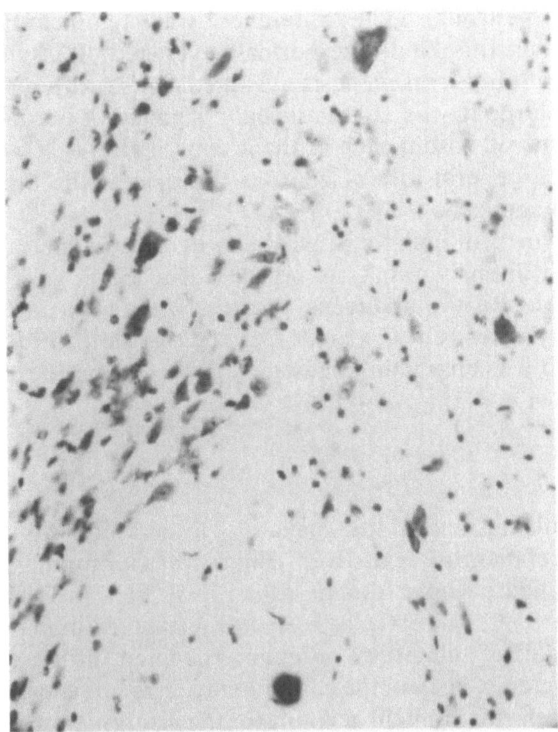

Fig. 6-5. A section through nucleus Y in a different pallid mouse (#5) having no otoconia within either labyrinth. These Y cells also appear shrunken, ventrolaterally displaced, and tightly packed.

3. The grid was placed 130 μm (one-half the total grid width and height) dorsally and laterally to place it in position to take the ventral 259 × 259 μm sample shown in Fig. 6-2b and c.

Neurons within the medial vestibular nucleus were counted within a 6 × 10 cell grid measuring 259 μm wide and 431.7 μm high. The ventrolateral edge of the grid overlapped the dorsomedial edge of the lateral vestibular nucleus (Fig. 6-2b). The block extended 100 μm rostral to the caudal edge of the abducens nucleus and 50 μm caudal to that landmark. The region in which cell loss had been detected in the pilot studies corresponded to the posterior 100 μm and the ventral 259 μm (Fig. 6-2b and c).

Results

The results are summarized in the central six rows in Table 6-1 marked *A* through *F*. *A* through *F* refer to the blocks of sample shown in Fig. 6-2c. As in the pilot studies, we found that in mice with unilateral otoconial agenesis the cell reduction was apparently confined to the side on which agenesis was found (mean: 416.9 cells) rather than to the opposite or contralateral side (mean: 475.9 neurons). The difference was significant ($t = 2$; $n = 4$; $p < 0.05$). Within this block the normals averaged 483.5 neurons per side and the mice with bilateral agenesis averaged 444.3. Although the difference was relatively large it was not statistically significant because of the variation which occurred within each of the groups. Table 6-1*D–F* confirms that there was not a general loss of neurons throughout the brainstem of mice with otoconial agenesis.

Right dominance in cell count was also evident in the medial vestibular nucleus and an intensive analysis revealed it to be largely centered in the sections posterior to the abducens nucleus. In this 50-μm block including the whole 6 × 10 grid (C + F, Fig. 6-2c), the right-left ratio was 1.21 for all ten animals and was characteristic of all groups. Again, the effect was statistically reliable ($t = 3.31$; $p < 0.01$).

Discussion

Finding a significant neuron loss ipsilateral to otoconial loss in our unilaterally otoconia-deficient mice fits well with our earlier findings. Why then did we fail to replicate those findings of 17.7% cell loss in mice without otoconia? The 8.8% fewer cells seen in medial vestibular nucleus region $B + C$ (Fig. 6-2c) of bilaterally otoconia-free mice in this study is half the percentage difference seen in the earlier study. The important differences between the studies which might account for the discrepancy are the use of heterozygote controls instead of pallid controls, 11.5% smaller sample size in the present study due to microscope magnification factors, or differences in ages of the animals (to name a few possibilities). We believe that cells are lost in the lateral part of the medial vestibular nucleus due to loss of utricular afferent input.

The reader should bear in mind that results presented up to this point have been measured in a fixed volume of tissue. The data are measures of cell packing. In the following section the data are numbers of cells within rostrocaudal divisions of fastigial nucleus without regard to volume of tissue sampled.

FASTIGIAL NUCLEUS

Procedure

Both microscopes were used in counting cells in the fastigial nucleus. Because all the cells of the nucleus were counted in each section field size differences did not affect count totals. We counted neuron nucleoli in a block of tissue beginning at the anterior edge of the fastigial nucleus and extending caudally for 210 μm. This represents roughly the anterior half of the nucleus. For purposes of analysis we considered this to be divided into three 70-μm-thick slabs (Fig. 6-2c). The borders of the nucleus could be readily identified throughout the 210 μm and all neuron nucleoli within these borders were counted.

Results

The results are summarized in rows *L*, *M*, and *N* of Table 6-1 which respectively refer to the rostral, middle, and caudal 70-μm-thick slabs into which the sample was divided for data analysis (Fig. 6-2c).

In the anterior 70-μm slab, the bilaterally deficient mice averaged 602 neurons as compared to 825 in the normal mice. Almost identical numbers of neurons were found in both groups for the middle 70-μm slab of the fastigial nucleus, but in the posterior slab the positions of the two groups were reversed; i.e., the normal mice averaged 1321 neurons whereas the bilaterally defective animals averaged 1490. Interestingly, these groups differed by less than 5% for the total number of cells in the entire 210-μm sample of tissue. The difference thus appears to be in the location of the neurons along a rostrocaudal axis. Thus, mice with bilateral agenesis appear to have a cell reduction in the rostral fastigial nucleus and an increase in neurons in a central portion of the nucleus, neither of which reached statistical significance in our small sample.

Right-left asymmetry also was found in the fastigial nucleus, where it was confined to the anterior portion. In the fastigial nucleus, however, the asymmetry favored the left side rather than the right. In the anterior slab the average cell count per animal was 331.1 for the right side and 382.3 for the left, a difference of over 15.5%. The right-left difference was reliable at the 5% level ($t = 2.85$). In contrast, cell counts were almost equal for the right and left sides in a combination of the posterior and middle slabs. The asymmetry thus appears to be confined largely to the anterior or more "vestibular" part of this nucleus.

Discussion

We chose to examine the fastigial nucleus for a variety of reasons. Tilt-responsive neurons are located in the rostral half of the fastigial nucleus in cats (22, 23). Moreover, these neurons were responsive to electric stimulation of proprioceptive afferents from the neck and limb musculature (13) and to contralateral macular input, following section of one VIIIth nerve (14). First order vestibular axons have been reported to synapse in and around the fastigial nucleus (5) and certainly synapse in the vermis (34). The vermis, in turn, massively projects to the fastigial nucleus. The medial vestibular nucleus and possibly nucleus Y also project on the fastigial nucleus ipsilaterally (48). Electrical stimulation of the anterior fastigial nucleus has been found to produce a variety of emotional-motivational responses (43), and we found mice with bilateral otoconial agenesis to display gross deviance from normality in terms of emotionality (12). In the latter study we also found the deficient mice to behave on several tests as if the hippocampus were not functioning properly (e.g., poor habituation and alternation plus excessive activity). Harper and Heath (26) produced lesions in the fastigial nucleus in cats and traced a degenerating pathway to the hippocampus and septal region.

It is highly suggestive that we did in fact find a great reduction in the number of fastigial neurons, specifically in the rostral fastigial nucleus, in mice with bilateral otoconial agenesis. The difference did not attain statistical significance largely because of great variability within the bilateral group. We expect that this effect of this high variability will be reduced by the addition of several subjects in the future. We also have a number of "suggestive but inconclusive" findings related to large versus small neurons. Such size distinctions unfortunately introduce inter-rater variability, a problem that will be solved when we turn to automated cell counting.

GENERAL DISCUSSION

Animals without otoconia are unable to apply threshold shearing force to the stereocilia of macular hair cells. We believe that the static organs of mice without otoconia are totally unable to respond to appropriate stimuli and that the present experimental animals were thus "sensory deprived" (44). Findings of decreased neuron number of packing in sensory-deprived animals are rare in the literature except for neurons within the chimpanzee retina (25,42).

However, there is one study actually involving hair cell deprivation in mice. Webster and Webster (59) found that mice raised by avocal mothers to 45 days of age in a sound-attenuated chamber, and mice whose external auditory meatus was surgically closed on the third day of life, had fewer cells per 10,000 μm^2 of dorsal cochlear nucleus than did controls. These two experimental groups also had significantly smaller cells than did controls in the medial nucleus of the trapezoid body and the globular cell group of the ventral cochlear nucleus, both of which are known to have bilateral cochlear input. Interestingly, the two experimental groups were not significantly different from each other on any measures taken.

Webster and Webster (59) also report on a comparison between a 9-year-old congenitally sensorineuronal deaf girl and an 8-year-old hearing girl. Within the cochlear nuclei of the deaf girl, all neurons were "shrunken and unrecognizable by their cytological appearance," and only the globular cell group of the ventral cochlear nucleus could definitely be identified. This brain, when compared to the brain of the hearing girl, showed deficits very similar to those seen in the deprived mice. They observed in the deaf girl a reduction in cell number in the globular cell group and the medial nucleus of the trapezoid body, and also a size reduction in cells of the medial superior olivary nucleus, an area not measured in the mouse. Other histologic studies of the brains of congenitally deaf men (39) and mice (46,47) have demonstrated grossly underdeveloped cochlear nuclei. The Websters conclude, "... even incomplete conductive hearing losses can result in abnormal central auditory structures. One must wonder whether there is a critical development period in which proper meaningful sound must be received for the central auditory system to mature normally."

Lateralization of brain function is an established fact, although the physical basis for functional differences between the two hemispheres remains obscure. In some cases biochemical differences may underlie lateralization. Rats, for example, have been found to have right-left differences in the dopamine content of the nigrostriatal tract, and this has been reported to be strongly related to turning preferences (24). In other cases, however, functional asymmetries may be related to right-left differences in size or number of neurons. Gross neuroanatomic asymmetries have been found in both humans and animal brains (60,61). Although these studies have concentrated largely upon cortex, there are also reports of subcortical asymmetries. In *Rana temporaria*, for example, there are two habenular nuclei in the left hemisphere but only one in the right (3). To our knowledge, however, there have been no previous reports of asymmetries, whether anatomic or chemical, at levels below the midbrain.

We report here the discovery of three hindbrain regions which display right-left differences in cell number and/or distribution in pallid mice. Two of these are within vestibular nuclei whereas the other is in the fastigial nucleus of the cerebellum.

Our finding that the right medial vestibular nucleus and Y area sampled contained significantly more cells than the left was not predicted, and would probably have gone unnoticed had we not been making bilateral comparisons within our unilateral subjects. While we know that several investigators have localized spatial orientation in the right hemisphere of the human brain, we are not the first to report asymmetry within the vestibular system. A number of authors have reported that tilting the head to the right elicits more counter rolling of the eyes than does equal tilting of the head to the left (11,18,33,38). Kompanejetz cites Barnay (1) as having said that "in rotary stimulation the right labyrinth preponderates over the left physiologically" (p. 344). If that were so it might help explain why we found significant asymmetry only in those areas that showed static organ effects. In light of the relationship between vestibular functioning and spatial orientation, and that between spatial orientation and the right hemisphere in man, it is noteworthy that in rats the cortex of the right hemisphere is consistently thicker than that of the left (10).

ACKNOWLEDGMENTS

This work was carried out under the auspices of the Lauretta Bender–Paul Schilder Memorial Project and was supported by a grant from the Leland Fikes Foundation to the University of Texas at Dallas, subcontracted through the Aberrant Behavior Center, Dallas, Texas. We wish to thank Margaret Pfeil, Paul Myers, Tom Gardner, and Mejun Chen for their help with histology and cell counting. We wish to thank Sara Clark for her help with photography and manuscript preparation and for her moral support.

References

1. Barnay, R.: Uber die vom ohr labyrinth ausgeloste gegenrollung der augen bei Normalharenden, ohrenkranken u. Taubstummen. Arch. Ohrenheilk. 68(1): 1, 1906.
2. Bhatara, V., Clark, D.L., and Arnold, L.E.: Behavioral and nystagmus response of a hyperkinetic child to vestibular stimulation. Am. J. Occup. Ther. 32(5):311, 1978.
3. Braitenberg, V. and Kemali, M.: Exceptions to bilateral symmetry in the epithalamus of lower vertebrates. J. Comp. Neurol. 138(2):137, 1970.
4. Brodal, A. and Hoivik, B.: Site and mode of termination of primary vestibulo-cerebellar fibres in the cat. An experimental study with silver impregnation methods. Arch. Ital. Biol. 102:1, 1964.
5. Carpenter, M.B.: Experimental anatomical-physiological studies of the vestibular nerve and cerebellar connections. In Rasmussen, G.L. and Windle, W.F., (eds.): Neural Mechanisms of the Auditory and Vestibular Systems. Springfield, Ill., Thomas, 1960, pp. 297–323.
6. Chan, Y.S., Hwang, J.C., and Cheung, Y.M.: Crossed sacculo-ocular pathway via the Deiters' nucleus in cats. Brain Res. Bull. 2(1):1, 1977.
7. Chee, F.K.W., Kreutzberg, J.R., and Clark, D.L.: Semicircular canal stimulation in cerebral palsied children. Phys. Ther. 58:1071, 1978.
8. Clark, D.L., Kreutzberg, J.R., and Chee, F.K.W.: Vestibular stimulation influence on motor development in infants. Science 196:1228, 1977.
9. Clark, G.M. and Douglas, R.J.: Neuronal loss in the vestibular nuclei of mice lacking otoconia. Presented at Southwestern Psychological Association Meeting, Fort Worth, Texas, April, 1977.
10. Diamond, M.C., Johnson, R.E., and Ingham, C.A.: Morphological changes in the young, adult, and aging rat cerebral cortex, hippocampus, and diencephalon. Behav. Biol. 14:163, 1975.
11. Diamond, S., Markham, C.H., Simpson, N.E. and Carthoys, I.S.: Binocular counter rolling in humans during dynamic rotation. Acta Otolaryngol. 87:490, 1979.
12. Douglas, R.J., Clark, G.M., Erway, L.C., Hubbard, D.G., and Wright, C.G.: The effects of genetic vestibular defects upon behavior related to spatial orientation and emotionality. J. Comp. Physiol. Psychol. 93(3):467, 1979.
13. Erway, L.C., Ghelarducci, B., Pompeiano, O., and Stanojevic, M.: Responses of cerebellar fastigial neurons to afferent inputs from neck muscles and macular labyrinthine receptors. Arch. Ital. Biol. 116:173, 1978.
14. Erway, L.C., Ghelarducci, B., Pompeiano, O., and Stanojevic, M.: Responses of cerebellar fastigial neurons to stimulation of contralateral macular labyrinthine receptors. Arch. Ital. Biol. 116:205, 1978.

15. Erway, L.C., Hurley, L.S., and Fraser, A.: Neurological defect: Manganese in phenocopy and prevention of a genetic abnormality of inner ear. Science 152:1766, 1966.

16. Erway, L.C., Hurley, L.S., and Fraser, A.: Congenital ataxia and otolith defects due to manganese deficiency in mice. J. Nutr. 100:643, 1970.

17. Erway, L.C., Hurley, L.S., and Fraser, A.: Prevention of congenital otolith defect in pallid mice by manganese supplementation. Genetics 67:97, 1971.

18. Fisher, M.H.: Untersuchungen uber die Gegenrollung der Augen und die Lokalisation der scheinbaren vertikalen bei seitlicher Neigung des Gesamtkorpers bis zu 360° II. Mitteliung. v. Graefes. Arch. Ophthalmol. 123:476, 1930.

19. Frank, J. and Levinson, H.: Dysmetric dyslexia and dyspraxia. J. Am. Acad. Child Psychiat. 12:690, 1973.

20. Fredrickson, J.M., Schwarz, D., and Kornhuber, H.H.: Convergence and interaction of vestibular and deep somatic afferents upon neurons in the vestibular nuclei of the cat. Acta Otolaryngol. (Stockh.) 61:168, 1966.

21. Gacek, R.R.: The course and central termination of first order neurons supplying vestibular endorgans in the cat. Acta Otolaryngol. (Stockh.) 254:1, 1969.

22. Ghelarducci, B.: Responses of the cerebellar fastigial neurons to tilt. Pflugers Arch. 334:195, 1973.

23. Ghelarducci, B., Pompeiano, O., and Spyer, K.M.: Distribution of the neuronal responses to static tilts within the cerebellar fastigial nucleus. Arch. Ital. Biol. 112:126, 1974.

24. Glick, S.D., Jerussi, T.P., and Zimmerberg, B.: Behavioral and neuropharmacological correlates of nigro-striatal asymmetry in rats. In Harnad, S. (ed.): Lateralization in the Nervous System. New York, Academic Press, 1977, pp. 213–249.

25. Globus, A.: Brain morphology as a function of presynaptic morphology and activity. In Riesen, A.H. (ed.): The Developmental Neuropsychology of Sensory Deprivation. New York, Academic Press, 1975, pp. 9–91.

26. Harper, J.W. and Heath, R.G.: Anatomic connections of the fastigial nucleus to the rostral forebrain in the cat. Exp. Neurol. 39:285, 1973.

27. Henkel, C.K. and Martin, G.F.: The vestibular complex of the American opossum, *Didelphis virginiana*. J. Comp. Neurol. 172:299, 1977.

28. Hoffman, R.B., Salinas, G.A., and Baky, A.A.: Behavioral analyses of killifish exposed to weightlessness in the Apollo-Soyuz Test Project. Aviat. Space Environ. Med. 48:712, 1977.

29. Hoffman, R.B., Salinas, G.A., Baky, A.A., Boyd, J.F., and von Baumgarten, R.J.: Effect of pre-hatching weightlessness on adult fish behavior in dynamic environments. Aviat. Space Environ. Med. 49(4):576, 1978.

30. Kantner, R.M., Clark, D.L., Allen, L.C., and Chase, M.F.: Effects of vestibular stimulation on nystagmus response and motor performance in the developmentally delayed infant. Phys. Ther. 56:414, 1976.

31. Kimm, J., Hassul, M., and Cogdell, B.: Fastigial neuronal responses to sinusoidal horizontal rotation. Exp. Neurol. 50(3):579, 1976.

32. Klosovskii, B.N. (ed.): The Development of the Brain and Its Disturbance by Harmful Factors. Oxford, Pergamon Press, 1963.

33. Kompanejetz, S.: Investigation on the counter rolling of the eyes in optimum head positions. Acta Otolaryngol. (Stockh.) 12:332, 1928.

34. Kotchabhakdi, N. and Walberg, F.: Primary vestibular afferent projections to the cerebellum as demonstrated by retrograde axonal transport of horseradish peroxidase. Brain Res. 142:142, 1978.

35. Kovac, W. and Denk, H.: Der Hirnstamm der Maus. New York, Springer-Verlag, 1968.

36. Lyon, M.F.: Hereditary absence of otolith in the house mouse. J. Physiol. 114:410, 1951.

37. Mason, W.A.: Early social deprivation in non-human primates: Implications for human behavior in environmental influences. In Glass, D.C. (ed.): Environmental Influences. New York, Rockefeller University Press and Russell Sage Foundation, 1968.

38. Miller, E.F., II: Counter rolling of the human eyes produced by head tilt with respect to gravity. Acta Otolaryngol. 54:479, 1962.

39. O'Leary, J.: Abnormalities in the brainstem associated with malformations of the ear. Anat. Rec. 55:223, 1933.

40. Ornitz, E.M.: Vestibular dysfunction in schizophrenia and childhood autism. Compr. Psychiatry 11:159, 1970.

41. deQuiros, J.B.: Diagnosis of vestibular disorders in the learning disabled. J. Learn. Dis. 9:39, 1976.

42. Rasch, E., Swift, H., Riesen, A.H., and Chow, K.L.: Altered structure and composition of retinal cells in dark reared mammals. Exp. Cell Res. 25:521, 1961.

43. Reiss, D.J., Doba, N., and Nathan, M.A.: Predatory attack, grooming, and consummatory behaviors evoked by electrical stimulation of cat cerebellar nuclei. Science 182:845, 1973.

44. Riesen, A.H. (ed.): The Developmental Neuropsychology of Sensory Deprivation. New York, Academic Press, 1975.

45. Ritvo, E., Ornitz, E., Eviatar, A., et al.: Decreased nystagmus in early infantile autism. Neurology 19:653, 1969.

46. Ross, H.D.: Auditory pathway of the epileptic waltzing mouse. I. A comparison of the acoustic pathways of the normal mouse with those of the totally deaf epileptic waltzer. J. Comp. Neurol. 119:317, 1962.

47. Ross, H.D.: The auditory pathway of the epileptic waltzing mouse. II. Partially deaf mice. J. Comp. Neurol. 125:141, 1965.

48. Ruggiero, D., Balton, R.R., III, Jayaraman, A., and Carpenter, M.B.: Brain stem afferents to the fastigial nucleus in the cat demonstrated by transport of horseradish peroxidase. J. Comp. Neurol. 172:189, 1977.

49. Scheld, H.W., Baky, A., Boyd, J.F., et al.: Killifish hatching and orientation MS-161. In: Apollo-Soyuz Test Project Final Science Report. NASA Special Publications SP-412, 1977.

50. Schilder, P.: The vestibular apparatus in neurosis and psychosis. J. Nerv. Ment. Dis. 78:137, 1933.

51. Smith, D.E.: The effect of deafferentation on the development of brain and spinal nuclei. Prog. Neurobiol. 8:349, 1977.

52. Steinberg, M. and Rendle-Short, J.: Vestibular dysfunction in young children with minor neurological impairment. Dev. Med. Child Neurol. 19:639, 1977.

53. Torok, N. and Peristein, M.A.: Vestibular findings in cerebral palsy. Ann. Otol. Rhinol. Laryngol. 71:51, 1962.

54. Vinnikov, Y.A.: The evolution of the gravitation receptor and its investigation under conditions of acceleration and weightlessness. Arkh. Anat. 66:10, 1974.

55. Vinnikov, Y.A., et al.: Struktur ana i funktsional' naya organizatsiya vestibulyarnogo apparata hrys, prebyvavshikh y usloviyakh nebesosti 19.5 dney na sputnikye "Kosmos-782." Arkh. Anat. Cistol. Embriol. 1:22, 1978.

56. Vinnikov, Y.A., et al.: The development of the vestibular apparatus (the labyrinth) of the frog *Rana temporaria* under conditions of weightlessness. Z. Evol. Biokhim. Fiziol. 7:343, 1972.

57. Vinnikov, Y.A., et al.: The development of the vestibular apparatus under conditions of weightlessness. Arkh. Anat. Gistol. Embriol. :11, 1976.

58. Von Baumgarten, R.J., Simmonds, R.C., Boyd, J.F., and Garriott, O.K.: Effects of prolonged weightlessness on the swimming pattern of fish aboard Skylab 3. Aviat. Space Environ. Med. 46:902, 1975.

59. Webster, D.B. and Webster, M.: Neonatal sound deprivation affects brain stem auditory nuclei. Arch. Otolaryngol. 103:392, 1977.

60. Webster, W.G. and Webster, I.H.: Anatomical asymmetry of the cerebral hemisphere of the cat brain. Physiol. Behav. 14:867, 1975.

61. Witelson, S.F.: Anatomical asymmetry in the temporal lobes: its documentation, phylogenesis, and relationship to functional asymmetry. In Diamond, S. (ed.): Evolution and lateralization of the brain. Ann. N.Y. Acad. Sci. 299:328, 1977.

Discussion

MEHLER: Maybe I didn't hear your explanation. Why did you pick this area for sample?

CLARK: The question is why did we choose to sample this particular portion of the medial vestibular nucleus. The answer is that while we expected from Dr. Gacek's work that the lateral portion of this nucleus was receiving utricular input, in our first study we sampled from both medial and lateral parts of the medial vestibular nucleus, but found group differences only in the lateral part of the nucleus.

MEHLER: That's a good point. Because it looks like the point, the place in the midline, starting midline between descending and medial nucleus. It's good to look at it.

7

Central Projections to the Vestibular Nuclei From the Subparafascicular Region of the Rabbit

N.H. BARMACK, C.K. HENKEL, AND V.E. PETTOROSSI

Recent neurophysiologic experiments have implicated the vestibular complex of mammals as a relay site for nonvestibular sensory information. In particular, it has been demonstrated that both neck proprioceptive and visual stimulation can modify the activity of secondary vestibular neurons (17,20). The vestibulocerebellum undoubtedly constitutes an indirect source of neck proprioceptive and visual information to the vestibular nuclei (3,22). However, neck proprioceptive information also ascends to the vestibular nuclei via a more direct spinal pathway (10). We have investigated the contribution of the vestibulocerebellum to the visual modulation of the activity of neurons in the vestibular complex. We have found that lesions of the dorsal cap of the inferior olive, the source of visual climbing fibers to the vestibulocerebellum, cause changes in the dynamics of visual modulation of secondary vestibular neurons. Nevertheless, neither dorsal cap lesions nor total cerebellectomies abolish the visual modulation of secondary vestibular neuronal activity (Barmack and Pettorossi, unpublished observations). These findings are in agreement with behavioral observations which stress that the vestibular nuclei but not the cerebellum are essential for the preservation of optokinetic reflexes (2,8). Recent preliminary electrophysiologic observations by others support the idea that there is an extracerebellar pathway by which visual information may reach the vestibular nuclei (11).

Therefore, using the horseradish peroxidase (HRP) tracer technique, we have undertaken an investigation of alternative pathways by which visual information could reach the vestibular nuclei.

We have discovered a group of cells in the subparafascicular complex of the mesodiencephalic region which fill retrogradely when HRP injections are restricted to the vestibular complex. The afferent projections to this subparafascicular area suggest that these cells might relay visual information to the vestibular complex.

METHODS

Eight albino and Dutch-belted rabbits, weighing 1 to 2 kg, were used for unilateral and, in two cases, bilateral injections of HRP into the vestibular complex. In a preparatory operation, rabbits were anesthetized with ketamine hydrochloride (50 mg/kg intramuscularly) and halothane. The rabbit's head was aligned in a stereotaxic apparatus. Two inverted stainless steel 10-32 screws and four 2-56 anchor screws were fixed to the closed calvarium with dental cement. The screws mated with holes in a steel rod which was used subsequently to immobilize the head at the center of rotation of a rate table. At least two days after implantation of the fixation screws the rabbits were anesthetized temporarily with halothane and placed on the rate table. The body was encased in "egg carton" foam rubber, and the head was bolted to the steel rod, placing it in the center of the rate table for rotations about the vertical and longitudinal axes. A tracheal cannula was inserted and the tissues surrounding the wound margin were infiltrated with 1% lidocaine-epinephrine and coated with 2.5% Xylocaine ointment. The same local anesthetics were applied to a 1-cm wound margin overlying the dorsal surface of the calvarium. A small hole was drilled through the calvarium exposing the left cerebellar area medullaris. Following completion of these surgical procedures the halothane anesthetic was discontinued. The rabbit was allowed to recover for 1 h before microelectrode recording with tungsten microelectrodes from the vestibular nuclei was attempted. The end-tidal carbon dioxide was monitored during spontaneous respiration. The rabbit was paralyzed with gallamine triethiodide (5 to 10 mg/kg/h) and artificially respired to maintain the end-tidal carbon dioxide at the spontaneous level (4.0 to 4.5%).

Extracellular microelectrode recordings from the vestibular nuclei were obtained while the rabbit was rotated in the horizontal or vertical planes. Single medial vestibular neurons were identified by their increased rate of discharge in phase with ipsilateral velocity of the rate table, and by their increased rate of discharge evoked by posterior-anterior optokinetic stimulation of the ipsilateral eye. Following physiologic identification of the medial vestibular nucleus, the tungsten microelectrode was withdrawn and replaced with a glass micropipette containing 25% HRP (Sigma IX). The pipette had a tip diameter of about 15 μm. It was connected electrically to an amplifier and hydraulically to a 1-μl Hamilton syringe. The pipette was capable of recording multiple-unit activity and was driven into the previously identified medial vestibular nucleus. Its position was confirmed by observing multiple-unit activity during horizontal oscillation of the rate table. Pressure injections of HRP (50 to 150 nl) were confirmed physiologically by observing a transitory depression in multiple-unit activity as the HRP was ejected.

The rabbit was allowed to recover from the gallamine triethiodide. When spontaneous breathing returned, halothane anesthesia was reintroduced and the wound margins were sutured. After a survival period of 24 h, the rabbit was deeply anesthetized with sodium pentobarbital (60 mg/kg) and sacrificed by intracardiac perfusion. The brains were processed according to the HRP procedures of de Olmos (9), using *o*-dianisidine as the chromagen.

RESULTS

In five rabbits, the HRP injections were confined primarily to the rostral medial vestibular nucleus. The injection in one case spread to the superior vestibular nucleus, and in most experiments light reaction product was found in the descending vestibular nucleus and the perihypoglossal complex. The perihypoglossal complex was spared, however, in one case. Two of the eight rabbits used in this study had bilateral injections. Following HRP injections which were restricted to the medial vestibular nucleus, retrogradely labeled cells were found in the cerebellum, the contralateral vestibular complex, and the ipsilateral vestibular ganglion. The extent and distribution of these projections will not be considered in this report. Rather, we will focus our attention on a large population of labeled neurons found in the rostral midbrain and mesodiencephalic border, primarily on the side of the injection, but occasionally bilaterally. This population of marked neurons can be divided into three subgroups on the basis of histologic criteria.

The most rostral labeled neurons, group 1 (Fig. 7-1), were small and medium-sized cells capping the medial border of the cerebral peduncle. These cells were located mainly within the zona incerta, but in some cases extended into the territory of the fields of Forel and the lateral hypothalamus. Although in some cases labeled cells of this group were near the medial terminal nucleus of the optic tract, they were never clearly a part of this nucleus. The zona incerta group of labeled cells was especially small in animals with injections restricted to the medial vestibular nucleus, and it was most heavily labeled in the one case in which the injections spread into the reticular formation.

The most extensively and consistently labeled neurons, group 2 (Figs. 7-1 and 7-3), were found in a position caudal and dorsal to group 1, and were associated intimately with the course of the fasciculus retroflexus. We have chosen Bodian's (5) designation for this area as a subparafascicular region. It is not to be confused with the thalamic nucleus parafascicularis in the rabbit (13). The subparafascicular region, as we have called it, is confined to the

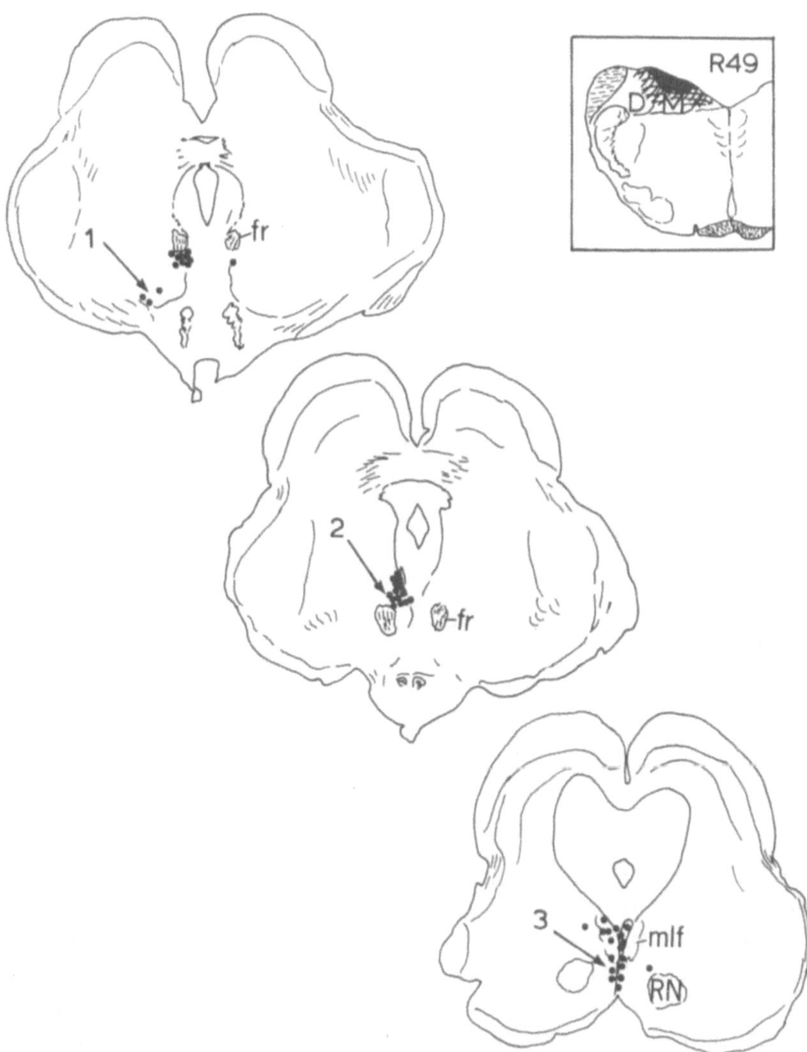

Fig. 7-1. Distribution of retrogradely labeled neurons in the mesodiencephalic region following an HRP injection (100 nl) into the left medial vestibular nucleus. Labeled cells were found in three regions which are numbered in a rostrocaudal direction. Insert illustrates the maximal spread of the reaction product at the injection site. *M*, medial vestibular nucleus; *D*, dorsal vestibular nucleus; *fr*, fasciculus retroflexus; *mlf*, medial longitudinal fasciculus; *RN*, red nucleus.

region rostral to and merging near the fasciculus retroflexus with the interstitial nucleus of Cajal. It contains a mixture of very small and medium-sized neurons. The small neurons (10 to 15 μm) are generally spindle-shaped or round with only a narrow rim of lightly staining cytoplasm. The medium-sized neurons (16 to 30 μm) are multipolar, have a slightly eccentric nucleus,

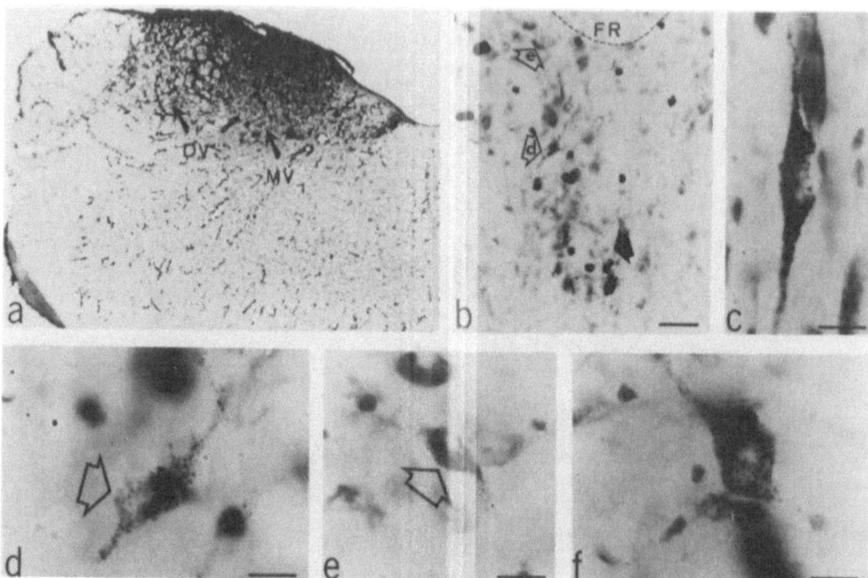

Fig. 7-2. Photomicrographs of HRP injection site and retrogradely labeled cells. **a** HRP injection site. **b** Subparafascicular region. **c** Labeled cell in nucleus linearis. **d** and **e** Higher magnification view of medium (**d**) and large (**e**) labeled neurons. The locations of these cells are illustrated by block arrows in **b**. **f** Labeled neuron located in the interstitial nucleus of Cajal. Calibration bars: **b**, 50 μm; **c**–**f**, 15 μm. *DV*, dorsal vestibular nucleus; *MV*, medial vestibular nucleus; *FR*, fasciculus retroflexus.

and contain a greater volume of cytoplasm that is either lightly stained or filled with darkly stained Nissl substance. The labeled cells in the subparafascicular region eventually merge dorsal to the fiber bundle with slightly larger cells in the interstitial nucleus of Cajal that were also labeled.

The HRP-labeled, subparafascicular region is shown in Fig. 7-2 alongside the injection site in the medial vestibular nucleus. The solid arrow points to a cluster of labeled neurons surrounding a couple of stained erythrocytes. The block arrows *d* and *e* in Fig. 7-2b indicate a medium-sized and a small neuron, respectively, that contain granules of reaction product in their cytoplasm. The same cells are shown at a higher magnification in Fig. 7-2d and e. In addition, Fig. 7-2f shows one of the larger labeled cells found in the interstitial nucleus of Cajal.

Labeled neurons found most caudally, group 3, lie along the midline of the midbrain tegmentum (Figs. 7-1 and 7-3). This midline group consisted mainly of cells in the nucleus linearis, but also included a few labeled cells interspersed between the oculomotor nuclei and not clearly distinguishable from the nucleus of Edinger-Westphal. The labeled cells in nucleus linearis were spindle-shaped (Fig. 7-2c), with their long axes aligned dorsal to ventral.

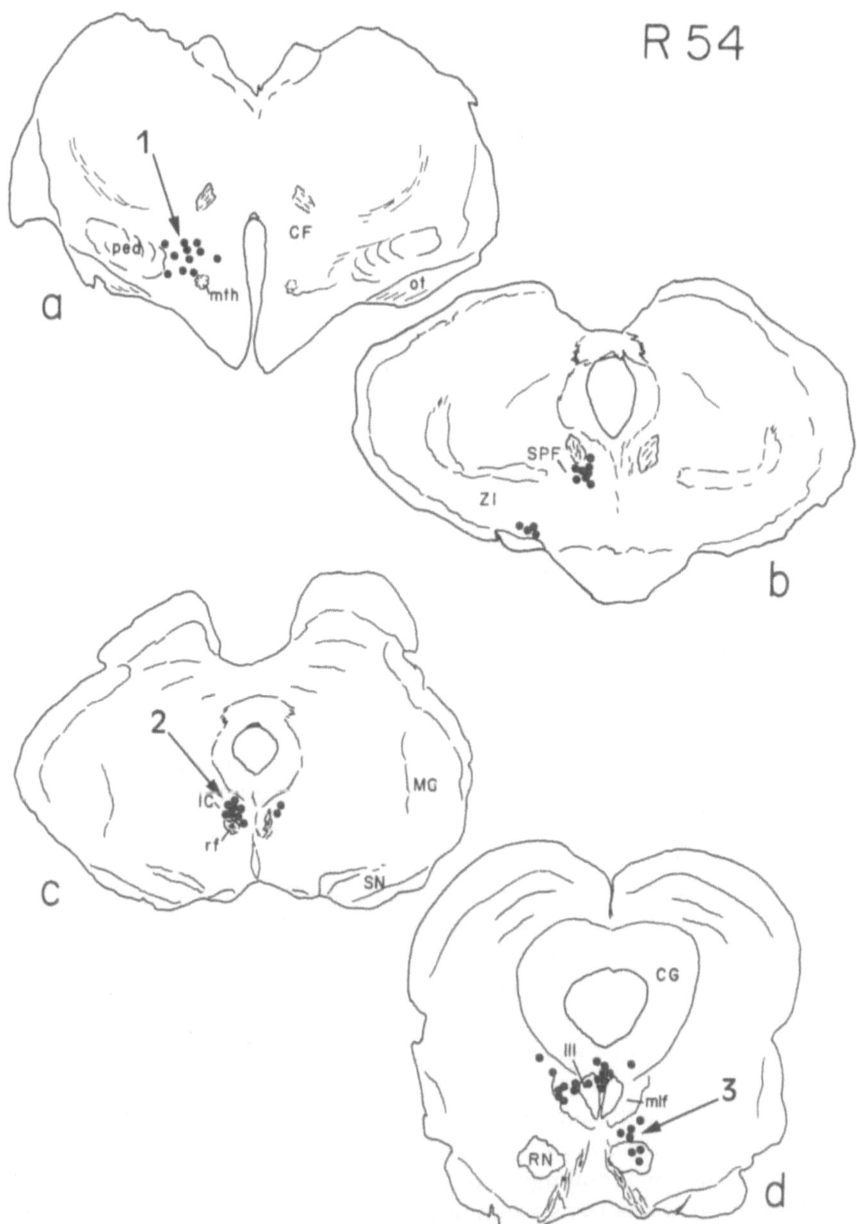

Fig. 7-3. Rostrocaudal distribution of retrogradely labeled neurons following HRP injection (150 nl) into the left medial vestibular nucleus. The three labeled regions are illustrated in rostrocaudal sequence. *ot*, optic tract; *ped*, cerebral peduncle; *mth*, mamillothalamic tract; *CF*, field of Forel; *ZI*, zona incerta; *SPF*, subparafascicular region; *IC*, interstitial nucleus of Cajal; *rf*, fasciculus retroflexus; *SN*, substantia nigra; *MG*, medial geniculate nucleus; *CG*, central gray; *III*, oculomotor nucleus; *mlf*, medial longitudinal fasciculus; *RN*, red nucleus.

It should be pointed out that although labeling was present in all three of the aforementioned areas in each rabbit, labeled cells were most numerous and densely labeled in the subparafascicular area (group 2). Since the vestibular complex was entered from a dorsal approach through the cerebellum, there was of necessity in all cases some leakage along the electrode track of HRP into the cerebellum. Therefore, some caution should be used in interpreting these data. Although it has been reported that cells in the midline of the mesodiencephalic region project to the cerebellum (18), our control injections of HRP in the cerebellum did not reproduce the pattern of labeling seen after injections in the vestibular complex.

DISCUSSION

Our data have demonstrated a projection from the subparafascicular region to the medial vestibular nucleus of the rabbit. We do not know the extent of the projections from the subparafascicular region to other vestibular nuclei, since we have attempted to confine our HRP injections to the medial vestibular nucleus. Our findings are in agreement with the results obtained by another group using silver degeneration techniques following the placement of large electrolytic lesions in the vicinity of the subparafascicular region of the cat (15). These workers found projections to both the vestibular complex and the medial accessory division of the inferior olive. Other studies using degeneration techniques have demonstrated projections from the subparafascicular region to the dorsal cap, β-nucleus, and other regions of the medial accessory division of the inferior olive (12,21). These projections have been confirmed recently by both orthograde (14) and retrograde (6,19) tracer studies.

The subparafascicular region receives projections from both the pretectum and the superior colliculus (1,4,16). These afferent connections suggest that the subparafascicular region may relay visual information to the medial vestibular nucleus and the medial accessory division of the inferior olive. Thus, the medial vestibular nucleus may receive visual information from the cerebellum *and* from the subparafascicular area. The output of the subparafascicular region may influence the activity of the medial vestibular nucleus directly via the pathway demonstrated in the present experiment, or indirectly via a subparafascicular-olivo-cerebellar-vestibular pathway. Our supposition that the tectal input to the subparafascicular region conveys visual information needs to be examined directly using physiologic techniques. Recordings from what appears to be an homologous region in the monkey have demonstrated cells whose activity is modulated in association with vertical eye movements (7). Future experiments should determine the extent of additional sensory influences on the discharge characteristics of cells in this region, as well as the possibility of projections from this region to other vestibular nuclei.

SUMMARY

The possible central origin of pathways conveying visual information to the medial vestibular nucleus was examined using the HRP technique in rabbits. A projection to the medial vestibular nucleus was discovered which originates primarily from the ipsilateral subparafascicular region. The cells in this region extend from the ipsilateral nucleus of Cajal, near the anterior border of the red nucleus, rostrally to the region of the mamillothalamic tract, near the medial borders of the zona incerta and cerebral peduncle. This subparafascicular region is of particular interest because it is known to receive afferents from the superior colliculus as well as to project to the dorsal cap of the inferior olive. Therefore, this region might constitute an important source of extracerebellar afferents by which visual information in particular could influence the discharge of secondary vestibular cells.

ACKNOWLEDGMENTS

This research was supported by N.I.H. Grant EY-00848 and by the Oregon Lions Sight and Hearing Foundation. V.E.P. is the recipient of a C.N.R.-N.A.T.O. Fellowship and Oregon Lions Sight and Hearing Foundation Fellowship.

References

1. Altman, J. and Carpenter, M.B.: Fiber projections of the superior colliculus in the cat. J. Comp. Neurol. 116:157, 1961.
2. Azzena, G.B., Azzena, M.T., and Marini, R.: Optokinetic nystagmus and the vestibular nuclei. Exp. Neurol. 42:158, 1974.
3. Barmack, N.H.: Immediate and sustained influences of visual olivo-cerebellar activity on eye movement. In Talbott, R.E. and Humphrey, D.R. (eds.): Control of Posture and Movement. New York, Raven Press, 1979.
4. Berman, N.: Connections of the pretectum in the cat. J. Comp. Neurol. 174:227, 1977.
5. Bodian, D.: Studies on the diencephalon of the Virginia opossum. I. The nuclear pattern in the adult. J. Comp. Neurol. 71:259, 1939.
6. Brown, J., Chan-Palay, V., and Palay, S.L.: A study of afferent input to the inferior olivary complex in the rat by retrograde axonal transport of horseradish peroxidase. J. Comp. Neurol. 176:1, 1977.
7. Büttner-Ennever, J.A. and Büttner, U.: A cell group associated with vertical eye movements in the rostral mesencephalic reticular formation of the monkey. Brain Res. 151:31, 1978.
8. Collewijn, H.: Dysmetria of fast phase of optokinetic nystagmus in cerebellectomized rabbits. Exp. Neurol. 28:144, 1970.
9. de Olmos, J.S.: An improved method for the study of central nervous connections. Exp. Brain Res. 29:541, 1977.

10. Hikosaka, O. and Maeda, M.: Cervical effects on abducens motoneurons and their interaction with vestibulo-ocular reflex. Exp. Brain Res. 18:512, 1973.
11. Keller, E.L. and Precht, W.: Persistence of visual response in vestibular nucleus neurons in cerebellectomized cat. Exp. Brain Res. 32:591, 1978.
12. King, J.S., Hamos, J.E., and Maley, B.E.: The synaptic terminations of certain midbrain-olivary fibers in the oposum. J. Comp. Neurol. 182:185, 1978
13. Kuhlenbeck, H. and Miller, R.N.: The pretectal region of the rabbit's brain. J. Comp. Neurol. 76:323, 1942.
14. Linauts, M. and Martin, G.F.: An autoradiographic study of midbrain-diencephalic projections to the inferior olivary nucleus in the opossum (*Didelphis virginiana*). J. Comp. Neurol. 179:325, 1978.
15. Mabuchi, M. and Kusama, T.: Mesodiencephalic projections to the inferior olive and the vestibular and perihypoglossal nuclei. Brain Res. 17:133, 1970.
16. Martin, G.F.: Efferent tectal pathways of the opossum (*Didelphis virginiana*). J. Comp. Neurol. 135:209, 1969.
17. Rubin, A.M., Young, J.H., Milne, A.C., Schwarz, D.W.F., and Fredrickson, J.M.: Vestibular-neck integration in the vestibular nuclei. Brain Res. 96:99, 1975.
18. Sugimoto, T., Itoh, K., and Mizuno, N.: Direct projections from the Edinger-Westphal nucleus to the cerebellum and spinal cord in the cat: an HRP study. Neurosci. Lett. 9:17, 1978.
19. Takeda, T., and Maekawa, K.: The origin of the pretecto-olivary tract. A study using the horseradish peroxidase method. Brain Res. 117:319, 1976.
20. Waespe, W., and Henn, V.: Neuronal activity in the vestibular nuclei of the alert monkey during vestibular and optokinetic stimulation. Exp. Brain Res. 27:523, 1977.
21. Walberg, F.: Descending connections from the mesencephalon to the inferior olive: An experimental study in the cat. Exp. Brain Res. 21:145, 1974.
22. Wilson, V.J., Maeda, M., and Franck, J.I.: Inhibitory interaction between labyrinthine, visual, and neck inputs to the cat flocculus. Brain Res. 96:357, 1975.

Discussion

BÜTTNER: I thought your pictures were very nice, and I would like to comment on a cell group that we've described. This is a cell group that we have named the rostral interstitial nucleus of the medial longitudinal fasciculus. It lies actually directly beneath the subparafascicular nucleus, and is divided from it by a large blood vessel. After various sets of anatomic and physiologic studies, we've shown that it's a cell group involved in controlling vertical eye movements. The location of your labeled neurons fitted very well with our studies. Although our studies have been in the monkey, and we've identified the same nucleus in the cat. I think the area you're describing is the same nucleus, or the equivalent nucleus, in the rabbit. There is quite a lot of evidence now that this rostral interstitial does send descending projections down into the vestibular nuclei and into the perihypoglossal nuclei. Would you like to comment?

HENKEL: Yes, I've looked at your work, and I think that we're talking about comparably the same areas. Although I haven't been able to look at the cell types in some of the other forms. I would say that we're in the same general vicinity.

8

Vestibular Projections to the Monkey Thalamus and Rostral Mesencephalon: An Autoradiographic Study

J.A. BÜTTNER-ENNEVER AND W. LANG

The existence of vestibular nucleus fibers which terminate rostral to the oculomotor complex is still a matter for dispute (17,42). A vestibulothalamocortical pathway is proposed on the basis of physiologic results (6,13,14,16,27,28,32,40). Convincing anatomic proof for a direct vestibulothalamic projection in the monkey is still missing, since Tarlov (41) found no evidence for thalamic projections in his study using the Nauta-Gygax and the Fink-Heimer techniques. The increasing electrophysiologic evidence and the recent development of highly sensitive, autoradiographic methods encouraged us to reinvestigate this problem. Our report will deal with the vestibular pathways terminating rostral to the oculomotor complex. It will present data on the exact location of vestibular projection sites in the thalamus (24) and regions of the rostral mesencephalon involved in the premotor control of vertical eye movements (9).

METHODS

The experiments were carried out on macaque monkeys. After localizing the vestibular nuclear complex by single unit recordings during rotational stimuli in the alert monkey, unilateral injections of radioactive amino acids (L[5-^{3}H]proline, sp. act. 18 Ci/mmol; and L[4,5-^{3}H]leucine, sp. act. 38 Ci/mmol) were placed in different vestibular nuclei of the anesthetized animal, using a thermosyringe or a pressure syringe attached to a glass micropipette (tip diameter 30–100 μm). The injected volume was 0.1 to 0.4 μl; the total amount of injected radioactivity ranged between 3 and 20 μCi. All injections were displayed on the right side of the monkey. After 5 h to 7 days survival time, the animals were perfused under deep Nembutal anesthesia with 0.4% paraformaldehyde and 1.25% glutaraldehyde, phosphate buffered to pH 7.4, following an intravenous injection of 2 ml 1% sodium nitrate and 2 ml heparin, and a preliminary intracranial infusion of 100–200 ml 6% Macrodex, a plasma expander. The brain was left in fixative overnight and then transferred into a 0.1 M phosphate-buffered 30% sucrose solution for 48 to 72 h. Frozen sections of 40 μm were taken in the coronal plane, and at least every sixth section was prepared for autoradiography. This procedure consisted of dipping the sections in NTB 2 emulsion diluted 1:1 with water and exposing them for 8 weeks at 4°C. After this period, the sections were developed in Dektol at 18°C and fixed in 30% sodium thiosulfate solution. Finally, the sections were stained with cresyl violet and examined under dark- and bright-field illumination. For the description of thalamic and vestibular structures, the nomenclatures of Olszewski (33) and Brodal and Pompeiano (3), respectively, were used.

RESULTS

All the injection sites are illustrated in Fig. 8-1. The injection site of exp. 22 was almost exclusively confined to the brachium conjunctivum. Only those experiments in which the rostral and more ventral parts of the vestibular nucleus were involved in the uptake area (exp. 19, 21, 24, 33) gave rise to projections rostral to the oculomotor complex (Table 8-1). As expected in these four "positive" experiments, labeling was found in the vestibular nuclei contralateral to the injection site, the abducens, trochlear, and oculomotor nuclei (Fig. 8-2a) (29). If the injection site was located mainly in the superior vestibular nucleus, labeled fibers ascending in the medial longitudinal fasciculus were ipsilateral, whereas after injections into the medial, and ventral part of the lateral, vestibular nuclei ascending fibers in the Mlf were labeled predominantly contralaterally. This is in agreement with earlier studies (42).

Rostral Mesencephalon

Labeling was observed in the interstitial nucleus of Cajal (Fig. 8-2a), and further rostrally, in the area lying dorsomedial to the red nucleus, the *rostral* interstitial nucleus of the medial longitudinal fasciculus, as described by Büttner-Ennever and Büttner (Figs. 8-2b, 8-4, 8-5 and Table 8-1)(9). Labeling in the latter cell group never filled the whole nucleus, and was often weaker than that seen in the interstitial nucleus of Cajal. No labeling was found in either nucleus of experiments 22, 27, and 36; the labeling in the

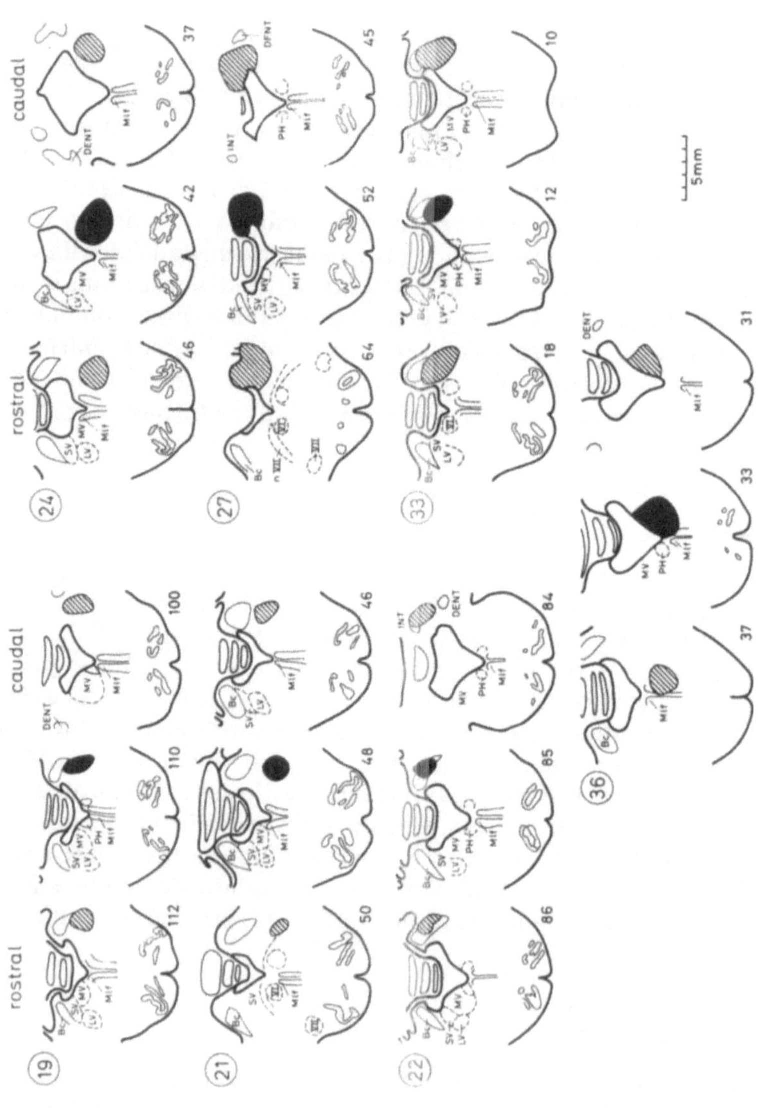

19
21
22
24
27
33
36
rostral
caudal
rostral
caudal
5mm

other four experiments is summarized in Table 8-1. Again the silver grain density was heavier on the ipsilateral sides in cases involving mainly the superior vestibular nucleus, and contralateral labeling was observed if the border zone of the lateral and medial vestibular nuclei was included in the injection site (Table 8-1). An important observation was that no significant labeling was found in the nucleus of Darkshevich.

Thalamus

At thalamic levels, projection sites were located predominantly in the ventroposterior lateral nucleus pars oralis (VPL_0), covering only small regions scattered within the nucleus (Figs. 8-3 and 8-4). This terminal labeling was arranged in clusters, lying mainly in the caudal and ventral half of VPL_0, in the lateral, intermediate, and medial portions. A small number of grain patches was also found in regions of the inferior ventroposterior nucleus (VPI) and in the ventrolateral nucleus pars caudalis (VL_c) adjacent to VPL_0. Every single patch of terminal labeling covered an area of approximately

Fig. 8-1. The location of the amino acid injection sites in the vestibular nuclei, and the size of the uptake area in all the cases used in this study (case number is circled). The first and third section of each individual case contain the rostral and caudal limits of the (hatched) uptake area. The middle section contains the point of injection of the amino acids and the uptake area is shaded black. Consecutive section numbers represent a rostrocaudal separation of 320 μm. The following abbreviations are used in Figs. 8-1–8-5. *a*, artifact; *Atd*, ascending tract of Deiters; *Bc*, brachium conjunctivum; *Bp*, brachium pontis; *CD*, nucleus caudatus; *CL*, nucleus centralis lateralis; *CM*, centrum medianum; *DBc*, decussatio brachii conjunctivi; *DENT*, nucleus dentatus cerebelli; *FAST*, nucleus fastigii cerebelli; *GL*, geniculatum laterale; H_1, H_1 field of Forel; H_2, H_2 field of Forel; *IC*, nucleus interstitialis Cajal; *INT*, nucleus interpositus cerebelli; *LD*, nucleus lateralis dorsalis; *Ll*, lateral lemniscus; *LP*, nucleus lateralis posterior; *LRF*, formatio reticularis lateralis; *LV*, nucleus vestibularis lateralis; *Lvst*, tractus vestibulospinalis lateralis; *MD*, nucleus medialis dorsalis; *Mlf*, fasciculus longitudinalis medialis; *MV*, nucleus vestibularis medialis; *n III*, nervus oculomotorius; *n IV*, nervus trochlearis; *n VII*, nervus facialis; *ND*, nucleus Darkshevich; *NR*, nucleus ruber; *NRTP*, nucleus reticularis tegmenti pontis; *PH*, nucleus praepositus hypoglossi; *R*, nucleus reticularis; *RE*, nucleus reuniens; *RI*, rostral interstitial nucleus of the medial longitudinal fasciculus (*rostral* iMLF); *SN*, substantia nigra; *STH*, nucleus subthalamicus; *SV*, nucleus vestibularis superior; *TR*, tractus retroflexus; *VLc*, nucleus ventralis lateralis pars caudalis; *VP*, nucleus ventralis posterior; *VPI*, nucleus ventralis posterior inferior; VPL_c, nucleus ventralis posterior lateralis pars caudalis; VPL_0, nucleus ventralis posterior lateralis pars oralis; *VPM*, nucleus ventralis posterior medialis; *VPMpc*, nucleus ventralis posterior medialis parvocellularis; *X*, cell group *x* of the vestibular complex; *Y*, cell group *y* of the vestibular complex; *ZI*, zona incerta; *III*, nucleus nervi oculomotorii; *IV*, nucleus nervi trochlearis; *VI*, nucleus nervi abducentis; *VII*, nucleus nervi facialis.

Table 8-1.
THE EXTENT OF AUTORADIOGRAPHIC LABELING IN INDIVIDUAL STRUCTURES OF EACH CASE IN WHICH THE *ROSTRAL* INTERSTITIAL NUCLEUS AND/OR VPL_0 WERE LABELED.

Exp.	Interstitial n. Cajal		n. Dark-shevich		*rostral* interstitial n.		VPL_0		1		2	3	4	5
	i	c	i	c	i	c	i	c	i	c				
19	++	(+)	—		+	—	(+)	++	+	(+)	(+)	+	+	+
21	++	(+)	—		+	—	++	+	++	++	+	—	+	(+
24	—	++	—		?	+	++	+	(+)	+	+	—	(+)	—
33	++	+	(+)		+	(+)	+	++	++	(+)	+	+	+	+

i = ipsilateral
c = contralateral

1) medial longitudinal fasciculus (Mlf), 2) ascending tract of Deiters (Atd),
3) brachium conjunctivum (Bc), 4) 'lateral lemniscus' (L1)
5) lateral pontine reticular formation (LRF)

The labeling is estimated as strong ++, significant +, weak (+), and absent —. In case 24 the extent of fiber labeling is hard to estimate, since the survival time was 5 h.

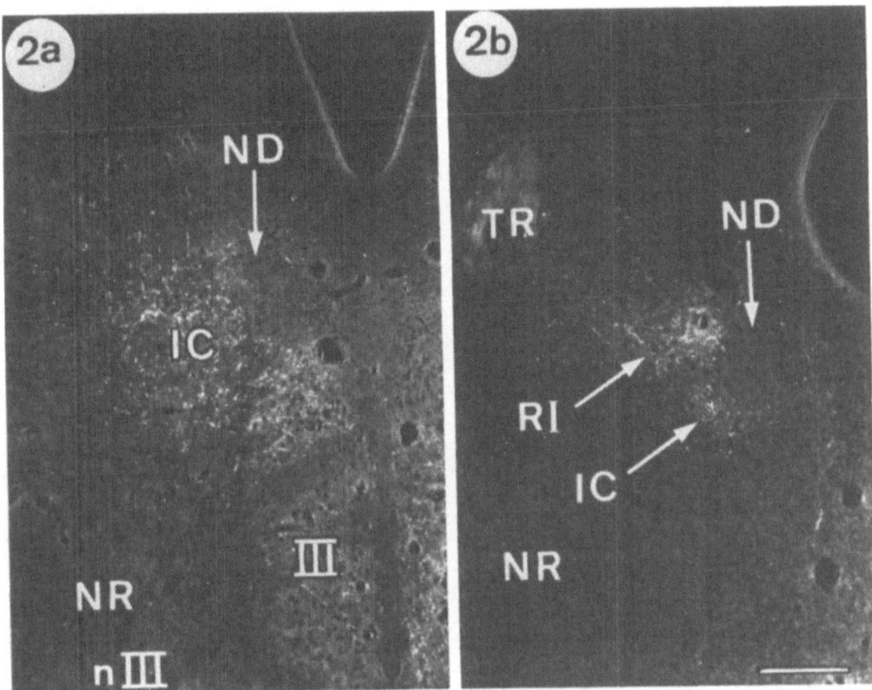

Fig. 8-2. **a** Darkfield photograph of silver grain deposits lying over the interstitial nucleus of Cajal (*IC*) and parts of the oculomotor nucleus (*III*). This terminal labeling arises from an ipsilateral injection of radioactive amino acids into the rostral portions of the superior and lateral vestibular nuclei (exp. 33). Note that nucleus Darkshevich (*ND*) is not labeled. **b** Darkfield photograph taken from the same experiment, 2 mm further rostral to **a**. It shows silver grain deposits lying over the caudal tip of the *rostral* interstitial nucleus of the MLF (*RI*), whereas the rostral extreme of IC, lying just ventral to RI, is still visible. This section was selected to demonstrate the relative location, but distinct separation, of RI and IC. ND is not labeled. The calibration represents 0.5 mm.

0.02 mm^2. Fig. 8-3b shows such a patch of grains arising from the branch of a fiber ascending through the lateral part of VPL_o.

The projection sites were bilateral and showed an ipsi- or contralateral predominance, according to the location of the injection site (Table 8-1). If the injection affected mainly the lateral and medial vestibular nucleus the ipsilateral thalamic projections were more numerous, whereas if injections were located mainly in the superior vestibular nucleus a greater number of terminal fields on the contralateral side was observed. No labeling was found in the caudal ventroposterior lateral nucleus (VPL_c), the medial ventroposterior nucleus (VPM), and the posterior group (i.e., the magnocellular part of the medial geniculate body, the posterior nucleus, and the suprageniculate nucleus) in any of the experiments.

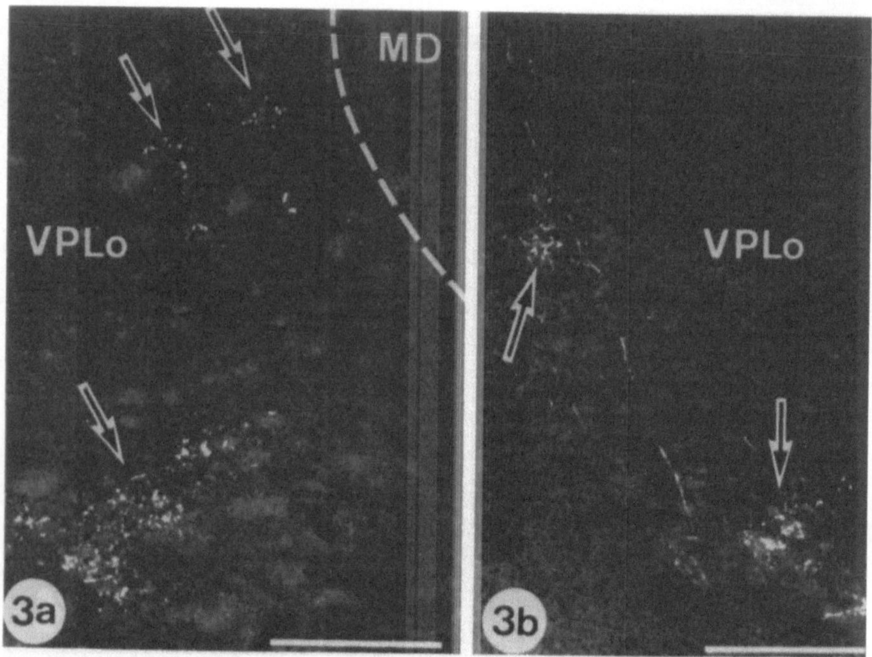

Fig. 8-3. **a** Darkfield photograph of the quadrant shown in exp. 33, section 68, of Fig. 8-4. The arrows indicate patches of labeling in VPL_0 of the thalamus. **b** Darkfield photograph to show terminal deposits of silver grains and a labeled fiber in the contralateral thalamus of exp. 24, after injection of radioactive amino acids into the medial and lateral vestibular nuclei. The fiber is ascending along the lateral border of the thalamus and gives rise to small branches and terminal clusters. Calibrations represent 0.5 mm.

Ascending Fibers

All together five ascending fiber tracts were visible in the four "positive" cases at the level of trochlear nucleus: (1) the medial longitudinal fasciculus; (2) the ascending tract of Deiters; (3) the brachium conjunctivum; (4) fibers lying medial to the brachium pontis in the ventral part of the lateral lemniscus; and (5) a compact bundle of fibers (and endings) in the lateral pontine reticular formation (4,17,25,31,43). The location of these groups is shown in Fig. 8-5.

The contributions of groups 2 and 4 appear to be the most important with respect to the thalamic projections. However, our results do not permit us to exclude groups 1, 3, and 5 as possible routes for vestibulothalamic fibers. The extent of the labeling of each tract in individual cases is shown in Table 8-1.

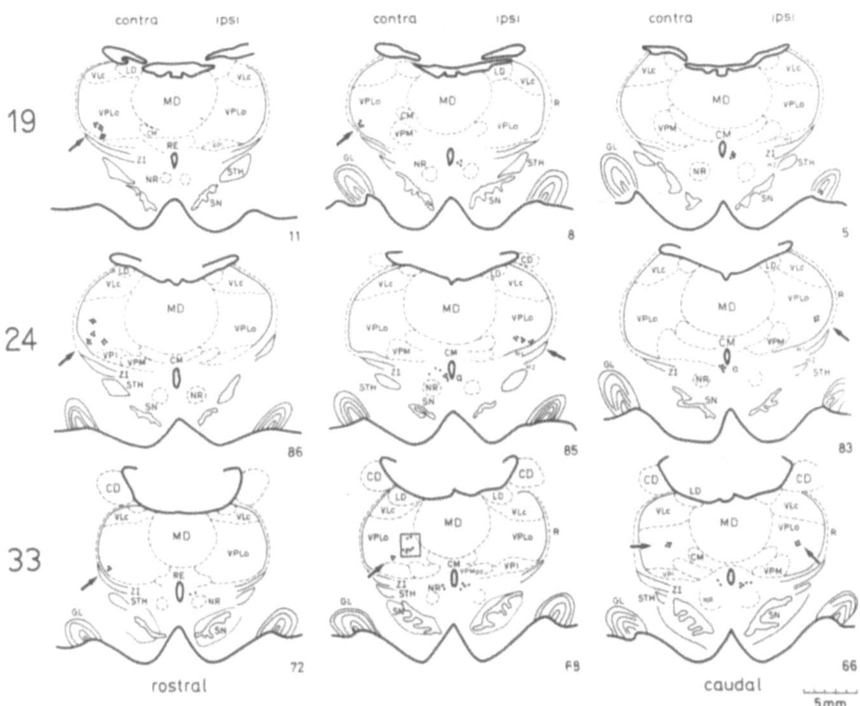

Fig. 8-4. The sites of terminal labeling (small dots) in the thalamus of three selected cases (exp. 19, 24, 33), after injection of radioactive amino acids into the vestibular nuclei. Some patches of the scattered thalamic labeling are indicated by arrows. The area encased in section 68, exp. 33 is the region photographed for Fig. 8-3a. *a*, in section 83, exp. 24, marks an artifact, which prevented analysis of this region. Consecutive section numbers represent a rostrocaudal separation of 320 μm.

DISCUSSION

The results of this autoradiographic study describe two new vestibular pathways terminating rostral to the oculomotor nuclei: a direct vestibulothalamic pathway which has not previously been demonstrated by anatomic methods, and a more medial projection site which is confined to a newly defined region, designated as the *rostral* interstitial nucleus of the Mlf by Büttner-Ennever and Büttner (9). The partial labeling of the *rostral* interstitial nucleus of the Mlf is of current interest since it has only recently been shown that the area is part of the direct premotor system controlling vertical eye movements (5,9,18,23). Its possible role in the control of head movements has long been known, see n. praestitialis (15,21,22). The *rostral* interstitial nucleus of the Mlf is immediately adjacent to the nucleus of Darkshevich and ventral to the nucleus subparafascicularis; thus, it could easily be con-

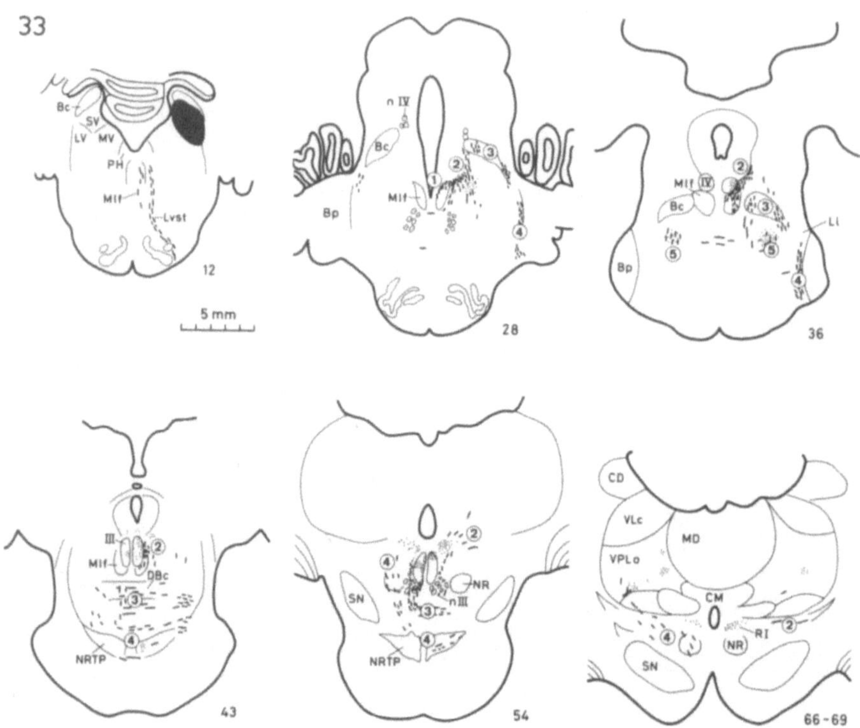

Fig. 8-5. Drawings from exp. 33 to demonstrate the route taken by the ascending pathways labeled with an injection of radioactive amino acids in the vestibular nuclei (shaded black). Five separate pathways are labeled; *1*, MlF; *2*, ascending tract of Deiters; *3*, brachium conjunctivum; *4*, "lateral lemniscus;" and *5*, fibers running in the lateral pontine reticular formation. The pathways which appear to give rise to the ipsilateral and contralateral thalamic projections are the ascending tract of Deiters (*2*) and the lateral lemniscus (*4*), respectively. Fibers are represented by short lines and terminal labeling as fine stippling.

fused with either of these two nuclei, especially in animals such as the cat and rabbit, where the cytoarchitecture is not as clear as in the monkey. The functional implications of the vestibular projection are important because they suggest that the *rostral* interstitial nucleus of the Mlf can be subdivided into a caudomedial and a rostrolateral portion, where only the former is under direct vestibular control. The origin and route taken by the ascending vestibular fibers are very similar to those of the vestibular afferents supplying the oculomotor nucleus and interstitial nucleus of Cajal (42), which supports the supposition that the vestibulo–*rostral* interstitial nucleus of the Mlf pathway also takes part of the vestibular regulation of eye and head movements.

The direct vestibulothalamic pathway in the monkey ends *bilaterally* in mainly VPL_o and to a lesser extent in the adjacent regions of VPI and VL_c.

This result is supported by additional evidence from retrograde tracer experiments in the study of Lang et al. (24). Several authors have claimed the existence of such a pathway in the monkey on the basis of electrophysiologic observations (13,14,27), whereas the anatomic studies with degeneration techniques, reviewed by Tarlov (41), are contradictory. In cats, the concept of a vestibulothalamic link is based on electrophysiologic (30,37,38) and anatomic results (10,20,35,36).

The thalamic projection in the monkey, although definite, is diffuse and weak compared with the vestibular projection to the oculomotor nuclei. The fact that no degeneration in the thalamus was observed by Tarlov (41) with the Fink-Heimer and Nauta-Gygax techniques could be explained by the small number and size of the scattered terminal sites (Figs. 8-3 and 8-4), which could well exceed the resolution of the degeneration techniques.

Most remarkable is the distribution of projection sites within the thalamus. In all instances, small, compact patches of terminals were widely scattered throughout the ventral VPL_0, with some clusters arising from branches of one axon (Fig. 8-3b). The sparse distribution is in good agreement with electrophysiologic (14,27) and neurophysiologic (6,7,28) studies in the monkey, in which neurons with vestibular-related activity were always intermingled with other, nonvestibular neurons. This pattern of distribution is not typical of a specific vestibular projection area, but it resembles the patchy distribution of spinothalamic afferents in the primate (2). The vestibular information probably interacts here with other proprioceptive inputs, which are known to provide most of the afferents to the VPL_0(34). In agreement with this hypothesis are findings that most of the vestibular thalamic neurons are also activated by proprioceptive (14,27) and optokinetic (6) stimuli.

In an earlier anatomic study in the monkey (41), no terminal degeneration was found rostral to the posterior commissure, following lesions of the vestibular nuclear complex. It was argued that degeneration found in the thalamus by other authors was due to concomitant lesions of either the brachium conjunctivum or the dorsal column nuclei. Our results do not appear compatible with this interpretation, since first there was no uptake in the dorsal column nuclei in the cases with vestibulothalamic projections; and second, the contribution of fibers of passage at the injection site to the labeled terminals (e.g., the brachium conjunctivum) can be neglected in the autoradiographic technique, since radioactive amino acids are rarely taken up by fibers (12). This is confirmed by exp. 22 where the injection site lies within the brachium conjunctivum itself but no endings in VPL_0 were found.

Other regions outside the vestibular nuclei included in the uptake areas are the mesencephalic nucleus of the Vth nerve in exp. 33 and the medullary reticular formation in exp. 24. Neither of these sources could account for the typical thalamic labeling also seen in exp. 19 and 21, in which only the vestibular nuclei were labeled.

Most anatomic (19,35) and electrophysiologic (1,30,37,39) studies conclude that the ascending vestibular thalamic fibers run outside the Mlf. It was also found in our study that the majority of terminals originate from fibers outside the Mlf, in group 2 (ipsilateral) and 4 (contralateral) of Fig. 8-5. However, we cannot exclude the possibility that a few vestibulothalamic fibers, of particularly the ipsilateral component, ascend in the Mlf, which would account for the thalamic degeneration found after Mlf lesions (11). The contribution of fibers ascending within the brachium conjunctivum (44) and the pontine reticular formation is thought on the basis of our results to be negligible.

Responses evoked by electric stimulation of the vestibular nerve have been reported in several regions in addition to VPL_0 in the monkey (13,27): the pulvinar, centrum medianum, nucleus centralis lateralis, zona incerta, nucleus lateralis posterior, nucleus reticularis, the posterior group, and VPM. Direct vestibular connections to these nuclei could not be found in our study. The lack of afferents to the posterior group was particularly surprising (35). In the posterior group of the squirrel monkey, short latency responses comparable to those in the VPL_0 were found (27), which would indicate two thalamic relays for ascending vestibular information, and thus two separate, parallel vestibular pathways to the cortex (8,26,27). The source of vestibular afferents to the posterior group remains then a puzzle.

SUMMARY

Vestibular projections rostral to the oculomotor nucleus were studied in the monkey (*Macaca mulatta*) by injecting anterograde tracer substances (radioactive leucine and proline) into the vestibular nuclear complex. Terminal labeling in the thalamus was found bilaterally, mainly in the nucleus ventroposterior lateralis pars oralis (VPL_0). The labeling was sparse, and scattered over wide areas. No labeling was found in the posterior group. Other vestibular nucleus afferents, ascending within the medial longitudinal fasciculus, terminated in the *rostral* interstitial nucleus of the medial longitudinal fasciculus (*rostral* iMLF) and the interstitial nucleus of Cajal, but not within the nucleus of Darkshevich.

ACKNOWLEDGMENTS

This work was supported by the Swiss National Science Foundation, grant No. 3.636.75, and the Dr. Eric Slack-Gyr Foundation in Zürich.

References

1. Abraham, L., Copack, P.B., and Gilman, S.: Brain stem pathways for vestibular projections to cerebral cortex in the cat. Exp. Neurol. 55:436, 1977.
2. Boivie, J.: An anatomical reinvestigation of the termination of the spinothalamic tract of the monkey. J. Comp. Neurol. 186:343, 1979.

3. Brodal, A. and Pompeiano, O.: The vestibular nuclei in the cat. J. Anat. (Lond.). 91:438, 1957.
4. Buchanan, A.R.: The course of the secondary vestibular fibers. J. Comp. Neurol. 67:183, 1937.
5. Büttner, U., Büttner-Ennever, J.A., and Henn, V.: Vertical eye movement related unit activity in the rostral mesencephalic reticular formation of the alert monkey. Brain Res. 130:239, 1977.
6. Büttner, U. and Henn, V.: Thalamic unit activity in the alert monkey during natural vestibular stimulation. Brain Res. 103:127, 1976.
7. Büttner, U., Henn, V., and Oswald, H.P.: Vestibular related neuronal activity in the thalamus of the alert monkey during sinusoidal rotation in the dark. Exp. Brain Res. 30:435, 1977.
8. Büttner, U. and Lang, W.: The vestibulo-cortical pathway: neurophysiological and anatomical studies in the monkey. Reflex control of posture and movements. Prog. Brain Res. 50:581, 1979.
9. Büttner-Ennever, J.A. and Büttner, U.: A cell group associated with vertical eye movements in the rostral mesencephalic reticular formation of the monkey. Brain Res. 151:31, 1978.
10. Carpenter, M.B. and Hanna, G.R.: Lesions of the medial longitudinal fasciculus in the cat. Am. J. Anat. 110:307, 1962.
11. Carpenter, M.B. and Strominger, N.L.: The medial longitudinal fasciculus and disturbances of conjugate horizontal eye movements in the monkey. J. Comp. Neurol. 125:41, 1965.
12. Cowan, W.M. and Cuénod, M. (eds.): The Use of Axonal Transport for Studies of Neuronal Connectivity. Amsterdam, Elsevier, 1975.
13. Deecke, L., Schwarz, D.W.F., and Fredrickson, J.M.: Nucleus ventroposterior inferior (VPI) as the vestibular thalamic relay in the rhesus monkey. I. Field potential investigation. Exp. Brain Res. 20:88, 1974.
14. Deecke, L., Schwarz, D.W.F., and Fredrickson, J.M.: Vestibular responses in the rhesus monkey ventroposterior thalamus. II. Vestibulo-proprioceptive convergence at thalamic neurons. Exp. Brain Res. 30:219, 1977.
15. Duensing, F., Schaefer, K.-P., and Trevisan, C.: Die raddrehenden vermittelnden Neurone in der centralen Funktionsstruktur der Labyrinthstellreflexe auf Kopf und Augen. Arch. Psychol. Z. Ges. Neurol. 204:113, 1963.
16. Fredrickson, J.M., Figge, U., Scheid, P., and Kornhuber, H.H.: Vestibular nerve projection to the cerebral cortex of the rhesus monkey. Exp. Brain Res. 2:318, 1966.
17. Gacek, R.R.: Anatomical demonstration of the vestibulo-ocular projections in the cat. Acta Otolaryngol. (Suppl.) 293:1, 1971.
18. Graybiel, A.M.: Organization of oculomotor pathways in the cat and rhesus monkey. In Baker, R. and Berthoz, A. (eds.): Control of Gaze by Brain Stem Neurons. Developments in Neuroscience, Vol. 1. Amsterdam, Elsevier, 1977, pp. 79–88.
19. Hassler, R.: Forels Haubenfaszikel als vestibuläre Empfindungsbahn mit Bemerkungen über einige andere sekundäre Bahnen des Vestibularis und Trigeminus. Arch. Psychol. Nervenkr. 180:23, 1948.
20. Hassler, R.: Hexapartition of inputs as a primary role of the thalamus. In

Frigyesi, T.L., Rinvik, E., and Yahr, M.D. (eds.): Corticothalamic Projections and Sensorimotor Activities. New York, Raven Press, 1972, pp. 551–579.

21. Hassler, R.: Supranuclear structures regulating binocular eye and head movements. Cerebral control of eye movements and motion perception. Bibl. Ophthalmol. 82:207, 1972.

22. Hess, W.R.: Das Zwischenhirn: Syndrome, Localisation, Funktionen. Basel, Schwabe, 1954.

23. Kömpf, D.: Vertikale Augenbewegungen. Nervenarzt 49:337, 1978.

24. Lang, W., Büttner-Ennever, J.A., and Büttner, U.: Vestibular projections to the monkey thalamus: an autoradiographic study. Brain Res. 177:3, 1979.

25. Lewandowski, M.: Untersuchungen über die Leitungsbahnen des Truncus cerebri und ihren Zusammenhang mit denen der Medulla spinalis und des Cortex cerebri. In Vogt, O. (ed.): Neurobiologische Arbeiten, 2. Weitere Beiträge zur Hirnanatomie. Jena, Fischer, 1904, p. 80.

26. Liedgren, S.R.C., Kristensson, K., Larsby, B., and Oedkvist, L.M.: Projections of thalamic neurons to cat primary vestibular cortical fields studied by means of retrograde axonal transport of horseradish peroxidase. Exp. Brain Res. 24:237, 1976.

27. Liedgren, S.R.C., Milne, A.C., Rubin, A.M., Schwarz, D.W.F., and Tomlinson, R.D.: Representation of vestibular afferents in somatosensory thalamic nuclei of the squirrel monkey. J. Neurophysiol. 39:601, 1976.

28. Magnin, M. and Fuchs, A.F.: Discharge properties of neurons in the monkey thalamus tested with angular acceleration, eye movement, and visual stimuli. Exp. Brain Res. 28:293, 1977.

29. McMasters, R.E., Weiss, A.H., and Carpenter, M.B.: Vestibular projections to the nuclei of the extraocular muscles. Am. J. Anat. 118:163, 1966.

30. Mickle, W.A. and Ades, H.W.: Rostral projection pathway of the vestibular system. Am. J. Physiol. 176:243, 1954.

31. Muskens, L.J.J.: An anatomico-physiological study of the posterior longitudinal bundle in its relation to forced movements. Brain 36:352, 1913–1914.

32. Oedkvist, L.M., Schwarz, D.W.F., Fredrickson, J.M., and Hassler, R.: Projection of the vestibular nerve to the area 3a arm field in the squirrel monkey. Exp. Brain Res. 21:97, 1974.

33. Olszewski, J.: The Thalamus of the *Macaca mulatta*. An Atlas for Use with the Stereotaxic Instrument. New York, Karger, 1952.

34. Poggio, G.F. and Mountcastle, V.B.: The functional properties of ventrobasal thalamic neurons studied in unanaesthetized monkeys. J. Neurophysiol. 26:775, 1963.

35. Raymond, J., Demêmes, D., and Marty, R.: Voies et projections vestibulaires ascendantes emanant des noyaux primaires: étude radioautographique. Brain Res. 111:1, 1976.

36. Raymond, J., Sans, A., and Marty, R.: Projections thalamiques des noyaux vestibulaires. Etude histologique chez le chat. Exp. Brain Res. 20:273, 1974.

37. Roucoux-Hanus, M. and Boisacq-Schepens, N.: Ascending vestibular projections: further results at cortical and thalamic levels in the cat. Exp. Brain Res. 29:283, 1977.

38. Sans, A., Raymond, J., and Marty, R.: Réponses thalamiques et corticales à la stimulation électrique du nerf VIII chez le chat. Exp. Brain Res. 10:265, 1970.

39. Sans, A., Raymond, J., and Marty, R.: A vestibulothalamic pathway: electrophysiological demonstration in the cat by localized cooling. J. Neurosci. Res. 2:167, 1976.

40. Schwarz, D.W.F. and Fredrickson, J.M.: Rhesus monkey vestibular cortex: A bimodal primary projection field. Science 172:280, 1971.

41. Tarlov, E.: The rostral projections of the primate vestibular nuclei: An experimental study in macaque, baboon, and chimpanzee. J. Comp. Neurol. 135:27, 1969.

42. Tarlov, E.: Anatomy of the two vestibulo-oculomotor projection systems. Basic aspects of central vestibular mechanisms. Prog. Brain Res. 37:471, 1972.

43. Winkler, C.: The central course of the nervus octavus and its influence on motility. Verh. Akad. Wet. Amst. 14:31, 1907.

44. Yamamoto, M., Shimoyama, I., and Highstein, S.M.: Vestibular nucleus relaying excitation from the anterior canal to the oculomotor nucleus. Brain Res. 148:31, 1978.

Discussion

PRECHT: Did you ever look at the magnocellular part of the medial geniculate body when you injected retrograde tracers into 2V?

BÜTTNER: There are some labeled cells in the medial geniculate after HRP injections into 2V, but on the other hand the highest proportion of cells lie in the suprageniculate nucleus. However, the injection into 2V is a small one. We can't conclude that the magnocellular area does not project there. In fact, there's a lot of evidence that it does.

PRECHT: We have done some experiments last year showing that the magnocellular part of the medial geniculate body is a vestibulocortical relay area. Injecting horseradish peroxidase in the magnocellular part of the medial geniculate body, after having recorded there, and after having found very short latency responses following vestibular nerve stimulation, one finds retrograde labeling of vestibular nucleus neurons. This particular area of the medial geniculate body seems to project to the anterior suprasylvian gyrus. So there may be another parallel pathway. Maybe it's not as clear in the monkey, or differently organized as in the cat.

BÜTTNER: Yes, I think that may be the case. But in our studies on the monkey, although there are some quite large injections in the vestibular nuclei, and we've looked very carefully in this region expecting, in fact, to find something, we never found any labeling there.

9

Ionic Mechanisms in the Vestibular Apparatus: The Resting State

HUMBERTO BRACHO, RUBEN BUDELLI, AND FRANK GALEY

In the vestibular apparatus, the sensory elements that inform the central nervous system about the position and movements of the head are part of the epithelium that covers the inner ear. In this organ the sensory elements are found in specific regions where connections with afferent terminals are made. These regions include the macula sacculi, the macula utriculi, and the crista ampullaris of each semicircular canal. The sensory areas consist of two types of epithelial cells: hair cells and supporting cells. The former have been identified as the mechanoreceptors, in which it is thought, the mechanoelectric transduction takes place (15). Morphologically different from the supporting cells, these receptors have developed a ciliary apparatus on the endolymphatic (apical) side and a synaptic region on the perilymphatic (basolateral) side (Fig. 9-1). The vestibular epithelium separates two fluids of different ionic composition: the endolymph, which has a high potassium concentration (above 100 mM), and the perilymph, with a low potassium content (below 5 mM) (20,22).

Because of their epithelial character and their excitable properties, the cells in the sensory areas of the inner ear should be identified as neuroepithelial cells and studied as such. This means that both transepithelial transport and membrane excitability should be considered in studying the participation of this tissue in the mechanoelectric transduction.

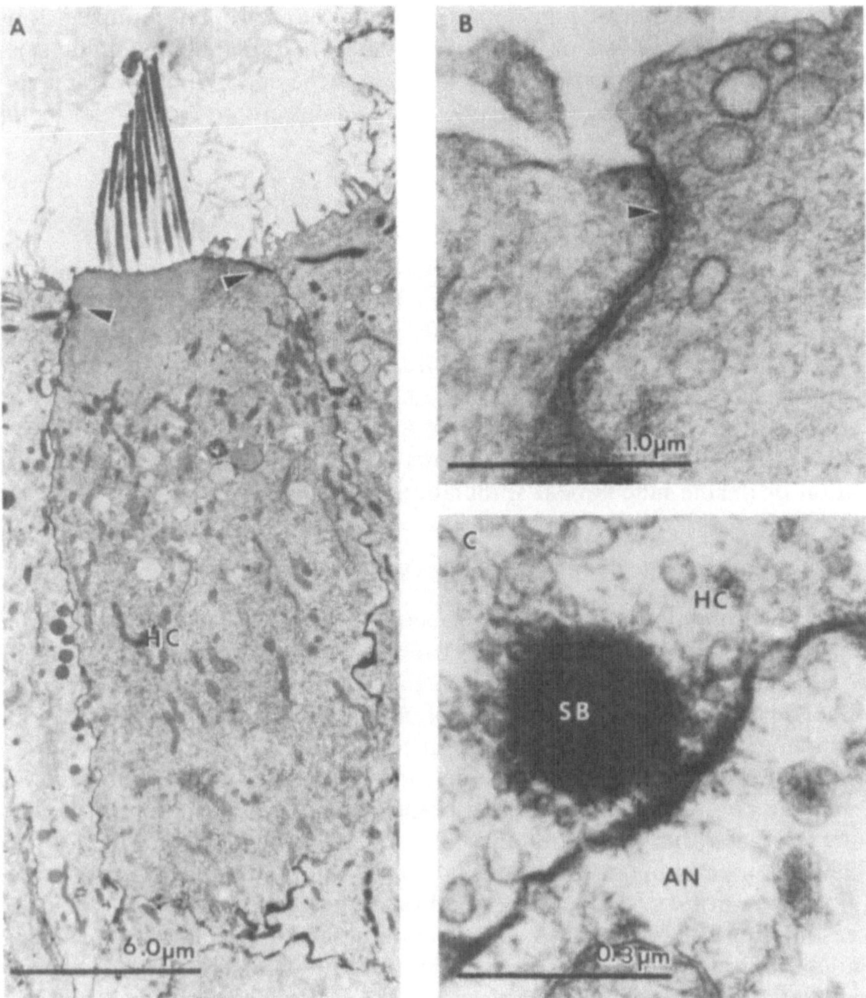

Fig. 9-1. Electron micrograph of the macula sacculi of the mudpuppy showing morphological features of excitable and secretory epithelial cells. The features corresponding to excitable cells are: ciliary apparatus at the apical membrane (in **A**) and synaptic body (*SB*) in hair cells (*HC*) associated with an afferent nerve (*AN*) ending (in **C**). Note the typical granules associated with the synaptic body. The features corresponding to secretory epithelial cells are: zonulae occludentes (arrows in **A** and **B**), microvilli, basolateral membrane infoldings and a tortuous extracellular space (in **A**).

Despite remarkable advances in the elucidation of the cellular mechanisms involved in excitable processes and transepithelial transport in other organs, little is known about these mechanisms in the inner ear.

This chapter summarizes our knowledge of the ionic mechanisms which operate in the vestibular neuroepithelium in the absence of mechanical stimulation. Included are a description of the inner ear preparation used in our experiments, an analysis of the neuroepithelial cell membranes' ionic selectivity, an appraisal of the function and origin of the endolymph, and an evaluation of inner ear resting electrogenesis linked directly to the cellular metabolism.

THE AMPHIBIAN MACULA SACCULI

One of the problems involved in the determination of the ionic mechanisms in the inner ear was the availability of a biologic preparation suitable for the study of individual cells. We used the macula sacculi from the mudpuppy (*Necturus maculosus*) (5,6,8) to record intracellular potentials from identifiable cells, perfuse independently each side of the neuroepithelium with solutions of known ionic composition, and perform in vitro and in vivo experiments (Fig. 9-2). These experiments can be done in the absence of stimulation or during mechanical stimulation of the hair cells.

IONIC SELECTIVITY OF THE MEMBRANE

Initial experiments on the macula sacculi were done to elucidate the ionic dependence of membrane potentials from the neuroepithelial cells present in the sensory area. The preparation was placed in a single bath surrounded by artificial perilymph and with the endolymphatic surface facing upward. The resting membrane potentials of several cells were measured by impaling the cell membranes with glass microelectrodes connected with the appropriate electronic equipment. Resting membrane potentials from different preparations fluctuated between −60 and −80 mV.

The cell membrane of these cells is selective to potassium, as are other excitable membranes (1,3). The membrane potential's absolute value decreases when the potassium concentration in the bath is increased. The potential–potassium concentration relationship, which is similar to relationships found in other excitable cells, fits a modified version of the constant field equation that includes a resting sodium permeability (Fig. 9-3) (18):

$$e^{VF/RT} = \frac{[K]_o}{[K]_i} + b\frac{[Na]_o}{[K]_i} + b\frac{[Na]_o}{[K]_i}$$

The symbol b is a constant given by the Na/K permeability ratio. By using the potential values obtained for different potassium concentrations, the internal potassium ($[K]_i$) and the Na/K permeability ratio (b) can be estimated. These values are 136 mV and 0.05, which are similar to the ones found in other excitable cells (16).

ELECTRODIFFUSION ACROSS THE NEUROEPITHELIUM

Although the described experiments suggest that the membrane potassium selectivities of neuroepithelial cells and other excitable tissues are strikingly

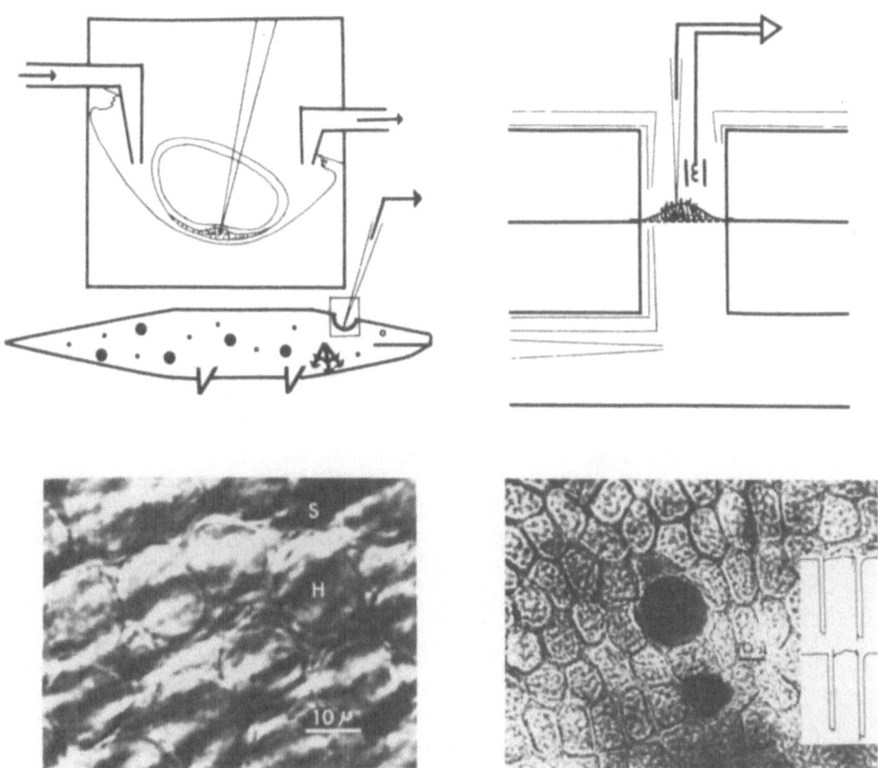

Fig. 9-2. Top: Experimental arrangements for the in vivo (left) and in vitro (right) experiments. Bottom: endolymphatic surface of the center of the macula (left). Circular and polygonal cells have been identified as hair cells (*H*) and supporting cells (*S*) , respectively. Microphotograph of the endolymphatic surface taken at the edge of the macula (right), where the melanocyte network is less dense. Two cells have been stained by intracellular ionophoresis. The records show the membrane potential obtained from the two cells, before and after the injection of the dye (Chicago blue). In this particular experiment the circular cell had a potential of −76 mV and the polygonal one of −78 mV. (After Bracho and Budelli, 1978)

similar, we must consider the epithelial character of the inner ear mechanoreceptors. The membrane potential in these cells depends on the ionic sensitivity of two membranes, apical and basolateral, which often have different permeabilities (19). Since in our experiments the same solution was perfused on both sides of the epithelium, they did not reveal whether only one side of the cell or both sides had a potassium-selective membrane. To characterize each side of the epithelial cells, it was necessary to perfuse each side independently. When this was done, changes in potassium concentration on each side of the epithelium produced changes in the membrane potentials of the cells. This indicates that both membranes (endolymphatic and perilymphatic) are potassium selective. However, when the concentration was modified in the perilymphatic side, the potential changes were larger than

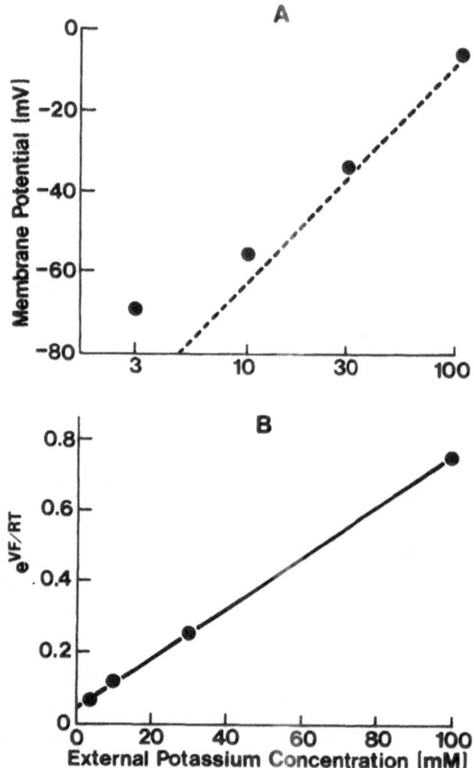

Fig. 9-3. **A** The membrane potentials were plotted against the logarithm of the external potassium concentration. The interrupted line represents the Nernst equation for a potassium electrode at an internal potassium concentration of 136 mM. **B** The data from **A** were plotted using the Moreton equation. The value $\exp(VF/RT)$ was plotted against the potassium concentration in a linear scale. The straight line is the least-square fitting of the experimental results. Each point represents the mean of 15 experiments; in each experiment the membrane potential was obtained in at least 10 cells for each experimental condition. (From Bracho and Budelli, 1978)

those produced with concentration changes in the endolymphatic side (Fig. 9-4). This does not necessarily mean that the ionic selectivities of these membranes are different. Since the perilymphatic membrane area is larger than the endolymphatic area (Fig. 9-1), it is possible that the ionic selectivity of both cell membranes is the same but the overall permeability is different. To clarify the interpretation of the above results, the electrodiffusion equations developed for a two-compartment system were modified to consider three compartments and applied to the experimental results. It was assumed that the membrane ionic selectivity is uniform over the whole cell.

The equation for a two-compartment system is as follows:

$$M_x = P_x\,(X_o - X_i \cdot e^{VF/RT}) \cdot f(v)$$

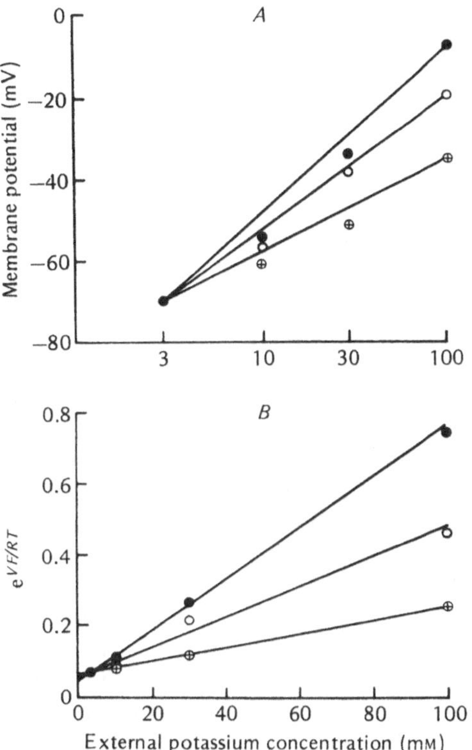

Fig. 9-4. **A** Dependence of membrane potentials on the external concentrations of potassium under three different experimental conditions: when the potassium concentration was changed in the perilymph and the endolymph kept constant at 3 mM (○), when the potassium concentration was changed in the endolymph and the perilymph kept constant at 3mM (⊕), and when the changes were made symmetrically (●). (Data obtained from Fig. 9-3) **B** The same data from **A** were plotted after calculating $\exp(VF/RT)$ and plotting this value against the external potassium concentration in a linear scale. The straight lines represent the least-square linear fitting to the experimental results. (From Bracho and Budelli, 1978)

In the macula sacculi mounted between two chambers, three compartments are separated by two membranes. The net flux of ions during the steady state would be:

$$M_K + M_{Na} + M'_K + M'_{Na} = 0$$

Since the changes in potassium concentration in the experimental solutions were accomplished by substituting Na for K in equimolar amounts, the total amount of cations remained constant (118 mM).

$$K_p + Na_p = K_e + Na_e = 118$$

Combining the above three equations and considering that there is no difference of potential between the endolymphatic and the perilymphatic side. the following equation is obtained:

$$e^{VF/RT} = \frac{(1-b)}{[K]_i} \cdot \frac{P_k}{P_k + P'_k} \cdot K_e + \frac{(1-b)}{[K]_i} \cdot \frac{P'_k}{P_k + P'_k} \cdot K_p + \frac{118}{[K]_i} \cdot b$$

This equation states that the slope of the potassium potential relationship should be different if the overall membrane permeability is different on each side but the ionic selectivity (b) is the same. As shown in Fig. 9-4, the electrodiffusion equation for a three-compartment system fits the experimental results closely, supporting the hypothesis of uniform ionic selectivity in the neuroepithelial cell membrane.

ROLE OF THE ENDOLYMPH

With this analysis the functional significance of the endolymphatic-perilymphatic potassium gradient becomes apparent: the ciliated membrane (endolymphatic) separates two fluids with a similar potassium concentration (endolymph and intracellular), and the innervated membrane (perilymphatic) separates two fluids with a drastically different potassium content (perilymph and intracellular). Since both membranes have the same high potassium selectivity, each would generate an electric potential according to its potassium gradient across the membrane. Electrically, this means that two potassium batteries of different value are present in the inner ear neuroepithelial cells (Fig. 9-8). The first battery (V_e) has a low absolute value given by the potassium concentration difference between the endolymph and the interior of the cell.

$$V_e = \frac{RT}{F} \cdot log\frac{100}{136} = -\ 8mV$$

The second battery (V_p) has a potential close to:

$$V_p = \frac{RT}{F} \cdot log\frac{3 + 115b}{136} = -\ 70\ mV$$

If these potassium batteries were connected through a low transepithelial resistance (Rs), a continuous flux of ionic current would be present without stimulation (Fig. 9-8). In other words, the most likely physiologic role of the endolymph is to provide a source of electric current during resting conditions.

THE SHUNT RESISTANCE

An important aspect of this hypothesis is the existence of a shunt resistance connecting the two ionic batteries to allow for the flux of ionic current. In epithelial tissues, in addition to the contribution of cellular membranes to the overall transepithelial resistance, an electric resistance is localized in the intercellular junctions (zonulae occludentes) which hold the cells together on the apical side. These junctions have different ionic permeabilities and therefore different electric resistances. "Leaky" epithelia have low-resistance intercellular junctions ($< 500\ \Omega \cdot cm_2$), and "tight" epithelia have high resistance ($>1500\ \Omega \cdot cm_2$). In the presence of an ionic gradient, "tight" epithelia may develop a transepithelial potential (up to 100 mV) and "leaky" epithelia will not (24).

Observation of the movements of electron-opaque tracers, such as lanthanum, under the electron microscope reveals that these tracers do not permeate tight zonulae occludentes but do move freely through leaky junctions (7,24). Evidence on vestibular epithelium indicates that the zonulae occludentes belong to the leaky type. This is based on the following experimental observations: the transepithelial potential in the vestibule has a value around 0 mV (6,21); the transepithelial resistance in the amphibian macula sacculi is about $200\ \Omega \cdot cm_2$ (6); and, as seen in Fig. 9-5, lanthanum appears to permeate the zonulae occludentes of the vestibular neuroepithelium in the mudpuppy. The existence of a low-resistance transepithelial pathway makes it possible to close the circuit connecting the two potassium batteries and therefore establish a continuous flux of ionic current. Interestingly, the cochlear neuroepithelium has the characteristics of a tight epithelium: large transepithelial potentials (about 100 mV) (4) and impermeability of zonulae occludentes to lanthanum (13).

In general, leaky and tight epithelia have not only different transepithelial resistances and potentials but also different transport properties. Among them is the suitability of leaky epithelia for the isotonic transport of fluids (10), which apparently occurs not only across the cellular membranes but also through the intercellular junctions. This property of leaky epithelia may be of clinical interest. The inner ear is covered by a mixture of tight (cochlear) and leaky (vestibular) epithelia. In some pathologic conditions, such as Meniere's disease and syphilis, the transport of fluids across the inner ear neuroepithelium is disrupted. As a result, the endolymphatic space enlarges, producing severe impairment of inner ear function (2). It would not be surprising to find that the initial disturbance in these diseases is a change in the permeability of the intercellular junctions in the leaky epithelial region of the inner ear neuroepithelium.

METABOLIC ELECTROGENESIS

The analysis of ionic movements in vitro shows that all the forces involved are passive and that participation of cellular metabolism is not necessary to generate membrane potentials directly. However, inner ear electrogenesis has been linked to cellular metabolism in several animal species. The existence of a steep ionic gradient across the neuroepithelium, linked with the high number of mitochondria in the hair cells (23), strongly suggests that this epithelium is involved in active transport of ions (14). Since the macula

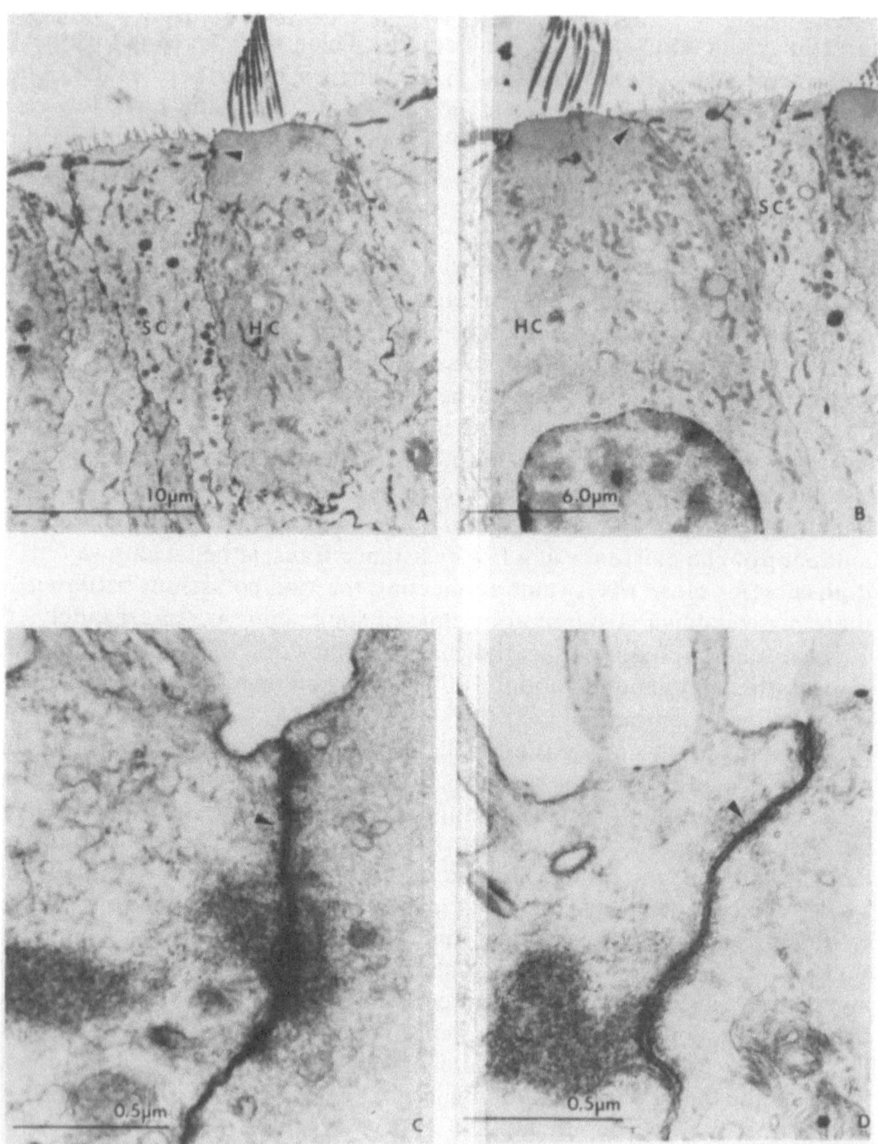

Fig. 9-5. Electron micrographs of sensory epithelium of mudpuppy macula. **A** and **B** are low-power electron micrographs of hair cells (*HC*) and supporting cells (*SC*). **A** and **B** are of colloidal lanthanum-treated and -untreated maculae respectively. Note the higher density of extracellular space in the lanthanum-treated epithelium. **C** and **D** Treated and untreated intercellular junctional complexes (zonulae occludentes, arrows in **A** and **B**), respectively. The junctional complex of the treated macula is filled with electron-opaque material. This difference from untreated maculae is interpreted as due to the presence of colloidal lanthanum in the zonula occludens.

sacculi can be exposed to record neuroepithelial cell membrane potentials in vivo, we performed some experiments on an anesthetized mudpuppy with a window open in the skull to expose the inner ear (Fig. 9-2). The inner ear blood supply was intact, as was therefore the continuous supply of oxygen and metabolites which are necessary to preserve aerobic metabolism. The perfusion of solutions with different potassium concentrations yields a potassium-potential relationship different from the one obtained in vitro. In vivo, membrane potentials recorded at different potassium concentrations lie on a line parallel to the theoretical line generated by the Nernst equation and are consistently hyperpolarized with respect to their in vitro counterparts (Fig. 9-6). The value of $[K]_i$, estimated with the use of the electrodiffusion equation, is too large (200 mM) to be compatible with the osmotic equilibrium of the cell. We interpret these results as evidence that, in vivo, the neuroepithelial cell membrane potentials cannot be explained by passive diffusion of ions alone; the contribution of the metabolism must be considered.

In other excitable cells which demonstrate metabolic electrogenesis, it has been shown that a membrane-bounded enzyme, the Na-K adenosinetriphosphatase (ATPase), is involved in the electrogenic exchange of ions across the membrane (16). This enzyme is activated by extracellular potassium and inactivated by glycosides, such as ouabain. If the in vivo–in vitro difference of membrane potentials recorded at a given potassium concentration is generated by the Na-K ATPase activity, this difference should increase as the external potassium concentration increases. This was the case when the membrane potentials obtained in vitro at a given potassium concentration were subtracted from the ones obtained at the same potassium concentration in vivo.

Another indication of the Na-K ATPase involvement in the neuroepithelial cell resting electrogenesis is the depolarizing effect of ouabain and ethacrynic acid, which are Na-K ATPase inhibitors (Fig. 9-7) (8).

EQUIVALENT CIRCUIT

The described experimental analysis of vestibular resting electrogenesis can be summarized by using an equivalent circuit of the system (Fig. 9-8). This circuit includes three potassium-dependent batteries: one generated by the endolymphatic-intracellular potassium gradient, the second generated by the intracellular-perilymphatic potassium gradient, and the third by the metabolic activity of the cell. This circuit is closed through leaky intercellular junctions, providing a continuous flow of current.

The modulation of a resting current by a specific stimulus has been proposed to constitute the basis of the transduction mechanism in several sensory receptors (9,23,26). At this time, only the resting current generated by the vertebrate photoreceptors has been directly measured (25). Although we have not measured directly the resting current generated by the inner ear

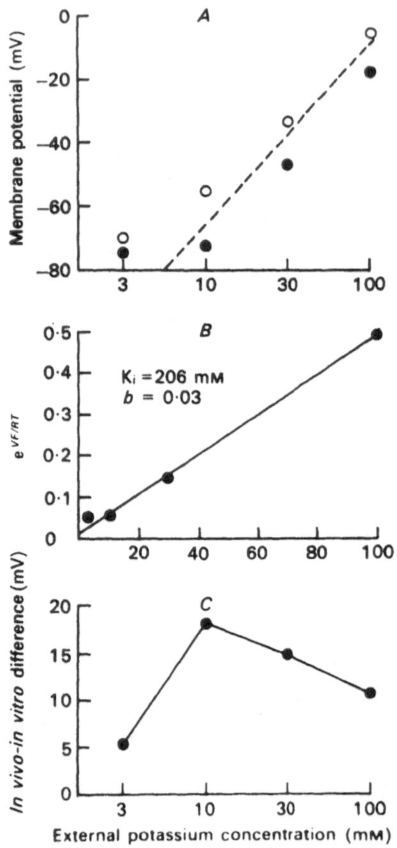

Fig. 9-6. Dependence of membrane potentials on the external potassium concentration in the in vivo preparation. **A** The membrane potentials were plotted against the logarithm of the external potassium concentration. (●) in vivo experiments. (○) in vitro experiments. The dashed line represents the Nernst equation assuming an internal potassium concentration of 136 mM. **B** exp(VF/RT) was plotted against the external potassium concentration. The estimated value of b and K_i are 0.03 and 206 mM. **C** The in vivo–in vitro differences were plotted against the logarithm of the external potassium concentration using the values from **A.** This graph shows that the in vivo–in vitro potential difference is potassium dependent. (From Bracho and Budelli, 1978)

mechanoreceptors, our results strongly support the existence of such a mechanism. The generation of a resting current may well be a ubiquitous characteristic of sensory receptors.

THE ORIGIN OF THE ENDOLYMPH

Besides their participation in the transduction process, the vestibular inner ear neuroepithelial cells are probably involved in the production of

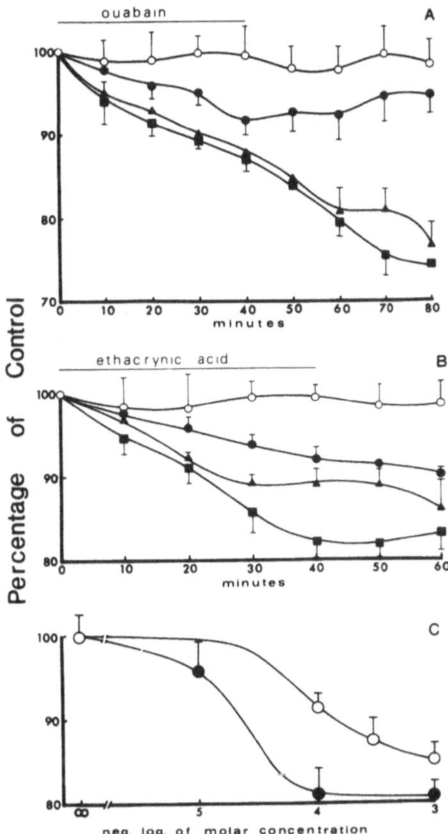

Fig. 9-7. Depolarizing effect of different concentrations of ouabain and ethacrynic acid on the neuroepithelial cell resting potentials. **A** Ouabain at 0 (○), 10^{-5} (●), 10^{-4} (▲), and 10^{-3}M(■). **B** Ethacrynic acid at 0 (○), 10^{-4} (●), 3 x 10^{-4} (▲), and 10^{-3} M (■) **C** Dose-effect curve taken 1 h after the preparation was exposed to ouabain (●) and ethacrynic acid (○). In all three graphs the points represent the average resting potential of the neuroepithelial cells, measured as the percentage of the control value obtained at zero time. The vertical bars represent the standard error of the mean when this is larger than the symbol. The values were obtained from 3 to 7 different animals for each concentration of the drugs. (From Budelli et al., 1979)

endolymph. This involvement is more likely in animals like the mudpuppy which do not possess a specialized epithelium (the stria vascularis) to perform such a function. In the light of the described experimental analysis, let us speculate about the origin of the high endolymphatic potassium. To produce an accumulation of potassium in the endolymph, this ion has to be transported through the neuroepithelium from the perilymph. The forces that determine the passive transcellular potassium movements across the inner ear epithelium are the potassium concentration gradient and the electric potentials across the epithelial membranes. The estimated intracellular

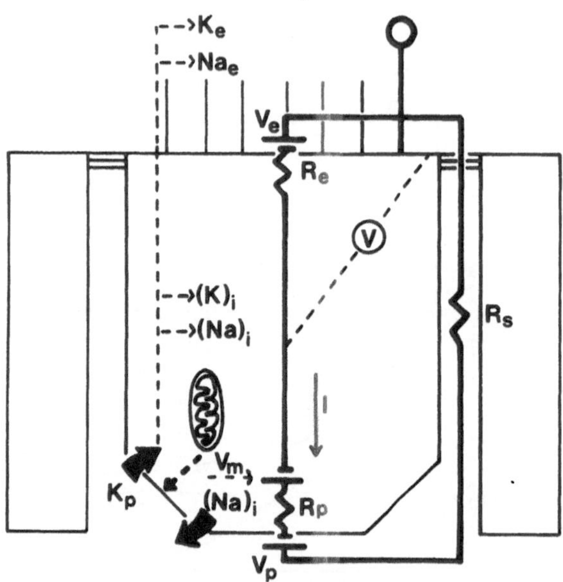

Fig. 9-8. Schematic representation of the two-gradient hypothesis. The membrane potential is the result of the interaction of three batteries: V_e which is generated by the endolymphatic-intracellular potassium gradient; V_p, produced by the perilymphatic-intracellular potassium gradient, and V_m, established by the active exchange of sodium and potassium which was mediated by the activity of the sodium-potassium adenosine triphosphatase. The dotted lines represent the unknown pathway followed by sodium and potassium ions. (From Bracho and Budelli, 1978)

potassium concentration is 136 mM, and the endolymphatic and perilymphatic potassium content is 100 and 3 mM, respectively. By substituting these values in the Nernst equation, we can determine the equilibrium potential for potassium at each membrane:

$$V_{ke} = \frac{RT}{F} \cdot log\frac{100}{136} = -8\ mV$$

$$V_{kp} = \frac{RT}{F} \cdot log\frac{3}{136} = -96\ mV$$

Since the actual neuroepithelial cell membrane potential in vivo is −70 mV (6), potassium is off its equilibrium across the perilymphatic membrane by 26 mV and across the endolymphatic membrane by 62 mV. To transport potassium from the perilymph into the endolymph, it would be necessary for an epithelial cell to actively incorporate potassium from the perilymph into the intracellular fluid and to actively extrude this ion from the cell into the endolymph. These processes require two active transport mechanisms (Fig. 9-9). The one that incorporates potassium into the cell is most likely the Na-K active exchange mediated by the Na-K ATPase, which apparently is present in the perilymphatic membrane (17). The second mechanism may be

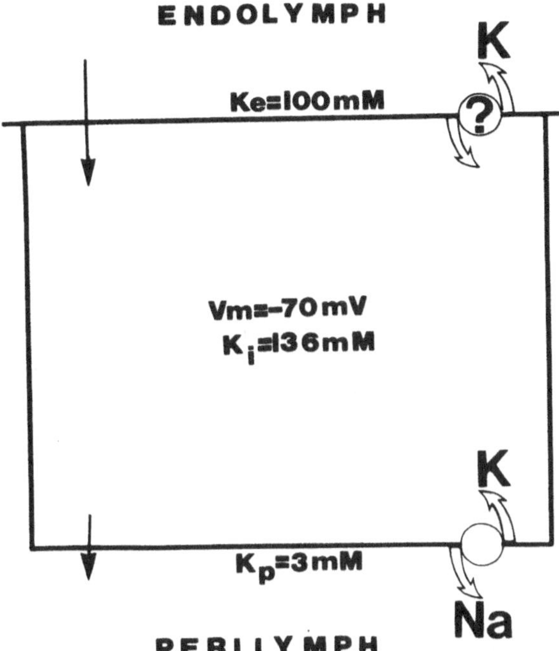

Fig. 9-9. Hypothetical movements of potassium across the neuroepithelial cells. The arrows on the left represent the electrochemical gradients acting upon the potassium ion at each epithelial membrane. The pumps necessary to overcome the passive movements of potassium ions are included on the right.

a potassium extrusion process. This type of active transport has been shown to be present in the insect midgut (12) and in the gastric mucosa of vertebrates (11), but has not been investigated in the inner ear. A clue to the existence of this process was found in the experiments using the in vitro preparation of the amphibian macula sacculi. When the membrane potentials were recorded periodically in the absence of continuous perfusion, the absolute value of the potentials decayed, reaching 50% of the control value in about 30 min. Membrane potentials returned to control values after perfusing the preparation with a fresh solution for 5 min. This may be indirect evidence of the existence of a potassium secretory process at the endolymphatic membrane: in the absence of continuous perfusion, the secreted potassium accumulates in the vicinity of the endolymphatic membrane, modifying the potassium gradient and therefore the transmembrane potential.

PROSPECTS

The results obtained from an amphibian inner ear preparation are encouraging. Now that a reasonable amount of information about the resting state in the vestibule is available, efforts should be directed to clarify the membrane phenomena occurring when the hair cells are stimulated mechanically, and to clarify the relation of mechanoelectric events to the release of synaptic transmitter. More research is also needed on the origin of the endolymph. We believe this preparation can be used to obtain important information about all aspects of hair cell function in the inner ear, leading to the understanding of both mechanisms in inner ear physiology and the pathophysiology of some inner ear diseases.

ACKNOWLEDGMENTS

We thank Diane Foster for editorial help; Sheila Odnert, Cynthia Smith, Ibrahim Hernandez, and Michel Walker for technical assistance; and Janet Lewis, Carl Muller, and Manuel Don for critically reading this manuscript.

References

1. Adrian, R.H.: The effect of internal and external potassium concentration on the membrane potential of frog muscle. J. Physiol. 133:631, 1956.
2. Antoli-Candela, F., Jr.: The histopathology of Meniere's disease. Acta Otolaryngol. (Suppl.) 340:1, 1976.
3. Baker, P.F., Hodgkin, A.L., and Shaw, T.I.: The effect of changes in internal ionic concentration on the electrical properties of perfused giant axons. J. Physiol. 164:355, 1962.
4. Bekesy, G.V.: DC resting potentials inside the cochlear partition. J. Acoust. Soc. Amr. 24:72, 1952.
5. Bracho, H.: Potassium dependence of membrane potentials in the *Necturus maculosus* vestibular system. Brain Res. 136:366, 1977.
6. Bracho, H. and Budelli, R.: Generation of resting membrane potentials in an inner ear cell system. J. Physiol. (Lond.) 281:445, 1978.
7. Bracho, H., Erlij, D., and Martinez-Palomo, A.: The site of permeability barriers in frog skin epithelium. J. Physiol. 213:52P, 1970.
8. Budelli, R., Canalis, R., and Bracho, H.: Effects of ouabain and ethacrynic acid on the resting potentials of neuroepithelial cells in the inner ear. Gen. Pharmacol. 10:335, 1979.
9. Davis, H.: A model for transducer action in the cochlea. Cold Spring Harbor Symp. Quant. Biol. 30:181, 1965.
10. Diamond, J.: The epithelial junction: bridge, gate, and fence. Physiologist 20:10, 1977.
11. Ganser, A.L. and Forte, J.G.: K^+ - stimulated ATPase in purified microsomes of bullfrog Oxyntic cells. Biochem. Biophys. Acta 307:169, 1973.
12. Harvey, W.R. and Wood, J.L.: Route of cation transport across skilworm midgut. Benzon Symp. 5:342, 1973.
13. Iurato, S., Frank, K., Luciano, L., Wermbter, G., Pannese, E., and Reale, E.: Intercellular junctions in the organ of Corti as revealed by freeze fracturing. Acta Otolaryngol. 82:57, 1976.

14. Konishi, T. and Mendelsohn, M.: Effects of ouabain on cochlear potentials and endolymph composition in guinea pigs. Acta Otolaryngol. 69:192, 1970.

15. Lowenstein, O.: Comparative morphology and physiology. In Kornhuber, H.H. (ed.): Handbook of Sensory Physiology, Vol. 6/1. Vestibular System. Berlin, Springer-Verlag, 1974, pp. 75–120.

16. Marmor, M.F.: The membrane of giant molluscan neurons: electrophysiological properties and origin of the resting potentials. Prog. Neurobiol. 5:169, 1975.

17. Matsura, S., Ikeda, K., and Furokawa, T.: Effects of Na, K, and ouabain on microphonic potentials of the goldfish inner ear. Jp. J. Physiol. 21:563, 1971.

18. Moreton, R.B.: An investigation of the electrogenetic sodium pump in snail, using the constant field theory. J. Exp. Biol. 51:181, 1969.

19. Nagel, W.: The dependence of the electrical potentials across the membranes of the frog skin upon the concentration of sodium in the mucosal solution. J. Physiol. 269:777, 1977.

20. Rauch, S. and Rauch, I.: Physico-chemical properties of the inner ear, especially ionic transport. In Keidel, W.D. and Neff, W.D. (eds.): Handbook of Sensory Physiology, Vol. 5/1. Auditory System. Berlin, Springer-Verlag, pp. 647–682.

21. Smith, C.A., Davis, H., Deatherage, B.H., and Gessert, C.F.: DC potentials of the membrane labyrinth. Am. J. Physiol. 59:311, 1958.

22. Smith, C.A., Lowry, O.H., and Wu, M.L.: The electrolytes of the labyrinthine fluids. Laryngoscope 64:141, 1954.

23. Thurm, U.: Mechanisms of electrical responses in sensory receptors illustrated by mechanoreceptors. In Jaenicke, L. (ed.): Biochemistry of Sensory Function. Berlin, Springer-Verlag, p. 367.

24. Ussing, H.H., Erlij, D., and Lassen, V.: Transport pathways in biological membranes. Annu. Rev. Physiol. 36:17, 1974.

25. Yau, K.W., Lamb, T.D., and Baylor, D.A.: Light-induced fluctuations in membrane current of single toad rod outer segments. Nature 269:78, 1977.

26. Zuckerman, R.: Ionic analysis of photoreceptor membrane currents, J. Physiol. 235:333, 1973.

10

Centrally Originating Efferent Terminals on Hair Cells: Fact or Fancy?

M. D. ROSS

The second puzzle has to do with the function of the efferent fibers that pass to the cochlea, vestibular apparatus, and other sensory organs . . . . Perhaps the reason we have not succeeded in experimentally defining [the function of Rasmussen's efferent cochlear bundle] as yet reflects... a misconception of what its function actually is. Obviously what is needed is a fresh set of ideas and some clean new experiments.

R. Galambos, 1960[1]

The title of this article is not meant to be inflammatory. Its purpose is, rather, to call attention to what I consider to be one of the most pressing problems in auditory and vestibular research: the continued domination of the theory of olivocochlear and vestibular efferent modulation of the end organs of the inner ear in spite of the fact that more than 30 years of research has failed to produce agreement concerning the location of the cells of origin of the efferent bundles or their functional significance. The apparent unwillingness to consider alternative points of view, or to deal objectively

[1]Quoted by R.B. Livingston: A discussion of research potentialities. In Rasmussen, G.L. and Windle, W.F. (eds.): Neural Mechanisms of the Auditory and Vestibular Systems. Springfield, Ill., Thomas, 1960, p. 367.

with conflicting evidence, reminds one of T.H. Huxley's comment in regard to Herbert Spencer. Huxley said that Spencer's idea of a tragedy was a deduction killed by a fact. This chiding remark was prompted by Spencer's continued, uncritical attachment to theories of his own for which he sought support by looking for and emphasizing findings that tended to be confirmatory, while ignoring or reinterpreting (to his own satisfaction) conflicting data.

Heavy responsibility for the current state of disarray in our knowledge of the efferent fibers of the eighth nerve must be shouldered by anatomists, who have most consistently failed to give sufficient weight to conflicting or contradictory evidence, or who have misinterpreted the earlier literature. It should be recognized from the outset that there is no positive proof that the olivocochlear or the vestibular efferent bundles exist as currently described. Even recent application of the horseradish peroxidase (HRP) histochemical method (35,97) has failed to resolve the issue. The results are partly in conflict, and Warr's (97) findings emerged from studies carried out largely on early postnatal kittens in which the efferent-type terminals on the hair cells and the organ of Corti itself are both incompletely developed (88). Moreover, no one has yet demonstrated uptake of HRP by efferent terminals on hair cells at either the light or the electron microscopic level, and current ultrastructural evidence indicates that uptake does not occur (25,50,60).

Neither does degeneration of vesiculated terminals in the organ of Corti after lesions have been made in the brain stem prove that the fibers interrupted centrally and the terminals are parts of the same neurons, although the finding is often interpreted that way (49). Proof of physical continuity between the two would require a direct demonstration that no one has yet accomplished. Moreover, the degeneration has unique features (for example, rupture of the surface membranes) not characteristic of either central or peripheral terminal degeneration occurring elsewhere (8,16,23,41,42,63). An alternative explanation for the enlargement of the mitochondria and the disruption of the membranes of the preterminal and terminal segments is that section of specific fibers centrally results in fluid, ion, and/or nutritive imbalances peripherally detrimental to the nerve endings in question. That a local calcium ion imbalance alone can cause mitochondrial swelling and other degenerative changes in motor nerve axons and terminals that are comparable to those observed after axotomy has already been shown (46,90).

This second explanation would require that the fibers interrupted centrally be autonomic. The viewpoint has already been expressed that the efferent fascicles extending from the brain stem into the cochlear and vestibular nerves are parasympathetic (29–31,84). This viewpoint is not based solely upon the results of the studies noted, but also upon solid neuroanatomic and experimental findings that go back to the last century. It is with this latter evidence that I shall largely deal in this article, in order to demonstrate the wide basis of support that exists for stating that the centrifugal fibers of the eighth nerve are autonomic and for urging that some other origin be sought for the efferent terminals on the hair cells.

THE AUTONOMIC INNERVATION OF THE INNER EAR

Centrifugal fibers in the cochlear nerve were identified at the light microscopic level by Held in 1897 (44), who speculated that they were of central origin. Much earlier, Arnold (4,5) had described two anastomoses between the facial and the vestibular nerves. One of these (sometimes double) was a communication between nervus intermedius and the vestibular nerve that was said to have been found by Wrisberg in the 18th century. Fibers of the second communication were traced by Arnold from the carotid plexus along the major superficial petrosal nerve to the external genu of the facial nerve, and then to the vestibular nerve. Arnold described the fibers of these two communications as sympathetic in type (ortho- and parasympathetic subdivisions of the autonomic nervous system had not yet been recognized), and their existence was verified by numerous succeeding anatomists (45,72,76,82). Arnold postulated that the autonomic fibers were essential to the nutrition of the labyrinth and to the production of inner ear fluids. He also thought that disturbances in the sympathetic innervation resulted in pathologic phenomena.

It is unfortunate that these early findings of the anatomists somehow became lost as later authors described the facial and vestibular nerves in their anatomy texts and other publications. The first person in recent history to redescribe the orthosympathetic fibers anastomosing with the facial and then with the vestibular nerve, so far as I can determine, was Lorente de Nó (67), who detailed his findings in the embryonic mouse. It was not until 1966, however, when Spoendlin and Lichtensteiger (93) carried out their fluorescence studies of the cochlear nerve in the cat, that the vast number of sympathetic fibers existing in the eighth nerve became apparent. Moreover, this was the first work to demonstrate that not all of the fibers were related to blood vessels, but many were blood vessel independent and extended up to the foramina nervosa, where they terminated in close relationship to the auditory nerve fibers. Subsequently, Spoendlin and Lichtensteiger's work was confirmed by a number of investigations and similar innervation patterns of the sympathetic fibers in the vestibular nerve were recognized (20–22,85,86,94).

The communication between nervus intermedius and the vestibular nerve described by Arnold was observed by Ramon y Cajal in 1911 (77) and was redefined embryologically by Hovelacque in 1927 (48) and more recently by Eyries et al. (32), Zelenka and Subrt (102), and Obrebawski and Skornicki (70). Two or more fascicles are often involved. Eyries and Chouard (29,30) and Chouard (15) stressed that these anastomoses, which they called the acousticofacial, carried the parasympathetic fibers to the inner ear. In the meantime, I demonstrated both the central course of the parasympathetic fibers and the anastomoses between the facial and vestibular nerves, with a component continuing into the auditory nerve, on the basis of my acetylcholinesterase findings (84). I pointed out that the intramedullary course of the parasympathetic component corresponded to that of both the

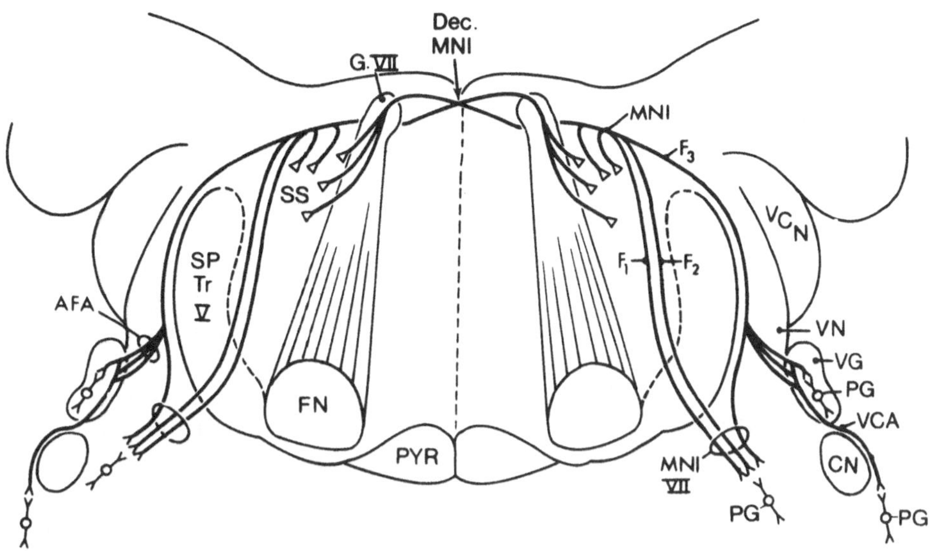

Fig. 10-1. This diagram illustrates the general location of the cells of the superior salivatory nucleus (*SS*) in the brain stem of the rat and the course of their axons in motor nervus intermedius (*MNI*). All the axons appear to have a genu, as shown, and some of the axons cross the midline (*Dec. MNI*) at the rostral end of the internal facial genu (*G. VII*). Compare this diagram with Figs. 10-3 and 10-4. Note that, unlike the situation in the dog and the cat in which only one intramedullary root of motor nervus intermedius occurs (corresponding in course to F_3 in this diagram), motor nervus intermedius of the rat has three or more intramedullary branches (F_1, F_2, and F_3). Commonly, two such branches leave the brain stem medial to the spinal tract of the trigeminal nerve (*Sp. Tr V*). Another major branch, often double (F_3), passes across the dorsolateral tegmentum to reach the lateral border of the spinal tract of V and turns ventrolateralward medial to the sensory root of nervus intermedius (not shown) and the vestibular nerve (*VN*). This branch supplies one small twig to motor nervus intermedius of the facial nerve (*MNI VII*) and two or more larger fascicles to the inner ear through the acousticofacial anastomoses (*AFA*). The anastomosing fibers carry the parasympathetic preganglionic nerve supply to the inner ear. Postganglionic neurons (*PG*) are located in the vestibular (*VG*) and spiral ganglia. Compare the crossed and uncrossed origin, the decussation, and the course of the component of motor nervus intermedius to the inner ear with the pathway ascribed to an "olivocochlear" bundle by Rasmussen (ref. 80, Fig. 2). Except for assigned cells of origin and ultimate sites of termination, the fiber pathways are identical. *CN*, cochlear nerve; *FN*, motor nucleus of the facial nerve; *PYR*, pyramid; *VCA*, vestibulocochlear anastomosis; *VCN*, ventral cochlear nucleus.

olivocochlear and the vestibular efferent bundles; only the origin, in the superior salivatory nucleus, was different (Fig. 10-1). The opinion was offered that the efferent fascicles to the inner ear were, in reality, parasympathetic and branches of nervus intermedius. This same viewpoint was expressed independently by Eyries and Chouard (30) and by Eyries et al. (31), and is supported by Chouard's (15) description of the parasympathetic nerve supply to the inner ear in man.

Although several investigators have demonstrated that the motor component of nervus intermedius sends one or more branches into the eighth nerve, the matter of the location of the required postganglionics has seldom been explored. Such postganglionic neurons would most likely, but not necessarily, be multipolar (75). Among the early investigators who described multipolar neurons in the spiral ganglion are Ayers (7) and Retzius (83), who used Golgi preparations, and Cannieu (12). Bovero (10) found similar type neurons in the vestibular ganglion. Retzius and others thought that the multipolar neurons might be primitive bipolar cells, which possibly explains the minor impression the finding made on students of the inner ear. However, we know from embryologic studies (98) that multipolar neurons represent a more advanced stage of development than the bipolar configuration.

More recently, Ross (84) reported that certain spiral and vestibular ganglion cells were acetylcholinesterase-positive and appeared to have terminals of the preganglionic parasympathetic fibers on or near them. The spiral ganglion cells identified as possibly postganglionic were located chiefly in or near the intraganglionic spiral bundle, which was considered to be comparable to an autonomic plexus. Later, Maw (62) provided ultrastructural evidence that the intraganglionic spiral bundle resembled an autonomic ground plexus and described axodendritic synapses in that bundle. Ehrenbrand and Wittemann (26) showed the presence of synaptic terminals on some vestibular ganglion cells. In further ultrastructural investigations, Ross and Burkel (87) found that the unmyelinated spiral ganglion cells were distributed largely in or near the intraganglionic spiral bundle and likely corresponded to the postganglionic neurons described earlier by Ross (84). Study of serial sections showed that at least some of these cells were multipolar.

The unmyelinated neurons are ultrastructurally similar to the postganglionic neurons of the rabbit otic ganglion (24) and of the otic, ciliary, and pterygopalatine ganglia of the golden hamster (101). Moreover, the relatively constant proportion of the unmyelinated spiral ganglion cells across species (generally, 8 to 10%) speaks for a common, generalized function for them. However, these same neurons have been identified by Spoendlin (92) as the source of the entire afferent innervation of the outer hair cells of the organ of Corti in the cat. This conflict must be resolved by further study, and the postganglionic neurons identified conclusively.

In spite of the problem of identification of the parasympathetic postganglionic neurons, it is clear that a great number of the centrifugal fibers in the auditory and vestibular nerves can be accounted for on the basis that they are autonomic. Additionally, there appear to be afferent fibers of sensory nervus intermedius in the eighth cranial nerve. In the mouse (Ross, unpublished observations) and in man (15) their cell bodies are located close to Scarpa's ganglion in a displaced portion of the geniculate ganglion. The question is whether still other populations of fibers, olivocochlear and vestibular efferent, occur in the eighth nerve, or whether the communicating fibers of nervus intermedius and of the orthosympathetic system together account for all the fibers originating outside the cochlear and vestibular ganglia.

RASMUSSEN'S EARLY OBSERVATIONS

It was in 1946 that Rasmussen first published his findings in Marchi preparations after placing lesions in the basal part of the brain stem in, or in proximity to, the superior olivary complex (78). He found Marchi granules along a fascicle out of the superior olive that had been called the olivary peduncle, but also noted that this peduncle continued past the abducens nucleus (where most of its fibers had been thought to terminate), across the midline and the dorsolateral tegmentum to reach the medial border of the vestibular nerve root (Fig. 10-2). He traced this crossed fascicle to the cochlear nerve, through the vestibulocochlear anastomosis. It was not until 1953, however, that Rasmussen succeeded in applying the Marchi technique to the cochlear tissues, and then traced the fibers of the fascicle through the intraganglionic spiral bundle and into the osseous spiral lamina. After this finding, the bundle of fibers was called "olivocochlear." Later, in 1960, Rasmussen documented an uncrossed component of the bundle (80).

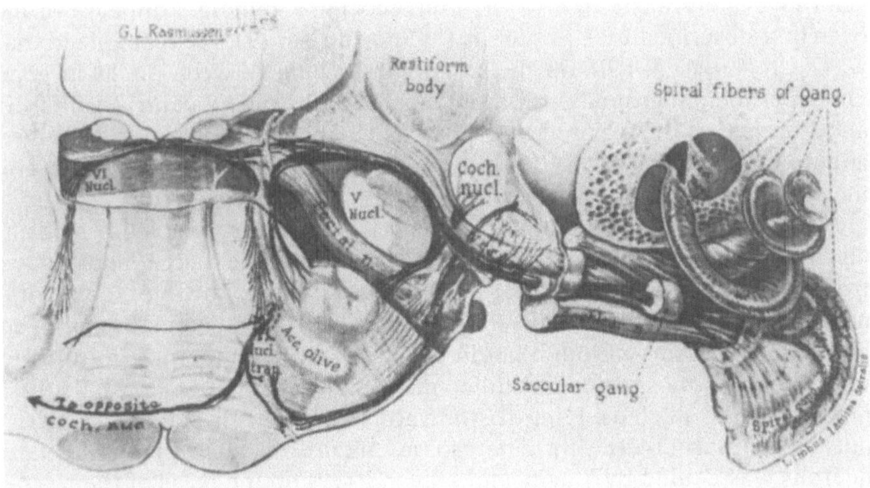

Fig. 10-2. This diagram illustrates the origin and course of the crossed and uncrossed fibers of the olivocochlear bundle as described by Rasmussen (80). Compare with Figs. 10-1 and 10-3. *Acc. olive*, accessory superior olive; *nucl. trap.*, trapezoid nucleus; *V Nucl.*, nucleus of the spinal tract of the trigeminal nerve; *VI Nucl.*, nucleus of the abducens nerve. From Rasmussen, G.L. and Windle, W.F. (eds.): Neural Mechanisms of the Auditory and Vestibular Systems. Springfield, Ill., Thomas, 1960.

Rasmussen's early observations were not entirely novel, because Papez had presented a similar viewpoint regarding the central course of the olivary peduncle in 1930 (71). However, the findings of these two investigators stood in marked contrast to a great volume of work on the connections of the superior olivary complex that had preceded them. The efferent projections of the complex most commonly described were (1) into the lateral lemniscus; (2) into the medial longitudinal fasciculus; and (3) to the abducens nucleus (2,43,56,57,100).

None of the early investigators described a connection out of the superior olivary complex that extended lateralward across the midline to exit with the vestibular root. This is remarkable because the existence of such a fascicle, associated with both the facial and the vestibular nerve roots, was well known and a matter of some controversy. The main difference of opinion seemed to concern whether the fascicle in question was an efferent component of the facial nerve or an aberrant, vestibular afferent bundle of fibers. However, Leidler (58) proved experimentally that the bundle was centrifugal and indicated his belief that it corresponded to motor nervus intermedius of the facial nerve as described earlier by Kohnstamm (55).

THE CENTRAL ORIGIN AND COURSE OF MOTOR NERVUS INTERMEDIUS

The fascicle anatomically related to both the facial and the vestibular roots centrally, lately attributed to the "olivocochlear bundle," appears to have been first described by Bischoff in 1899 for the cat (9). He recognized that it was partly crossed, with the decussation occurring between the facial genua. On experimental grounds, Bischoff decided that it was a centrifugal fascicle and a branch of the facial nerve; whereas Kohnstamm (55) later described it as autonomic, partly crossed and partly uncrossed, and as having its origin in the superior salivatory nucleus (Fig. 10-3).

According to Kohnstamm, the superior salivatory nucleus is located chiefly in the dorsolateral tegmentum, between the facial nucleus ventromedially and the lateral vestibular nucleus (which, however, some of the neurons invade) dorsolaterally. This location for the nucleus has been supported in the main, although not in every detail, by numerous investigators working on many species, including man (6,18,19,27,53,66,84,86,95,99). A portion of the nucleus lying rostralward (3) has been called the lacrimal nucleus in man (14,18,28). The crossed nature of some of the salivatory neurons was denied by Yagita and Hayama (99) on the basis of their experimental results, but Kaida (51) later showed that a few of the fibers of the chorda tympani in the rabbit were of crossed origin.

Shute and Lewis (91) have placed the salivatory neurons of the rat in two nuclei, medial and lateral to the facial genu. Brown and Howlett (11), while disagreeing with many of the findings of Shute and Lewis, concurred that the cells closely apposed to the lateral border of the genu were salivatory. The results of these two groups of investigators are in conflict with those of numerous workers who have placed the bulk of the salivatory neurons more

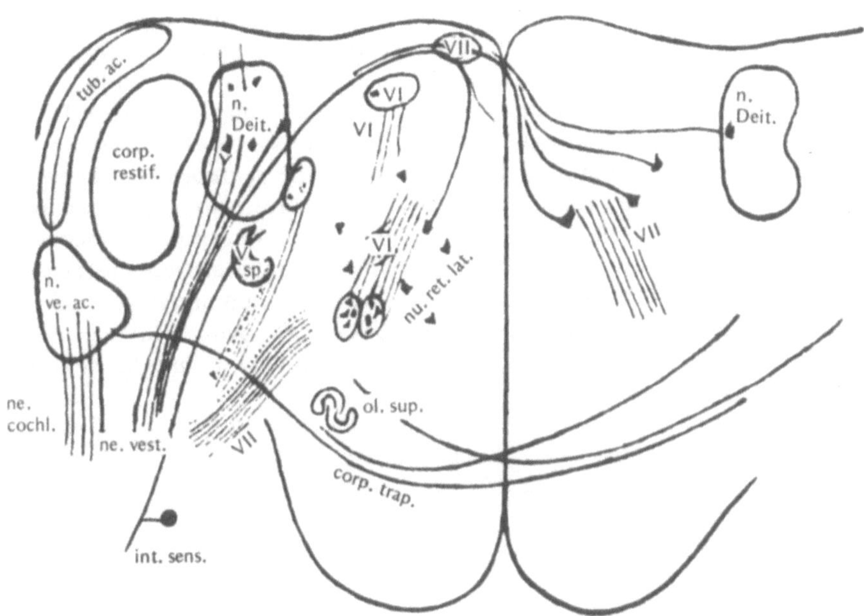

Fig. 10-3. This figure illustrates the origin and course of motor nervus intermedius. The cells of the superior salivatory nucleus are represented by solid triangles and their axons are shown streaming dorsalward, forming a genu, with some axons crossing the midline. Crossed and uncrossed fibers join lateral to the internal facial genu (*VI*) (*left center*) to traverse the dorsolateral tegmentum, pass over the spinal tract of V (*V sp*), and leave the brain stem medial to the vestibular nerve (*ne. vest.*). Compare with Figs. 10-1 and 10-2. *corp. restif.*, restiform body; *corp. trap.*, trapezoid body; *int. sens.*, sensory portion of the intermediate nerve; *n. Deit.*, Deiters' nucleus; *n. ve. ac.*, ventral acoustic nucleus; *ne. cochl.*, cochlear nerve; *nu. ret. lat.*, lateral reticular nucleus; *ol. sup.*, superior olive; *tub. ac.*, tuberculum acusticum; *VI*, abducens nucleus (also nerve); *VII*, motor root of VII and its internal genu. From Kohnstamm, O.: Verh. Deutsch. Kongr. Inn. Med. 20:361, 1902.

caudalward and lateralward in the tegmentum. They are also difficult to reconcile with the findings of Windle (98) in the fetal cat, who documented the lateral migration of the salivatory neurons from the primitive, dorsomedial motor cell column.

The central course of the parasympathetic fibers in the cat, dog, and rabbit has been described as across the dorsolateral tegmentum, over or through the spinal tract of the trigeminal nerve, and medial to the vestibular root. Lorente de Nó (68) found the parasympathetic fibers grouped into several fascicles in the mouse, with some following the course just outlined and others traversing the brain stem ventrolaterally to leave the brain stem independently of the vestibular nerve. These species differences will be dealt with further below.

The anatomic findings on the location of the salivatory neurons and fibers have been supported by physiologic experimental results. Stimulation of well-circumscribed "points" at the level of the facial motor nerve and its nucleus, and slightly caudalward, resulted in copious salivary outflow that was, however, mostly ipsilateral to the side of stimulation (13,61,64,96). The trails marked by the distribution of the reactive points as a whole correspond to the course of motor nervus intermedius or of the secretomotor fibers of the glossopharyngeal nerve, or else lie within the boundaries of the salivatory nuclei (Figs. 10-4 and 10-5). Thus, the salivatory response appears to be a direct result of stimulation of the salivatory nuclei or their fibers (61) and not an indirect, coordinated reaction indicative of an integrative salivatory center. Analysis of the results of Corbin et al. (17), who obtained salivation, dilatation of the blood vessels of the tongue, and sometimes lacrimation

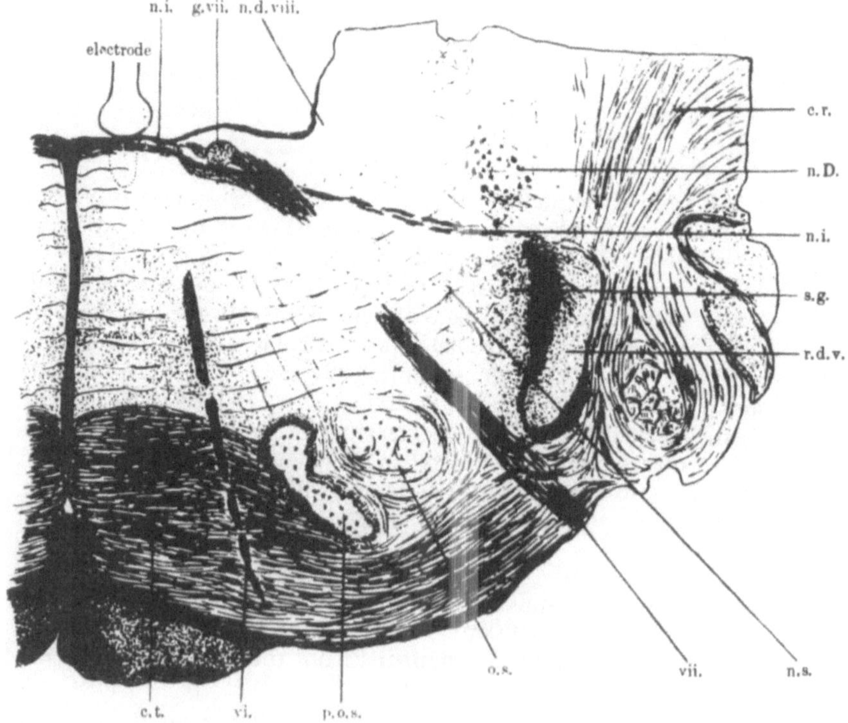

Fig. 10-4. This drawing shows the place at which the fibers of motor nervus intermedius (*n.i.*) were stimulated (*electrode*) in the cat with resulting copious (ipsilateral) salivatory outflow. Note that Miller obtained salivatory outflow upon stimulation of the fascicle described later as olivocochlear by Rasmussen. *c.r.*, restiform body; *c.t.*, trapezoid body; *n.D.*, Deiters' nucleus; *n.s.*, salivatory nucleus; *o.s.*, superior olive; *p.o.s.*, medial superior olive; *r.d.v.*, spinal tract of the trigeminal nerve; *s.g.*, nucleus of the spinal tract of the trigeminal nerve; *vi.*, abducens nerve root; *vii.*, facial motor root. From Miller, F.R.: Q.J. Exp. Physiol. 6:57, 1913.

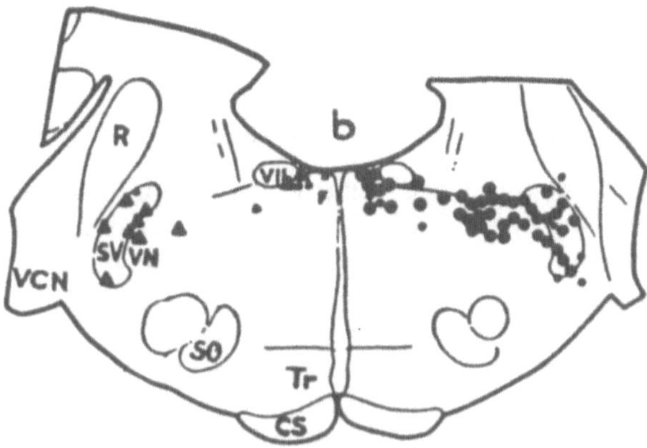

Fig. 10-5. Some of the results obtained by Wang (96) upon stimulation of the brain stem in a series of 35 cats are illustrated here. Large solid circles or triangles represent responses of salivatory flow of more than 0.25 ml/min; and small solid circles or triangles, 0.1–2.5 ml/min. Circles represent submaxillary outflow from the right side; triangles represent parotid outflow from the right side but transposed to the left side of the diagram. Note the correspondence between the points shown here and those in Fig. 10-4. *CS*, corticospinal tract; *R*, restiform body; *SO*, superior olive; *SV*, spinal tract of the trigeminal nerve; *Tr*, trapezoid body; *VCN*, ventral cochlear nucleus; *VN*, nucleus of the spinal tract of the trigeminal nerve; *VII*, internal genu of the facial nerve. From Wang, S.C.: J. Neurophysiol. 6:197, 1943.

in the cat upon stimulation of the intramedullary course of the seventh cranial nerve and the reticular formation close to it, leads to a similar conclusion.

Rasmussen was aware of much of the work on the origin, decussation, and course of the autonomic branch of the facial nerve that had already been carried out by 1946, and perhaps on that basis he originally suggested that the fascicle he described might be autonomic. The paths of motor nervus intermedius and of the olivocochlear bundle differ little, except for their predicated origins, as already stated. Rasmussen was unsuccessful in finding the cells of origin for the bundle either in the superior olivary complex or in the superior salivatory nucleus on the basis of his own retrograde degeneration studies and later (in 1953; ref. 79) abandoned the viewpoint that the fibers might be parasympathetic. It should be stressed, however, that Rasmussen thought that the superior salivatory neurons were located dorsal to the superior olivary complex, whereas the majority of the cells occur more caudalward and lateralward. Moreover, neither he nor Papez took into account the fact that their lesions interrupted the blood supply to the dorsal part of the tegmentum. Thus, it is possible that both the olivary peduncle and the motor nervus intermedius were affected, leading to the conclusion in Marchi preparations that they formed one continuous bundle. There is the further problem that Rasmussen (78) noted a great consistency in the size of the decussating peduncular fibers in diverse species, from opossum to man, whereas it is well known that the superior olivary complex itself is extremely variable in mammals (65).

THE PHYLOGENETIC HISTORY OF THE SEVENTH CRANIAL NERVE

The early experimental work carried out on motor nervus intermedius utilized species of animals in which the entire autonomic component leaves the brain stem ventrolaterally with the sensory root, medial to the vestibular nerve. However, in the mouse (68,84) and the rat (86,91) many of the parasympathetic fibers take a deviant, ventral course centrally to exit through the ventromedial part of the spinal tract of the trigeminal nerve (Fig. 10-1). These fibers, which form two or more fascicles, unite at the surface of the brain stem and join a part of the lateral branch more distally, after other fibers of that branch have already communicated with the vestibular nerve. This has led to some confusion in the literature so that the more lateral fibers of motor nervus intermedius have sometimes gone unrecognized and have instead been assigned partly (91) or entirely (11) to the olivocochlear bundle. There has been a concomitant failure to recognize that *all* of the motor fibers of nervus intermedius course lateralward in some species, as in the cat. In order to understand these variations, one must turn to the phylogenetic history of the facial motor root.

In all vertebrates with the exception of the placental mammals, the autonomic fibers of the facial nerve cannot be distinguished from those supplying the facial musculature by their position in the brain stem. In nearly all these vertebrates (amphibians are exceptions), all of the motor fibers of the facial nerve pass through the dorsolateral part of the tegmentum, over or through the dorsal part of the spinal tract of the trigeminal nerve, and leave the brain stem in close proximity to the vestibular root (Figs. 10-6 and 10-7). More than one investigator has deplored the impossibility of distinguishing between fibers of the seventh and eighth cranial nerves near the brain stem in many of these vertebrates.

It is only in placental mammals that some of the facial motor fibers, those supplying the muscles of facial expression, consistently leave the brain stem medial to the spinal tract of the trigeminal nerve. When this happens, however, the autonomic fibers either keep the phylogenetically old position of the motor facial root entirely, as in cat, dog, and rabbit, or else part of them do, as in the mouse (68,84) or rat (86,91). The intramedullary divergence of the two kinds of fascicles, medial and lateral, occurs just lateral to the facial motor root, ventral to its genu.

Thus, the pathway across the brain stem attributed to the olivocochlear and vestibular efferents belongs to the facial nerve phylogenetically and, in many vertebrates, is still traveled by the entire parasympathetic component of that nerve. The comparative neuroanatomic findings, then, support the concept that the efferents to the inner ear described by Rasmussen (78–80), Rasmussen and Gacek (81), Petroff (73), Gacek (34), Rossi and Cortesina

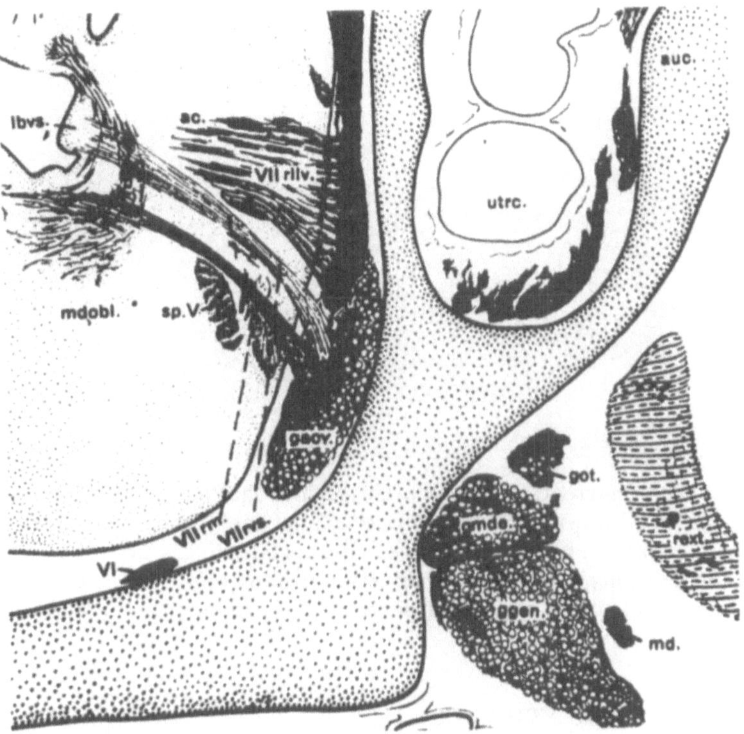

Fig. 10-6. This diagram shows the level of the facial nerve in the dogfish. It illustrates the phylogenetically old intramedullary course of the entire motor component of the facial nerve (*VII rm*) and the close anatomic relationship of the facial nerve to the vestibular ganglion (*gacv*) in vertebrates below placental mammals. Note that the intramedullary course of the motor fibers of the facial nerve (*VII rm*) corresponds to that of motor nervus intermedius in the dog (Fig. 10-3) and the cat (Fig. 10-4), and to the pathway taken by fibers of motor nervus intermedius that supply the inner ear in the rat (Fig. 10-1). This suggests that the autonomic components of the facial nerve, particularly those fibers supplying the inner ear, are phylogenetically old. The course pursued by facial motor root fibers is identical to the one assigned to the olivocochlear bundle (Fig. 10-2). *ac.*, acusticum (tuberculum acusticum); *auc.*, auditory capsule; *ggen.*, geniculate ganglion; *gmde.*, ganglion of the ramus mandibularis externus VII; *got.*, ganglion of the ramus oticus VII; *lbvs.*, lobus visceralis; *md.*, ramus mandibularis; *mdobl.*, medulla oblongata; *rext.*, m. rectus lateralis; *sp. V*, spinal tract of the trigeminal nerve; *utrc.*, utricle; *VI*, abducens nerve; *VIIrllv.*, ventral lateral line root of the facial nerve; *VIIrvs*, visceral sensory root. × 32.5 From Norris, H.W. and Hughes, S.P.: J. Comp. Neurol. 31:293, 1920.

(89), and many others are part of the facial nerve and autonomic. They suggest, moreover, that the parasympathetic neurons, including those to the inner ear, are phylogenetically very old and functionally related to basic, homeostatic matters.

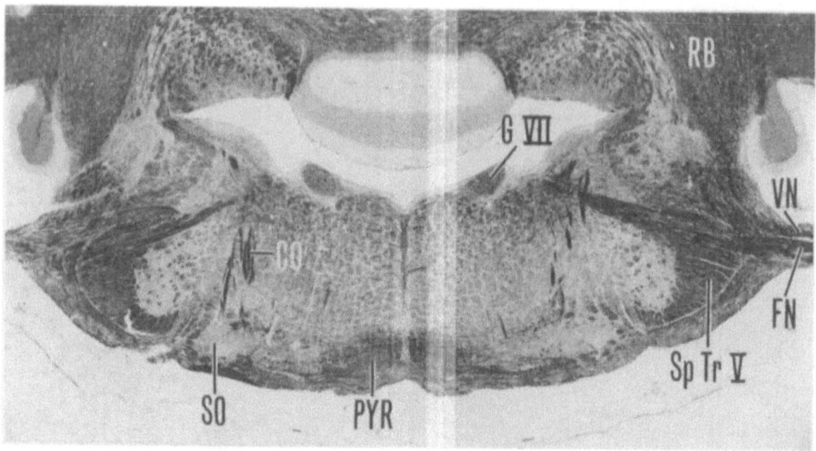

Fig. 10-7. Photomicrograph of the brain stem of the opossum at the level of the facial nerve (*FN*) showing the trajectory of its root fibers across the dorsolateral tegmentum. Compare with Figs. 10-1–10-6. *CO*, cochleoolivary tract; *G. VII*, internal genu of the facial nerve; *PYR*, pyramid; *RB*, restiform body; *SO*, superior olive; *Sp. tr. V*, spinal tract of the trigeminal nerve; *VN*, vestibular nerve. Weil stain, × 15 Courtesy of Dr. Nathan Gross.

THE EMBRYOLOGIC EVIDENCE

That ontogeny repeats phylogeny is an old, but still useful, saw. If the branch of the facial nerve to the inner ear is autonomic and phylogenetically very old, it should appear relatively early embryologically and be present at a fetal age when the superior olivary complex, a late-evolving, auditory center, has not yet differentiated. Such is indeed the case, as we learn from study of cross sections of the cat fetus illustrated by Windle (Fig. 10-8) (98). The motor facial root and the motor nervus intermedius are present in the 8-mm embryo, at which time the facial motor neurons have not completely migrated from their primitive position near the midline, the vestibular nerve is incompletely developed, the cochlear nerve is not defined, and neither the primary vestibular nor the secondary superior olivary centers are differentiated. It is highly unlikely that the axons could be so well developed if the cells of origin have not yet differentiated in the brain stem!

Figure 10-8 also demonstrates that the decussation attributed to the olivocochlear bundle is present at this same early stage in the fetal cat, and for reasons already stated this crossing of nerve fibers cannot be olivocochlear. Windle indicated that he traced axons of facial motor neurons into this decussation and across the midline. The decussation belongs to the seventh cranial nerve, specifically to its autonomic component as first documented by Kohnstamm (55).

Windle's work alone is sufficient to disprove the concept of olivocochlear and vestibular efferents. It is unfortunately true that Rasmussen knew of Windle's findings in the cat and did not interpret them correctly. On the

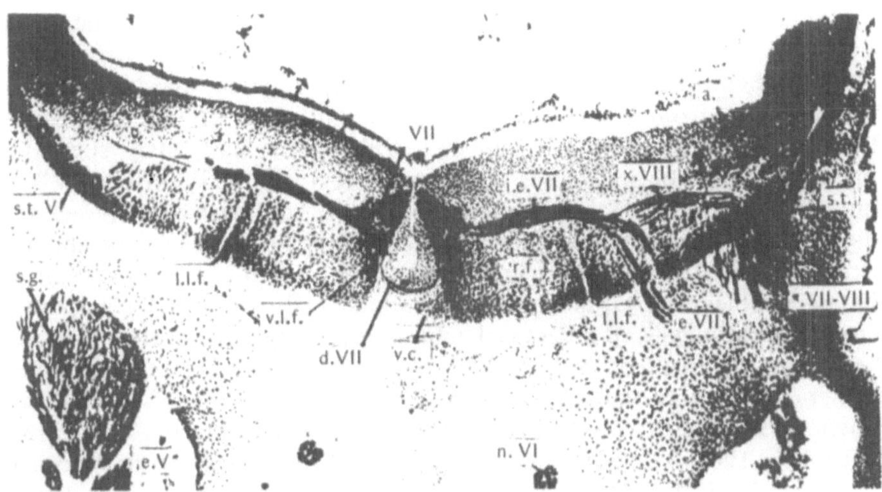

Fig. 10-8. This illustration shows the intramedullary course of the motor rootlets of the facial nerve *(i.e. VII)* and the exiting motor facial root *(i.e. VII)* in a 10-mm cat embryo. The fascicle marked *VIII* was indicated by Windle to consist of fibers from the inferior portion of the vestibular ganglion that joined motor fibers of the facial nerve centrally; this fascicle corresponds to motor nervus intermedius in the cat (Fig. 10-4). The fibers labeled *d. VII* are decussating motor axons of the facial nerve, and possibly vestibular nerve fibers, according to Windle. They correspond to the decussation of motor nervus intermedius (Figs. 10-1, 10-3, and 10-4). This same decussation and the fibers marked *VIII* were attributed to the olivocochlear bundle by Rasmussen (78); however, observe that the superior olivary complex has not differentiated at this stage of development and cannot, therefore, be the origin of the decussating fibers or of *VIII* of Windle. *a.*, association and commissural neuroblasts; *l.l.f.*, lateral longitudinal fascicle; *r.f.*, reticular formation; *s.t.*, composite sensory tract containing facial, vestibular, and, in some cases, possibly trigeminal fibers; *s.t.V*, sensory tract of trigeminal nerve; *s.g.*, semilunar ganglion; *s. VII-VIII*, sensory fibers of facial and vestibular nerves; *v.c.*, ventral commissure; *v.l.f.*, ventral longitudinal fascicle; *VII*, facial motor nucleus; *n. VI*, abducens nerve. × 70. From Windle, W.F.: J. Comp. Neurol. 58:643, 1933.

other hand, Windle himself seems to have been peculiarly unaware of the work of Bischoff, Kohnstamm, and the others who had already traced the central course of motor nervus intermedius, and could not decide what to name the fascicle connecting the facial motor and vestibular sensory roots.

THE FUNCTIONS OF THE COCHLEAR AND VESTIBULAR EFFERENTS

Just as the anatomic results have been largely in conflict, so there has been little agreement concerning the functional significance of the vestibular and cochlear efferents. Physiologic findings up to 1974 were discussed by Klinke and Galley in their excellent article (54), and no attempt will be made here to review all of the conflicting opinions.

First indications of possible functions for the cochlear and vestibular efferents came from Galambos' laboratory (39). He and his colleagues described a cochleocochlear influence which, they cautiously suggested, might reflect a function of the cochlear efferent bundle described by Rasmussen (78). Later, Galambos (36) found that stimulation of the crossed olivocochlear bundle reduced the amplitude of the auditory nerve compound action potential (recorded at the round window), while Fex (33) reported that the cochlear microphonics were simultaneously increased in amplitude. These results suggested a feedback inhibitory role for the cochlear efferents.

Close on the heels of this later work, however, Galambos (37) carried out further experiments on the cochlear efferents and stated that he could find no function for them in awake, mobile cats. In contrast to his 1956 paper (36), Galambos' 1960 article (37) is seldom quoted. Rose and his colleagues routinely sectioned the vestibular nerve in their investigations of the suppression of one low-frequency tone by another in the squirrel monkey. Although this procedure eliminated the cochlear efferents, it was without effect on the suppression observed (47). Pfalz pointed out (74) that prior physiologic experiments conducted on cochlear efferents had employed unnatural levels of electric stimulation. He could find no function for the olivocochlear bundle under physiologic experimental conditions. In the meantime, Kiang (52) reported that two-tone inhibition and normal spontaneous activity of auditory nerve fibers were preserved in an animal 9 days after complete transection of crossed and uncrossed olivocochlear tracts. Histologic verification of complete transection was provided by Gacek, and Kimura confirmed the loss of vesiculated terminals on the hair cells ultrastructurally (52). Liff and Goldstein (59) studied peripheral inhibition in auditory nerve fibers of the frog and discounted a role for the centrally originating efferents in this process. They stated that the minimum latency of the inhibitory effect appeared to be too short to involve a feedback mechanism through the central nervous system. Chouard (15) reported a slight improvement in hearing in three patients in which the efferent bundle, interpreted, however, as the parasympathetic preganglionic nerve supply to the inner ear, had been cut. Iurato et al. (49) recently found that transection of the crossed olivocochlear bundle had no effect upon either auditory nerve action potentials (AP) or cochlear microphonics (CM), which would appear to contradict a role for the efferents in feedback inhibition of the peripheral receptor organ. Klinke and Galley (54) reviewed the studies centering on a feedback inhibitory function for the efferents of the eighth nerve, and while agreeing that there is little evidence to support that function, suggested that the efferent systems may operate in a feed-forward way, at least in the case of the vestibular apparatus.

It is clear that physiologic experiments have not been successful in uncovering the function of the efferent-type endings on the hair cells. This is not surprising, however, because the origin of these terminals is a matter of dispute. Like Galambos in 1960, we can only plead again today for fresh new ideas and clean new experiments to resolve the problem.

CONCLUSION

Although the concept of a fiber system of central origin terminating on the hair cells of the inner ear is not currently considered to be controversial, the findings considered above provide evidence that there are good anatomic and physiologic grounds for questioning the correctness of this concept as presently formulated. The problem is not whether or not efferent fascicles reach the inner ear. They do, as has been recognized since the work of Arnold and of Held in the last century. It is, rather, whether the centrifugal fibers are entirely autonomic, or whether they are a unique, somatic efferent component ending on receptor cells, without analogy anywhere else in the body. This neuroanatomic uniqueness must be emphasized. Gamma motor neurons modulating the activity of the muscle spindle (40) are, after all, a type of somatic motor neuron innervating a special kind of somatic muscle, the intrafusal fibers. Centrifugal influences on the retina (40) or on the olfactory bulb (1) are central both in origin and in termination, comparable with other feedback systems known to exist in the central nervous system.

The concept that the centrifugal fibers of the vestibular and cochlear nerves are autonomic fits well with the existing comparative and embryologic neuroanatomic information. The concept of efferent olivocochlear and vestibular fiber systems does not. As long as the theory of central neuronal modulation of hair cell activity remains unproven, alternative opinions concerning the origin of the efferent-type terminals on the hair cells should be considered. To do less is intellectually suppressive and the resulting narrowing of vision but delays the proper functional interpretation of the centrifugal fibers of the eighth nerve and of the vesiculated terminals on the hair cells. Certainly, the important possibility that the vesiculated terminals are branches of the peripheral processes of bipolar spiral and vestibular ganglion cells is presently being neglected. This hypothesis for their origin was presented in an earlier article (86), but is very similar to a suggestion offered by Galambos and Davis in 1944 (38), before the theory of centrally originating efferents terminating on hair cells became dominant. It is time to test ideas like this for degree of fit to physiologic data, and to turn our attention toward other explanations for degeneration of the efferent-type endings on hair cells when lesions are made centrally. I am convinced that when we do this, we shall begin to have new and wonderful insights as to how the inner ear really functions.

ADDENDUM 1

Since presenting this paper, Dr. C.H. Chouard has kindly made me aware of an article that was published in 1976 entitled "Le faisceau olivo-cochléaire de Rasmussen existe-t-il?"[1] In this article Chouard and his colleagues critique the classic description of the efferent fascicles to the inner ear. They point out paradoxes in findings said to support the concept of distinct vestibular and cochlear efferents and stress the fact that the trajectories of fascicles of motor nervus intermedius and of the presumed olivocochlear and vestibular efferents across the brain stem are the same. The acousticofacial anastomoses also correspond to the communications between the efferent fascicles and the vestibular and cochlear nerves. Chouard et al. conclude that the origin of the efferent fibers to the inner ear is not in the region of the superior olive or in Deiters' nucleus, but in the parasympathetic nucleus of the facial nerve. They further reason that the acousticofacial aggregate (ensemble) is of great functional significance, with the parasympathetic contingent acting as an organ-protector of the inner ear. I am delighted to acknowledge this article, of which I was previously unaware, and hope that this statement will help explain the relative similarity between the title of the article by Chouard et al. and the title of this paper.

ADDENDUM 2

A stimulating new era in our understanding of the inner ear may already be beginning. Very recently, the presence of multipolar neurons in the spiral ganglion was corroborated by Kimura et al. (1979)[2] in their study of human material. These same authors described axosomatic and axodendritic synapses on some of the small spiral ganglion cells that would be consistent with and are strongly supportive of the concept of a parasympathetic innervation of the inner ear. Moreover, Nadol (1980)[3] has reported that there are reciprocal synapses at the bases of the outer hair cells in man, with afferent and efferent synapses within the same nerve terminal. This observation would be in keeping with the hypothesis (Ross, 1973) that both afferent and efferent endings in the organ of Corti are branches of peripheral processes of bipolar spiral ganglion cells and are involved in peripheral integration and/or analysis of auditory signals, or in lateral inhibition. Much further work will be required to resolve all of the issues these and other observations have raised about inner ear organization and function, but there can be little doubt that the future promises to be exciting.

ACKNOWLEDGMENTS

I thank Dr. Elizabeth C. Crosby for her advice and encouragement during my work on the efferents to the inner ear. I am grateful to her and to Dr. Joseph E. Hawkins, Jr., for their helpful comments during preparation of this manuscript. I thank all the publishers and authors who allowed me to reproduce their copyrighted figures for this article. This research was supported by Grants # NB-07306 and NS-13428, USPHS.

[1]Chouard, C.-H., Eyries, C., and Meyer, B.: Le Faisceau olivo-cochléaire de Rasmussen existe-t-il? C.R. Seances du 73° Congrés Francais ORL, October 1976. Paris, Libraire Arnette, 1976, pp. 428–437.

[2]Kimura, R.S., Ota, C.Y., and Takahashi, T.: Nerve fiber synapses on spiral ganglion cells in the human cochlea. Ann. Otol. 88, No. 6, Part 3. Supplement 62:1-17, 1979.

[3]Nadol, J.B., Jr.: Reciprocal synapses at the base of outer hair cells in the organ of Corti of man. Ann. Otol. 1980.

References

1. Arduini, A. and Moruzzi, G.: Sensory and thalamic synchronisation in the olfactory bulb. EEG Clin. Neurophysiol. 5:234, 1953.
2. Ariëns Kappers, C.U.: Vergleichende Anatomie des Nervensystems. Haarlem, Bohn, 1920.
3. Ariëns Kappers, C.U.: The Evolution of the Nervous System in Invertebrates, Vertebrates and Man. Haarlem, Bohn, 1929, p. 168.
4. Arnold, F.: Observationes nonnullas neurologicas de parte cephalica nervi sympathici in homine. Med. diss. Heidelberg, 1826.
5. Arnold, F.: Handbuch der Anatomie des Menschen, Vol. 2. Freiburg im Breisgau, Herdersche Verlagshandlung, 1851.
6. Åström, K.E.: On the central course of afferent fibres in the trigeminal, facial, glossopharyngeal, and vagal nerves and their nuclei in the mouse. Acta Physiol. Scand. 29(Suppl. 106):209, 1953.
7. Ayers, H.: Ueber das peripherische Verhalten der Gehörnerven und den Wert der Haarzellen des Gehörorganes. Anat. Anz. 8:435, 1893.
8. Birks, R., Katz, B., and Miledi, R.: Physiological and structural changes at the amphibian myoneural junction, in the course of nerve degeneration. J. Physiol. (Lond.) 150:145, 1960.
9. Bischoff, E.: Ueber den intramedullären Verlauf des Facialis. Neurol. Zentralbl 18:1014, 1899.
10. Bovero, A.: Sulla fine strutture e sulle connessioni del ganglio vestibolare del nervo acustico. Mem. Roy. Accad. Sci. Torino (Ser. II) 64/10:1, 1914.
11. Brown, J.C. and Howlett, B.: The facial outflow and the superior salivatory nucleus: an histochemical study in the rat. J. Comp. Neurol. 134:175, 1968.
12. Cannieu, A.: Note sur les cellules des ganglions de l'oreille. Rev. Laryngol. (Bordeaux) 19:97, 1899.
13. Chatfield, P.O.: Salivation in response to localized stimulation of the medulla. Am. J. Physiol. 133:637, 1941.
14. Chouard, C.-H.: Recherches sur l'organisation intra-axiale des formations motrices et parasympathiques du nerf facial. Thèse Méd. Paris, 1962.
15. Chouard, C.-H.: Acousticofacial anastomoses in Ménière disorder. Arch. Otolaryngol. 101:296, 1975.
16. Colonnier, M. and Gray, E.G.: Degeneration in the cerebral cortex. In Breese, S.S., Jr. (ed.): International Congress for Electron Microscopy, Vol. 2. New York, Academic Press, 1962.
17. Corbin, K.B., Harrison, F., and Wigginton, C.: Elicitation of the "pseudomotor contracture" in the tongue by intramedullary stimulation. Arch. Neurol. Psychiat. 45:271, 1941.
18. Crosby, E.C. and DeJonge, B.R.: Experimental and clinical studies of the central connections and central relations of the facial nerve. Ann. Otol. 72(3):735, 1963.
19. Crosby, E.C., Humphrey, T., and Lauer, E.W.: Correlative Anatomy of the Nervous System. New York, Macmillan, 1966.

20. Densert, O.: Adrenergic innervation in the rabbit cochlea. Acta Otolaryngol. (Stockh.) 78:345, 1974.

21. Densert, O.: A fluorescence and electron microscopic study of the adrenergic innervation in the vestibular ganglion and sensory areas. Acta Otolaryngol. (Stockh.) 79:96, 1975.

22. Densert, O. and Flock, Å.: An electron microscopic study of adrenergic innervation in the cochlea. Acta Otolaryngol. (Stockh.) 77:185, 1974.

23. De Robertis, E.: Submicroscopic changes of the synapse after nerve section in the acoustic ganglion of the guinea pig. An electron microscope study. J. Biophys. Biochem. Cytol. 2:503, 1956.

24. Dixon, J.: The fine structure of parasympathetic nerve cells in the otic ganglion of the rabbit. Anat. Rec. 156:239, 1966.

25. Duvall, A.J. and Sutherland, C.R.: Cochlear transport of horseradish peroxidase. Ann. Otol. 81:705, 1972.

26. Ehrenbrand, F. and Wittemann, G.: Ueber synaptische Strukturen im Ganglion vestibulare der Maus. Ant. Anz. 126:300, 1970.

27. Eyries, C. and Chouard, C.-H.: Le noyau lacrymo-muco-nasal chez l'homme. XLVIIIe reunion des anatomistes. Toulouse, 1962.

28. Eyries, C. and Chouard, C.-H.: Les origines réelles du nerf facial. Ann. Otolaryngol. (Paris) 80(9):775, 1963.

29. Eyries, C. and Chouard, C.-H.: Les anastomoses acoustico-faciales. Ann. Otolaryngol. (Paris) 87(6):321, 1970.

30. Eyries, C. and Chouard, C.-H.: Les voies cochléaires bulbo-protubérentielles. In Maduro, R., et al. (eds.): Problèmes Actuels d'O.R.L. Paris, Librairie Maloine, 1970.

31. Eyries, C., Chouard, C.-H., and Peytral, C.: Systematisation des voies cochléaires. Encycl. méd. chir., Oto-Rhino-Laryngologie, 3.24.10, 20020 A-20. Paris, 1974, pp. 1–16.

32. Eyries, C., Perles, B., and Pineau, H.: Developpement de l'oreille interne humaine du 26e au 70e jour de l'embryon. Ann. Otolaryngol. (Paris) 77:877, 1960.

33. Fex, J.: Augmentation of cochlear microphonic by stimulation of efferent fibres to the cochlea. Acta Otolaryngol. (Stockh.) 50:540, 1959.

34. Gacek, R.R.: Efferent component of the vestibular nerve. In Rasmussen, G.L. and Windle, W.F. (eds): Neural Mechanisms of the Auditory and Vestibular Systems. Springfield, Ill., Thomas, 1960, pp. 276–284.

35. Gacek, R.R. and Lyon, M.: The localization of vestibular efferent neurons in the kitten with horseradish peroxidase. Acta Otolaryngol. (Stockh.) 77:92, 1974.

36. Galambos, R.: Suppression of the auditory nerve activity by stimulation of efferent fibers to the cochlea. J. Neurophysiol. 19:424, 1956.

37. Galambos, R.: Studies of the auditory system with implanted electrodes. In Rasmussen, G.L. and Windle, W.F. (eds.): Neural Mechanisms of the Auditory and Vestibular Systems. Springfield, Ill., Thomas, 1960, pp. 137–151.

38. Galambos, R. and Davis, H.: Inhibition of activity in single auditory nerve fibers by acoustic stimulation. J. Neurophysiol. 7:287, 1944.

39. Galambos, R., Rosenblith, W.A., and Rosenzweig, M.R.: Physiological evidence for a cochleo-cochlear pathway in the cat. Experientia 6:438, 1950.

40. Granit, R.: Receptors and Sensory Perception. New Haven, Yale Univ. Press, 1955.

41. Gray, E.G. and Guillery, R.W.: Synaptic morphology in the normal and degenerating nervous system. Int. Rev. Cytol. 19:111, 1966.

42. Gray, E.G. and Hamlyn, L.H.: Electron microscopy of experimental degeneration in the avian optic tectum. J. Anat. 96(3):309, 1962.

43. Held, H.: Die centrale Gehörleitung. Arch. Anat. Physiol. Abt. 201, 1893.

44. Held, H.: Zur Kenntnis der peripheren Gehörleitung. Arch. Anat. Physiol. Abt. 350, 1897.

45. Henle, F.G.J.: Handbuch der systematischen Anatomie des Menschen, Vol. 3, Part 2, 2nd ed. Braunschweig, Vieweg & Sohn, 1876.

46. Heuser, J., Katz, B., and Miledi, R.: Structural and functional changes of frog neuromuscular junctions in high calcium solutions. Proc. R. Soc [Biol.] 178:407, 1971.

47. Hind, J.E., Anderson, D.J., Brugge, J.F., and Rose, J.E.: Coding of information pertaining to paired low-frequency tones in single auditory nerve fibers of the squirrel monkey. J. Neurophysiol. 30:794, 1967.

48. Hovelacque, A.: Anatomie des nerfs crâniens. Paris, Doin et Cie, 1927.

49. Iurato, S., Smith, C.A., Eldredge, D.H., Henderson, D., Carr, C., Ueno, Y., Cameron, S., and Richter, R.: Distribution of the crossed olivocochlear bundle in the chinchilla's cochlea. J. Comp. Neurol. 182(1):57, 1978.

50. Jahnke, K.: Verteilung intrathekal applizierter Peroxydase in der Meerschweinchen-Cochlea. Arch. Ohr Nas Kehlkopfheilk. 202:418, 1972.

51. Kaida, Y.: Ueber den Ursprung und den peripheren Verlauf der sog. zentrifugalen Vestibularisnerven nach Leidler (Fasciculus vestibularis medialis nach Kaplan). Arch. Ohr Nas Kehlkopfheilk. 123:62, 1929.

52. Kiang, N.Y.: A survey of recent developments in the study of auditory physiology. Ann. Otol. 77:656, 1968.

53. Kimmel, D.L.: Differentiation of the bulbar motor nuclei and the coincident development of associated root fibers in the rabbit. J. Comp. Neurol. 72(1):83, 1940.

54. Klinke, R. and Galley, N.: Efferent innervation of vestibular and auditory receptors. Physiol. Rev. 54(2):316, 1974.

55. Kohnstamm, O.: Vom Centrum der Speichelsekretion, dem Nervus intermedius und der gekreuzten Facialiswurzel. Verh. Deutsch Kongr. Inn. Med. 20:361, 1902.

56. von Kölliker, A.: Handbuch d. Gewebelehre, 6 Aufl., Bd. 2., Leipzig, Engelmann, 1893, pp. 268, 400.

57. Larsell, O.: The differentiation of the peripheral and central acoustic apparatus in the frog. J. Comp. Neurol. 60(8):473, 1934.

58. Leidler, R.: Experimentelle Untersuchungen über das Endigungsgebiet des Nervus vestibularis. 2. Mitteilung. Arb. Neurol. Inst. Univ. Wien 21:151, 1914.

59. Liff, H.A., and Goldstein, M.H.: Peripheral inhibition in auditory fibers in the frog. J. Acoust. Soc. Am. 47:1538, 1970.

60. de Lorenzo, A.J., Shiroky, D.V., and Cohn, E.I.: Distribution of exogenous horseradish peroxidase in perilymphatic and endolymphatic spaces of the guinea pig cochlea. In de Lorenzo, A.J. (ed.): Vascular Disorders and Hearing Defects. Baltimore, University Park Press, 1973.

61. Magoun, H.W. and Beaton, L.E.: The salivatory motor nuclei in the monkey. Am. J. Physiol. 136:720, 1942.

62. Maw, A.R.: Further morphological modifications of the concept of efferent cochlear innervation at an ultrastructural level. J. Laryngol. 87:619, 1973.

63. Miledi, R. and Slater, C.R.: On the degeneration of rat neuromuscular junctions after nerve section. J. Physiol. 207:507, 1970.

64. Miller, F.R.: On the reactions of the salivary centres. Q. J. Exp. Physiol. 6:57, 1913.

65. Moore, J. and Moore, R.: A comparative study of the superior olivary complex in the primate brain. Folia Primatol. (Basel) 16:35, 1971.

66. Nó, R. Lorente de: Contribucion al conocimiento del nervio trigemino. Libro en honor Cajal, T.2. Madrid, 1922, pp. 13–30.

67. Nó, R. Lorente de: Etudes sur l'anatomie et la physiologie du labyrinthe de l'oreille et du VIIIe nerf. II. Quelques données au sujet de l'anatomie des organes sensoriels du labyrinthe. Trab. Inst. Cajal Invest. Biol. 24:53, 1926.

68. Nó, R. Lorente de: The central projection of the nerve endings of the internal ear. Anatomy of the eighth nerve. Laryngoscope 43:1, 1933.

69. Norris, H.W. and Hughes, S. P.: The cranial, occipital, and anterior spinal nerves of the dogfish, *Squalus acanthias*. J. Comp. Neurol. 31:293, 1920.

70. Obrebawski, A. and Skornicki, R.: Communications between nerves and the internal auditory meatus in the dog. Folia Morphol. (Warsz.) 26:197, 1967.

71. Papez, J.W.: Superior olivary nucleus—its fiber connections. Arch. Neurol. Psychiat. 24:1, 1930.

72. Penzo, R.: Ueber das Ganglion geniculi und die mit demselben zusammenhängenden Nerven. Anat. Anz. 8:738, 1893.

73. Petroff, A.E.: An experimental investigation of the origin of efferent fiber projections to the vestibular neuro-epithelium. Anat. Rec. 121:352, 1955.

74. Pfalz, R.K.J.: Absence of a function for the crossed olivocochlear bundle under physiological conditions. Arch. Klin. Exp. Ohr Nas Kehlkopfheilk. 193:89, 1969.

75. Pick, J.: The Autonomic Nervous System. Philadelphia, Lippincott, 1970.

76. Pieschel, C.: De parte cephalica nervi sympathici in equo prodromus. Leipzig, Stange, 1844. Quoted by Arnold (5).

77. Ramon y Cajal, S.: Histologie du systeme nerveux de l'homme et des vertébrés. Paris, Maloine, 1911.

78. Rasmussen, G.L.: The olivary peduncle and other fiber projections of the superior olivary complex. J. Comp. Neurol. 84:141, 1946.

79. Rasmussen, G.L.: Further observations of the efferent cochlear bundle. J. Comp. Neurol. 99:61, 1953.

80. Rasmussen, G.L.: Efferent fibers of the cochlear nerve and cochlear nucleus. In Rasmussen, G.L. and Windle, W.F. (eds.): Neural Mechanisms of the Auditory and Vestibular Systems. Springfield, Ill., Thomas, 1960, pp. 105–115.

81. Rasmussen, G.L. and Gacek, R.R.: Concerning the question of an efferent fiber component of the vestibular nerve of the cat. Anat. Rec. 130:361, 1958.

82. Rauber, A.A. and Kopsch, F.: Lehrbuch der Anatomie, Vol. 5, 8th ed. Leipzig, Thieme, 1909.

83. Retzius, G.: Biologische Untersuchungen, Vol. 6. Leipzig, Vogel, 1895.

84. Ross, M.D.: The general visceral efferent component of the eighth cranial nerve. J. Comp. Neurol. 135:453, 1969.

85. Ross, M.D.: Fluorescence and electron microscopic observations of the general visceral efferent innervation of the inner ear. Acta Otolaryngol. (Stockh.) (Suppl.) 286:1, 1971.

86. Ross, M.D.: Autonomic components of the VIIIth nerve. Adv. Otorhinolaryngol. 20:316, 1973.

87. Ross, M.D. and Burkel, W.: Multipolar neurons in the spiral ganglion of the rat. Acta Otolaryngol. (Stockh.) 76:381, 1973.

88. Ross, M.D., Nuttall, A.L., and Wright, C.G.: Horseradish peroxidase acute ototoxicity and the uptake and movement of the peroxidase in the auditory system of the guinea pig. Acta Otolaryngol. (Stockh.) 84:187, 1977.

89. Rossi, G. and Cortesina, G.: Research on the efferent innervation of the inner ear. J. Laryngol. 77:202, 1963.

90. Schlaepfer, W.W.: Calcium-induced degeneration of axoplasm in isolated segments of rat peripheral nerve. Brain Res. 69:203, 1974.

91. Shute, C.C.D. and Lewis, P.R.: The salivatory centre in the rat. J. Anat. 94:59, 1960.

92. Spoendlin, H.H.: Degeneration behaviour of the cochlear nerve. Arch. Klin. Exp. Ohr Nas Kehlkopfheilk. 200:275, 1971.

93. Spoendlin, H.H. and Lichtensteiger, W.: The adrenergic innervation of the labyrinth. Acta Otolaryngol. (Stockh.) 61:421, 1966.

94. Terayama, Y., Holz, E., and Beck, C.: Adrenergic innervation of the cochlea. Ann. Otol. 75:69, 1966.

95. Torvik, A.: Die Lokalisation des "Speichelzentrums" bei der Katze. Z. Morphol. Mikrosk. Anat. 63:317, 1957.

96. Wang, S.C.: Localization of the salivatory center in the medulla of the cat. J. Neurophysiol. 6:195, 1943.

97. Warr, W.B.: Olivocochlear and vestibular efferent neurons of the feline brain stem: their location, morphology, and number determined by retrograde axonal transport and acetylcholinesterase histochemistry. J. Comp. Neurol. 161:159, 1975.

98. Windle, W.F.: Neurofibrillar development in the central nervous system of cat embryos between 8 and 12 mm long. J. Comp. Neurol. 58:643, 1933.

99. Yagita, K. and Hayama, S.: Ueber das Speichelsekretionscentrum. Neurol. Zentralbl. 28:738, 1909.

100. Yoshida, I.: Ueber die funktionelle Bedeutung der oberen Olive nebst ihrer Faserbahnen. Folia Anat. Jap. 3:111, 1925.

101. Yoshida, M.: Vergleichende elektronmikroskopische Untersuchungen an sympathischen und parasympathischen Ganglien des Goldhamsters. Z. Zellforsch. 88:138, 1968.

102. Zelenka, J. and Subrt, J.: Contribution to the problem of vestibular facial anastomosis. Cesk Otolaryngol. 15(5):300, 1966.

Discussion

RICHARD GACEK: I really think that the strongest argument for a centrally originating efferent bundle both in the auditory and vestibular systems is the fact that following the appropriate central lesion, degenerated axons are consistently demonstrated by a number of techniques into the vestibular nerve, past the vestibular ganglion without making contact with ganglion cells, out to a peripheral terminus in the sense organs. This demonstration is based on an axon which degenerates because it has been separated from its centrally located cell body. If that's not a definition of an efferent neuron, well, I don't know what is. In addition, the vesiculated nerve endings degenerate very rapidly within a few days after a similar lesion. Retrograde degeneration of a ganglion cell would not have that time course. There's also been the evidence we heard today from two or three very closely agreeing studies describing the central location of the vestibular efferents with the HRP technique. I will grant that there are some similarities between the nervus intermedius and the efferent vestibular bundle. For example, the nervus intermedius does travel with the olivocochlear and efferent vestibular bundles after leaving the brain stem and travels in the facial-vestibular anastomoses to reach the geniculate ganglion region of the facial nerve. But the remaining portion of the efferent vestibular component can be followed out separately to the periphery. These are some of the arguments based on proven experimental anatomic principles which have not been dispelled and which continue to support the established fact that there is an efferent neuron located centrally and terminating peripherally in the labyrinth.

ROSS: I would just like to comment that I think the cells you described as the origin for the vestibular efferents fall within the confines of what has been described by many people as the superior salivatory nucleus. This would be a point of disagreement, obviously. Furthermore, I thought I made the point that you can have terminal degeneration by a fluid and ion imbalance alone. This has been shown by other people. Are you talking about Marchi preparations...what kind of preparations? I know that Rasmussen used Marchi preparations, and these are not fool-proof, as anyone knows.

GACEK: The central and peripheral portions of degenerated efferent axons have been demonstrated with the Nauta silver method and Sudan black technique. These methods demonstrate the axis cylinder and the myelin sheath portions of the axon, respectively. The techniques are based on discontinuity of the axons following a central lesion. Histochemically, the acetylcholinesterase activity in efferent axons permits the demonstration of their continuity. All these methods have overcome the limitations of the Marchi technique.

ROSS: Well, I believe that this is a problem we must come to grips with: Is there room out there for all these systems? I think we must also recognize that since the facial root is phylogenetically old and develops much before this other system would, the real question is whether or not an olivocochlear or any other such bundle of nerve fibers can grow into an area that is already occupied by facial root fibers. I think this embryologic problem is one that has to be addressed.

Part 2

Morphology and Function Interrelationship

INVITED LECTURE

11

Thick and Thin Mammalian Vestibular Axons: Afferent and Efferent Response Characteristics

JAY M. GOLDBERG

In recent years, there have been several studies of the discharge characteristics of mammalian vestibular afferents (1,3–7,11,12,17,20,22,28). On this basis, we now have a reasonably comprehensive picture of the kinds of information encoded by labyrinthine receptors. We shall review these studies and, where possible, draw inferences concerning the functional contributions of individual end organs. The physiology of the efferent vestibular system, which is only now being investigated (15), will also be considered. Throughout the paper, emphasis will be placed on the fact that the units supplying a particular end organ can show great variations in both their afferent and efferent response characteristics. We begin by presenting evidence that the differences in the physiologic properties of individual afferents are related to their fiber caliber and, by inference, to the innervation patterns they make within the sensory epithelium.

FIBER CALIBER AND AFFERENT RESPONSE CHARACTERISTICS

Mammalian vestibular nerve fibers range in size from 1 to 10 μm (10). Although the fiber-size spectrum has only one peak, it is customary to recognize three groups of axons (19,21,34). Thick fibers mainly supply the central part of the sensory epithelium, where each axon gives rise to calyceal formations enveloping a few neighboring type I hair cells. Medium-sized and thin fibers form the predominant innervation to peripheral parts of the epithelium and may also contribute to the central zone; the axons run in an intraepithelial plexus and individual afferents may distribute to several widely spaced receptors. Medium-sized fibers provide a mixed innervation consisting of calyceal endings to type I hair cells and bud-shaped endings to type II hair cells. Thin fibers furnish bouton-type endings solely to type II receptors.

The differences in innervation patterns of various fiber groups should be reflected in the physiology of the corresponding afferents. That afferents do vary in their response properties was shown in studies of both semicircular canal (1,3,12,17,22,28) and otolith afferents (5,7). One difference in discharge is illustrated in Fig. 11-1, which shows the resting activities of two

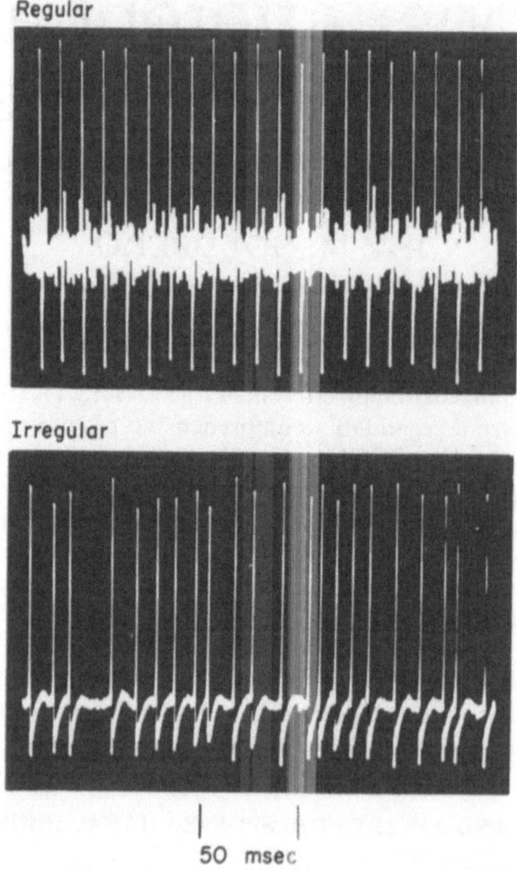

Fig. 11-1. Resting activity of two semicircular canal afferents in the squirrel monkey. Both afferents are firing at 98 spikes/s. From Goldberg, J.M. and Fernández, C.: J. Neurophysiol. 34:635, 1971.

Table 11-1.
SUMMARY OF DISCHARGE CHARACTERISTICS OF THICK AND THIN AXONS.

Thick Axons	Thin Axons
Irregular discharge pattern	Regular discharge pattern
Phasic-tonic response dynamics, including adaptation and velocity sensitivity	Tonic response dynamics, mirroring presumed mechanics of end organ
High sensitivity, particularly in bandwidth of physiological head movements (1−10 Hz)	Low sensitivity in the same bandwidth
Weak relation between resting discharge and sensitivity to natural stimulation	Strong positive relation between resting discharge and sensitivity to natural stimulation
Easily silenced by inhibitory accelerations	Not easily silenced by inhibitory accelerations
Large efferent responses	Small efferent responses

semicircular canal afferents. Although both afferents are firing at just under 100 spikes/s, there are obvious differences in their discharge patterns. The unit at the top has a regular spacing of action potentials; for the bottom unit, the spacing is irregular.

Afferents first classified as regularly or irregularly discharging differ in other respects as well. To ascertain whether the differences were related to fiber caliber, antidromic conduction times (CTs) were measured (Fig. 11-2E and F). Interspike-interval histograms and CTs are compared for two units, one with a regular spacing of action potentials (Fig. 11-2A and C), the other irregular (Fig. 11-2B and D). The regularly discharging afferent has the longer CT and, hence, has a thinner axon. This was a general finding and has been confirmed by Yagi et al. (36), who measured orthodromic CTs in the cat.

We conclude that thick afferents are characterized by irregular discharge patterns, thin afferents by regular patterns. Other differences between the two populations are summarized in Table 11-1. Thick afferents are phasic-tonic in the sense that their response dynamics deviate from the expected mechanical behavior of the cupula or otolithic membrane. They show high sensitivity, particularly in the bandwidth of naturally occurring head movements, which we take to be 1 to 10 Hz. There is only a weak relation between resting discharge and sensitivity. The units are readily silenced by inhibitory head rotations or head tilts. In all of these respects, thick axons are distinguishable from thin axons. Many of these points are best illustrated by considering the responses of units innervating particular end organs. We first turn to the semicircular canals.

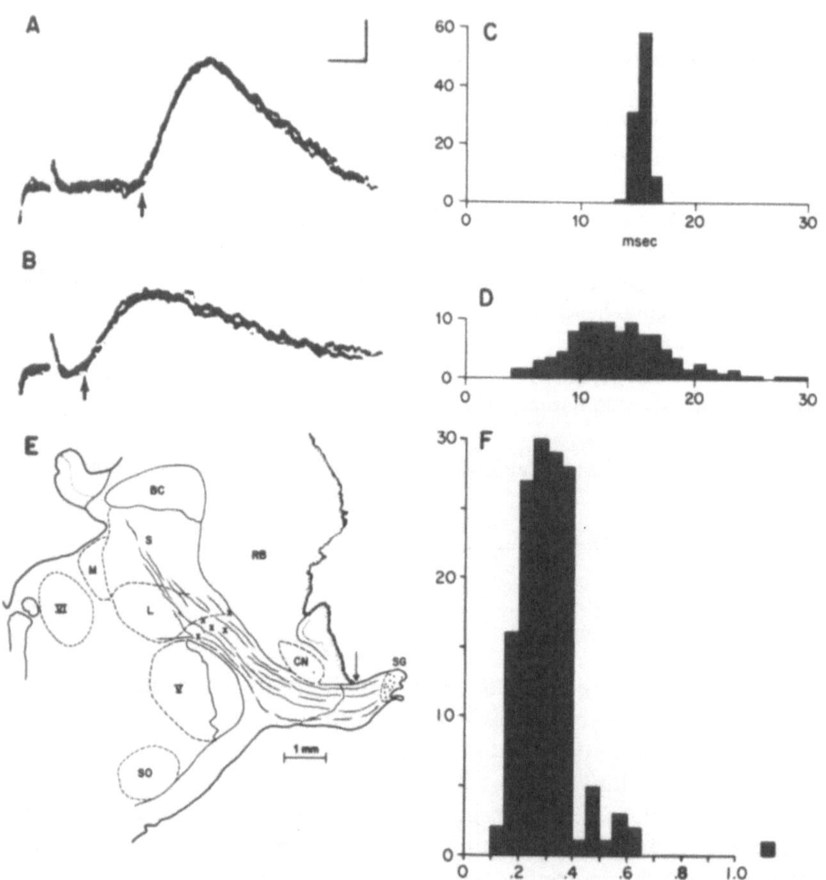

Fig. 11-2. **A** and **B** Antidromic action potentials for a regular and an irregular unit, respectively. Both units, superior canal neurons, recorded consecutively in same puncture. *Arrows*, foot of action potentials. Each record contains 10 superimposed sweeps. Calibrations: 0.1 ms, 1 mV. **C** and **D** Corresponding interspike-interval distributions. *Ordinate*, percentage of intervals encountered; *abscissa*, interval values in ms; *bins*, 1 ms. **E** Locations (*x*) of most effective cathodal stimulating electrode for each of five animals. *Arrow*, approximate microelectrode recording location. *V*, *VI*, spinal trigeminal and abducens nuclei, respectively; *BC*, brachium conjunctivum; *CN*, cochlear nuclei; *L, M, S,* lateral, medial, and superior vestibular nuclei, respectively; *RB,* restiform body; *SG,* Scarpa's ganglion; *SO,* superior olivary complex. **F** Conduction times, all units. *Ordinate,* number of units; *abscissa*, conduction time in ms. From Goldberg, J.M. and Fernández, C.: Brain Res. 122:545, 1977.

SEMICIRCULAR CANAL AFFERENTS

Canal neurons are sensitive to angular accelerations of the head. The afferents can also be affected by constant linear forces (3,23), although it may be doubted whether this would occur under entirely physiologic circum-

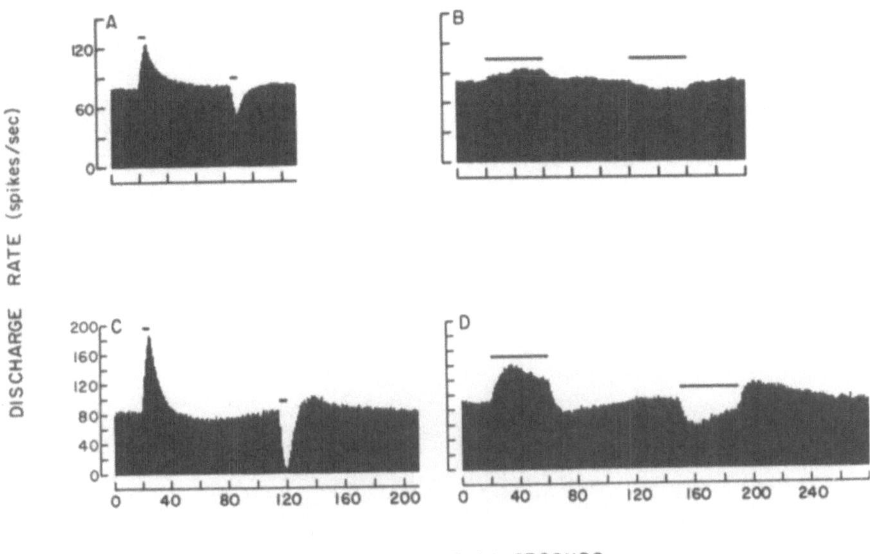

Fig. 11-3. Responses of semicircular canal afferents in the squirrel monkey to oppositely directed steps of constant angular acceleration separated by periods of no acceleration. Each stimulus is a velocity trapezoid consisting of the following five periods: (1) head stationary; (2) constant angular acceleration (*left-hand bar*); (3) constant angular velocity; (4) constant angular deceleration (*right-hand bar*); and (5) head stationary. **A** and **B** Responses of a regularly discharging afferent, respectively, to 5-s, 60 deg/s² and to 40-s, 7.5 deg/s² steps of acceleration and deceleration. **C** and **D** Responses of an irregularly discharging afferent to the same stimulus conditions. From Goldberg, J.M. and Fernández, C.: J. Neurophysiol. 34:635, 1971.

stances (13). The response to rotational stimulation is bidirectional (Fig. 11-3). The directions of excitatory and inhibitory accelerations are the same for all afferents innervating a single canal (11,24,25) and are consistent with the morphologic polarization of the hair cells (26). A particular canal is only affected by accelerations acting within its plane (3). Any acceleration which excites a canal on one side will inhibit the parallel canal of the opposite side. Since the three sets of parallel canals are almost orthogonal to one another, precise information is provided about the magnitude and orientation of any arbitrarily directed angular acceleration.

The mechanics of the semicircular canals may be likened to a heavily damped torsion pendulum (30,31). The response to constant angular accelerations should then be largely determined by a single first-order time constant. Some afferents behave in this way (Fig. 11-3A and B). The time constant of response is about 5 s. For other units, however, there are major discrepancies from the predictions of the torsion-pendulum model. One discrepancy involves a form of sensory adaptation, reflected by the presence of prestimulus response declines during long accelerations (Fig. 11-3D) and

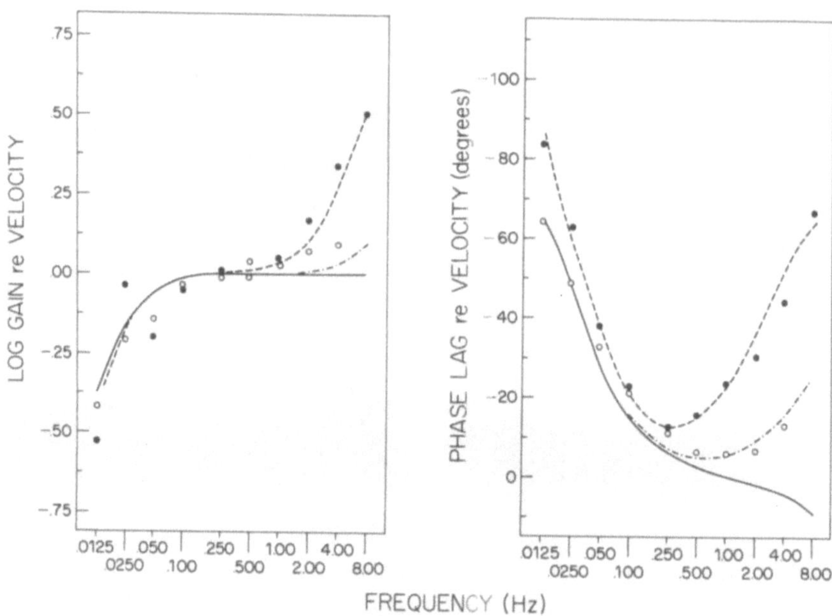

Fig. 11-4. Responses of semicircular canal afferents in the squirrel monkey to sinusoidal head rotations. Data are for a regularly discharging (○) and an irregularly discharging (●) unit. Log gains *re* velocity normalized to gain at 0.25 Hz. Phase leads are shown as negative phase lags. Solid lines are gains and phases predicted from the torsion-pendulum model; the other two curves are based on empirical transfer functions. From Fernández, C. and Goldberg, J.M.: J. Neurophysiol. 34:661, 1971.

poststimulus secondary responses following even short accelerations (Fig. 11-3C). These effects, which are most conspicuous in irregularly discharging afferents (1,12), bear a formal resemblance to the response declines and secondary responses seen in human studies of vestibular-induced sensation and nystagmus (16). Adaptation, it should be noted, would not come into play during naturally occurring head movements.

The afferents' response dynamics deviate in a second way from those of the torsion-pendulum model. The discrepancy is seen during sinusoidal stimulation. Figure 11-4 shows the expected response of the model (solid curves), as well as data from two afferents. Based solely on its predicted mechanical behavior, the canal should function as an angular-velocity transducer. The predicted gain *re* velocity is constant above 0.1 Hz and the expected phase is close to zero. In contrast, the observed gains, rather than remaining constant, begin to increase for frequencies greater than 1 Hz. Corresponding to the gain enhancement, the phase lead reaches a minimum and then starts increasing again. Both the gain enhancement and the phase lead are more conspicuous in irregular units (12,17,22,28). In the squirrel monkey, the two effects can be simulated by assuming that the afferents' response is linearly related to the velocity, as well as to the displacement, of the cupula (4). It has been suggested that the velocity sensitivity could compensate for the phase lags and gain attenuations characterizing the high-frequency response of various reflex pathways (4,29).

OTOLITH AFFERENTS

The saccular and utricular maculae are both sensors of linear forces acting on the head. Otolith afferents do not respond to even intense angular accelerations (13).

Functional Polarization and Response to Static Tilts

The two otolith organs are disposed at almost right angles to one another. The utricular macula lies in a horizontal plane, the saccular macula in a parasagittal plane (19). Each macula is divided in two by a morphologically distinctive zone, the striola. The hair cells to either side of the striola have opposing morphologic polarizations (Fig. 11-5, lower right). The disposition of the two maculae and the arrangement of the hair cells within them have implications for the directional properties of the afferents.

Most mammalian otolith neurons respond in a maintained manner to static tilts (5,8,20). Figure 11-5 shows how the discharge of otolith neurons can be modulated as the head is held in various positions. The modulation takes place around a resting (or zero-force) discharge, which for the unit in Fig. 11-5B amounts to 80 spikes/s. The response, taken as the difference between the discharge (d) and the resting discharge (d_o), is an approximately trigonometric function of tilt angle ϕ. This is consistent with a linear model. Suppose that each afferent is characterized by a functional polarization vector of unit length. Then were linearity assumed, the response would be

$$d - d_o = s \cos \theta = sF$$

θ is the angle between the polarization vector and the gravity vector; s is a sensitivity factor; and $F = \cos \theta$ is the effective force. From this equation, polarization vectors can be calculated from static-tilt data (for details, see ref. 8). The trigonometric relation for tilts around either the pitch or roll axes follows from simple geometric principles.

The eight units of Fig. 11-5A–H innervate the utricular macula. In each case, the discharge is close to its resting value when the animal is in a horizontal position and approaches a maximum or minimum when the animal is tilted 90° from the horizontal. In the horizontal position, gravity is directed perpendicular to the macula. A 90° tilt will bring the gravity vector parallel to the macula. The observations thus imply that gravity is effective only when it acts within the plane of the macula. In short, the utricular end organ is only sensitive to shearing forces. The conclusion is confirmed when variously directed centrifugal forces are used (6). Utricular units vary in

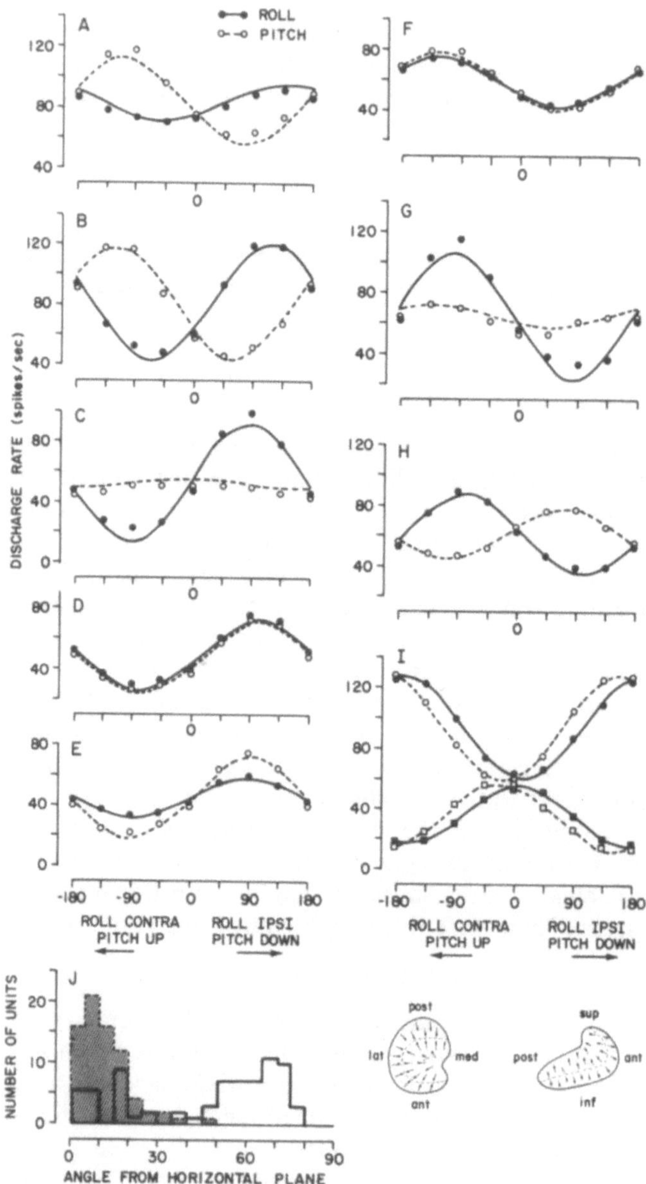

Fig. 11-5. Responses of otolith afferents in the squirrel monkey to static tilts. **A–H** Eight separate utricular afferents. **I** Two saccular afferents. Each point based on a 10-s sample of steady-state activity. Curves are best-fitting trigonometric functions. Abscissa shows tilt in degrees. **J** Distribution of functional polarization vectors for utricular afferents (*shaded bars*) and saccular afferents (*open bars*). Modified from Fernández, C. and Goldberg, J.M.: J. Neurophysiol. 39:970, 1976. **Lower right** Morphological polarization maps for utricular and saccular maculae of the squirrel monkey are seen, respectively, to the left and right. From Lindeman, H.H.: Ergeb. Anat. Entwickl. 42:1, 1969.

their directional properties. Some are excited by ipsilateral rolls, others by contralateral rolls, some by downward pitches, others by upward pitches. This diversity is entirely consistent with the morphologic polarization map (Fig. 11-5, lower right). The two units of Fig. 11-5I are related to the saccular macula. For these, the horizontal positions are points of maximum or minimum discharge, whereas 90° tilts bring the discharge near d_o. Hence, the parasagittally oriented saccular macula also responds only to shearing forces. The two saccular units have static-tilt curves which are mirror images of one another. This reflects the parallel orientation of the hair cells to either side of the striola (Fig. 11-5, lower right).

The differences in the directional properties of saccular and utricular afferents can be summarized in terms of their functional polarization vectors (Fig. 11-5J). Most utricular vectors lie within 15° of the horizontal plane, whereas most saccular vectors are disposed more than 50° from this plane (or equivalently, within 40° of the vertical). As a result, utricular units are sensitive to small tilts from the normal head position and to linear forces arising during locomotion in the horizontal plane. Saccular units, in contrast, are sensitive to dorsoventral accelerations, such as occur in jumps and falls. The utricular macula, given the broad distribution of its vectors, can signal information about any arbitrarily directed force acting in the horizontal plane. To the extent that the utricular vectors are confined to this plane, the sensory representation is two-dimensional. The saccular macula provides the third dimension.

Response Dynamics

Regularly discharging otolith afferents are tonic receptors (5,7,32). Their response parallels the applied force during prolonged stimulation (Fig. 11-6B) and also during abrupt force transitions (Fig. 11-7C and D). Irregularly discharging afferents behave in a more phasic manner. They adapt to maintained stimulation (Fig. 11-6A) and their response to rapid force transitions provides evidence of a velocity sensitivity (Fig. 11-7A and B).

It may be concluded that the otolith organs subserve both tonic and phasic functions. The two modes of otolith activation may be related to the dual role of these end organs, i.e., to monitor static head position and also the dynamic accelerations arising during movement.

EFFERENT VESTIBULAR SYSTEM

Peripheral Action

In all other hair-cell systems studied, including the mammalian cochlea (35) and lateral lines (27), electric stimulation of efferent pathways results in an inhibition of afferent activity. The recent localization of efferent vestibular (EVS) neurons (9,33) has permitted us to do comparable experiments in this system (15). Remarkably, electric stimulation of the EVS almost always

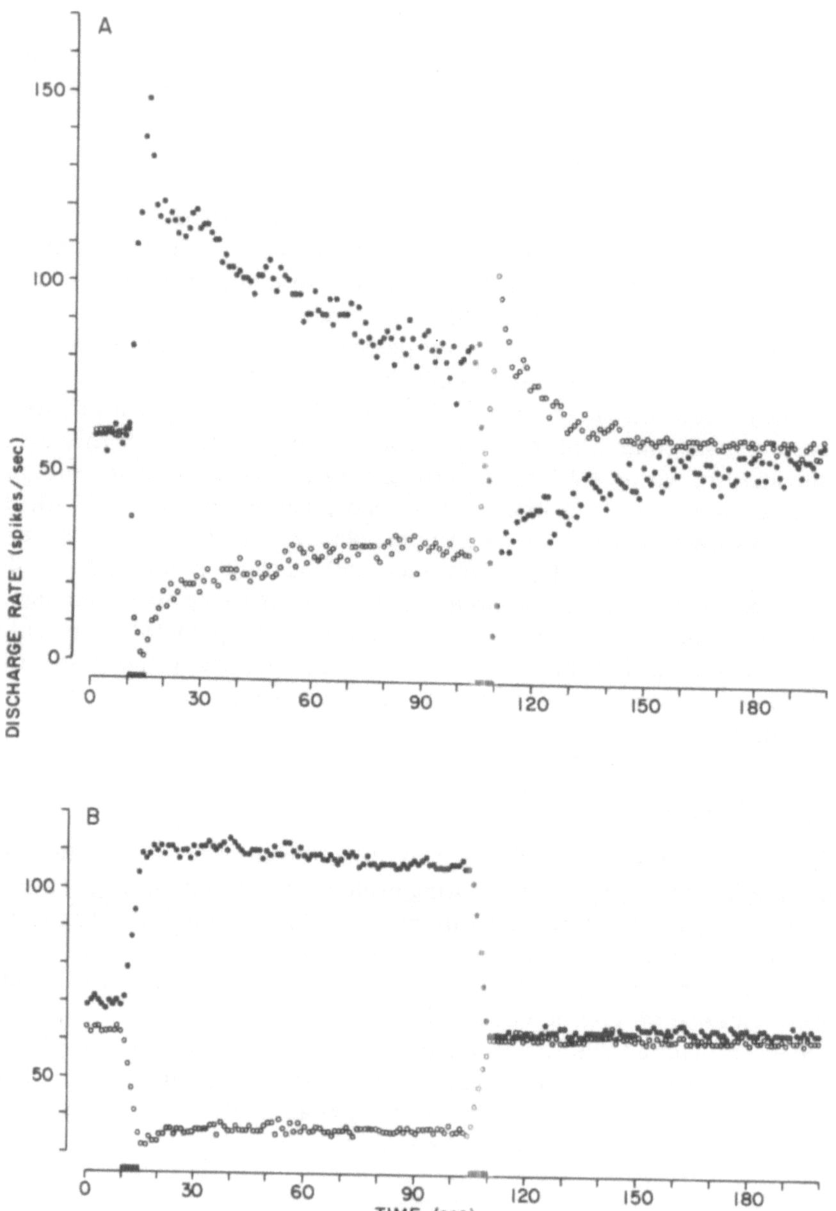

Fig. 11-6. Responses of otolith afferents in the squirrel monkey to centrifugal-force trapezoids of long duration. **A** Irregularly discharging afferent. **B** Regularly discharging afferent. Each stimulus consists of the following five periods: (1) background force (0.077 *g*); (2) linear force transition (*left-hand bar*); (3) constant force (1.23 *g*); (4) linear force transition (*right-hand bar*); and (5) background force. Responses to excitatory (●) and inhibitory (○) forces directed, respectively, parallel and antiparallel to each unit's functional polarization vector. Based on Fernández, C. and Goldberg, J.M.: J. Neurophysiol. 39:970, 1976.

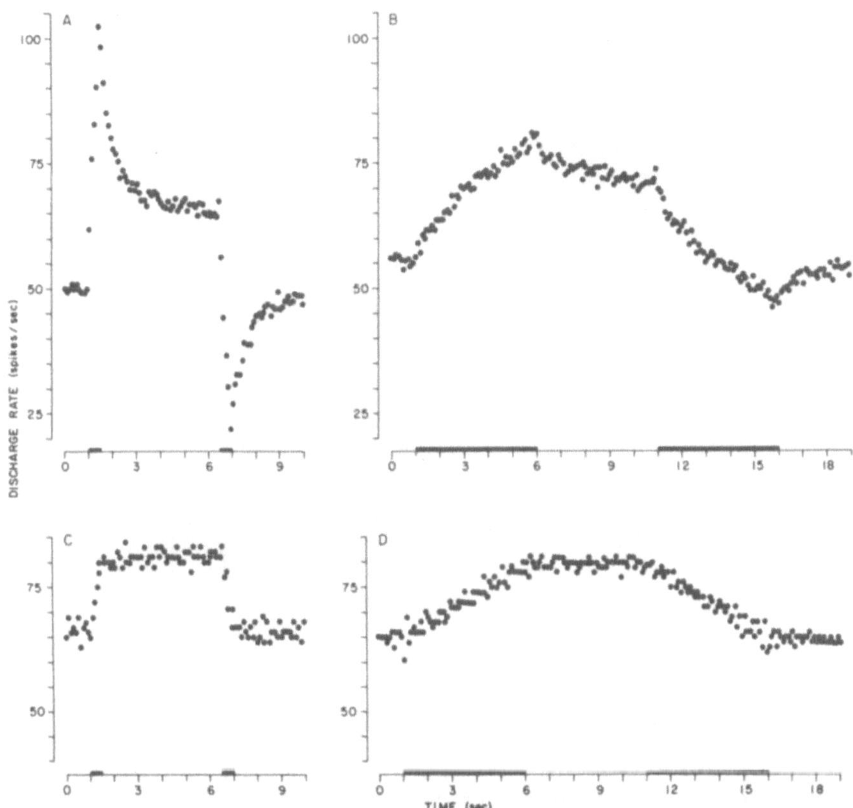

Fig. 11-7. Response to short-duration centrifugal-force trapezoids, same units as in Fig. 11-6. **A** and **B** The irregularly discharging afferent shows a phasic response, which is larger during 0.5-s force transitions (**A**) than during 5-s transitions (**B**). The background and peak forces are each identical for the two trapezoids. **C** and **D** The response of the regularly discharging afferent is determined by the instantaneous force, not by the rate of force application. Based on Fernández, C. and Goldberg, J.M.: J. Neurophysiol. 39:996, 1976.

results in an increase in background discharge. Excitation is seen when the ipsilateral and contralateral efferent pathways are stimulated simultaneously or separately and is observed in neurons innervating each of the five end organs.

The response of an irregularly discharging afferent to high-frequency electric stimulation of the EVS is seen in Fig. 11-8A. Abrupt changes in activity occur at the onset and termination of the shock train. During stimulation, there is a gradual build-up of activity. Following the train, the discharge does not immediately return to background levels. Rather, there is a persistent excitatory response lasting some 30 s. The rapid changes in discharge reflect a *fast* response component. The gradual prestimulus build-up and poststimulus decline indicate the presence of a *slow* response component. Ex-

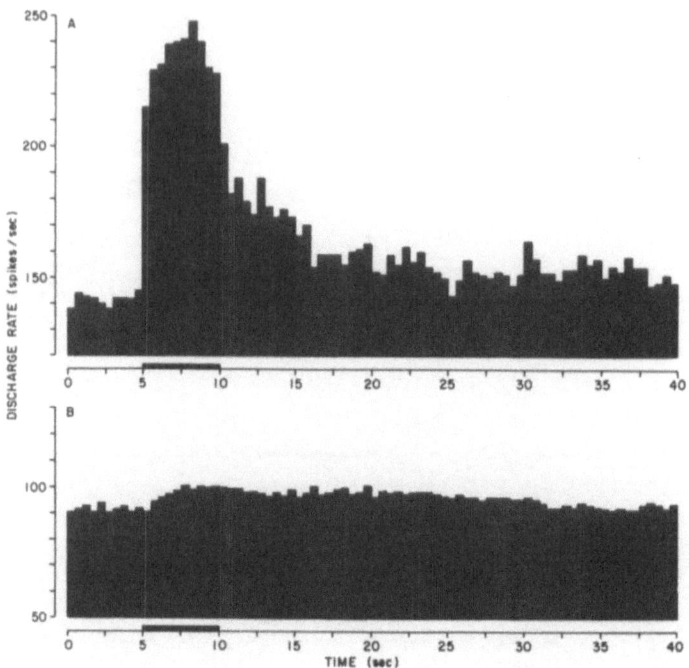

Fig. 11-8. Response of two vestibular afferents to electrical stimulation of brainstem efferent pathways in the squirrel monkey. Stimulus (*bar*) consists of 5-s trains of 0.1-ms shocks, 40 μA, delivered at 333/s. **A** Irregular afferent. **B** Regular afferent. Both units obtained in same preparation. Based on Goldberg, J.M. and Fernández, C.: J. Neurophysiol. 43:986, 1980.

cept for the fact that it is excitatory, the *fast* component resembles the efferent responses seen in other hair-cell systems. The *slow* response has a number of unusual features and these would suggest that it is not mediated by a conventional synaptic process.

There is a relation between the magnitude of the EVS responses and the afferents' regularity of discharge. The response of a regularly discharging unit is shown in Fig. 11-8B. As is typical, the response is 10 to 20 times smaller than that usually observed in irregular units (Fig. 11-8A) and, further, it is almost entirely composed of a *slow* component. The results may reflect differences in the actions of the postsynaptic efferent innervation of afferent chalices and the presynaptic efferent innervation of type II hair cells. Presumably the irregular units receive a postsynaptic innervation, the regular units a presynaptic innervation.

Functional Considerations

Any discussion of the function of the mammalian EVS must be speculative. One reason is that, given the unphysiologic nature of electric stimulation, only limited deductions can be made from our own studies. Moreover, virtually nothing is known concerning the discharge characteristics of mammalian EVS neurons or about possible efferent modifications of afferent discharge under normal conditions.

In this circumstance, it is helpful to consider the lateral lines of lower vertebrates, since this is the one hair-cell system in which there is some understanding of efferent control (27). Lateral lines are exteroceptors, which are potentially excited by the animal's own movement. The lateral line efferents fire during movement and they presumably function to prevent self-excitation by inhibiting afferent discharge.

It is unlikely that the EVS works in precisely the same way since the vestibular end organs are proprioceptors, whose function is to monitor the animal's own head movements. Klinke and Schmidt (18) did propose that the EVS suppresses afferent discharge during movement. That this is not the case would seem proved by the observation that the vestibuloocular reflex and, by inference, the vestibular end organs have normal gains during voluntary head movements (2). As an alternative, it may be suggested that during movements the EVS activates the labyrinths on the two sides. Such a bilateral activation, while it would not lead to vestibular reflex movements or sensation, could improve the performance of the receptors.

The vestibular afferents have a potential dynamic range of 0–400 spikes/s (11). Although the afferents have a high resting discharge, they usually work in the lower part of the range (5,6,11). The result is that it is easier to silence units than it is to drive them into excitatory saturation. A silencing or inhibitory saturation is of no concern during the small, passive head accelerations that occur when an animal is standing still and the vestibular system is functioning to maintain postural stabilization. But the large accelerations accompanying intended movements could easily lead to afferent silencing and, hence, render the operation of the vestibular system nonlinear. A bilateral activation of the EVS, presumably triggered from premotor centers, would reduce or eliminate the nonlinearity. At the same time, there would be a symmetric increase in the excitatory drive on vestibular reflex pathways and this should ensure that they respond appropriately.

Stated somewhat more generally, it is envisioned that the EVS functions to modulate the dynamic range of the afferents to match expected accelerations. The alternative solution of raising the afferents' resting discharge may simply be too costly in terms of the metabolic requirements of the labyrinth. This notion, although speculative, has two attractions. First, it provides a rationale for the excitatory action of the EVS; and second, it offers a functional explanation as to why the EVS exerts most of its influence on irregularly discharging units. These afferents, being more sensitive than regular units, are more readily driven into inhibitory saturation.

SUMMARY

Thick and thin vestibular axons have different discharge characteristics. Thin afferents, which innervate type II hair cells, are characterized by a regular spacing of action potentials and are tonic receptors. Thick afferents supply type I hair cells; the spacing of action potentials is irregular and the receptors are more phasic in their behavior. Thick (irregular) units are also more sensitive to natural stimuli, particularly to those whose spectrum falls within the bandwidth of naturally occurring head movements. *Semicircular canal afferents* respond to angular accelerations acting within the plane of the corresponding canal. The three sets of parallel canals provide precise information about the magnitude and direction of any arbitrarily directed angular acceleration. The response dynamics of the afferents deviate from the expected mechanical response of the canals. The afferents show a sensory adaptation and a velocity sensitivity. *Otolith afferents* respond to linear forces. The utricular macula is potentially sensitive to any linear force acting within the horizontal plane, whereas the saccular macula is sensitive to dorsoventrally oriented forces. The otolith organs subserve both tonic and dynamic functions, which may be related to their dual role of monitoring static head position and also the dynamic accelerations arising during movement. Electric stimulation of the *efferent vestibular system* (EVS) results in an excitation, which is particularly prominent in irregularly discharging (thick) afferents. It is proposed that the EVS functions to modulate the dynamic range of the afferents to match the expected accelerations occurring during movement.

ACKNOWLEDGMENTS

The author's research is supported by Grant NS 01330 from the National Institutes of Health and by Grant NGR-14-001-225 from the National Aeronautics and Space Administration.

References

1. Blanks, R.H.I., Estes, M.S., and Markham, C.H.: Physiologic characteristics of vestibular first-order canal neurons in the cat. H. Response to constant angular acceleration. J. Neurophysiol. 38:1250, 1975.
2. Dichgans, J., Bizzi, E., Morasso, P., and Tagliasco, V.: Mechanisms underlying recovery of eye-head coordination following bilateral labyrinthectomy in monkeys. Exp. Brain Res. 18:548, 1973.
3. Estes, M.S., Blanks, R.H.I., and Markham, C.H.: Physiologic characteristics of vestibular first-order canal neurons in the cat. I. Response plane determination and resting discharge characteristics. J. Neurophysiol. 38:1232, 1975.
4. Fernández, C. and Goldberg, J.M.: Physiology of peripheral neurons innervating semicircular canals of the squirrel monkey. II. Response to sinusoidal stimulation and dynamics of peripheral vestibular system. J. Neurophysiol. 34:661, 1971.
5. Fernández, C. and Goldberg, J.M.: Physiology of peripheral neurons innervating otolith organs of the squirrel monkey. I. Response to static tilts and to long-duration centrifugal force. J. Neurophysiol. 39:970, 1976.

6. Fernández, C. and Goldberg, J.M.: Physiology of peripheral neurons innervating otolith organs of the squirrel monkey. II. Directional selectivity and force-response relations. J. Neurophysiol. 39:985, 1976.

7. Fernández, C. and Goldberg, J.M.: Physiology of peripheral neurons innervating otolith organs of the squirrel monkey. III. Response dynamics. J. Neurophysiol. 39:996, 1976.

8. Fernández, C., Goldberg, J.M., and Abend, W.K.: Response to static tilts of peripheral neurons innervating otolith organs of the squirrel monkey. J. Neurophysiol. 35:978, 1972.

9. Gacek, R.R. and Lyon, M.: The localization of vestibular efferent neurons in the kitten with horseradish peroxidase. Acta Otolaryngol. 77:92, 1974.

10. Gacek, R.R. and Rasmussen, G.L.: Fiber analysis of statoacoustic nerve of guinea pig, cat, and monkey. Anat. Rec. 139:455, 1961.

11. Goldberg, J.M. and Fernández, C.: Physiology of peripheral neurons innervating semicircular canals of the squirrel monkey. I. Resting discharge and response to constant angular accelerations. J. Neurophysiol. 34:635, 1971.

12. Goldberg, J.M. and Fernández, C.: Physiology of peripheral neurons innervating semicircular canals of the squirrel monkey. III. Variation among units in their discharge properties. J. Neurophysiol. 34:676, 1971.

13. Goldberg, J.M. and Fernández, C.: Responses of peripheral vestibular neurons to angular and linear accelerations in the squirrel monkey. Acta Otolaryngol. 80:101, 1975.

14. Goldberg, J.M. and Fernández, C.: Conduction times and background discharge of vestibular afferents. Brain Res. 122:545, 1977.

15. Goldberg, J.M. and Fernández, C.: Efferent vestibular system in the squirrel monkey: Anatomical location and influence on afferent activity. J. Neurophysiol. 43:986, 1980.

16. Guedry, F.E., Jr.: Psychophysics and vestibular sensation. In Kornhuber, H.H. (ed.): Handbook of Sensory Physiology. VI. Vestibular System. Part 2: Psychophysics, Applied Aspects and General Interpretations. Berlin, Springer-Verlag, 1974, pp. 3–154.

17. Keller, E.L.: Behavior of horizontal semicircular canal afferents in alert monkey during vestibular and optokinetic stimulation. Exp. Brain Res. 24:459, 1976.

18. Klinke, R. and Schmidt, C.L.: Efferent influence on the vestibular organ during active movements of the body. Arch. Ges. Physiol. 318:325, 1970.

19. Lindeman, H.H.: Studies on the morphology of the sensory regions of the vestibular apparatus. Ergeb. Anat. Entwickl. 42:1, 1969.

20. Loe, P.R., Tomko, D.L., and Werner, G.: The neural signal of angular head position in primary afferent vestibular axons. J. Physiol. (Lond.) 230:29, 1973.

21. Lorente de Nó, R.: Études sur l'anatomie et la physiologie du labyrinthe de l'oreille et du VIIIe nerf. II. Trav. Lab. Rech. Biol. Univ. Madrid 24:53, 1926.

22. Louie, A.W. and Kim, J.: The response of 8th nerve fibers to horizontal sinusoidal oscillations in the alert monkey. Exp. Brain Res. 24:447, 1976.

23. Lowenstein, O.: Physiology of vestibular receptors. Progr. Brain Res. 37:19, 1972.

24. Lowenstein, O. and Sand, A.: The individual and integrated activity of the

semicircular canals of the elasmobranch labyrinth. J. Physiol. (Lond.) 99:89, 1940.

25. Lowenstein, O. and Sand, A.: The mechanism of the semicircular canal. A study of the responses of single-fibre preparations to angular accelerations and to rotation of constant speed. Proc. R. Soc. Lond. B 129:256, 1940.

26. Lowenstein, O. and Wersäll, J.: A functional interpretation of the electron-microscopic structure of the sensory hairs in the cristae of the elasmobranch *Raja clavata* in terms of directional sensitivity. Nature 184:1807, 1959.

27. Russell, I.J.: Amphibian lateral line receptors. In Llinás, R. and Precht, W. (eds.): Frog Neurobiology. A Handbook. Berlin, Springer-Verlag, 1976, pp. 513–550.

28. Schneider, L.W. and Anderson, D.J.: Transfer characteristics of first and second order lateral canal vestibular neurons in gerbil. Brain Res. 112:61, 1976.

29. Skavenski, A.A. and Robinson, D.A.: Role of abducens neurons in vestibulo-ocular reflex. J. Neurophysiol. 36:724, 1973.

30. Steinhausen, W.: Über den Nachweis der Bewegung der Cupula in der intakten Bogengansampulle des Labyrinthes bei der natüralichen rotatorischen und calorischen Reizung. Arch. Ges. Physiol. 228:322, 1931.

31. Steinhausen, W.: Über die Beobachtung der Cupula in den Bogengansampullen des Labyrinthes des levenden Hechts. Arch. Ges. Physiol. 232:500, 1933.

32. Vidal, J., Jeannerod, M., Lifschitz, W., Levitan, H., Rosenberg, J., and Segundo, J.P.: Static and dynamic properties of gravity-sensitive receptors in the cat vestibular system. Kybernetik 9:205, 1971.

33. Warr, W.B.: Olivocochlear and vestibular efferent neurons of the feline brain stem: their location, morphology, and number determined by retrograde axonal transport and acetylcholinesterase histochemistry. J. Comp. Neurol. 161:159, 1975.

34. Wersäll, J.: Studies on the structure and innervation of the sensory epithelium of the cristae ampullaris in the guinea pig. A light and electron microscopic investigation. Acta Otolaryngol. (Suppl.) 126:1, 1956.

35. Wiederhold, M.L. and Kiang, N.Y.S.: Effects of electrical stimulation of the crossed olivocochlear bundle on single auditory-nerve fibers in the cat. J. Acoust. Soc. Am. 48:950, 1970.

36. Yagi, T., Simpson, N.E., and Markham, C.H.: The relationship of conduction velocity to other physiological properties of the cat's horizontal canal neurons. Exp. Brain Res. 30:587, 1977.

Discussion

LOWENSTEIN: How would you feel if what you just told us might be compatible with our very early speculation that the function of the efferent is to set the working point of the afferent along an S-shaped characteristic?

GOLDBERG: That is precisely the idea we would like to put forward concerning efferent function, i.e., that the efferents set the operating point of the input-output curve so as to match expected accelerations accompanying intended head movement. At the same time, it must be emphasized that all considerations of efferent function will remain speculative until we know more about the efferents' discharge characteristics and about the efferent modifications of afferent activity under normal circumstances.

YOUNG: I'd like to return to the otolith afferent part of the story. It is well known that there are behavioral and oculomotor asymmetries in which there are preferred axes of stimulation; particularly in the human, sensitivity to changes in orientation from the normally erect position is much greater than for movements from an initially reclining position. These effects have been interpreted as some sort of ambiguity in the saccular information, whether in the source of the information or its treatment. Do you see anything in the recordings of presumed saccular units which would distinguish them from utricular units in terms of threshold, irregularities of discharge, or other dynamic characteristics?

GOLDBERG: Two classes of saccular units can be distinguished on the basis of their directional properties. One set has its maximum discharge when the head is erect; whereas for the other set, the discharge is minimum in this head position. In the terminology of polarization vectors, the first kind of afferent has a downwardly directed vector and should innervate the inferior part of the saccular macula. The second kind of afferent is characterized by an upwardly directed vector and so should supply the superior part of the macula. As illustrated in Fig. 11-5I, the two groups of afferents differ in their resting discharges and, to a lesser extent, in their sensitivities. None of the afferents had an apparent threshold. Moreover, there was no obvious relation between regularity of discharge and directional properties.

YOUNG: Were there any tests to determine if there were thresholds in the afferents at the level corresponding to behavioral thresholds, or tilts of fractions of a degree?

GOLDBERG: No attempt was made in our experiments to use near-threshold stimuli. What can be said, however, is that there is no obvious discontinuity in the input-output curve as it passes through zero force. Because of this, the concept of a threshold has not proved particularly relevant in describing the discharge of peripheral afferents.

FLOCK: One of the big differences here, apart from the excitation that you find, is the time constant of the effect that you see. The inhibition that we find in the lateral line organ lasts about a couple of hundred milliseconds at the most and your slow component is apparently several tens of seconds. This brings to mind the recent findings by Kuffler and others that synapses can put out two transmitters, where one component is apparently a peptide which has a very slow time constant. Also relevant is the finding that the lateral line, after blocking the efferent inhibition with Flaxedil, sometimes showed a slow excitation instead.

GOLDBERG: The efferent responses consisted of a *fast* and a *slow* component. The *fast* component had time constants on the order of 10–100 ms, not dissimilar to those characterizing efferent actions in the lateral lines or in the mammalian cochlea. The time constant of the *slow* response, in contrast, was 5–20 s. Because of its unusually long time course, it is difficult to see how the *slow* response could be mediated by a conventional synaptic process. Two alternatives may be suggested. One is a slow EPSP similar to that described in the cervical sympathetic ganglion. The other would involve nonsynaptic mechanisms. It will be recalled that the efferents form a profuse, unmyelinated plexus in the lower part of the sensory epithelium. Conceivably, impulse traffic in the efferent plexus might increase extracellular K^+ levels and this could lead to an afferent excitation by depolarizing the hair cells or the afferent terminals.

GUALTIEROTTI: Do you have any evidence that the efferents exert a tonic effect, which might be blocked by prolonged electrical stimulation?

GOLDBERG: Dr. Gualtierotti raises an important issue. We have interpreted the excitatory responses as due to an activation of the efferent vestibular system (EVS). Another possibility is that our stimulating electrodes, instead of activating the EVS, somehow blocked it. According to this view, the EVS would be inhibitory and the excitatory response would really be a form of disinhibition. Two observations effectively rule out the latter explanation. First, the explanation requires that the EVS exert a tonic inhibitory influence in our deeply anesthetized preparations. We studied this possibility by acute section of the vestibular nerve between our stimulating and recording electrodes. Now if there were a tonic inhibition, the nerve section should have resulted in a perceptible increase in the afferents' background discharge. No such effect was observed. The lesion can be assumed to have severed the efferent fibers since the responses to brainstem electric stimulation were eliminated. Second, the auditory and vestibular efferent pathways travel in close proximity within the brainstem and both of them should have been stimulated by our brainstem electrodes. Consistent with this, we found that the same shock train which excites vestibular afferents also inhibits auditory afferents. It is difficult to see how an identical shock train could activate auditory efferents and yet block vestibular efferents. A final point is that the excitatory EVS responses in no way depend upon antidromic activation of afferents.

O'LEARY: I am curious about your correlation between the two types of hair cells and irregular versus regular firing activity. We have found in elasmobranchs the same basic distribution of discharge patterns, and response dynamics in a preparation that has only one type of hair cell.

GOLDBERG: Your work in elasmobranchs is important in establishing that differences in the regularity of firing and in response dynamics can be observed in a vestibular end organ only containing type II hair cells. It is of interest to note that Cajal (1908) described thick and thin vestibular afferents in fish. The thick fibers, as in mammals, supply a few, closely spaced hair cells by way of large terminal expansions. The main difference between this expansion and the calyceal terminal of mammals is that the former only surrounds the base of the hair cell and presumably consists of a cluster of boutons. The thin fibers in the fish were described as entering an intraepithelial plexus and presumably give rise to a more diffuse innervation. What is suggested is that the distinctive physiology of an afferent is more closely related to its innervation pattern than to the kinds of hair cell (type I or type II) innervated or the precise morphology of the afferent synapse. Other differences in hair cells, more subtle than their classification into type I and type II varieties, must also be considered.

GUALTIEROTTI: You've demonstrated that the distinction between afferents showing variable frequencies and afferents showing a constant one is impossible and the frequency spectrum is a continuum. Is it also possible to make a correlation between frequency of discharge of the activity at rest and variability? It has been demonstrated, and we have seen it ourselves in the frog, that the coefficient of variation changes inversely with the rate of firing. We might find a situation in which by increasing the frequency of a certain afferent a regular discharge appears and by decreasing it the firing rate becomes irregular.

GOLDBERG: We, too, have observed that the coefficient of variation (CV) increases in an afferent as its discharge rate decreases. For this reason, we have always compared CVs at a standard mean interval. Hence, the differences we have described in regularly and irregularly discharging afferents cannot be ascribed to the two populations having different discharge rates.

PRECHT: A comment regarding the tonic activity of the efferent system and its possible effect on afferents. We have recently done a study in the cat and compared the resting rate of afferents of the deeply anesthetized cat (40 mg/kg Nembutal) to that in the entirely alert cat with efferents tonically active and found absolutely no difference. Statistically the means of the two populations were completely overlapping. This would indicate that there is not much of a contribution of the efferents to resting rate. However, I know of data of Dr. Haupt Hartman of Frankfurt who managed in the goldfish to occasionally hold an afferent fiber over a long time. He applied Flaxedil or curare and he saw a slight increase of the resting rate in all the afferents which he could hold over a long period following the injection. He explained this effect of disinhibition, i.e., blocking of the inhibitory transmitter by curare. Finally, it was reported at a meeting in Florence, by Rossi, that in the frog inhibition, but also excitation, of afferents is found. They report that approximately 10% of the afferents are excited; curare blocks only the inhibition but never affects the excitation. Chronic denervation of the eighth nerve, i.e., abolishing the efferent system, abolishes both effects and indicates that both excitation and inhibition seem to be really efferent effects.

GOLDBERG: Dr. Precht has raised an important point about the comparative physiology of efferents. It is certainly possible that the efferent action in lower vertebrates might be inhibitory, whereas our results suggest that in mammals the main efferent effect is excitatory. Were such a difference documented it might have profound implications for our thinking about the evolution of the vertebrate labyrinth.

WIEDERHOLD: You got a continuum in the coefficient of variation and a continuum in fiber diameter. Is there also a continuum in innervation patterns? Are there afferent fibers that innervate both type I and type II hair cells?

GOLDBERG: This appears to be the case. The medium-sized fibers are usually described as innervating both type I and type II hair cells. In most respects, the physiology of medium-sized fibers can be inferred to be intermediate between that of thick and thin fibers.

FREE PAPER

12

Semicircular Canal Morphology and Function in Crabs

PETER J. FRASER

All crabs have a well-developed balancing organ or statocyst which has long been considered analogous to the vertebrate vestibular system. The foundations for our understanding of the statocyst were laid by Hensen (8) and others, and further developed by Dijkgraaf (1). However, it was 1972 before two independent lines of research indicated the closeness of the analogy between the crab statocyst and the vertebrate semicircular canal system. Sandeman and Okajima (11) in Canberra demonstrated that the statocyst of the Australian mud crab *Scylla* consisted of two toroids at right angles. They also showed by introducing dye into the canals that during angular accelerations, fluid was displaced by inertial forces differentially in the separate canals, and they were able to investigate the responses of the various receptor hairs to fluid displacement. At the same time, I discovered a set of large interneurons each with input from one statocyst in the green shore crab *Carcinus* which illustrated the orthogonal arrangement of the statocyst. Each interneuron responded optimally to rotation in one direction about either the vertical axis or one of the two horizontal axes midway between pitch and roll axes (2,3). Using these interneurons or the sensory cells innervating the thread hairs and free hook hairs to monitor the working of the statocyst, it has been possible to investigate many properties of the statocyst and its associated central pathways in *Scylla* (4,7,9,10,12), in *Carcinus* (6; unpublished observations; Campbell, unpublished observations), and to a lesser extent in other crabs.

In this paper I intend to draw on the sources quoted above together with some new data in order to demonstrate the working of the crab statocyst as an angular acceleration detector. Hence, I shall largely deal with the fluid-dependent components.

MORPHOLOGY

The crab statocyst consists of an invagination of the base of the first antenna which is folded in a complex way. It can best be visualized by producing an invagination of the basal joint in the shape of a football, and then bursting the football and pushing inward on either side to produce a toroid. By pushing inward in two places one can produce two linked toroids at right angles to one another (Fig. 12-1 a), thus forming a horizontal canal and a vertical canal. In *Scylla*, the in-pushing is almost complete, whereas in *Carcinus* it is incomplete. *Scylla* has a closed canal system, *Carcinus* has an open system (Figs. 12-1 b, and 12-2). Statocysts of *Leptograpsus, Ocypode* (which is extremely agile), and *Eriphia* (which is less agile), from tropical climates, have closed canals, whereas statocysts from the sluggish edible crab *Cancer* and *Macropipus* species, which are agile swimming crabs from temperate zones, have open canals. The giant crab *Geryon* from deep Atlantic waters (275–1300 m) has a closed canal system (Fig. 12-2). No correlation between activity or agility and open or closed canals is apparent. However, detailed differences in shape, e.g., between *Carcinus*, where the depth of the vertical

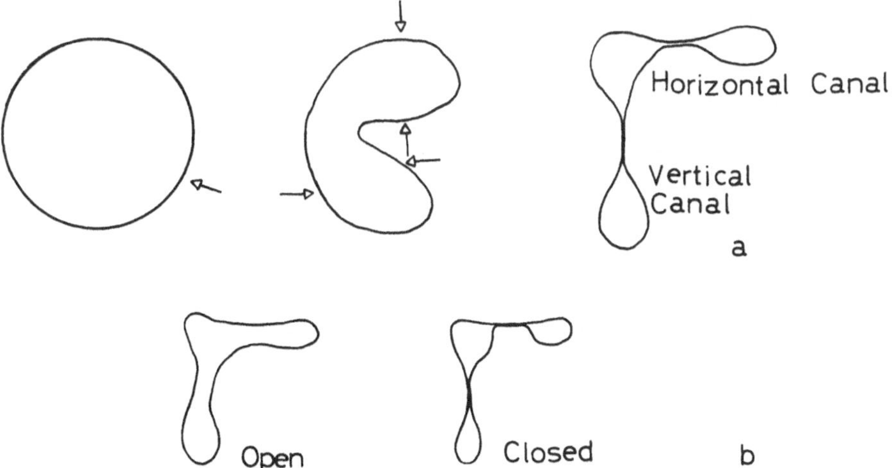

Fig. 12-1. a The crab statocyst is best visualized by bursting an imaginary football and squashing it as shown by the arrows to form two orthogonal toroids. **b** Diagrammatic cross sections showing open and closed canal systems.

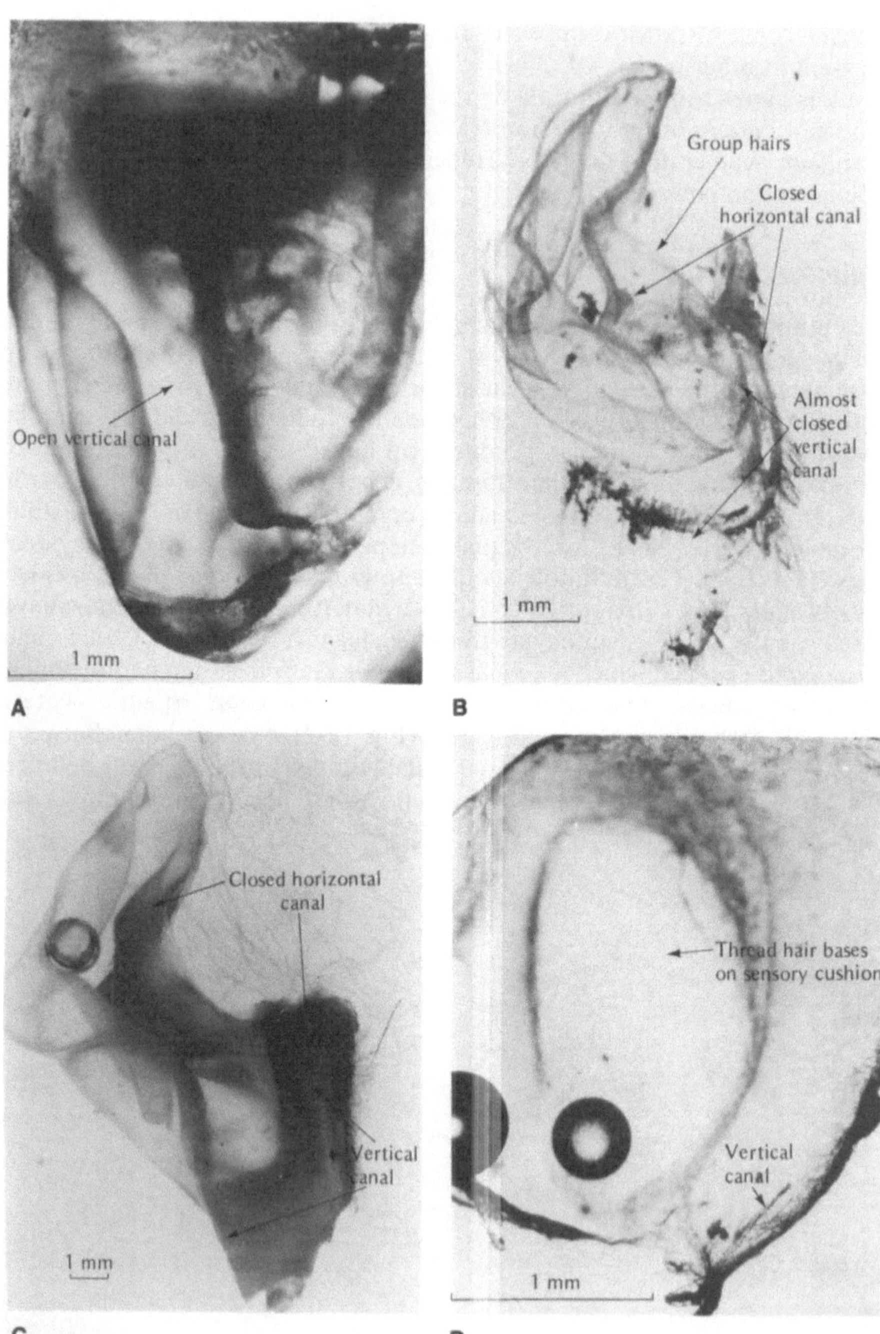

Fig. 12-2. Statocysts of *Carcinus* (**A**), *Scylla* (**B**), and *Geryon* (**C**) looking perpendicular to the planes of the two canals; and *Carcinus* (**D**), *Scylla* (**E**), and *Geryon* (**F**) looking at the sensory cushion of the vertical canal.

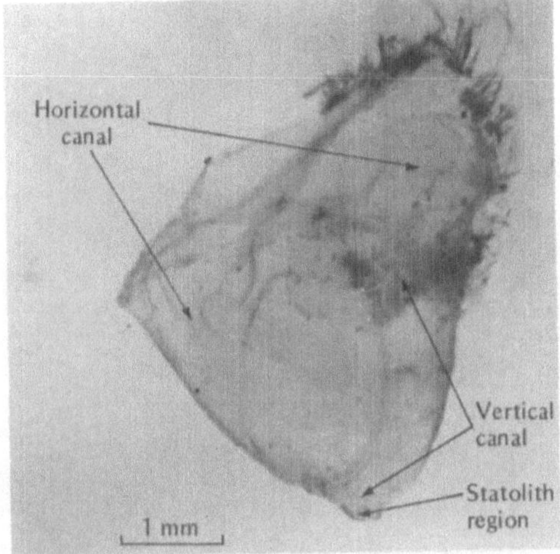
Horizontal
canal
Vertical
canal
Statolith
region
1 mm

E

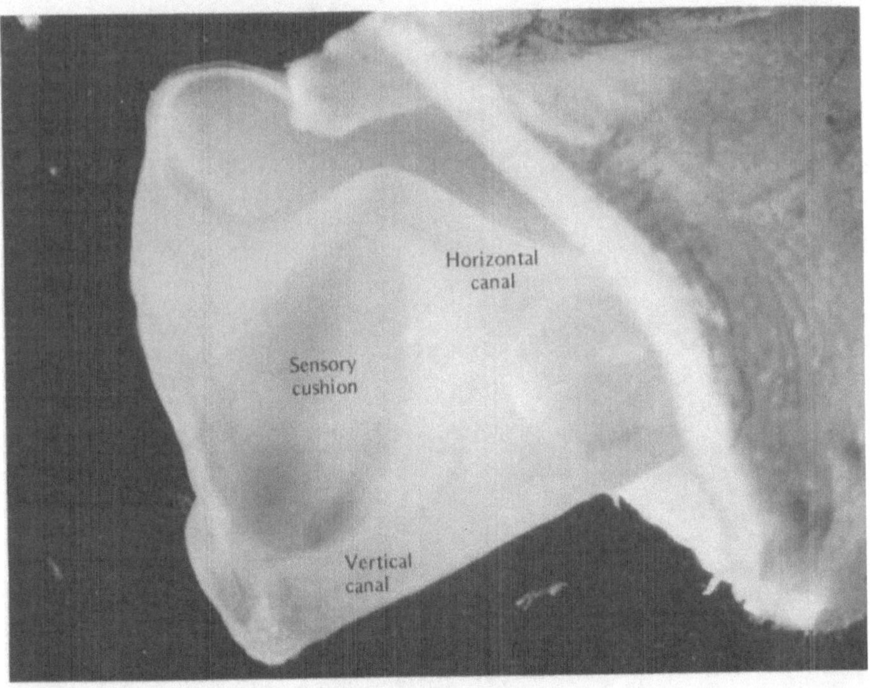
Horizontal
canal
Sensory
cushion
Vertical
canal

F

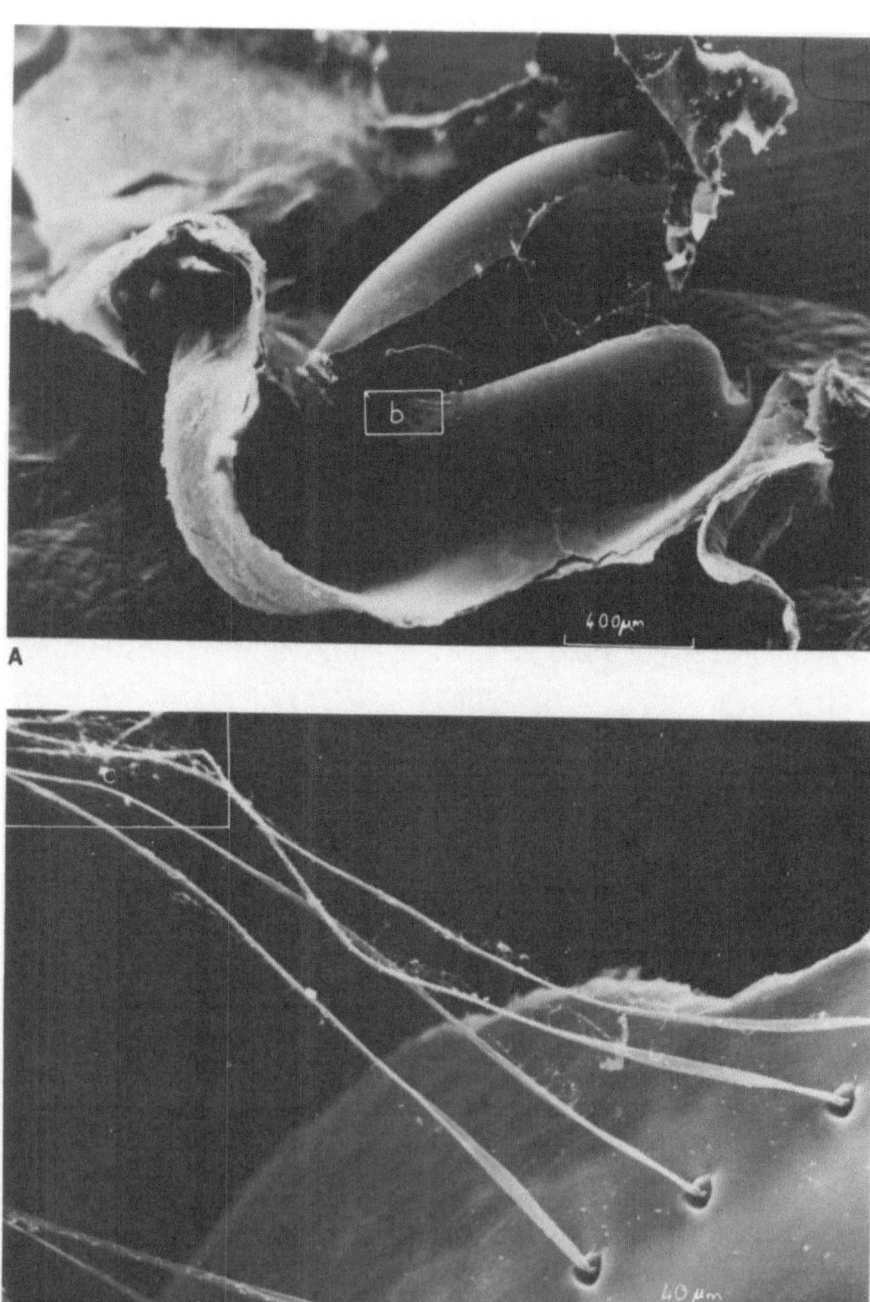

Fig. 12-3. Scanning electron micrographs (**A**–**C**) of the inside of *Carcinus* statocyst showing the line of thread hairs and the feathering on the hairs. **D** *Carcinus* free hook hair.

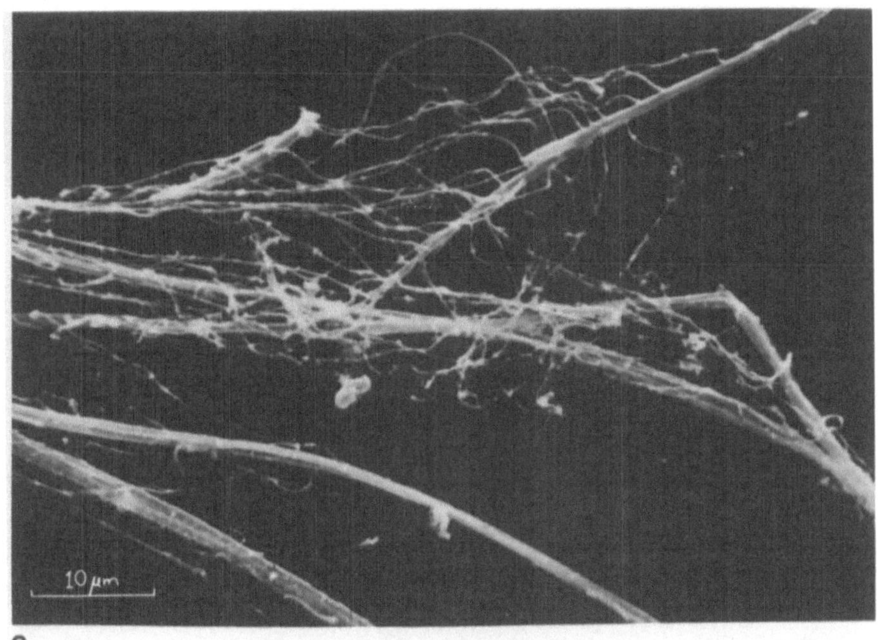

C

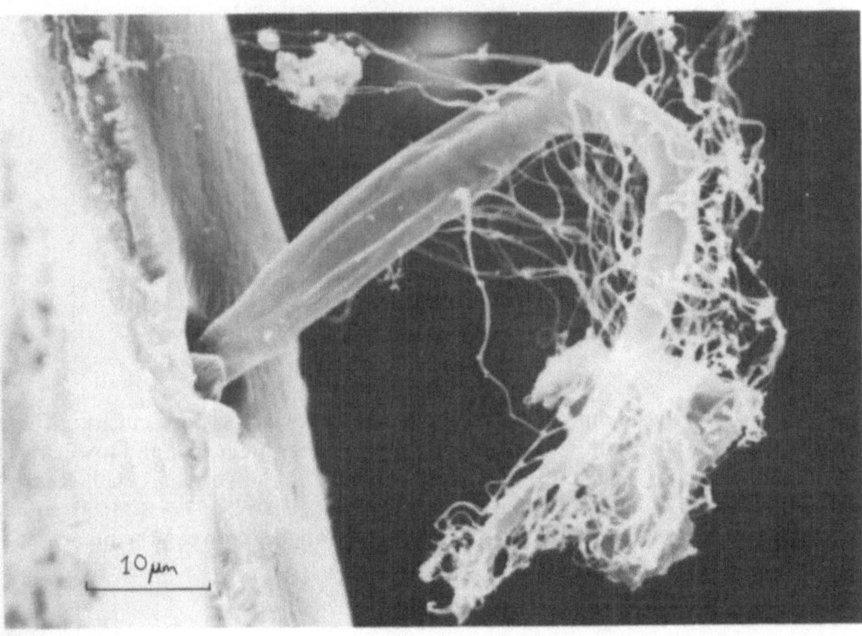

D

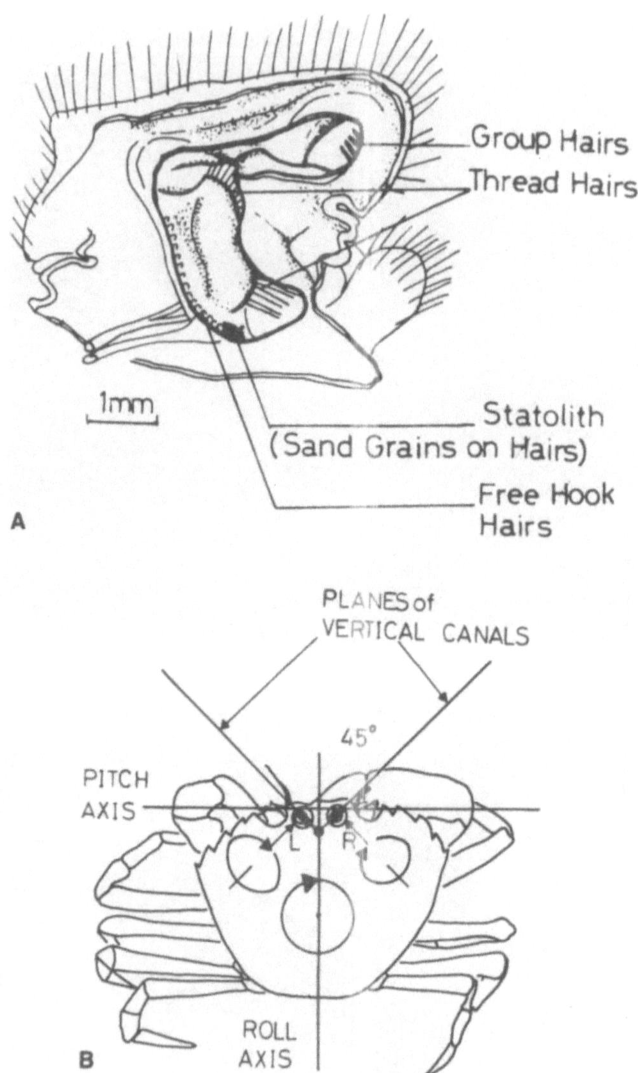

Fig. 12-4. **A** The basal joint of the first antennule of *Scylla* has been cut transversely to show group hairs and upper thread hairs in the horizontal canal and lower thread hairs, free hook hairs, and statolith on statolith hairs in the vertical canal. Redrawn from Sandeman, D.C. and Okajima, A.: J. Exp. Biol. 57:187, 1972. **B** Diagrammatic representation of position of statocysts showing the angles between the planes of the vertical canals and the median plane of the crab.

canal is less than the breadth, and *Ocypode* where the depth of the vertical canal is greater than the breadth, may yet prove significant. There is a tendency for large crabs (e.g., *Scylla* and *Geryon*) to have closed canals, but large *Cancer* have an open statocyst. However, the size of the statocyst is greater relative to the size of the crab in *Scylla* and *Geryon*. The best correlations are between high environmental temperature and closed canals, and low temperature and open canals.

Inside the statocyst are four types of hair which have best been described in *Carcinus* and *Scylla*. Firstly, thread hairs are arranged in a single line over the sensory cushion, which is the name given to the in-pushing which helps form the vertical canal (Figs. 12-3 and 12-4). The hairs are about 400 μm long in *Scylla*, 300 μm long in *Carcinus*, and 600 μm long in *Geryon*. They are between 2 and 4 μm wide. Hairs near the center of the sensory cushion are slightly shorter than those from the edges. In *Scylla* there are about 96 hairs. In *Carcinus*, *Macropipus*, *Cancer*, and *Geryon* there are about 48 hairs. Hairs are often stuck together by means of their feathering. The hair base is polarized. In fresh unfixed preparations the hairs can be seen to bend together mainly about an axis defined by the line joining the bases. Secondly, the free hook hairs lie along the posterior arm of the vertical canal and are 40 to 60 μm long in *Scylla*, 5 μm wide at their widest point and about 2 μm wide at their bases. In *Carcinus* the hairs are slightly smaller, but similar in shape and appearance (Fig. 12-3). In *Geryon*, the hairs are 150 μm long. Thirdly, the statolith hairs are confined to a patch at the bottom of the vertical canal and are surrounded by two arrays of free hook hairs in *Scylla*. They are unfeathered, short (3 to 4 μm), and less than 1 μm in diameter. They support a small statolith which consists of a bundle of sand grains (Fig. 12-4). The fourth type of hair, the group hairs, are 15 μm thick, 800 μm long, and feathered in *Scylla*. They are found along the extreme lateral part of the horizontal canal (Fig. 12-4). No innervation has been found for these hairs, and they do not appear to have a role in regulating fluid flow in the canals. They are most likely a relic from the hairs covering the opening of the statocyst sac which is open to the outside water in many crustacea. They might even be used when the crab is replenishing the sand grains which form the statolith following a molt at which the whole lining of the statocyst and all the hairs are discarded.

Vertical and horizontal canals do not lie exactly in the vertical and horizontal planes of the crab. However, in the normal (slightly elevated head) position of the crab, the horizontal canal is held earth horizontal and the vertical canal vertical. Right and left vertical canals lie at angles of approximately 45° and 315° to the longitudinal plane of the crab (Fig. 12-4).

RESPONSES OF SENSORY NERVES FROM THREAD HAIRS AND FREE HOOK HAIRS

Thread hairs are innervated by two bipolar nerve cells. Free hook hairs are innervated either by one or perhaps more bipolar cells. These nerve cells are normally either silent, or they show a spontaneous level of firing of about 10 to 20 spikes/s. The responses of a small number of thread hair units from *Carcinus* during sinusoidal oscillation are shown in Fig. 12-5. A given unit fires to one direction of displacement in the plane of one canal and is silent during displacement in the opposite direction (Fig. 12-5A). Poststimulus

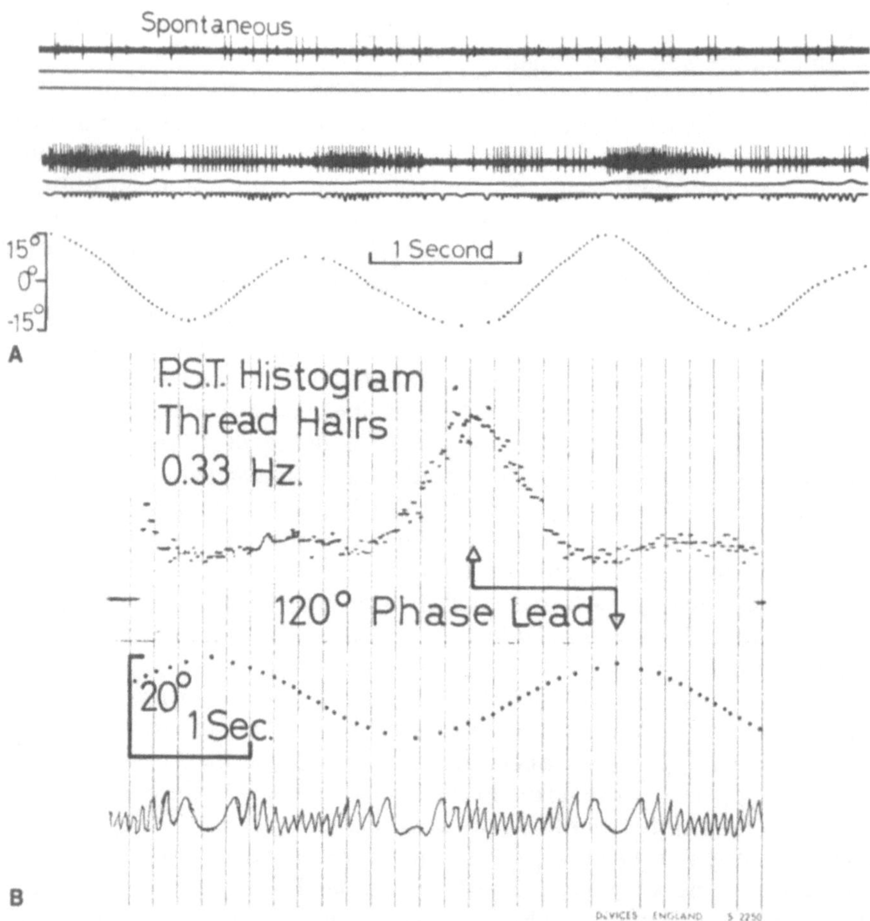

Fig. 12-5. **A** Oscilloscope record from a small group of thread hair units from an isolated statocyst of *Carcinus*, showing spontaneous activity and activity evoked by oscillating the crab at about 0.5 Hz. Middle and lower traces for each record represent a photocell monitor of angular displacement (1° steps) in the plane of the horizontal canal (*middle*) and vertical canal (*lower*). **B** Poststimulus time histogram relating frequency of spikes (*upper trace*) to angular position for approximately 250 cycles of sinusoidal oscillation in the plane of the vertical canal. Isolated statocyst of *Carcinus*.

time histograms show that peak response precedes peak position by about 120° (Fig. 12-5B). Similar records from thread hairs in *Scylla* show a similar picture, but the phase of the response relative to the oscillation cycle differs. In *Scylla*, peak response precedes peak position by about 90°. The relationship between oscillation amplitude and the response is shown in Fig. 12-6. Typically, thresholds of approximately 1 to 10° apply for oscillations in the range 0.5 to 2.0 Hz. Thread hairs are most sensitive and adapt least to oscillation frequencies in the range 0.5 to 10.0 Hz, and adapt out rapidly at

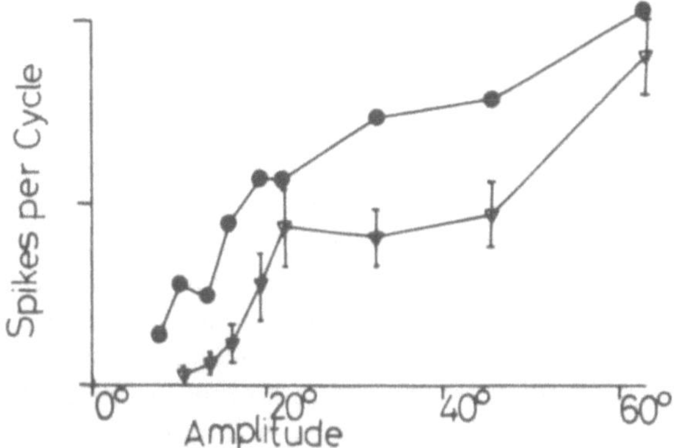

Fig. 12-6. Number of spikes/cycle plotted against oscillation amplitude for two different thread hair units oscillated in the plane of the vertical canal above 0.5 Hz. The curve is linear up to about 25° amplitude. From E.A. Campbell and P.J. Fraser, unpublished results.

higher frequencies although they do respond initially. Free hook hairs respond reliably above 0.5 Hz and follow frequencies up to 70 Hz, firing 1 spike/cycle for long periods. In general free hook hairs are vibration receptors responding to 60 to 70 Hz vibration at low amplitude, whereas thread hairs monitor large amplitude oscillations at lower frequencies. Poststimulus time histograms show that peak response of free hook hairs is in phase with peak angular acceleration. Spikes are sharply phase locked at higher frequencies (Fig. 12-7).

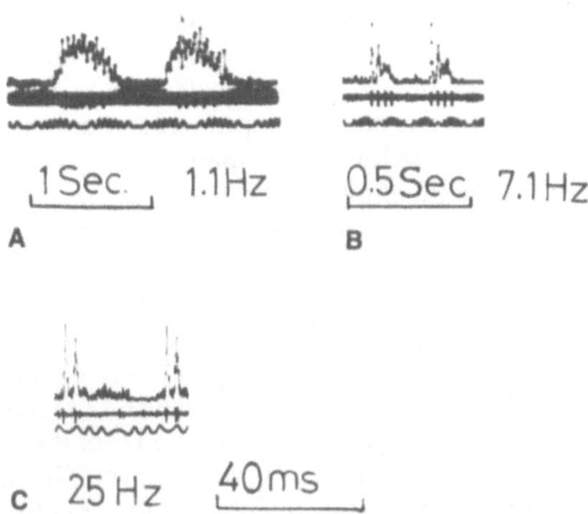

Fig. 12-7. Response from a free hook hair unit to various frequencies of stimulation. Note phase locking of spikes at higher frequencies. **A** Poststimulus time histogram, approximately 250 cycles. **B** Single free hook hair unit. **C** Photocell record of position (1° steps).

STATOCYST INTERNEURONS

Afferent nerve axons run in various branches of the antennulary nerve to the brain where they are known to synapse directly with oculomotor neurons and various interneurons, including a set of large interneurons found in all crabs so far investigated, which send axons down the esophageal connectives from the brain to the thoracic ganglia. These cells, which are premotor command cells, can be used as convenient monitors of the working of the statocyst, provided that conditions are adjusted to minimize other inputs to the cells. Additional inputs include particular combinations of leg joint proprioceptors from all walking legs and inputs from complex neuropils in the optic lobes (which in crustacea have important nonvisual functions). Each of the interneurons receives thread hair input from the group of thread hair units which responds to one particular direction of rotation in one particular plane. The units are paired (bilateral) cells and right and left homologs respond to angular displacement of homologous hairs in homologous canals. Different cells code different directions of rotation about different axes. Thus consider cell B, which receives input from the ipsilateral horizontal canal. The right cell B responds on anticlockwise rotation around the vertical axis. The left cell B responds on clockwise rotation around the vertical axis. Right and left cells A, which are the largest statocyst interneurons (with 50-μm diameter axons), respond to rotation about different axes. Each cell receives input from the contralateral vertical canal, and left and right vertical canals are mutually orthogonal (Fig. 12-4). Cells C and D each respond to rotation around a horizontal axis in a direction opposite to cell A.

PROPERTIES AND NONLINEARITIES IN THE WHOLE SYSTEM

The outputs of these various interneurons which are easily monitored can be used to illustrate the properties of the thread hair receptors and the statocyst canals. Figure 12-8 shows the response of cell A in *Scylla* during decaying oscillation on a pendulum. The cell is firing about a spontaneous level (under special circumstances, see below). The plot of instantaneous frequency against time is sinusoidal and shows a phase lead of 90° over peak displacement. Figure 12-9 illustrates the lack of adaptation and the linearity of response amplitude against oscillation amplitude over a limited amplitude range. *Scylla* exhibits a threshold of approximately 1° at about 1 Hz. (This increased sensitivity in *Scylla* correlates with the greater number of thread hairs.) The best response is to about 6 Hz oscillation frequency, but both *Scylla* and *Carcinus* show a second peak of sensitivity at about 50 or 60 Hz (Fig. 12-10). This remains after cutting the nerve from the thread hairs, but is much reduced by cutting the free hook hair nerve. This is an interesting finding, i.e., the thread hair pathways (analogous to semicircular canal pathways in vertebrates) are combined with the vibration receptor pathways in the same interneurons.

It can be shown by means of ablation experiments that the main thread

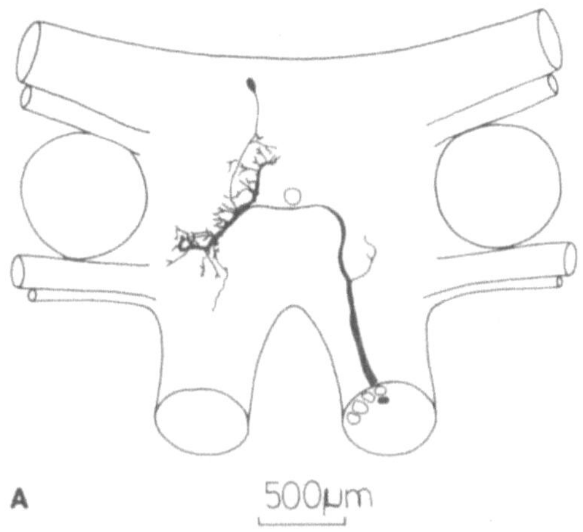

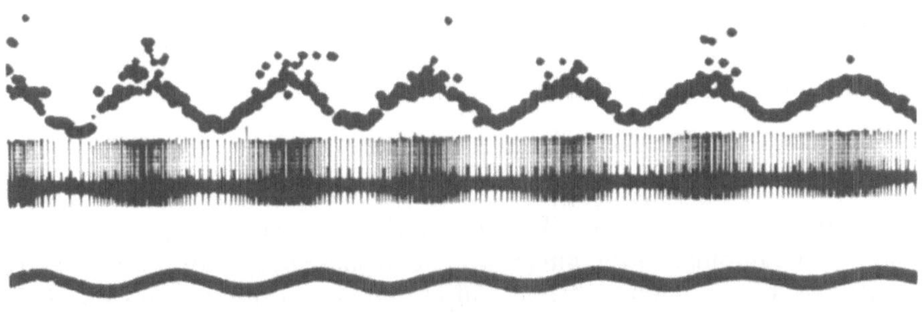

Fig. 12-8. **A** Morphology of large statocyst interneuron A in *Carcinus*. The cell sends its axon from the brain to lower ganglia which control the legs (3). **B** Response of cell A to decaying sinusoidal oscillation in the pitching plane. Oscillation frequency is 0.63 Hz. Peak-to-peak oscillation amplitude is 13° initially. Peak instantaneous frequency is about 50 spikes/s.

hair input to cells A, C, and D which respond to oscillations around horizontal axes is from the lower group of thread hairs. It is easy to see that this group will be preferentially displaced by fluid displacement in the vertical canal. As shown by Fig. 12-11, the interneurons monitor the component of angular acceleration in the plane of the canal. However, it is less easy to see how the upper group of hairs distinguishes between fluid displacement in the two canals.

From the above results we can imagine that displacement of fluid by inertial forces depending on the component of angular acceleration in the plane of one canal is monitored by the displacement-sensitive thread hairs. The small diameter of the canals should produce damping of fluid, causing fluid displacement and hence hair displacement to be more in phase with angular

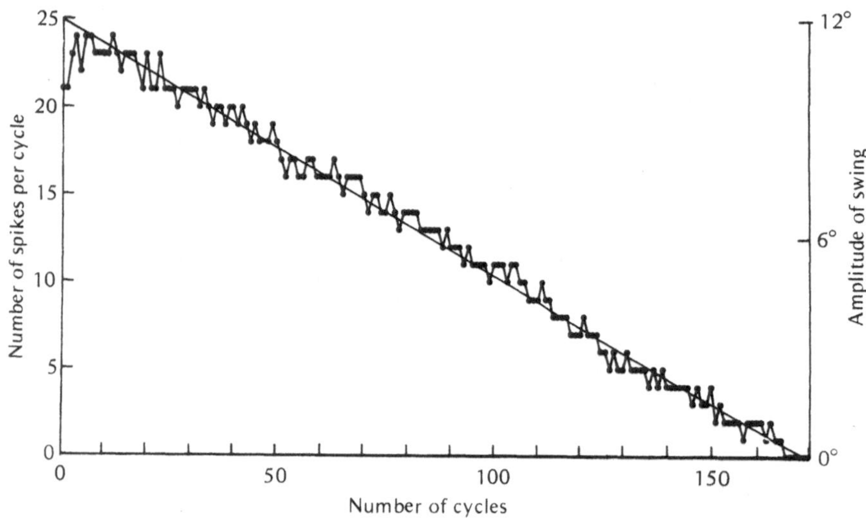

Fig. 12-9. Number of spikes/cycle from fiber 5 to a slowly decaying 1-Hz oscillation. The decay of amplitude was linear as far as could be determined by photocell monitoring and is represented by the straight line.

velocity than with angular acceleration. In the open canal system there will be less fluid damping, so the integration of angular acceleration into angular velocity will be incomplete. Open and closed canals differ in terms of the phase of peak response relative to peak position during an oscillation, but not in terms of the accuracy of abstraction of angular acceleration components. However, the real situation is more complicated than implied by the above, and the convergence with the vertebrate system may be due not so much to similarities in the mechanical system, but to differences in the properties of the displacement transducer. Qualitative observations on *Ocypode* and *Carcinus* thread hairs, and quantitative filming of the *Scylla* system through the transparent walls of an otherwise intact statocyst, have recently suggested that during oscillation, fluid displacement is quite large and the peak displacement of the hairs is closer to being in phase with angular acceleration than with angular velocity (9). The fluid is hence not damped so well as in the vertebrate system where cupula displacement is on the order of a few microns. It is obvious that the output of afferents cannot be directly proportional to the angular displacement of the thread hairs (Figs. 12-5 and 12-8).

Furthermore, the statocyst cum thread hair cum receptor nerve system as monitored by the interneurons, or by the receptor units directly, shows important nonlinearities. If a crab is oscillated about different positions, and the output to a series of oscillations monitored, then position dependence is marked, as shown in Fig. 12-12. When the output of cell A is enhanced, the

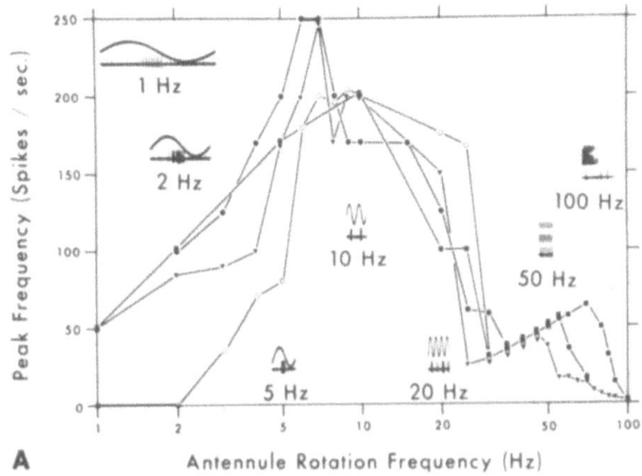

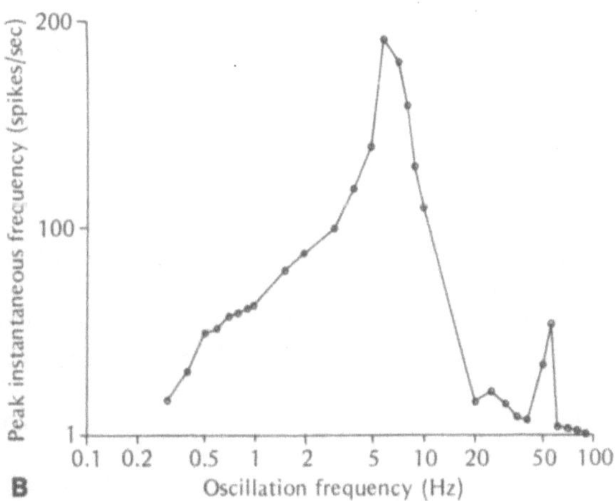

Fig. 12-10. Peak instantaneous frequency of cell A in *Scylla* (**A**) and *Carcinus* (**B**) to oscillation of the statocyst at different frequencies. Note at high frequencies oscillation amplitude was reduced. See ref. 4 for full gain curves and experimental details. **A:** From Fraser, P.J.: J. Comp. Physiol. 103:291, 1975.

output of cells C and D fed from the same vertical canal is diminished, and vice versa. Following a displacement in one direction, the units sensitive to that direction of displacement show tonic firing which increases rapidly and then declines exponentially. Following this the sensitivity of the unit to oscillation is diminished. The output of the system can be affected by relatively small changes in mean oscillation position. Consider the following experiment (4,7).

A crab, *Scylla*, was mounted on a pendulum so that the plane of one ver-

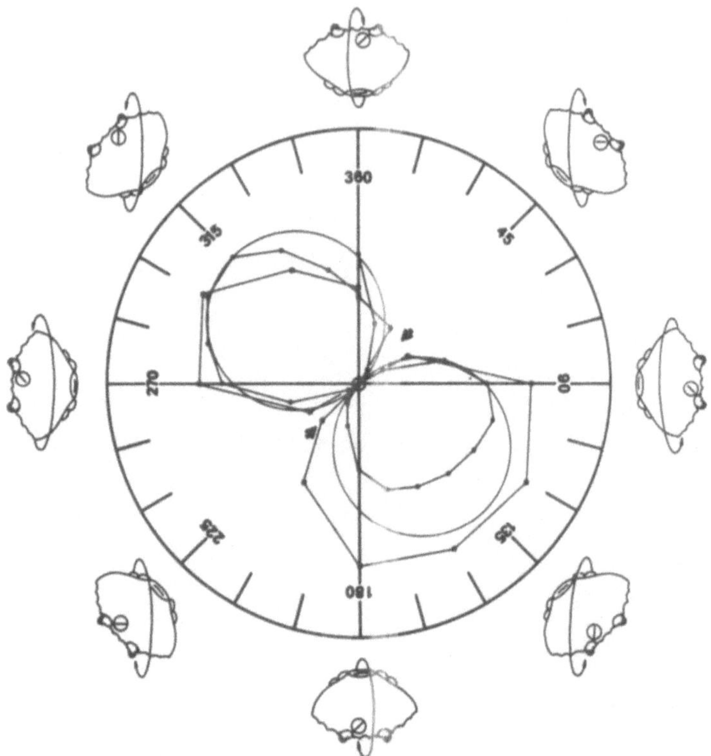

Fig. 12-11. Responses from cell A in *Scylla* which receives input from the right statocyst vertical canal to oscillation at different angles relative to the pitching plane of the crab. Distances from the center of the circle represent numbers of spikes/cycle. The circles show the cosine functions which would be expected if the cells coded the component of angular acceleration in the plane of the vertical canal. *Carcinus* curves follow the theoretical curves very well (4,6). Redrawn from Fraser, P.J. and Sandeman, D.C.: J. Comp. Physiol. 96:205, 1975.

tical canal coincided with the plane of oscillation. Initially the crab on the pendulum is displaced to an angle of 6.5°, head down. On release from this position one fiber, A, fires during the head-ascending part of the oscillation. Surprisingly, although the amplitude of oscillation is decreasing, the instantaneous frequency of firing and the number of spikes/cycle of fiber A increase for a period of about 10 s before decreasing. After a time the crab is displaced to 6.5°, head up. This evokes spikes on the head-ascending part of the cycle, and the firing level then decreases. Then the frequency increases for 1 or 2 s to a level higher than any level previously attained. This then declines exponentially over 1 or 2 min. Release from this head-up position evokes only a small response from the cell during the head-ascending part of the cycle (Fig. 12-13). Cells C and D, which fire on head-descending rotation, show exactly the same pattern of behavior but to opposite displace-

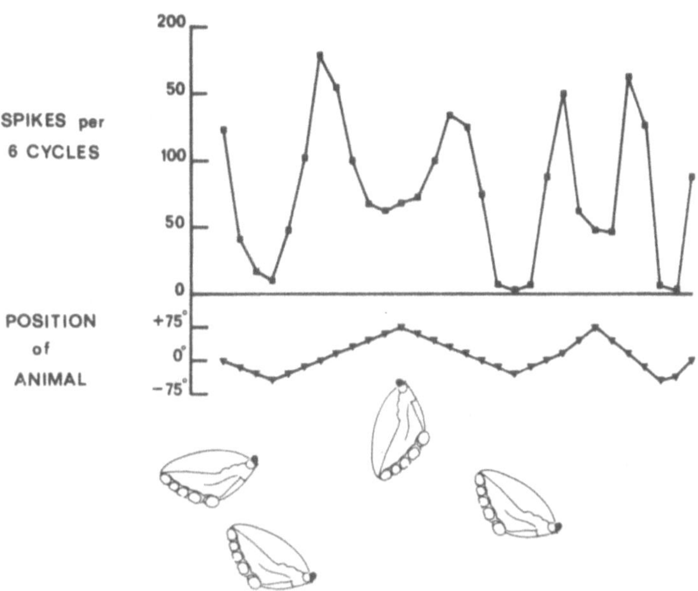

Fig. 12-12. Output of cell A in *Scylla* to a constant magnitude, small amplitude oscillation around different mean oscillation angles in the pitch plane. Optimal oscillation angle is when the crab is horizontal (position 0°). Adapted from Fraser, P.J. and Sandeman, D.C.: J. Comp. Physiol. 96:205, 1975.

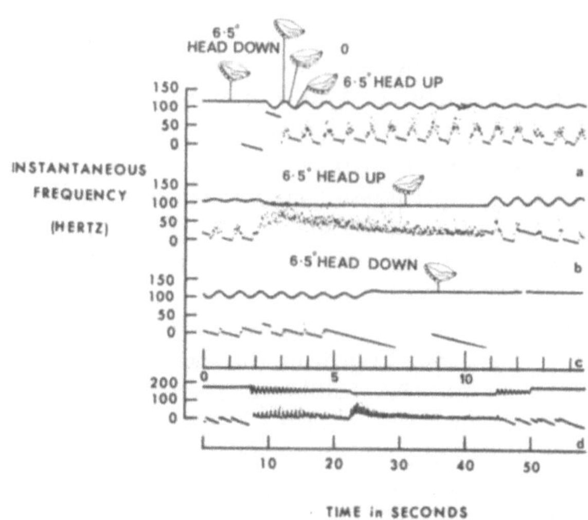

Fig. 12-13. The effect of initial displacement on the response of cell A in *Scylla* to oscillation around the horizontal axis in the plane of the vertical canal, providing input. **Top trace** Position of crab. **Lower trace** Instantaneous frequency of cell A. **a–c** Continuous recording. The crab begins in the head-down position (**a**), is allowed to oscillate and again arrested in the head-up position (**b**), allowed to swing from this position (**b** and **c**), and is arrested again in the head-down position. **d** A similar sequence on a different time scale. The ratemeter sets a voltage proportional to instantaneous frequency at each spike, and then decays slowly until the next spike.

ment. When cell A output is enhanced, output from cells C and D is diminished, and vice versa.

Three components of the thread hair system can now be recognized and can be seen at the level of receptor afferents and interneurons in *Carcinus* and *Scylla*. These are (1) a component sensitive to low frequency oscillation of fairly large amplitude; (2) a component sensitive to high frequency, small amplitude oscillation; and (3) a long duration effect which seems to work in antiphase to (1) but causes a sustained discharge following an angular displacement as described above. The detailed linkage between the two bipolar cells and the thread hair has still to be worked out. When this is known it may be possible to explain to some extent the above properties. Inspection of Fig. 12-13 reveals that the phase of peak response to peak displacement varies with time and the displacement parameters. Thus, immediately following head-down displacement fiber 5 peak response has a phase lead of 170° over peak position. This gradually reduces to about a 90° phase lead. Following head-up displacement phase lead is about 160 to 170°.

The effects above can be partly explained if the hairs show different sensitivities to oscillation about different positions, because following displacement the thread hairs will oscillate about different positions although the mean oscillation position of the crab is the same (Fig. 12-14). Following any large displacement gravitational forces will displace the hairs. Recordings from units during direct displacement of thread hairs confirms this sort of behavior and shows that in central positions the afferent neurons are position coders. In other positions they are more likely velocity coders (10).

The thread hairs are sensitive to tilts out of the plane of rotation which excite them optimally. Figure 12-15 illustrates an experimental recording from a bundle of thread hair units from the crab *Geryon* which respond to rotation around the vertical axis. Using an isolated statocyst, the turntable was rotated at constant amplitude and frequency around the plane perpendicular to the horizontal canal with this canal horizontal. The turntable was then tilted 15° alternately in either direction.

Following displacement to −15°, there was a brief high level of firing followed by a rapid decline. Following displacement to +15° there was a gradual increase in the number of spikes/cycle followed by a decrease back to a lower level than that obtained at 0°.

In summary, all crabs have a well-developed balancing organ or statocyst which has long been considered analogous to the vertebrate labyrinth. Similarities are greatest in agile tropical crabs and swimming crabs where the statocyst forms two orthogonal canals. In many temperate crabs the folding of the statocyst into canals is less complete. In the best-developed canal systems, recordings from thread hair afferent cells and identified interneurons show that each canal abstracts the component of angular acceleration in its plane and integrates it to give a signal proportional to angular velocity. No central convergence of information from separate canals has been found, although recent experiments indicate some interference between signals in horizontal and vertical canals of the same statocyst of some crabs. In

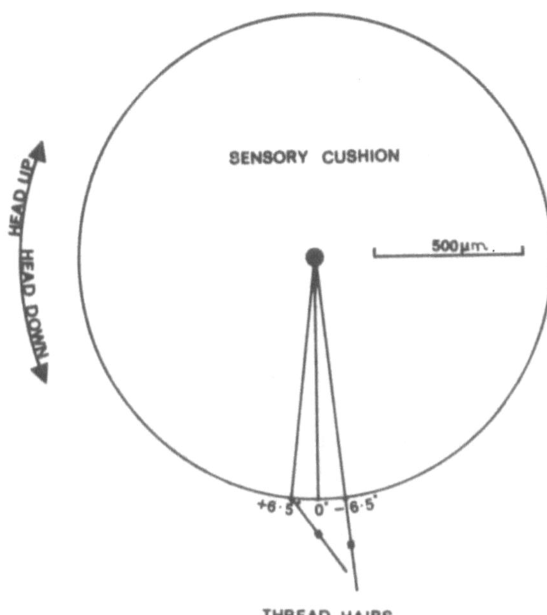

Fig. 12-14. Model to illustrate that the mean oscillation position of the thread hair will depend on the initial displacement of the crab. If the crab is displaced initially to $-6.5°$ and then released, the hair will adopt the position shown at $+6.5°$. Over the whole cycle the hair will oscillate around positions more anticlockwise as we see it than the initial position. Conversely, when displaced initially to $+6.5°$ the hair will oscillate around positions on the clockwise side of the initial position.

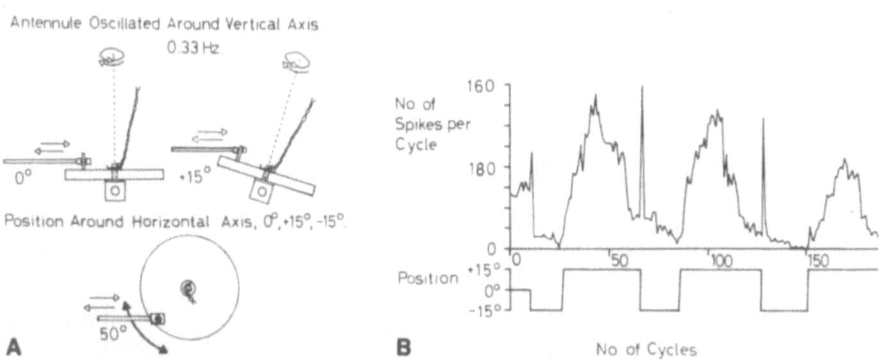

Fig. 12-15. Thread hairs are affected by displacement outwith the plane of normal sensitivity to oscillation. **A** A small bundle of thread hair units recorded from an isolated statocyst of *Geryon* during oscillation in the plane of the horizontal canal at different angles relative to a horizontal axis. **B** Output in terms of number of spikes/cycle was greatly altered. Output following positive displacement (equivalent to head up with statocyst in the whole animal) increased over a period of 30 s and then decreased.

crabs with a less complete canal system, angular acceleration components are abstracted, but there is less fluid damping in the statocyst and the signal is not fully integrated into angular velocity.

The statocyst also contains vibration receptors (free hook hairs) whose information converges on the same interneurons which carry the signal proportional to angular velocity. Positional effects on afferent sensitivity are also known.

References

1. Cohen, M.J. and Dijkgraaf, S.: Mechanoreception. In Waterman, T.H. (ed.): The Physiology of Crustacea, Vol. 2. New York, Academic Press, 1961, pp. 65–108.
2. Fraser, P.J.: Giant fibers and directional statocyst fibres: A study of interneurones between the brain and thoracic nervous system in the shore crab, *Carcinus maenas*. Ph.D. Thesis, Aberdeen University, 1973.
3. Fraser, P.J.: Interneurones in crab connectives (*Carcinus maenas*): directional statocyst fibres. J. Exp. Biol. 61:615, 1974.
4. Fraser, P.J.: Free hook hair and thread hair input to fibre 5 in the mud crab, *Scylla serrata*, during antennule rotation. J. Comp. Physiol. 103:291, 1975.
5. Fraser, P.J.: Three classes of input to a semicircular canal interneuron in the crab, *Scylla serrata*, and a possible output. J. Comp. Physiol. 104:261, 1975.
6. Fraser, P.J.: How morphology of semicircular canals affects transduction, as shown by response characteristics of statocyst interneurones in the crab *Carcinus maenas*. J. Comp. Physiol. 115:135, 1977.
7. Fraser, P.J. and Sandeman, D.C.: Effects of angular and linear accelerations on semicircular canal interneurones of the crab *Scylla serrata*. J. Comp. Physiol. 96:205, 1975.
8. Hensen, U.: Studien uber das Gehororgan der Decapoden. Z. Wiss. Zool. 13:317, 1863.
9. Janse, C. and Sandeman, D.C.: The role of the fluid-filled balance organs in the induction of phase and gain in the compensatory eye reflex of the crab *Scylla serrata*. J. Comp. Physiol. 130:95, 1979.
10. Janse, C. and Sandeman, D.C.: The significance of canal-receptor properties for the induction of phase and gain in the fluid-filled balance organs of the crab *Scylla serrata*. J. Comp. Physiol. 130:101, 1979.
11. Sandeman, D.C. and Okajima, A.: Statocyst induced eye movements in the crab, *Scylla serrata*. I. The sensory input from the statocyst. J. Exp. Biol. 57:187, 1972.
12. Silvey, G.E., Dunn, P.A., and Sandeman, D.C.: Integration between statocyst sensory neurons and oculomotor neurons in the crab, *Scylla serrata*. II. The thread hair sensory receptors. J. Comp. Physiol. 108:45, 1976.

Part 3

Vestibular Function

INVITED LECTURES

13

Functional Characteristics of Central Vestibular Neurons

WOLFGANG PRECHT

In the preceding papers the morphology of the afferent vestibular pathway from the receptor (Chap. 1) through the primary neurons to the vestibular nuclei (Chap. 2) has been summarized, and a review of the functional properties of the primary afferents has been given (Chap. 11). Since, with few exceptions, e.g., the direct projection to the cerebellum (62), vestibular afferents synapse in the vestibular nuclei, knowledge of the response properties of vestibular secondary and higher order neurons is an important prerequisite in the understanding of the information processing occurring between receptors and effectors. In this paper the functional properties of the neurons in the vestibular nuclei will be reviewed with emphasis on their response to vestibular (canal and otolith) stimulation. As far as data allow a comparative physiologic approach will be taken describing the vestibular neuron properties of frog, cat, and monkey.

Of the other sensory systems having access to the vestibular nuclei, such as somatosensory and visual systems, the visual or optokinetic influences have been most extensively studied. Since visual-vestibular interaction is of great importance for performance of vestibular neurons under conditions where the vestibular input is not sufficient to assure proper reflex movements, the main results of this work will be summarized. The behavioral and psychophysical aspects of visual-vestibular interaction will be treated in Chap. 24.

FUNCTIONAL SYNAPTOLOGY AND TOPOGRAPHY OF VESTIBULAR PROJECTION TO THE VESTIBULAR NUCLEI

Ipsilateral Input

Stimulation of the whole VIIIth nerve with brief electric pulses evokes in the vestibular nuclei of cat (73), frog (65), monkey (39), and rabbit (4) typical field potentials that serve as precise guides for electrode location within the vestibular nuclei during physiologic experiments. When single units are recorded, correlation of their response latencies with the various components of the field potential may be used for identification of the units as pre- or postsynaptic (mono- and polysynaptic) in nature (39,65,66).

As regards the termination sites of the various primary afferents in the vestibular nuclei, the results obtained with electric or natural stimulation of *cat* semicircular canals and otolith organs (12,31,50,58,66,72) agree well with those obtained from anatomic work (24). Fibers innervating the cristae of the *semicircular canals* terminate mainly in the superior and rostral part of the medial and descending vestibular nuclei as well as the interstitial nucleus of the VIIIth nerve; whereas *utricular* fibers end, for the most part, in the rostral descending and medial nuclei and rostroventral parts of the lateral nuclei; *saccular* afferents terminate mainly in cell group "y" and also in lateral and descending nuclei. With respect to termination of canal afferents in the *monkey*, both anatomic (76) and physiologic (1,2,22,39) studies have revealed distributions that appear to be similar to those described in the cat and rabbit at least as far as the canal representation is concerned. The anatomy and physiology of otolith projections are not known in the monkey. Finally, in the *frog* the various subdivisions of the vestibular nuclei described in mammals are too indistinct to allow a precise anatomic-functional correlation (25,65).

Anatomic studies (24,78) demonstrated that various areas in the vestibular nuclei did not receive primary vestibular afferents. The functional correlate to this finding is the exclusively polysynaptic activation (Fig. 13-1, neuron Ib) of many tonically active vestibular neurons after VIIIth nerve stimulation (66). Part of these neurons are located in areas that are conspicuously free of primary afferents; but others are intermingled with those that possess afferent input. Besides the pure polysynaptic (higher order neuron; Fig. 13-1, unit Ib) and pure monosynaptic units (second order neuron; Fig. 13-1, unit Ia), other neurons were found in the horizontal canal system: type I neurons which fired with polysynaptic latencies after weak stimuli and had additional monosynaptic discharge with strong stimulation (predominantly higher order), and neurons which fired monosynaptically and polysynap-

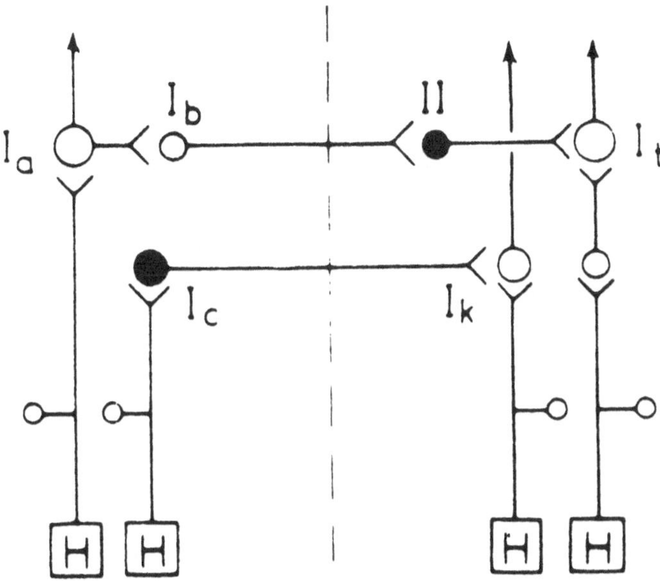

Fig. 13-1. Diagram showing neuronal circuitry of commissural vestibular pathways. Abbreviations: I and II, type I and II vestibular neuron; H., horizontal canal; filled circles, inhibitory neurons; open circles, excitatory neurons.

tically after weak and strong stimulation, respectively (predominantly second order units). Finally, some neurons located in the vestibular nuclei were not driven at all. Similar connections may exist in vertical canal (50) and otolith projections (58). Polysynaptic responses are, at least in part, mediated by intranuclear interneurons and/or axon collaterals of directly driven cells (for function, see below).

Intracellular studies of cat vestibular neurons show that the vestibular input may generate mono- or polysynaptic EPSPs or both (32,37). The double firing seen in many vestibular neurons after VIIIth nerve stimulation (66) is thus caused by a sequence of mono- and polysynaptic EPSPs. The synaptic and action currents generated by these EPSPs produce the N_1 and N_2 field potentials typically recorded in the vestibular nuclei (72). It is interesting to note that the unitary components of the vestibular EPSPs recorded in different vestibular neurons vary. Thus, non-Deiters' neurons have significantly larger unitary EPSPs as compared to Deiters' neurons. This would suggest that primary vestibular impulses activate some vestibular neurons more effectively than others. In general, however, the transmission from primary afferents is very powerful: many neurons can follow double stimuli with intervals of less than 1 ms with full action potentials. Also of interest is the fact that the earliest latencies of vestibular-evoked EPSPs and action potentials measure 0.6 and 0.7 ms, respectively. Occasionally, polysynaptic

IPSPs have also been evoked in vestibular neurons. They may be generated by collaterals of inhibitory secondary neurons.

In the frog (65), fish (42), pigeon (80), and rat (43) there is convincing evidence for electrotonic coupling between vestibular neurons and/or electric transmission between primary afferents and central neurons. These electric events have not been demonstrated in cats. Frog vestibular neurons are also different from their mammalian counterparts in that they frequently show partial action potentials or dendritic spikes after ipsi- or contralateral VIIIth nerve stimulation (16,57,65). The functional implications of these species differences in the properties of vestibular neurons are presently not well understood.

Contralateral Input

The crossed connections between the bilateral vestibular nuclei have been studied in the cat (35,36,46,49,64,73,74), frog (16,57), reptiles (68), and monkey (2). The results of these studies will be briefly summarized. In the *cat* the commissural neurons may be excitatory (Fig. 13-1, unit Ib) as well as inhibitory in nature (Fig. 13-1, unit Ic). Crossed excitation is found in neurons (type II) that mediate inhibition of higher order canal neurons (Fig. 13-1, unit It). On the other hand, crossed inhibitory fibers (Fig. 13-1, unit Ic) contact true second order neurons (Fig. 13-1, unit Ik). Thus, in the canal system the inhibitory type II neurons are excited and, in turn, inhibit the main sensory type I neurons as shown in detail in the circuit diagram of Fig. 13-1. The latter, however, may also be inhibited directly by commissural fibers (Fig. 13-1). It was found that this crossed inhibition exists only between functionally synergistic pairs of canals, i.e., bilateral horizontal canals, left anterior and right posterior and right anterior and left posterior canals.

It appears that crossed inhibition is specific to canal neurons; central otolith neurons are always excited by contralateral stimulation (64,74). Comparative physiologic studies apparently showed that frogs have a very poorly developed commissural inhibition, and instead show mainly crossed excitation (57). In reptiles, however, crossed inhibition was frequently found (68). It has been hypothesized that absence of commissural inhibition of canal neurons in the frog may be related to differences in the organization of eye movements (57). Some of the functional aspects of crossed pathways will be discussed in the following sections.

RESPONSES OF VESTIBULAR NEURONS TO NATURAL LABYRINTHINE STIMULATION

Since the pioneering work of Adrian (3), who was the first to record from central vestibular neurons during natural vestibular stimulation, a great number of studies have confirmed and substantially expanded his work. I have reviewed these studies in various monographs and shall, therefore, only briefly summarize the major points (59–63).

Responses to Semicircular Canal Input

When a functionally synergistic pair of canals is brought into the plane of rotation of a turntable, two major types of responses may be recorded from vestibular neurons of various species. In the horizontal canal system they have been designated type I and type II by Duensing and Schaefer (18). They appear to have approximately the same frequency of occurrence. Similarly, in the vertical canal system two major groups of neuronal responses were found (19,50). In addition, two other types of canal responses (types III and IV), implying bilateral excitation and inhibition, respectively, were found but they occurred much less frequently than type I and II responses (18,19,72) and are probably generated by canal-canal convergence. It became clear that horizontal type I neurons were the true second order (in the case of monosynaptic activation) or higher order (polysynaptic vestibular activation) neurons, i.e., had the same response polarities as the primary afferents of the corresponding ipsilateral canal in any given head position in space. On the other hand, type II neurons were demonstrated to be the inhibitory neurons receiving their input from the contralateral horizontal canal and projecting onto type I neurons on the same side (Fig. 13-1). Essentially similar results as in cat and rabbit were obtained in recordings from frog vestibular neurons (61), except that few type II neurons were encountered in this species (10% only versus approximately 40 to 50% in cat and monkey). They were, for the most part, derived from co-stimulation of the vertical canals. Since, as noted in the preceding section, frogs have a very poorly developed commissural inhibition, the dearth of inhibitory type II neurons seems reasonable.

Responses to Constant Angular Accelerations

In the time-domain studies which have been performed in vestibular neurons of the horizontal canal system of the lightly anesthetized rabbit (18), decerebrate cat (72,75), unanesthetized frog (61), and anesthetized monkey (2), the following major results have emerged (Table 13-1).

The mean *resting rates* of central vestibular neurons vary among species, being low (< 10 impulses/s) in the frog and relatively high in the alert monkey (22) and cat (41) (approximately 40 and 70 impulses/s in cat and monkey, respectively). Also, there are fewer silent neurons in the monkey than in the cat. When compared with the resting rates found in primary afferents (62), these values appear to be slightly smaller in central neurons.

Recent work on the postnatal development of vestibular neuron responses in the rat (47) has shown that the resting rate is very low and irregular at birth and then gradually increases in magnitude, reaching practically adult

Table 13-1.
RESPONSE CHARACTERISTICS OF CENTRAL TYPE I UNITS.

	Cat	Monkey	Frog
Resting rate (impulses/s)	40	70	<10
Threshold (°/s²)	<0.5	<0.5	<0.5
Input/output	nonlinear/linear	nonlinear/linear	nonlinear/linear
Adaptation	−/+	−/+	−/+
Sensitivity (step)	6(2.2) Sp/s	−0.3(−0.85)	4–6 (5–6)
Log. gain ≈ 0.25 Hz	−0.1(−0.4) per °/s²		±0.1 (0)
Time constant (s) (step)	3.7/8.1		
	3.0/7.2 (decerebrate)		3
(sinusoid)	3.0/11 (alert)	9–24	
Phase lag re. acceleration	55–60°/80–90° (decerebrate)		
	55–60°/105° (alert)	50–60°/120–170°	65°

levels by the end of the first postnatal month. During this time period the synapses between primary afferents and central vestibular neurons undergo maturational changes (34) and could probably not transmit very effectively activity arriving via primary afferents, even if the latter already had obtained the final level of resting rate. This, however, is not the case; primary afferent discharge shows a very similar progression toward higher and more regular values as do central neurons (13), and receptor-afferent synapses mature over about the same period of time (28).

The *thresholds* for frequency increase in all species are below 0.5°/s^2, and there exist units with linear or nonlinear *acceleration-frequency* increase/decrease (input-output) relationships. In general, the longer the acceleration lasts the more nonlinear the responses become, even in linear units. As with the resting rate, threshold values of rat secondary neurons undergo large changes from very high values at birth to essentially normal values by the end of the first month.

The distribution of *time constants* of the responses to constant angular acceleration in the cat is bimodal, showing short (approximately 3 s) and long (approximately 7 s) time constant groups; they have been called tonic and kinetic units, respectively, and were shown to have predominantly mono- and polysynaptic primary afferent connections, respectively (66). Frog vestibular neurons have mean time constants of approximately 3 s. All species have *adapting* as well as *nonadapting* units when tested with constant angular accelerations of long duration.

The mean *sensitivity* factor measured in cat central neurons is approximately 6 spikes/s per degree/s^2, which is significantly higher than the values found in primary afferents (9). Slightly higher values were found in the frog (Table 13-1). The sensitivity is also a function of time after birth and continues to increase even after the end of the first month.

In comparing the responses in the time domain of primary and secondary canal neurons, two findings deserve particular emphasis. Firstly, there is a group of secondary neurons with long time constants which is not encountered in the VIIIth nerve. This indicates that central mechanisms involving the neural integrator postulated by Robinson (69) modify the peripheral vestibular signal present in a particular group of central neurons. Secondly, the fact that the acceleration sensitivity of secondary neurons is much higher than that of the afferents has been shown to be due to commissural inhibition which increases the sensitivity of central neurons (2,52,72). As will be shown below, in animals with no commissural inhibition the sensitivity of primary afferents is much higher than in animals which have it (Table 13-1). Besides increasing the sensitivity of vestibular neurons, commissural inhibition introduces by way of saturation of disinhibitory effects nonlinearities in the input-output relationship (2). Finally, in chronically hemilabyrinthectomized cats (67) and frogs (17), commissural inhibition increases and assures sensitive type I responses on the deafferented side.

Responses to Sinusoidal Rotation

In the *frequency domain*, central vestibular neurons of the horizontal canals of frog (61); cat anesthetized (55), decerebrate (75), and alert (41); anesthetized rat (44), and alert monkey (11,22,39) have been measured. A summary of the mean phase and gain values of type I neurons obtained in these species is presented in Table 13-1 and Fig. 13-2. With respect to *phase*, secondary neurons of all three species have mean phases that are slightly larger than those of primary afferents; frog's phase values are slightly smaller than those of cat and monkey; and the shapes of the phase curves are similar in each species except that the values in the monkey diverge from the others at lower frequencies. The *gain* values increase from primary to secondary neurons in the cat and monkey and remain the same or slightly decrease in the frog. Primary afferents in the frog have a much higher gain than cat and monkey afferents; this high peripheral gain may be required to compensate for the lack of commissural inhibition in this species which, as mentioned above, increases the sensitivity of central neurons.

Interestingly, primary canal afferents of the pigeon have a similar high gain to that of frog afferents (Chap. 16). Whether this high gain likewise correlates with lack of commissural inhibition is not known.

The data obtained in *alert* cat (41) and monkey (22,39) deserve a closer inspection. In the cat (Figs. 13-3–13-5) as well as in the monkey, it was noted that central vestibular units may be divided into two groups: (1) a short time constant (mean of 3 s in cat), small phase lag (approximately 50 to 60° relative acceleration) group of units; and (2) a long time constant (11 s), large phase lag (approximately 105° in cat and 120 to 170° in monkey) group. It has also been shown that the second group receives disynaptic vestibular input and, therefore, is higher than second order, whereas the short time constant group is monosynaptically driven and truly second order. Correlation of the neural firing with eye movement revealed that the large phase lag units in the monkey had burst-tonic character very much like eye motoneurons and were, therefore, related to eye position. No such correlation was found in the small phase lag group; units, however, often paused during saccades. In the alert cat the large phase lag group was not burst-tonic in nature, but rather showed a weak eye position relationship also observed in some neurons in the monkey. It must be concluded that the large phase lag units receive signals other than those present in primary canal afferents. They presumably receive re-afference from the reticular integrator and contain a partially integrated signal, so that these phases tend more toward eye position than those of small phase lag units. It is clear, however, that the phases of even the large phase lag group (Fig. 13-5, filled circles) are still considerably different from those of abducens motoneurons (Fig. 13-5, dotted circles). Other neurons must provide the additional phase lag so that motoneurons receive an eye position signal.

In Fig. 13-4 a comparison of type I and II phases and gains of the alert cat

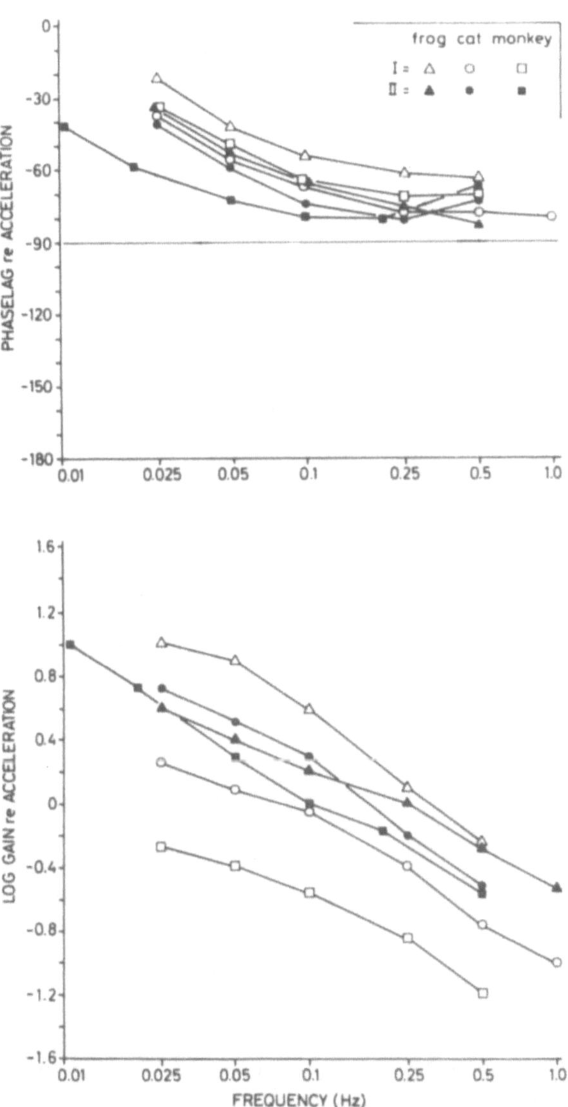

Fig. 13-2. Phase and gain of primary and secondary type I vestibular neurons of different species. Frog data are from Blanks and Precht (1976) and unpublished work; cat central and peripheral data were plotted from the work of Shinoda and Yoshida (1974) and Anderson et al. (1978), respectively; monkey peripheral and central data are from Fernandez and Goldberg (1971) and Buettner et al. (1979).

is given at one frequency (0.25 Hz). It can be seen that contrary to type I the phases of type II are unimodal and closely distributed around 95°, i.e., phases are similar to those of the large phase lag type I neurons. It should be remembered that type II neurons are inhibitory neurons mediating crossed inhibition to type I (Fig. 13-1), thereby enhancing the sensitivity of the lat-

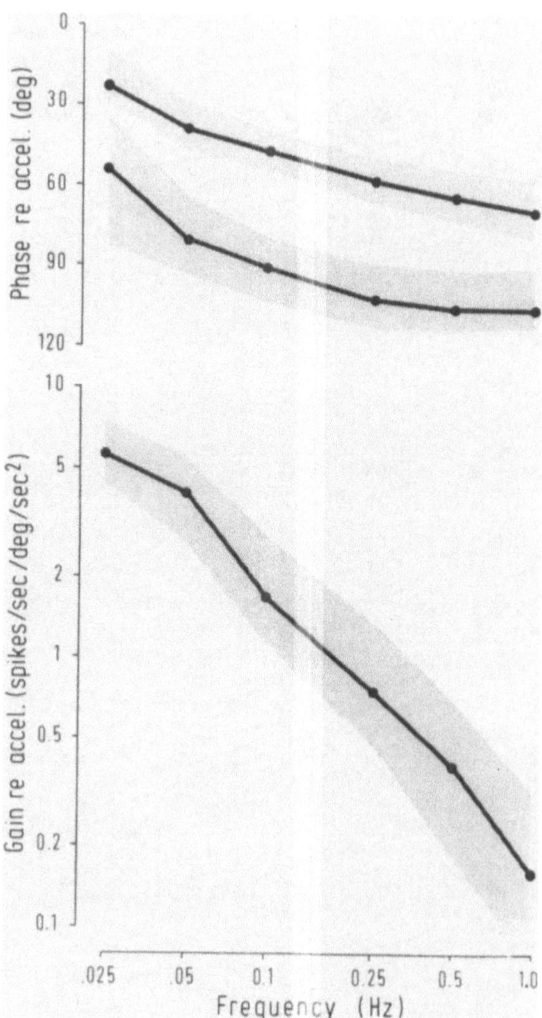

Fig. 13-3. Phase and gain values of central type I vestibular neurons in the alert cat. (Keller and Precht, 1979)

ter. In regard to the vestibular nerve input they are, of course, also higher order neurons.

Responses to Otolithic Input

Responses of central vestibular neurons to natural stimulation of the otoliths have been studied in rabbit (19), cat (8,23,29,53,54,58,71,74), and frog (61).

The major results obtained in these studies may be briefly summarized as follows:

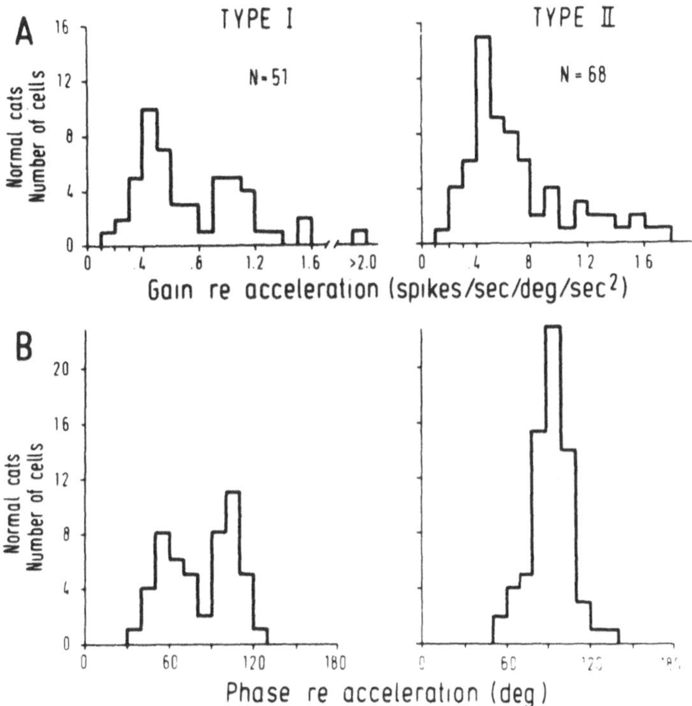

Fig. 13-4. Frequency distribution of phase and gain of type I and type II vestibular neurons of the alert cat at 0.25 Hz. (Keller and Precht, 1979)

1) Neurons responding to *static lateral tilt* (mainly utricular stimulation) have been divided into α- and β-neurons that are excited by tilting the ipsilateral or contralateral side down, respectively; their relative occurrence in the nuclei corresponds to that in the nerve (two-thirds α- and one-third β-responses) and reflects the polarization pattern of utricular hair cells. In addition, higher order γ- and δ-responses were noted which were excited or inhibited, respectively, by tilt in either direction. In general, α- and β-responses were driven monosynaptically by stimulation of the VIIIth nerve (true secondary neurons), whereas γ- and δ-responses (higher order neurons) were polysynaptically excited or not driven at all. The location of mono- and polysynaptic responses in the vestibular nuclei is in good agreement with the topography of otolith projections (see above). Also, monosynaptically driven cells were more sensitive than the polysynaptic units. However, in unanesthetized animals polysynaptic units may also be very sensitive. It has been shown that α- and γ-units project preferentially to the cervical and lumbar spinal cord, respectively.

2) According to the *time course* of the responses, neurons may be divided into *tonic, phasic-tonic,* and *phasic* units. Whereas in mammals tonic and phasic-tonic responses predominate, mainly phasic and phasic-tonic units

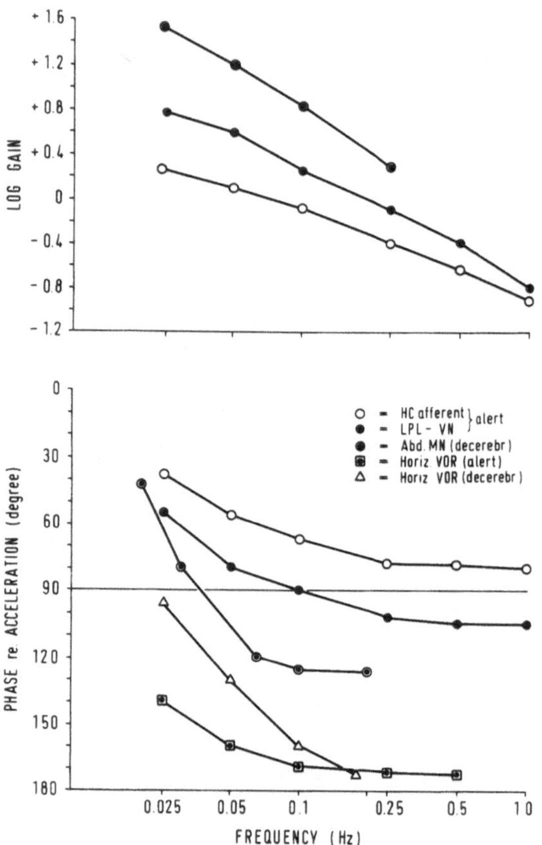

Fig. 13-5. Phase and gain values of cat primary horizontal canal afferents (Anderson et al., 1978), large phase lag type I units (Keller, Precht, 1979), abducens motoneurons (Shinoda, Yoshida, 1974), and horizontal VOR in decerebrate (Shinoda, Yoshida, 1974) and alert (Keller, Precht, unpubl. observ.) preparations. Note differences between decerebrate and alert condition.

are found in amphibians. Anesthesia and cerebellectomy appear to reduce the frequency of occurrence of phasic responses (56). In many units a given static head position was associated with different firing rates (multivaluedness), when tested at various times during a complete tilt protocol.

3) That the central otolith units have true *dynamic responses* has been shown in cats with all semicircular canals rendered nonfunctional. In this situation many cells showed greater gains as the frequency of sinusoidal tilt was increased. Frequency analysis of vestibular units with sinusoidally varying linear motion has shown that central responses differ from those recorded in primary afferents in that a frequency-dependent phase lag is introduced centrally. As in the case of canal neurons, these phase changes may result from the fact that primary otolith signals are passed through a leaky integrator (Chaps. 26 and 27).

VESTIBULAR RECEPTOR CONVERGENCE ON VESTIBULAR NEURONS

Canal-Canal Convergence

Electrophysiologic studies employing electric stimulation of individual canal nerves in the cat (36,51) and pigeon (79) revealed very little short latency convergence of ipsilateral (orthogonal) canals on central vestibular neurons. This finding indicates that there is selective projection of canal fibers on to vestibular neurons; and if there is significant orthogonal convergence under the condition of natural stimulation it can only be due to activation of polysynaptic paths, which are not activated by single shock stimulation and/or depressed by anesthesia. It should be remembered that coplanar (ipsilateral-contralateral) canal pairs converge via the vestibular commissure (Fig. 13-1). When natural stimulation was employed in the search for convergence in cat (14), rabbit (19), frog (61), and monkey (2), not only was coplanar canal convergence confirmed but also ipsi- and contralateral orthogonal canal convergence was often observed, in addition, of course, to pure single canal responses (approximately one-third). In fact, type III and IV responses may result from such interaction. Since there exists no significant canal-canal convergence at the receptor-primary afferent level the observed convergence must occur centrally. Whereas the function of in-plane canal-canal convergence is to increase the sensitivity of central neurons (see above), orthogonal canal convergence may be related to the fact that canal responses are a function of the cosine of the angle between the canal plane and plane of rotation. Such a relationship yields only small differences in response magnitude over a wide range of head positions and may require information from orthogonal canals to enable precise determination of plane of rotation.

Canal-Otolith Convergence

Of the horizontal type I and II vestibular neurons, about 50% responded also to otolith stimuli which consisted of constant velocity rotations or static head displacements in pitch and roll (14). Frequent occurrence of canal-otolith convergence was also reported in rabbits (19), frog (61), and rat (45). In a more recent study we recorded from a rather selected group of vestibular neurons, namely those projecting to the trochlear nuclei, during natural canal or otolith stimulation (8). It was found that approximately 50% of these vestibular axons were carrying both vertical canal and otolith signals. This finding implies that, e.g., in rotation in roll, which activates canals and otoliths, the rotating gravity vector is able to change the frequency response of the units in such a way as to compensate for the poor canal performance in the low frequency range.

VISUAL-VESTIBULAR CONVERGENCE IN VESTIBULAR NEURONS

In the preceding sections the responses of central vestibular neurons to vestibular stimulation in the dark have been described. From this summary it became apparent that their performance to low frequency stimulation (Fig. 13-5) is very poor and that the eye position signal required for proper functioning of the vestibuloocular reflex (VOR) cannot be provided by the semicircular canals alone. This deficiency is more pronounced in the cat than in the monkey (63). It should be remembered, however, that under biologic conditions, i.e., rotation against a stable visual world, optokinetic reflex mechanisms help the vestibular system in the low frequency or better, low velocity, range. This interaction of vestibular and visual systems results in a perfect VOR (i.e., 180° phase lag relative acceleration) over the whole frequency range.

Now, the question arises, where does this interaction occur? Recently it has been shown that visual-vestibular interaction occurs first at the level of central vestibular neurons, as demonstrated in fish (5), rabbit (15), cat (40), rat (11a), and monkey (27,77). At least in higher vertebrates there is no evidence for visual-vestibular interaction at the level of primary afferents (10,38). Figure 13-6 shows that the instantaneous firing rate of a semicircular canal afferent is modulated only during vestibular stimulation (Fig.

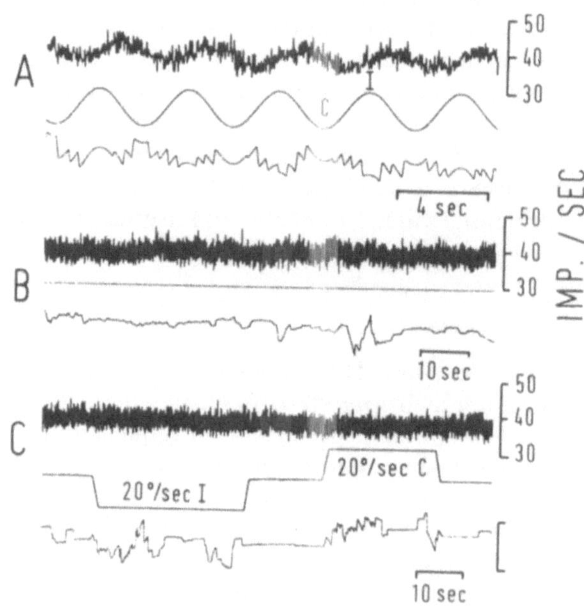

Fig. 13-6. **A–C** The activity of a single horizontal canal afferent during vestibular and optokinetic stimulation. The upper record is the instantaneous spike frequency; the middle trace, stimulus conditions; the lower trace, eye position. **A** Horizontal angular acceleration of the head (0.25 Hz, ±10°) (I, peak ipsilateral acceleration; C, peak contralateral acceleration). **B** Spontaneous resting activity in the absence of vestibular or optokinetic stimulation. **C** Unit response to optokinetic stimulation (20°/sec) to the ipsi- (I) and contralateral (C) side. Eye movement calibration is 15°. (From Blanks and Precht, 1978)

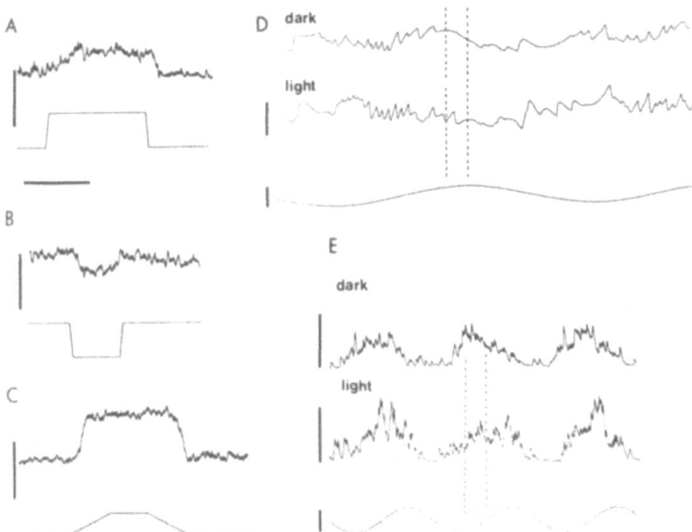

Fig. 13-7. **A** and **B** Response of a central vestibular neuron in cerebellectomized cat to pure visual stimuli. Smoothed unit discharge frequency shown in upper traces. Lower traces show velocity profile of visual pattern rotation (to the left in A and to the right in B). Constant velocity of 30°/sec in each. **C** Sustained response of another Vn (upper trace) to a velocity trapezoid of head rotation (lower trace), constant velocity of 60°/sec, normally lighted room. **D** Horizontal vestibuloocular reflex in cerebellectomized animal in response to sinusoidal head displacement (lower trace) (movement to the right is down). Upper trace shows compensatory eye movements in dark and middle trace with vision (movement to the right is up in both). Vertical dashed lines show the phase shift in compensatory eye movements that occurs from the case of rotation with vision to the dark. **E** Similar phase shifts occurring in Vn under light and dark conditions. Upper traces show smoothed discharge frequency. Lower trace shows head displacement. Vertical dashed lines show the trend from a unit response lagging maximum ipsilateral head velocity with vision to a response leading head velocity in dark. Time calibration 25 s except for D where it is 5 s. Unit frequency calibrations are 50 spikes/s with the foot of the mark at zero. Head displacement calibration 200° in D and E. Eye movement calibration is 20° for both traces in D. (Keller and Precht, 1978)

13-6A) and is unrelated to either spontaneous eye movements (Fig. 13-6B) or optokinetic stimuli (Fig. 13-6C).

Very much in contrast to the lack of any optokinetically induced changes in firing rate of primary afferents, secondary neurons display sustained responses to constant velocity head rotations (Fig. 13.7), and moreover respond in a direction-specific fashion to constant rotation of a large-field pattern in the absence of any vestibular stimulation (Fig. 13-7A and B). The input-output relationship between optokinetic drum velocity and increase in unit firing is shown in Fig. 13-8. It can be seen that in the paralyzed animal (optokinetic stimulus velocity = retinal slip velocity), the mean peak of fir-

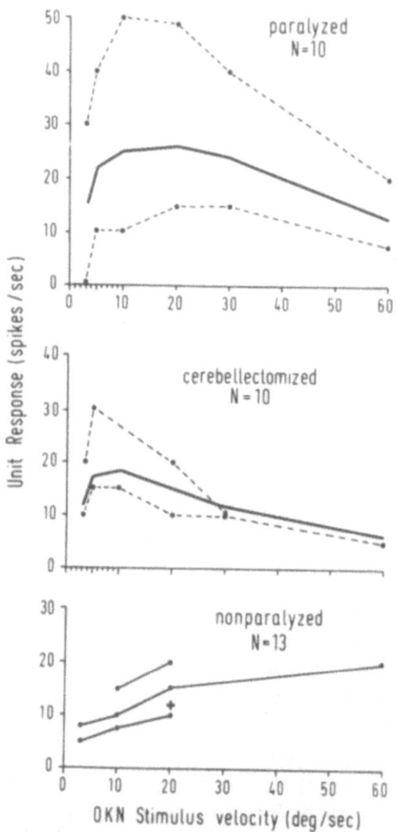

Fig. 13-8. Input-output relationship between optokinetic drum velocity and firing of central vestibular neurons in paralyzed, cerebellectomized and paralyzed and nonparalyzed cat. (Keller and Precht, unpubl. observ.)

ing occurs around 10°/s; lower and higher velocities give smaller responses. In the rat (11a) the peak occurs at about 1 to 2°/s, indicating large species differences. When the response of a given central unit is first studied with slow sinusoidal rotation in the dark and then in the light (against a stable world), clear shifts in phase and changes in sensitivity may be observed (Fig. 13-7E). Similar shifts are observed in eye position during vestibular stimulation in dark and light (Fig. 13-7D). The visual-induced changes in vestibular unit firing are found in almost all type I and type II horizontal canal neurons. In type I neurons firing increases and decreases on ipsilateral and contralateral table rotation, respectively, whereas optokinetic stimuli increase and decrease type I firing on contra- and ipsilateral rotation, respectively. In type II neurons the interaction is likewise synergistic, i.e., vestibular and visual stimuli leading to excitation or depression of firing cause eye movements in the same direction. Vision, therefore, will improve the vestibular performance, particularly at low velocities or during constant velocity of rotation. In the latter case unitary firing governed solely by semicircular canals will decay to resting levels, whereas with simultaneous presence of structured surround no such decrease occurs (Fig. 13-7C). In general, there is a linear relationship between unit firing and optokinetic nystagmus (OKN) or optokinetic afternystagmus (77) (OKAN) slow phase velocity. However, there are also several conditions in which dissociation between the two events occurred. This latter finding suggests that the vestibular nuclei, alone, do not control OKN. Finally, it should be emphasized that the above visual effects are, of course, initiated by retinal slip; but the signal present in the vestibular nuclei is most likely a kind of motor corollary signal probably related to an internal surround velocity signal.

At present very little is known regarding the pathways mediating the optokinetic effects onto vestibular neurons. Recent work in cerebellectomized cats (Fig. 13-8) indicates that the cerebellum is not necessary either for the optokinetic response (Figs. 13-7 and 13-8) or for the visual-vestibular interaction in vestibular neurons (Fig. 13-7A–C and E) and VOR (Fig. 13-7D), at least in the cat (40). It should be noted, however, that cerebellectomy had some effect on the mean peak discharge at all stimulus velocities (Fig. 13-8, middle graph). Cerebellectomy may have different effects in other species (26,33).

NECK PROPRIOCEPTIVE INFLUENCES ON VESTIBULAR NEURONS

In all animals except fish the head is free to move with respect to the body over a considerable range, and the labyrinths therefore signal head rather than body position. To monitor the relative position of head and body somatosensory-vestibular convergence is required. An important part of the somatosensory information arises in the neck between C_1 and C_4 (48). Natural stimulation of these neck receptors evokes the well-known tonic neck reflexes on the limbs as well as the eyes.

It is reasonable to assume that the vestibular nuclei, which are the first important central relay for vestibular information arising in the inner ear, are likewise informed by neck receptors on body-head position. Experimental evidence supporting this notion has recently been accumulated in several studies in the cat (21,30,70). As for nuchovestibular interaction in vestibuloocular reflex mechanisms, a synergistic interaction seems to occur in the vestibular neurons (30). Figure 13-9 illustrates the most simple pathways that may be involved. Inhibitory and excitatory type I neurons are excited by contralateral neck stimulation resulting in conjugate eye movements to the side of neck stimulation. Thus, under natural conditions of horizontal head rotation to the right, type I units in the right vestibular nucleus are activated by both labyrinthine stimulation (right) and probably stretching of left neck joints or muscles. Thoden (personal communication) has indeed shown that 90% of the vestibular-neck convergent neurons are synergistically modulated. The fact that he recorded mainly in the superior and medi-

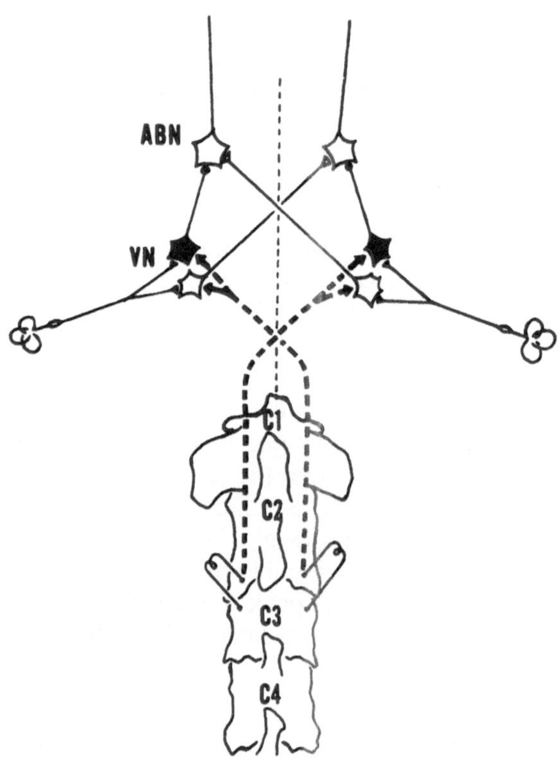

Fig. 13-9. Schematic of simplified pathways from neck joints to abducens motoneurons and their interaction with the vestibuloabducens reflex arc. Thick broken lines indicate cervical afferent pathways which converge on secondary vestibular neurons and facilitate them. *ABN*, abducens nucleus; *VN*, vestibular nucleus. Inhibitory neurons are filled in black and excitatory neurons are unfilled. To simplify the drawing commissural inhibitory pathways between right and left vestibular nuclei are not shown. From Hikosaka, O. and Maeda, M.: Exp. Brain Res. 18:512, 1973.

al nuclei where canal-ocular relays are located may explain the discrepancy with the results of Rubin et al. (70). The horizontal body against head movements, for the most part, activate vestibular neurons phasically; tonic neck position is only monitored by some type I or type II units (70). The phase of the response has its peak conjointly with neck velocity. A similar velocity signal was found in vestibular neurons during pure canal stimulation (see above).

Recent experiments in decerebrate cat (Thoden, personal communication) indicate that abducens motoneurons are activated by contralateral body movement (against the fixed head). One group of neurons was activated in phase with contralateral velocity, the other one with contralateral acceleration. Although these results seem to be in partial agreement with the data obtained in vestibular neurons presumably projecting to ocular motor nuclei (see above), they do not seem to fully support recent eye movement recordings in man suggesting that cervicoocular reflexes (COR) are antagonistic to the VOR (7). Recordings of COR are required in the cat to determine its direction and to allow further correlation with single unit work.

In addition to the predominantly synergistic interaction between neck and vestibular afferents at the level of vestibular neurons, antagonistic interaction may likewise be expected since in some conditions neck afferents should cancel the vestibular input to avoid vestibular-induced reflex movements. Such antagonistic effects on vestibular neurons have, indeed, been demonstrated (70). The next step should be to determine the projection of these various response types.

References

1. Abend, W.K.: Functional organization of the superior vestibular nucleus of the squirrel monkey. Brain Res. 132:65, 1977.
2. Abend, W.K.: Response to constant angular accelerations of neurons in the monkey superior vestibular nucleus. Exp. Brain Res. 31:459, 1978.
3. Adrian, E.D.: Discharges from vestibular receptors in the cat. J. Physiol. 101:389, 1943.
4. Akaike, T., Fanardjian, V.V., Ito, M., Kumada, M., and Nakajima, H .: Electrophysiological analysis of the vestibulospinal reflex pathway of rabbit. I. Classification of tract cells. Exp. Brain Res. 17:477, 1973.
5. Allum, J.H.J., Graf, W., Dichgans, J., and Schmidt, C.L.: Visual-vestibular interactions in the vestibular nuclei of the goldfish. Exp. Brain Res. 26:463, 1976.
6. Anderson, J.H., Blanks, R.H.I., and Precht, W.: Response characteristics of semicircular canal and otolith systems in cat. I. Dynamic responses of primary vestibular fibers. Exp. Brain Res. 32:491, 1978.
7. Barlow, D.E. and Freedman, W.: The cervico-ocular reflex in normal human adults. Soc. Neurosci. 4:912, 1978.
8. Blanks, R.H.I., Anderson, J.H., and Precht, W.: Response characteristics of semicircular canal and otolith systems in cat. II. Responses of trochlear motoneurons. Exp. Brain Res. 32:509, 1978.
9. Blanks, R.H.I., Estes, M.S., and Markham, C.H.: Physiological characteristics of vestibular first-order canal neurons in the cat. II. Response to constant angular acceleration. J. Neurophysiol. 38:1250, 1975.
10. Blanks, R.H.I. and Precht, W.: Response properties of vestibular afferents in alert cats during optokinetic stimulation. Neurosci. Lett. 10:225, 1978.
11. Buettner, U.W., Büttner, U., and Henn, V.: Transfer characteristics of neurons in vestibular nuclei of the alert monkey. J. Neurophysiol. 41:1614, 1978.
11a. Cazin, L., Precht, W, and Lannau, J.: Optokinetic responses of vestibular nucleus neurons in the cat. Pflügers Arch. 384:31, 1980.
12. Chan, Y.S., Hwang, J.C., and Chueng, Y.M.: Crossed sacculo-ocular pathway via the Deiters' nucleus in cats. Brain Res. Bull. 2:1, 1977.
13. Curthoys, I.S.: The development of function of horizontal semicircular canal primary neurons in the rat. Brain Res. 167:41, 1979.
14. Curthoys, I.S. and Markham, C.H.: Convergence of labyrinthine influences on

units in the vestibular nuclei of the cat. I. Natural stimulation. Brain Res. 35:469, 1971.

15. Dichgans, J. and Brandt, T.: Visual-vestibular interaction and motion perception. Bibl. Ophthalmol. 82:327, 1972.
16. Dieringer, N. and Precht, W.: Modification of synaptic input following unilateral labyrinthectomy. Nature 269:431, 1977.
17. Dieringer, N. and Precht, W.: Mechanisms of compensation for vestibular deficits in the frog. II. Modification of the inhibitory pathways. Exp. Brain Res. 36:329, 1979.
18. Duensing, F. and Schaefer, K.P.: Die Aktivität einzelner Neurone im Bereich der Vestibulariskerne bei Horizontalbeschleunigungen unter besonderer Berücksichtigung des vestibulären Nystagmus. Arch. Psychiat. Nervenkr. 198:225, 1958.
19. Duensing, F. and Schaefer, K.P.: Über die Konvergenz verschiedener labyrinthärer Afferenzen auf einzelne Neurone des Vestibulariskerngebietes. Arch. Psychiat. Nervenkr. 199:345, 1959.
20. Fernández, C. and Goldberg, J.M.: Physiology of peripheral neurons innervating semicircular canals of the squirrel monkey. II. Response to sinusoidal stimulation and dynamics of peripheral vestibular system. J. Neurophysiol. 34:661, 1971.
21. Fredrickson, J.M., Schwarz, D., and Kornhuber, H.H.: Convergence and interaction of vestibular and deep somatic afferents upon neurons in the vestibular nuclei of the cat. Acta Otolaryngol. (Stockh.) 61:168, 1966.
22. Fuchs, A.F. and Kimm, J.: Unit activity in vestibular nucleus of the alert monkey during horizontal angular acceleration and eye movement. J. Neurophysiol. 38:1140, 1975.
23. Fujita, Y., Rosenberg, J., and Segundo, J.P.: Activity of cells in the lateral vestibular nucleus as a function of head position. J. Physiol. (Lond.) 196:1, 1968.
24. Gacek, R.R.: The course and central termination of first order neurons supplying vestibular end organs in the cat. Acta Otolaryngol. (Stockh.) 254:1, 1969.
25. Gregory, K.M.: Central projections of the eighth nerve in frogs. Brain Behav. Evol. 5:70, 1972.
26. Hassul, M., Daniels, P.D., and Kimm, J.: Effects of bilateral flocculectomy on the vestibulo-ocular reflex in the chinchilla. Brain Res. 118:339, 1976.
27. Henn, V., Young, L.R., and Finley, C.: Vestibular nucleus units in alert monkeys are also influenced by moving visual fields. Brain Res. 71:144, 1974.
28. Heywood, P., Pujol, R., and Hilding, D.A.: Development of the labyrinthic receptors in the guinea pig, cat, and dog. Acta Otolaryngol. 82:359, 1976.
29. Hiebert, T.G. and Fernández, C.: Deitersian response to tilt. Acta Otolaryngol. (Stockh.) 60:180, 1965.
30. Hikosaka, O. and Maeda, M.: Cervical effects on abducens motoneurons and their interaction with vestibulo-ocular reflex. Exp. Brain Res. 18:512, 1973.
31. Hwang, J.C. and Poon, W.F.: An electrophysiological study of the sacculoocular pathways in cats. Jpn. J. Physiol. 25:241, 1975.
32. Ito, M., Hongo, T., and Okada, Y.: Vestibular-evoked postsynaptic potentials in Deiters' neurones. Exp. Brain Res. 7:214, 1969.

33. Ito, M., Shiida, T., Yagi, N., and Yamamoto, M.: Visual influence on rabbit's horizontal vestibulo-ocular reflex that presumably is effected via the cerebellar flocculus. Brain Res. 65:170, 1974.

34. Karhunen, E.: Postnatal development of the lateral vestibular nucleus (Deiters' nucleus) of the rat. Acta Otolaryngol. (Suppl.) 313:1, 1973.

35. Kasahara, M., Mano, M., Oshima, T., Ozawa, S., and Shimazu, H.: Contralateral short latency inhibition of central vestibular neurones in the horizontal canal system. Brain Res. 8:376, 1968.

36. Kasahara, M. and Uchino, Y.: Selective mode of commissural inhibition induced by semicircular canal afferents on secondary vestibular neurons in the cat. Brain Res. 34:366, 1971.

37. Kawai, N., Ito, M., and Nozue, M.: Postsynaptic influences on the vestibular non-Deiters nuclei from primary vestibular nerve. Exp. Brain Res. 8:190, 1969.

38. Keller, E.L.: Behavior of horizontal semicircular canal afferents in alert monkey during vestibular and optokinetic stimulation. Exp. Brain Res. 24:459, 1976.

39. Keller, E.L. and Kamath, B.Y.: Characteristics of head rotation and eye movement-related neurons in alert monkey vestibular nucleus. Brain Res. 100:182, 1975.

40. Keller, E.L. and Precht, W.: Persistence of visual response in vestibular nucleus neurons in cerebellectomized cat. Exp. Brain Res. 32:591, 1978.

41. Keller, E.L. and Precht, W.: Adaptive modification of central vestibular neurons in response to visual stimulation through reversing prisms. J. Neurophysiol. 42:896, 1979.

42. Korn, H., Sotelo, C., and Bennett, M.V.L.: The lateral vestibular nucleus of the toadfish *Opsanus tau:* Ultrastructural and electrophysiological observations with special reference to electrotonic transmission. Neuroscience 2:851, 1977.

43. Korn, H., Sotelo, C., and Crepel, F.: Electrotonic coupling between neurons in the rat lateral vestibular nucleus. Exp. Brain Res. 16:255, 1973.

44. Kubo, T., Matsunaga, T., and Matano, S.: Effects of sinusoidal rotational stimulation on the vestibular neurons of rats. Brain Res. 88:543, 1975.

45. Kubo, T., Matsunaga, T., and Matano, S.: Convergence of ampullar and macular inputs on vestibular nuclei unit of the rat. Acta Otolaryngol. 84:166, 1977.

46. Ladpli, R. and Brodal, A.: Experimental studies of commissural and reticular formation projections from the vestibular nuclei in the cat. Brain Res. 8:65, 1968.

47. Lannou, J., Precht, W., and Cazin, L.: The postnatal development of functional properties of central vestibular neurons in the rat. Brain Res. 175:219, 1979.

48. Magnus, R. Körperstellung. Berlin, Springer, 1924.

49. Mano, M., Oshima, T., and Shimazu, H.: Inhibitory commissural fibres interconnecting the bilateral vestibular nuclei. Brain Res. 8:378, 1968.

50. Markham, C.H.: Midbrain and contralateral labyrinth influences on brainstem vestibular neurons in the cat. Brain Res. 9:312, 1968.

51. Markham, C.H. and Curthoys, I.S.: Convergence of labyrinthine influences on units in the vestibular nuclei of the cat. II. Electrical stimulation. Brain Res. 43:383, 1972.

52. Markham, C.H., Yagi, T., and Curthoys, I.S.: The contribution of the contrala-

teral labyrinth to second order vestibular neuronal activity in the cat. Brain Res. 138:99, 1977.

53. Matsuoka, I., Fukuda, N., Takaori, S., and Morimoto, M.: Responses of single neurons of the vestibular nuclei to lateral tilt and caloric stimulation in the intact and hemilabyrinthectomized cats. Acta Otolaryngol. (Stockh.) 72:182, 1971.

54. Melvill Jones, G. and Milsum, J.H.: Neural response of the vestibular system to translational acceleration. In: Conference on Systems Analysis Approach to Neurophysiological Problems. Minnesota, Brainerd, 1969.

55. Melvill Jones, G. and Milsum, J.H.: Characteristics of neural transmission from the semicircular canal to the vestibular nuclei of cats. J. Physiol. 209:295, 1970.

56. Orlovsky, G.N. and Pavlova, G.A.: Response of Deiters' neurons to tilt during locomotion. Brain Res. 42:212, 1972.

57. Ozawa, S., Precht, W., and Shimazu, H.: Crossed effects on central vestibular neurons in the horizontal canal system of the frog. Exp. Brain Res. 19:394, 1974.

58. Peterson, B.W.: Distribution of neural responses to tilting within the vestibular nuclei of the cat. J. Neurophysiol. 33:750, 1970.

59. Precht, W.: The physiology of the vestibular nuclei. In Kornhuber, H.H. (ed.): Handbook of Sensory Physiology, Vol. VI. Heidelberg, Springer-Verlag, 1974, pp. 353–416.

60. Precht, W.: Vestibular system. In Goyton, A.C. and Hunt, C.C. (eds.): MTP International Review of Sciences, Neurophysiology. Physiology Series One, Vol. 3. London, Butterworths Univ. Park Press, 1975, pp. 82–149.

61. Precht, W.: Physiology of the peripheral and central vestibular systems. In Llinás, R. and Precht, W. (eds.): Frog Neurobiology. New York, Springer-Verlag, 1976, pp. 481–512.

62. Precht, W.: Neuronal operations in the vestibular system. In Braitenberg, V. (ed.): Studies of Brain Function, Vol. 2. New York, Springer-Verlag, 1978, pp. 226.

63. Precht, W.: Vestibular mechanisms. Ann. Rev. Neurosci. 2:265, 1979.

64. Precht, W., Grippo, J., and Wagner, A.: Contribution of different types of central vestibular neurons to the vestibulo-spinal system. Brain Res. 4:119, 1967.

65. Precht, W., Richter, A., Ozawa, S., and Shimazu, H.: Intracellular study of frog's vestibular neurons in relation to the labyrinth and spinal cord. Exp. Brain Res. 19:377, 1974.

66. Precht, W. and Shimazu, H.: Functional connections of tonic and kinetic vestibular neurons with primary vestibular afferents. J. Neurophysiol. 28:1014, 1965.

67. Precht, W., Shimazu, H., and Markham, C.H.: A mechanism of central compensation of vestibular function following hemilabyrinthectomy. J. Neurophysiol. 29:996, 1966.

68. Richter, A., Precht, W., and Ozawa, S.: Responses of neurons of lizard's, *Lacerta viridis,* vestibular nuclei to electrical stimulation of the ipsilateral and contralateral VIIIth nerves. Pfügers Arch. 355:85, 1975.

69. Robinson, D.A.: Oculomotor unit behavior in the monkey. J. Neurophysiol. 33:393, 1970.

70. Rubin, A.M., Young, J.H., Milne, A.C., Schwarz, D.W.F., and Fredrickson, J.M.: Vestibular-neck integration in the vestibular nuclei. Brain Res. 96:99, 1975.

71. Schor, R.H.: Responses of cat vestibular neurons to sinusoidal roll tilt. Exp. Brain Res. 20:347, 1974.

72. Shimazu, H. and Precht, W.: Tonic and kinetic responses of cat's vestibular neurons to horizontal angular acceleration. J. Neurophysiol. 28:991, 1965.

73. Shimazu, H. and Precht, W.: Inhibition of central vestibular neurons from the contralateral labyrinth and its mediating pathway. J. Neurophysiol. 29:467, 1966.

74. Shimazu, H. and Smith, C.M.: Cerebellar and labyrinthine influences on single vestibular neurons identified by natural stimuli. J. Neurophysiol. 34:493, 1971.

75. Shinoda, Y. and Yoshida, K.: Dynamic characteristics of responses to horizontal head angular acceleration in the vestibuloocular pathway in the cat. J. Neurophysiol. 37:653, 1974.

76. Stein, B.M. and Carpenter, M.B.: Central projections of portions of the vestibular ganglia innervating specific parts of the labyrinth in the rhesus monkey. Am. J. Anat. 120:281, 1967.

77. Waespe, W. and Henn, V.: Neuronal activity in the vestibular nuclei of the alert monkey during vestibular and optokinetic stimulation. Exp. Brain Res. 27:523, 1977.

78. Walberg, F., Bowsher, D., and Brodal, A.: The termination of primary vestibular fibers in the vestibular nuclei in the cat. An experimental study with silver methods. J. Comp. Neurol. 110:391, 1958.

79. Wilson, V.J. and Felpel, L.P.: Specificity of semicircular canal input to neurons in the pigeon vestibular nuclei. J. Neurophysiol. 35:253, 1972.

80. Wilson, V.J. and Wylie, R.M.: A short-latency labyrinthine input to the vestibular nuclei in the pigeon. Science 168:124, 1970.

Discussion

GOLDBERG: The data you presented on the cat vestibuloocular reflex were, if memory serves me correctly, partly based on observations in the decerebrate cat. I was particularly struck by the phase differences between abducens motoneuron discharge and the actual eye movement. Is there comparable information for the alert cat?

PRECHT: Yes. I think that if you look at the abducens in the monkey (David Robinson knows that better because he did these experiments), the phase lag is not as poor as in the cat in the lower frequency range. The VOR phase is also much closer to 180° and it doesn't seem to require so much additional optokinetic information to get to the figure. But the alert preparation showed that the animal in the dark has more of a phase lag, both in the VOR measurements as well as in the abducens nuclei recordings, than in the decerebrate condition. So the decerebrate preparation is really not ideal. I thought for some time it would be quite good because it abolishes the quick phases and you don't have the problem of the quick phase interfering, but it apparently changes the qualities of the integrator too.

GOLDBERG: Do you have information on the nature of the pathways from the visual system to the vestibular nuclei?

PRECHT: Yes, we are just working on this problem. When you shock the optic nerve or the optic chiasm, you get a field potential and single unit activation in the vestibular nuclei after latencies of about 15 to 20 ms. But it's very difficult to tell whether this is equivalent to what you see when you do an optokinetic stimulation. So what we are doing now is making various lesions, trying to find out which lesion is crucial for the optokinetic response, and at the moment we know that the cerebellum, as previously postulated, is not necessary. But it seems that the nucleus of the optic tract is quite essential for the optokinetic response. How the pathway continues from this nucleus is not clear yet. To our surprise, an enormous lesion in the central tegmental tract, which is one of the outputs of the nerves of the optic tract, does not abolish the optokinetic response of the vestibular neurons. So other, presumably polysynaptic, routes must carry the information.

14

The Influence of Duct and Utricular Morphology on Semicircular Canal Response

CHARLES M. OMAN

Recently a renewed interest has developed in the question of what constitutes the physiologic magnitude and bending mode of motion of the semicircular canal cupula. In 1972, Oman and Young (26) theoretically estimated that, regardless of bending mode, the motion of the middle of the human cupula ought to be limited to about 3 μm for maximal self-induced sinusoidal head motion. Experimental investigations of cupula motion in the frog (10,21,22) and in the skate (7,24,25) have supported the notion that the relatively large "swinging door" rotation described by early workers is likely artifactual, and have suggested that in the frog (and possibly also in the skate) the cupula is normally adherent not only to the crista, but also to the vault of the ampulla, and deforms as a diaphragm.

When calculating theoretical semicircular canal cupula volume displacement in response to head rotation, the usual point of departure for many workers has been Groen's formulation (6) of the "torsion pendulum" model of Steinhausen (29). A significant amount of morphologic data is now available documenting interspecies and intercanal differences in canal diameter, duct diameter and length, and cupula area. Several authors have noted that variations in some of these parameters appear systematically across species (14) or during ontogenetic development (12,13,15) when interpreted using analyses based on the model proposed by van Egmond et al. (6).

Although the experimental findings in frog and skate cited above appear generally consistent with theoretical predictions based on the van Egmond et al. model (21), several considerations now lead us to reexamine the extent to which theory and experiment can be quantitatively reconciled when comparing results from different species. For example, interspecies differences in utricular size and shape are known to exist, but have generally received little attention. Van Buskirk (31) has recently shown that the effect of the utricle on fluid flow has been improperly handled by previous workers. Using a theoretical model in which the canal duct and utricle were each represented as a separate segment of different circular cross section, each occupying about one-half of the circumference of a circle, Van Buskirk confirmed that the canal dynamics can be modeled as a heavily damped, second order system, and that flow entrance effects between the duct and the utricle may be ignored. He showed that the presence of a semicircular utricular section doubles the long time constant of the canal and also the magnitude of endolymph (and hence cupula) displacement in response to head movement, compared to the response of a canal with no utricle. However, in Van Buskirk's analysis, the functional variation of semicircular canal dynamics with utricular length and cross-sectional area were not analytically explored.

In recent years, data have become available (1,2,15) documenting earlier observations by Gray (8) and others that in many species, the membranous canal may deviate from the idealized form assumed by Groen in his model. In particular, the large radius of curvature of the canal torus may not be constant. Also, in some species, including man, the membranous narrow duct is known to be elliptic, rather than circular, in cross section. The difficulties posed by the former observation may be easily handled (9,15,17) by adopting a toroidal model of constant large radius whose enclosed area in the plane of acceleration is identical with that of the actual canal. With respect to the latter problem, Curthoys et al. (2) suggest that the narrow duct be modeled as a tube of circular cross section with the same cross-sectional area as that of the actual duct. In fact, the solution to this second problem is not so simple, as will be shown below.

In making theoretical interspecies comparisons of semicircular response dynamics, it is also necessary to consider possible differences in the physical characteristics of endolymph which result from differences in body temperature. Endolymph viscosity is expected to decrease with increasing temperature, since fluid viscosity is dependent upon cohesive forces between molecules, which decrease with temperature. As shown in Fig. 14-1, measurements taken by Money et al. (23) confirm that the temperature dependence of endolymph viscosity is similar to that of water and other weak solutions. Measurements taken on human endolymph using a similar method by Steer (27) are in reasonable agreement with those obtained from the pike (15) if temperature differences are taken into account, although some absolute difference from Money's pigeon data is indicated. Since cupula response to

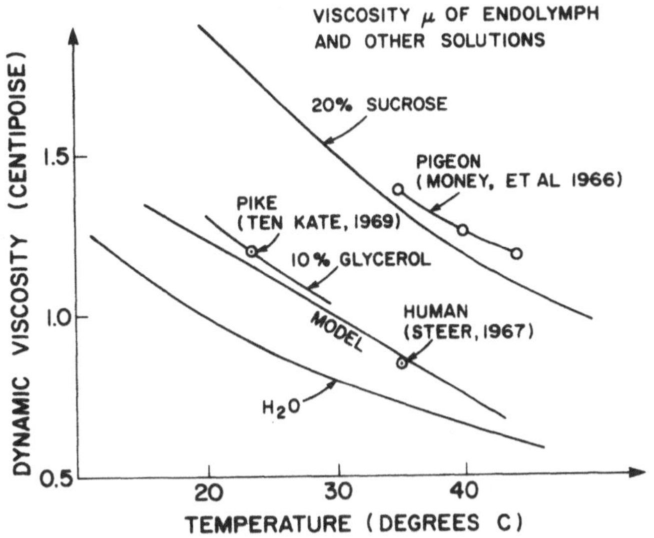

Fig. 14-1 Dynamic viscosity of endolymph, water, and solutions of glycerol and sucrose as a function of temperature.

normal head rotation is expected to be inversely proportional to endolymph viscosity, and viscosity may be expected to vary by 50% over the range from 15 to 37°C where many physiologic experiments are conducted, canal response is undoubtedly influenced.

In order to assess the combined effects of variations in length, cross-sectional area, and shape of both the semicircular canal narrow duct and utricle, as well as cupula area and temperature when making interspecies and intercanal comparisons, a revised model has been developed which specifically accounts for these various factors.

ANALYSIS

In the revised model, an individual semicircular canal is represented as a hollow rigid structure filled with endolymph of uniform density ρ and newtonian dynamic viscosity $\mu(T)$, where the temperature (T in °C) dependence of viscosity is approximated by

$$\mu(T) = 0.01 - 2.4 \times 10^{-4}(\text{T-30})\text{poise} \qquad [14\text{-}1]$$

as indicated in Fig. 14-1. (The implications of the newtonian assumption on semicircular canal models have been considered by Van Buskirk and Morse (32); the known dependence of the endolymph density on temperature, estimated to be −0.44%/°C for human endolymph by Steer (27), is here ignored.)

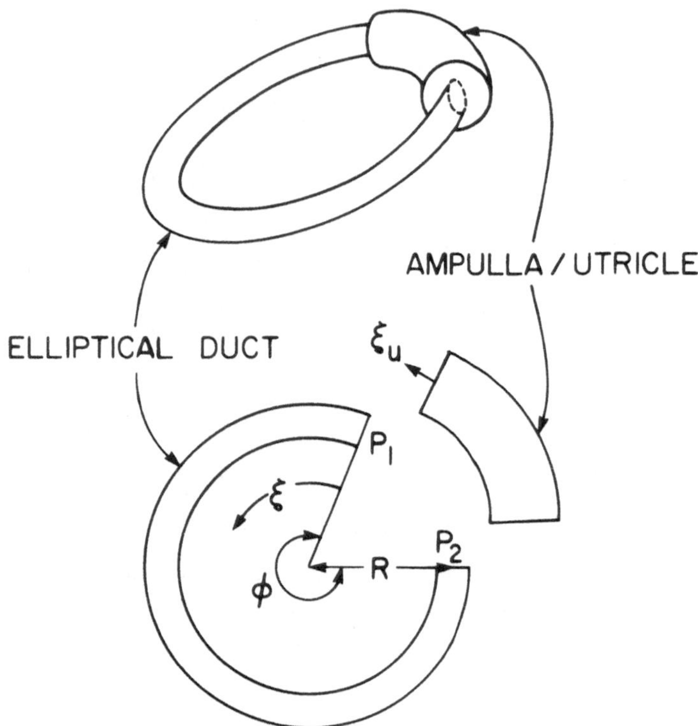

Fig. 14-2 Two-segment model for semicircular canal.

As shown in Fig. 14-2, the model canal is composed of two sections:

1) A segment representing the membranous canal narrow duct. This is idealized as a section of a toroidal ring of large radius R, subtending a central angle ϕ, a fraction $F = \phi/2\ \pi$ of a complete circle. The duct is elliptic in cross section; a and b are duct cross section major and minor semiaxes, respectively. Duct cross-sectional area $A_d = \pi\ ab$ is assumed constant around the duct segment.
2) A second toroidal segment with elliptic cross-sectional area A_u and large radius R, assumed to contain the fluid volume of the ampulla and utricle.

The duct segment is assumed to open into the ampullary/utricular segment (henceforth called the utricular segment) at each end through openings of cross-sectional area A_d. The mean angular displacement of endolymph flow in the duct segment, taken with respect to the center of the canal semicircle, is denoted by $\xi(t)$. The corresponding flow displacement in the utricular segment is denoted $\xi_u(t)$. As indicated in Fig. 14-2, P_1 and P_2 represent endolymph fluid pressures at the junctions between the duct and utricular segments, taken so that a positive pressure difference $\Delta P = P_1 - P_2$

produces a positive flow displacement ξ in the duct, and a negative flow displacement ξ_u in the utricular segment.

The objective of the present effort is to develop a single differential equation describing flow in the duct segment $\xi(t)$ in response to head angular accelerations $\alpha(t)$ in a form permitting comparison to the traditional formulation of the torsion pendulum model. This is accomplished by a method similar to the approach taken by Van Buskirk (31): one writes the equations of motion describing flow in the duct and utricular segments individually in terms of ΔP and the appropriate flow displacement variable for each segment. Since the two segments are physically connected, the desired equation may be directly obtained by equating expressions for ΔP, and by invoking the continuity equation to relate flow displacement variables in each segment.

The equation of motion for flow of endolymph in the duct segment is obtained by writing an expression for the time rate of change of fluid angular momentum with respect to a stationary coordinate system, and equating it to the torques on the fluid in the segment resulting from viscous drag on the duct walls and the pressure difference across the whole segment. If $a,b << R$, the moment of inertia of endolymph in the duct segment, Θ_d, can be approximated as:

$$\Theta_d \simeq 2\rho\pi^2\, ab\, FR^3 \qquad [14\text{-}2]$$

Since the total angular acceleration of the fluid in the duct segment is $\alpha + d^2\xi/dt^2$, then the equation of motion for endolymph flow can be written as:

$$\Theta_d\left(\alpha(t) + \frac{d^2\xi(t)}{dt^2}\right) = -\Pi_d\frac{d\xi(t)}{dt} + A_dR\Delta P \qquad [14\text{-}3]$$

Pressure drops due to fluid kinetic energy losses associated with changes in membranous canal cross-sectional area (entrance effects) are ignored in writing Eq. 14-3. Van Buskirk (31) has presented arguments that these effects are small.

In Eq. 14-3, Π_d is a coefficient expressing the relationship between mean angular flow velocity in the duct, $d\xi(t)/dt$, and the torque on the fluid segment produced by the pressure difference ΔP.

$$\Pi_d = \frac{\pi\, ab\, R\Delta P}{d\xi(t)/dt} \qquad [14\text{-}4]$$

The relationship between flow velocity and segment pressure difference may be estimated by analogy to the case where a newtonian fluid of viscosity $\mu(T)$, a function of temperature, T, has achieved low Reynolds number steady flow in a duct of elliptic cross section and length $2\pi RF$ in response to an imposed pressure gradient. We adopt a coordinate system in which x is taken along the center of the duct, and y and z directions are along the major semiaxis, a, and the minor duct semiaxis, b, respectively. The local endolymph flow velocity $u(y,z)$ must satisfy the Navier-Stokes equation for steady flow driven by a pressure gradient in the x direction:

$$\frac{\partial P}{\partial x} + \mu(T)\left(\frac{\partial^2 u}{\partial z^2} + \frac{\partial^2 u}{\partial y^2}\right) = \text{O} \qquad [14\text{-}5]$$

Solving this equation subject to a boundary condition of zero flow velocity at the duct wall, one obtains an expression for the endolymph flow velocity distribution in the duct:

$$u(y,z) = \frac{1}{2\mu(T)} \frac{a^2b^2}{a^2 + b^2}\left[1 - \left(\frac{z^2}{b^2} + \frac{y^2}{a^2}\right)\right]\frac{dP}{dx} \qquad [14\text{-}6]$$

As shown in Fig. 14-3, the velocity distribution is fundamentally parabolic in profile. Volume flow is obtained by integrating the velocity distribution over the duct cross section:

$$\iint u(y,z)dy\,dz = \frac{\pi\, a^3b^3}{4\mu(T)\,(a^2 + b^2)}\frac{dP}{dx} \qquad [14\text{-}7]$$

Hence, mean angular flow velocity in the duct is:

$$\frac{d\xi(t)}{dt} = \frac{a^2\,b^2}{4\mu(T)\,R\,(a^2 + b^2)}\frac{dP}{dx} \qquad [14\text{-}8]$$

Inspection of Eqs. 14-6–14-8 shows that the peak flow velocity, volume flow, and mean angular flow velocity $d\xi(t)/dt$ are smaller for an elliptic duct than for a circular duct with the same cross-sectional area driven by a given

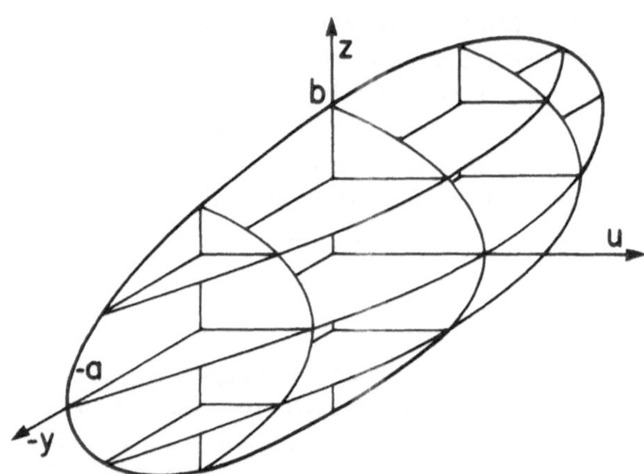

Fig. 14-3 Steady state endolymph flow velocity distribution in a canal duct of elliptic cross section.

pressure gradient dP/dx. The reduced flow in the elliptic duct results directly from the relatively greater duct wall surface area exposed to the viscous fluid. It is convenient to define a duct cross section shape factor S where

$$S = \frac{a^2 + b^2}{2ab} \qquad [14\text{-}9]$$

Note that $S = 1$ for a circular duct. Since the pressure gradient dP/dx is constant along the duct segment,

$$\Delta P = 2\pi RF \frac{dP}{dx} \qquad [14\text{-}10]$$

Combining Eqs. 14-4 and 14-8–14-10, one obtains

$$\Pi_d = 16F\ \pi^2\ \mu(T)\ R^3 S \qquad [14\text{-}11]$$

Notice that although a and b do not appear explicitly in this expression, duct ellipticity nonetheless exerts an influence through the shape factor S. For example, if $a = 2b$, as in the human duct (2), then Π_d is 25% larger than for a duct of equal area but circular cross section.

Equations 14-2, 14-3, and 14-11 collectively describe the motion of fluid in the duct segment. By analogy, for the utricular segment, one can write:

$$\Theta_u\left(\alpha(t) + \frac{d^2\xi_u}{dt^2}\right) = -\Pi_u \frac{d\xi_u}{dt} - A_u R \Delta P - KR^2 A_u^2 \xi_u \qquad [14\text{-}12]$$

where utricular segment moment of inertia, Θ_u, is given by:

$$\Theta_u = 2\rho\pi A_u(1 - F)\ R^3 \qquad [14\text{-}13]$$

and, assuming the utricular segment has a shape factor S_u, the utricular segment flow drag coefficient Π_u is:

$$\Pi_u = 16(1 - F)\ \pi^2\ \mu(T)\ R^3 S_u \qquad [14\text{-}14]$$

Recalling that the utricular segment must also reflect the presence of the ampulla, if the pressure versus endolymph volume displacement characteristic of the ampulla is assumed linear, cupula stiffness can be characterized by a constant $K = \Delta P/\Delta V$. Defining cupula stiffness in this way facilitates comparisons between canals of different utricular segment shape. The resulting torque on the utricular segment fluid is given by the last term in Eq. 14-12. Note also that the second term on the right side of this equation is written with a negative sign, reflecting the convention that a positive ΔP produces a negative flow in the utricular segment.

Given the equations of motion of the two segments, one can obtain the desired equation by invoking the continuity equation, which demands that the volume flow out of the utricular segment be equal to that flowing into the duct segment, i.e.,

$$A_u \xi_u(t) = A_d \xi(t) \qquad [14\text{-}15]$$

Using Eqs. 14-3, 14-12, and 14-15 to eliminate ΔP and ξ_u, substituting Eqs. 14-2, 14-11, 14-13, and 14-14, and defining kinematic viscosity $\vartheta(T) = \mu(T)/\rho$, one obtains the equation of motion of the semicircular canal model:

$$\frac{d^2\xi(t)}{dt^2} + \frac{8\vartheta(T)S}{ab}\frac{D_2}{D_1}\frac{d\xi(t)}{dt} + \frac{K\,ab}{2\rho RD_1}\xi(t) = -\frac{1}{D_1}\alpha(t) \quad [14\text{-}16]$$

where D_1 and D_2 are fluid inertia and drag distribution factors defined as:

$$D_1 = \left[F + \frac{A_d}{A_u}(1 - F)\right] \quad [14\text{-}17]$$

$$D_2 = \left[F + \frac{S_u}{S}\left(\frac{A_d}{A_u}\right)^2(1 - F)\right] \quad [14\text{-}18]$$

Laplace transforming equation 14-16, assuming zero initial conditions, and assuming the response is highly damped, one can write the system transfer function of the flow in the canal duct in response to head acceleration inputs as:

$$\frac{\xi(s)}{\alpha(s)} = \frac{-1/D_1}{s^2 + s/\tau_2 + 1/\tau_1\tau_2} \simeq \frac{-\tau_1\tau_2/D_1}{(\tau_1 s + 1)(\tau_2 s + 1)} \quad [14\text{-}19]$$

where

$$\tau_1 = \frac{16\mu(T)\,RSD_2}{a^2b^2K} \quad [14\text{-}20]$$

and

$$\tau_2 = \frac{abD_1}{8\vartheta(T)SD_2} \quad [14\text{-}21]$$

τ_1 and τ_2 are the familiar long and short time constants associated with cupula return and endolymph quasi-steady flow development, respectively.

Head movements are normally made in the frequency range $1/\tau_1 < \omega < 1/\tau_2$. Hence

$$\frac{\xi(j\omega)}{\alpha(j\omega)} = \frac{-\tau_1\tau_2/D_1}{(j\omega\tau_1 + 1)(j\omega\tau_2 + 1)} \simeq \frac{-\tau_2/D_1}{j\omega} \quad [14\text{-}22]$$

so endolymph flow displacement is normally in phase with head angular velocity, $\gamma(j\omega) = \alpha(j\omega)/j\omega$. Therefore,

$$\frac{\xi(j\omega)}{\gamma(j\omega)} = -\tau_2/D_1 \quad [14\text{-}23]$$

Since cupula deformation is proportional to the volume displacement, $\Delta V(j\omega) = \pi abR\xi(j\omega)$, of endolymph into the ampulla, a direct measure of the mechanical sensitivity of the canal cupula during normal head movements is given by the ratio of cupula volume displacement to head angular velocity in the frequency midrange defined above:

$$\frac{\Delta V(j\omega)}{\gamma(j\omega)} \simeq \frac{-\pi a^2 b^2 R}{8\vartheta(T)SD_2} \qquad [14\text{-}24]$$

Calculation of cupula linear displacement resulting from head movement requires an assumption about cupula bending mode, an issue discussed further below. However, a convenient index of the average cupula displacement to head velocity ratio is given by x/ω, where

$$\frac{x}{\gamma} = \frac{1}{A_c}\frac{\Delta V(j\omega)}{\gamma(j\omega)} \qquad [14\text{-}25]$$

and where A_c is the cross-sectional area of the lumen of the ampulla above the crista occupied by the cupula.

The functional dependence of τ_1, τ_2, and $\Delta V/\gamma$ on individual morphologic parameters and temperature is discussed in the next section. However, these parameters in fact covary between species, and effects on canal response produced by variations in one parameter may be compounded or offset by variations in other parameters. To evaluate these effects, calculations of τ_2, $\Delta V/\gamma$, and x/γ were made assuming the values for parameters a, b, R, F, A_c, and T for canals of five different species. Values for a, b, and R were obtained for the human, cat (1, 2), and pike (15) from the literature. Estimates of F were inferred from data on semicircular canal circumference and duct length in each species. Estimates for the frog (*Rana catesbeiana*) and the skate (*Raja erinacea*) are based on measurements made in our own material, and are believed representative. Results are shown in Table 14-1.

DISCUSSION

Comparison with Traditional Models

The fundamental analytic finding of the present analysis is that duct endolymph flow is described by a second order differential equation with three coefficients, $8\vartheta(T)SD_2/abD_1$, $Kab/2\rho RD_1$, and $1/D_1$. Assuming a small value for K, endolymph response to normal head movement is overdamped, and is described by two time constants, τ_1 and τ_2, given by Eqs. 14-20 and 14-21, and also a "gain factor," τ_2/D_1, given by Eq. 14-23.

By contrast, the 1949 "torsion pendulum" model formulation of van Egmond et al. (6) represented the canal simply as a torus of large radius R, and constant circular cross section πr^2. Van Egmond and co-workers recognized that substantial additional fluid volume was contained in the utricle and ampulla of an actual canal, but chose to represent the utricle and ampulla by a simple additional semicircular extension of the duct section, so that the moment of inertia of the entire endolymph ring was given by

$$\Theta \simeq 2\pi\rho^2 r^2 R^3 \qquad [14\text{-}26]$$

Table 14-1.
CALCULATED DATA

	Canal[1]	a (cm)	b (cm)	R (cm)	F	T (°C)	τ_2 (ms)	$\Delta V/\Omega$ (μm^3/deg/s)	A_c (mm^2)	x/γ ($m\mu$/deg/s)
Human (1,2)	H	0.0115	0.0230	0.280	0.7	37	3.179	18440	1.156	15.95
	A	0.0110	0.0230	0.368	0.7	37	2.959	22640	1.156	19.57
	P	0.0115	0.0225	0.358	0.7	37	3.152	22860	1.156	19.77
Frog	H	0.0175	0.0175	0.164	0.6	20	3.087	14170	0.3387	41.84
(*Rana*)	A	0.0163	0.0163	0.163	0.5	20	2.678	12720	0.4927	25.81
	P	0.0150	0.0150	0.159	0.5	20	2.268	8900	0.4531	19.64
Skate	H	0.0263	0.0263	0.380	0.6	14	6.240	150060	1.7090	87.80
(*Raja*	A	0.0370	0.0370	0.320	0.5	14	12.364	660000	2.7330	241.48
erinacea)	P	0.0350	0.0350	0.440	0.5	14	11.064	653960	2.1110	309.80
Cat (1, 2)	H	0.0110	0.0125	0.164	0.7	37	2.049	3619	0.4683	7.72
	A	0.0120	0.0150	0.220	0.7	37	2.638	8183	0.4683	17.48
	P	0.0110	0.0150	0.186	0.7	37	2.364	5683	0.4683	12.14
Pike (15)	H	0.0155	0.0155	0.284	0.6	20	2.420	15100	0.7070	21.35
	P	0.0195	0.0195	0.239	0.5	20	3.83	38200	0.7130	53.58

[1]H = Horizontal canal
A = Anterior canal
P = Posterior canal

In calculating Π, the net torque on the endolymph ring per unit angular velocity of flow in the duct, van Egmond et al. argued that the flow drag contribution due to the semicircular utricular segment should be ignored. By analogy to Poiseuille flow in a semicircular duct segment of circular cross section, they determined that

$$\Pi = 8\mu\pi^2 R^3 \qquad [14\text{-}27]$$

In the van Egmond model, the torque per unit angular flow in the duct, Δ, resulting from cupula stiffness was not evaluated in terms of the pressure/volume characteristics of the ampulla. However, for a toroidal duct of constant cross-sectional area, it may easily be shown that

$$\Delta = \pi^2 r^2 R^2 K \qquad [14\text{-}28]$$

The response of the van Egmond et al. "torsion pendulum" model is described by the second order differential equation

$$\frac{d^2\xi(t)}{dt^2} + \frac{\Pi}{\Theta}\frac{d\xi(t)}{dt} + \frac{\Delta}{\Theta}\xi(t) = -\alpha(t) \qquad [14\text{-}29]$$

with two coefficients, Π/Θ and Δ/Θ. If the response is highly damped, it is characterized by two time constants, $\tau_1^* = \Pi/\Delta$, $\tau_2 = \Theta/\Pi$, and a gain factor, $\xi/\gamma = \tau_2^*$. Because the model differential equation 14-29 has only two coefficients rather than three, the gain factor is determined by the short time constant.

It is instructive to describe the model developed by van Egmond et al. in terms of the parameters of the present model, and to examine the impact of their assumptions regarding utricular inertia and flow drag on the resulting model predictions. If Eqs. 14-26 through 14-28 are used to evaluate expressions for the time constants of the van Egmond model, one finds that

$$\tau_1^* = \frac{\Pi}{\Delta} = \frac{8\mu R}{Kr^4} \qquad [14\text{-}30]$$

$$\tau_2^* = \frac{\Theta}{\Pi} = \frac{r^2}{4\vartheta} \qquad [14\text{-}31]$$

As one might expect, τ_1^* and τ_2^* are respectively half and twice the values predicted by Eqs. 14-20 and 14-21 for the case of a toroidal duct of constant cross-sectional area (setting $F = 0.5$; $a = b = r$; $S = 1.0$; and $A_d = A_u$), because in developing Eqs. 14-20 and 14-21 Π_u was not neglected. However, it is perhaps more appropriate to also compare the van Egmond model time constants with those time constants predicted by Eqs. 14-20 and 14-21 for the case of a semicircular duct connected to a semicircular utricle of large cross-sectional area. Assuming $a = b = r$; $S = 1.0$; $A_d << A_u$; and $F = 0.5 \simeq D_1 \simeq D_2$; then

$$\tau_1 = \frac{8\mu R}{Kr^4} \qquad [14\text{-}32]$$

and

$$\tau_2 = \frac{r^2}{8\vartheta} \qquad [14\text{-}33]$$

Comparing the predictions made by the van Egmond model (Eqs. 14-30 and 14-31) with those in which utricular inertia and drag are specifically accounted for (Eqs. 14-32 and 14-33), it appears that the van Egmond et al. assumptions lead to significant overestimation of the short time constant of the canal ($\tau_2^* = 2\tau_2$), although the long time constants (and also the gain factor for the two cases, τ_2^* and τ_2/D_1, respectively) are identical. This results directly from the fact that the van Egmond et al. model describes the semicircular canal with a second order differential equation with two coefficients, whereas the analysis accounting for a larger utricular segment requires a differential equation of the same order but with three coefficients, each of which is a function of canal morphologic parameters. It is not possible to "adjust" the two-coefficient van Egmond model differential equation for the effects of different size utricles in various species simply by manipulating the length of the duct segment and calculating a different value of Π, as proposed by some workers (21,23), or by assuming a different value of Θ (2, 20,26). Increasing the value of Θ results in an overestimation of both time constants and also the response gain factor. These results were implicit in the analysis of Van Buskirk (31) who compared the response of a semicircular canal with a semicircular utricular segment of large cross-sectional area with that of a canal without a utricle. As noted earlier, Van Buskirk found that the short time constants were identical, but the long time constant and gain factor of the canal with the utricle were doubled. Clearly, if one seeks to model the dynamic response of a canal which has a dilated segment, the appropriate approach is to adopt a differential equation with three, rather than two, coefficients.

Influences of Utricular Segment Morphology

Van Buskirk's (31) analysis did not reveal the functional dependence of canal time constants and gain factor on the length, volume, or shape of the utricular and duct segments. Inspection of Eqs. 14-20, 14-21, and 14-24 shows that if one compares the response of several different canals, each of which has a large utricular segment of different length, if the large radii R and the ampulla mechanical characteristics of the canals are otherwise identical, and $A_u >> A_d$ for each canal, then the short time constant τ_2 of all the canals are the same, as the D_1 and D_2 terms cancel. It is the long time constant, τ_1, and gain factor, $\Delta V/\gamma$, which vary. For canals whose utricular segments are of relatively large cross section, $D_1 \simeq D_2 \simeq F$. Increased utricular length (decreased duct length) is associated with a decrease in canal long time constant τ_l (Eq. 14-20) and an increase in canal response gain factor $\Delta V/\gamma$ (Eq. 14-24) due to the associated decrease in endolymph drag distribution factor D_2. However, it is important to note that if A_d/A_u is small, the long time constant is dependent only on the angle subtended by the utricular segment about the center of the canal and not its absolute length, and is not much influenced by variations in utricular segment cross section if the cross-sectional area A_u remains large relative to A_d.

Because the analysis of van Egmond et al. predicted a dependence of canal time constants on canal large radius R and circular duct radius r, the ontogenetic and phylogenetic allometries of these two morphologic factors are well documented in the literature (1,2,12–15). Comparably systematic studies of utricular morphology, however, have not been made, perhaps because the theoretical dependence of canal time constants and gain factor on utricular segment morphologic factors has not until now been described. Estimates of semicircular canal circumference and duct length may be made, from which an estimate for utricular segment length may be inferred. For example, absolute utricular segment length $2\pi R(1 - F)$, for the five species described in Table 14-1, varies by more than a factor of four. Unfortunately, quantitative data on variations in utricular cross section A_u are generally not available in the literature. In calculating the results shown in Table 14-1, it has been assumed that $A_d/A_u << 1.0$. This appears to be a reasonable assumption. For most of the species listed, A_d/A_u probably falls in the range from 0.15 to 0.06. Although the posterior canal in the skate is a separate entity which is not joined to the utricle in a manner similar to the posterior canal of the other species tabulated, a portion of the skate posterior canal paralleling the endolymphatic duct is dilated with a cross-sectional area nearly three times that of the rest of the canal duct.

If results of greater accuracy are desired, the inertia distribution factors D_1 and D_2 may be explicitly calculated using the value of A_d/A_u and Eqs. 14-17 and 14-18. However, if A_d/A_u is relatively large along the utricular segment and there are also significant variations in A_u, it may be more appropriate in a given case to explicitly account for this by developing a model with a duct segment and two or more utricular segments. In such an analysis, each utricular segment would be assigned an appropriate length and cross-sectional area, a flow displacement variable, and a segment pressure difference. The present method of analysis can be directly extended by writing relationships corresponding to Eqs. 14-12 and 14-14. The resulting set of equations is solved by summing segment pressure difference terms around the canal, and invoking the continuity equation to relate flow displacement variables. However, such an approach is probably not justified until detailed morphologic measurements have been made on the species under consideration.

Impact of Duct Ellipticity

Duct segment ellipticity exerts an influence on canal response in a way which appears to have been overlooked in the past, probably because the conventional expression for the drag torque coefficient in a circular duct (Eq. 14-27) is independent of duct radius, r. However, in the present model, drag in both the utricular and duct segments depends on the corresponding segment shape factor. Utricular shape factor S_u enters the expression for D_2 (Eq. 14-18) in only a minor way, since $A_d < A_u$, whereas the duct shape factor S has a much more direct influence on both time constants and the gain factor (Eqs. 14-18, 14-19, and 14-24). Although the canal ducts of many animals are circular in cross section, in the human they are not. Values for a and b obtained from Curthoys et al. (2), and shown in Table 14-1, suggest that in the human, τ_l is 25% longer and τ_2 and $\Delta V/\gamma$ are 20% smaller than would be the case if the human canal had ducts of equal area but circular cross section. It is not appropriate to account for the effects of duct segment ellipticity by substituting the radius, $r_{eq} = \sqrt{ab}$, of a duct of equivalent cross-sectional area into the conventional models, as suggested in (2). To do so would ignore the relative increase in duct segment drag resulting from the correspondingly larger surface area exposed to the flow in the duct of elliptic cross section.

Temperature Effects

As noted earlier, the viscosity of endolymph may be expected to vary by 50% over the range of temperatures where physiologic experiments are conducted. Ten Kate and Kuiper (16) have noted that this temperature dependence would be expected to directly influence semicircular canal sensitivity. Equations 14-30–14-33 demonstrate that a 50% increase in endolymph viscosity due to decreased temperature would be expected to increase the canal long time constant, τ_1, by 50%, and reduce the canal short time constant, τ_2, and gain factor, $\Delta V/\gamma$, by a factor of two. This effect obviously must be taken into account when making interspecies comparisons of semicircular canal responses, particularly in cold-blooded animals.

Cupula Deformation Mode and Magnitude

In the past, the normal mode of cupula motion has been variously described as rotation about its base (4,28,30) in the manner of a swinging door; rotation about a point in the crista so that shearing at the surface of the crista also results (5,33); linear displacement as a plug (34); as angular rotation about an attachment at the cupula vault, with shearing taking place at the crista (35); or as deformation of a diaphragm fixed to the ampulla around its perimeter (3,10,11,19,22). This last interpretation has recently achieved considerable acceptance by workers in the field because a diaphragm mode of deformation has been carefully documented in the frog by the photographic studies of McLaren (21). Although a peripheral attachment of the type described by McLaren and Hillman (22) has not yet been conclusively demonstrated in other species, it is now generally believed that normal cupula movements—whatever the mode of deformation—are relatively small, and that the attachments of the cupula to the crista (and possibly also to the ampulla vault) may easily be traumatized. McLaren's (21) study in the frog provides experimental evidence that the mode of cupula deformation may be amplitude dependent. McLaren noted that the point of maximum displacement of the frog cupula is located near the crista for small stimuli, but moves toward the center of the cupula at higher stimulus levels.

Dohlman's (5) histologic description of the cupula suggests that the cupula substance might well behave in a nonlinear anisotropic fashion, since it is composed of fibrillar keratinous proteins arranged in a reticular structure, with small channels running through the cupula from above the receptor cells toward the ampulla vault. Fibrils of cupula ground substance are secreted by interstitial cells surrounding each hair cell, forming a loose reticulum at the level of the receptor cell cilia which coalesces at higher levels in the cupula above the receptor cell cilia. In general, reticular structures exhibit nonlinear elastic properties, particularly at large values of strain, as evidenced by the mechanical behavior of a knitted stocking. As indicated in Fig. 14-4, one could speculate that the effective Young's modulus, E_2, of the material in the region of the base of the cupula (given by the local slope of the unit stress/strain relation of the material at the working level of strain) may be substantially lower than the effective Young's modulus, E_1, of the remainder of the cupula. A constant angular acceleration stimulus to the canal produces a uniform virtual pressure loading of the cupula in the steady state (equal to $2\rho\pi R^2\alpha$, ref. 34). If the cupula is attached around its periphery, low frequency acceleration stimuli would be expected to produce maximum displacements at the top of the base region (E_1/E_2 interface) for small volume displacements, since the bottom of the E_1 region is effectively mechanically unsupported. If the cupula is thicker on its lateral sides adjacent to the plana semilunata, the point of maximum displacement would be expected to be above the center of the crista. For larger volume displacements, associated with larger working levels of strain, the effective Young's modulus of the base region may increase because of the nonlinear elastic properties postulated for the base region reticulum, so that effectively $E_2 \simeq E_1$. In this instance, the point of maximum displacement would be expected to move up toward the center of the cupula, as the effective mechanical properties of the cupula material become more uniform. Such a hypothesis is speculative, but could account for the observations of McLaren, who offers an explanation of a somewhat similar nature. It emphasizes the need for further studies of the mechanical properties of the cupula substance itself. To date the Young's modulus of the cupula has been estimated only by very indirect methods (2,15,26), and nonlinear aspects were not assessed. Because the mechanical properties of the cupula and its attachment are likely somewhat nonlinear, it remains entirely possible that the pressure/volume relationship of the ampulla K, assumed linear in Eq. 14-12, is also. However, it should be pointed out that unless K

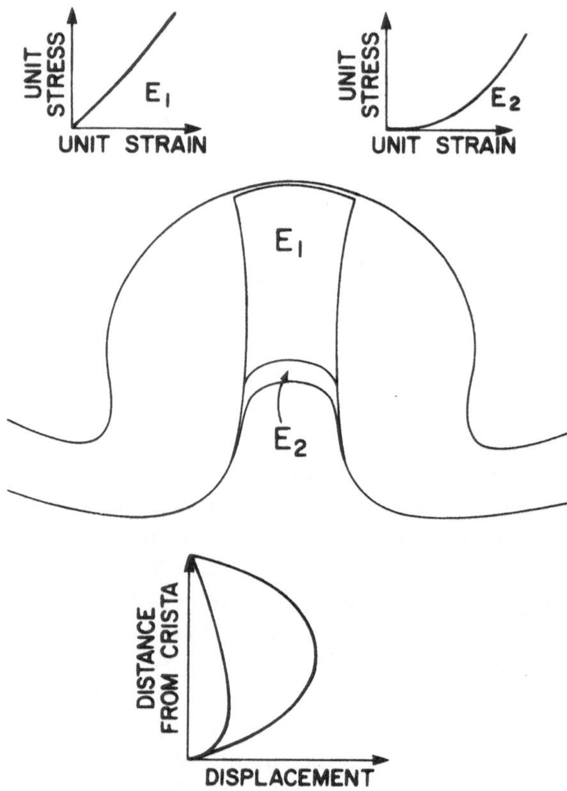

Fig. 14-4 Nonlinear elastic properties of cupula material make low frequency bending mode dependent on volume displacement.

increases dramatically with volume displacement, semicircular canal responses will likely remain overdamped. Although the long time constant τ_1 would be amplitude dependent, the short time constant τ_2 and the volume of endolymph accommodated by normal head movements in the frequency range $1/\tau_1 < \omega < 1/\tau_2$ are essentially unchanged. If this is the case, then the average displacement of all points on the face of the cupula is given by Eq. 14-25, even if K is nonlinear.

The hydrodynamic considerations implicit in Eq. 14-25 dictate only the total endolymph volume displacement into the ampulla, and say nothing about how this requirement is locally satisfied in terms of cupula displacement. However, the motion of the point of maximum displacement on the cupula cannot be less than the predicted average displacement. For all of the postulated modes of cupula displacement discussed earlier, the maximum displacement of the cupula does not much exceed twice the average value of cupula displacement. For example, if the cupula displaces as a solid plug, the maximum displacement is equal to the average displacement. If the cupula is attached about its periphery, and deforms parabolically, the maximum displacement is equal to twice the average displacement. Therefore, using Eq. 14-25, it is possible to estimate the maximum cupula displacement resulting from a given head velocity, although the location of this point on the cupula cannot be directly inferred. Average cupula displacement gain x/γ in mμ/degree/s of head velocity is estimated in Table 14-1. The range in these estimates comes directly from the assumed intercanal and interspecies differences in morphology and temperature. In the human, where the dynamic range of head motions is relatively well defined, head velocities occasionally exceed 1000°/ s, but most normal head motions (and vestibular testing) take place in the range from 2°/s (approximate subjective threshold) up to 200 to 300°/s in yaw. Hence, peak cupula displacement associated with subjective threshold motion is probably not less than 320 Å and reasonably may be estimated to lie in the range between 320 and 640 Å. This motion, of a dimension several times the thickness of unit membrane, is considerably larger than the estimated displacement of the basilar membrane at auditory threshold (18). At the other end of the head velocity dynamic range, peak cupula motion would be expected to be in the range from 3 to 10 μm for 200 to 300°/s head motions. These estimates are slightly larger than those made by Oman and Young in 1972 (26), and result directly from the revised model used. The calculations summarized in Table 14-1 suggest that roughly similar maximum displacements of the cupula should result from these head velocities in the frog, cat, and pike. Cupula displacement in the skate would be expected to be larger. However, it is important to note that the dynamic range of head velocities in nonhuman species is not well known, and in the case of the skate may well be lower. It therefore remains likely that the majority of head movements in many species are associated with cupula displacements so small that over much of the dynamic range they will be difficult or impossible to visualize through the wall of the ampulla using light microscopy. Although it may be possible to visualize cupula motion in some animals using large stimuli (7,21,22), such results should be interpreted cautiously, particularly since the bending mode of cupula deformation appears amplitude dependent. It may not be possible to infer the actual bending mode of the cupula for small volume displacements from measurements made at larger amplitudes.

It should also be emphasized that the particular bending mode exhibited by the cupula may be frequency dependent. In the absence of the cupula, flow velocities in the ampulla over the crista would be approximately parabolic for frequencies $\omega < 1/\tau_2$. The bending of the flow over the crista in the ampullary lumen, however, would tend to shift the peak flow velocity point toward the ampulla vault. As a limiting case, if the cupula material is assumed very weak, and sinusoidal endolymph displacements remain small, cupula deformation would be expected to conform to the profile of endolymph displacement. Whether or not this profile of displacement corresponds to the very low frequency, constant pressure difference deformation mode of the cupula depends, for reasons outlined earlier, on the properties of the cupula material and the attachment of the cupula to the crista and/or vault. Evidence obtained by McLaren in the frog, reviewed above, suggests that for small endolymph volume displacements, it probably does not. At low frequencies and small amplitudes, the point of maximum cupula deformation appears to be located near the crista, and not the center of the cupula. One would expect that for very low frequency accelerations, cupula deformation would be determined primarily by the equilibrium of strain dependent forces in the cupula with the inertial pressure load, whereas at some higher frequency, viscous shearing forces in the endolymph would be great enough to significantly limit cupula strain rate, and may, in fact, produce different, unanticipated, bending modes. To date, detailed measurements of cupula bending mode have been conducted only at a single frequency and at very large amplitudes. Tests at other frequencies in an animal preparation are clearly indicated. It may be ultimately necessary to resort to analytic or hydromechanical physical models to explore possible bending modes below the range of physical observation in vivo. However, such investigations may be necessary if workers in the field are to retain hope of predicting cupula displacement at the level of the hair cell cilia, and thereby improving our understanding of the detailed relationship between an angular acceleration stimulus and response in the peripheral afferent neuron.

CONCLUSIONS

1) Average endolymph flow displacement in the semicircular canal is more accurately described by a second order differential equation with three coefficients, rather than by the two-coefficient model advocated by van Egmond et al. (6) and adopted by most subsequent workers.

2) The values of the differential equation coefficients are shown to be functions of canal large radius and narrow duct area, as noted by van Egmond, and also utricular cross-sectional area, as first demonstrated by Van Buskirk (31). The dependence of the equation coefficients, the long and short response time constants, and the gain factors upon utricular length, cross-sectional area, narrow duct ellipticity, and endolymph temperature is here explicitly determined and numerically evaluated, using values for the morphologic parameters and temperatures which are believed representative of the three different canals in the human, cat, frog, skate, and pike.

3) In the past, investigators employing two-coefficient differential equation models for the semicircular canal have attempted to allow for the presence of the utricle and ampulla either by incorporating in their analysis only the drag of the narrow duct segment or by increasing the moment of inertia estimate above that for a simple toroidal ring of fluid. However, neither of these methods is totally appropriate. The former results in an overestimation of the short time constant of the canal, whereas the latter produces an overestimation of both time constants and gain factor. In fact, the short time

constant of the canal is shown to be independent of the relative (fractional) length of the utricle or narrow duct if the diameter of the utricular segment is significantly greater than that of the narrow duct. In such a case, the average displacement of the cupula in response to normal head movements is determined by the relative length of the utricle but not by its cross-sectional area.

4) If the utricle is relatively narrow, and there are significant variations on cross sectional area along its length, accuracy may be improved by extending the analysis employing a model containing multiple utricular segments.

5) The narrow canal duct in some animals, including man, is elliptic in cross section, resulting in a relative increase in duct inner surface area per unit volume of contained endolymph. As a result, viscous drag in elliptic ducts is greater, and is expected to produce a significant increase in the canal long time constant, as well as a concomitant decrease in the canal short time constant and the average cupula displacement resulting from normal head movements.

6) Over the years, several different modes of cupula deformation have been postulated. Experimental evidence obtained by McLaren and Hillman (22) in the frog indicates that the cupula may be fixed around its perimeter and function as a diaphragm whose mode of deformation is dependent on the magnitude of displacement. It is suggested that the low frequency dependence of cupula deformation mode on displacement magnitude may result from nonlinear elastic properties associated with different regions of the cupula reticulum. It is also noted that the bending mode of the cupula may be frequency dependent. Except at very low frequencies, viscous shearing forces in the endolymph could limit cupula strain rate. An understanding of the amplitude and frequency dependence of cupula bending mode will be necessary in order to assess the magnitude of the mechanical stimulus to underlying receptor cell cilia.

7) Although the cupula may exhibit nonlinear mechanical characteristics, during normal head movements these nonlinearities are expected to influence the bending mode but not the average displacement of the cupula, since the latter is determined only by the hydrodynamic characteristics of endolymph flowing in the membranous canal. Because the motion of the point of maximum displacement on the cupula cannot be less than the average cupula displacement, and probably does not much exceed twice the average cupula displacement for any of the presently known bending modes, the magnitude of the largest displacement of any point on the cupula can be estimated. For the human, the dynamic range of this motion appears to be from not less than 320 Å, for motions at the subjective threshold of rotation sensation, to upward of 10 μm. Consequently, it appears that the majority of head movements in many species probably produce maximum displacements so small that the motion should prove difficult or impossible to visualize through the wall of the ampulla. Therefore, results from experimental

studies of cupula motion should be interpreted with some caution: it may not be possible to infer the actual bending mode of the cupula for modest volume displacements from measurements made at larger amplitudes.

ACKNOWLEDGMENTS

The author is indebted to colleagues Lawrence S. Frishkopf and Sheila E. Widnall for their constructive criticism and useful suggestions during the course of this work, which was supported by NIH Grant NIH 5 R01 NS11080.

References

1. Curthoys, I.S., Blanks, R.H.I., and Markham, C.H.: Semicircular canal radii of curvature (R) in the cat, guinea pig, and man. J. Morphol. 151:1, 1977.
2. Curthoys, I.S., Markham, C.H., and Curthoys, E.J.: Semicircular duct and ampulla dimensions in cat, guinea pig, and man. J. Morphol. 151:17, 1977.
3. Dittrich, F.L., Extermann, R.E., and Greiner, G.F.: Biophysics of the Ear. Springfield, Ill., Thomas, 1962.
4. Dohlman, G.F.: Some practical and theoretical points in labyrinthology. Proc. R. Soc. Med. 28:1371, 1935.
5. Dohlman, G.F.: The attachment of the cupulae, otolith, and tectorial membranes to the sensory cell areas. Acta Otolaryngol. 71:89, 1971.
6. van Egmond, A.A.J., Groen, J.E., and Jongkees, L.B.W.: The mechanics of the semicircular canal. J. Physiol. 110:1, 1949.
7. Flock, A. and Goldstein, M.H.: Cupular movement and nerve impulse response in the isolated semicircular canal. Brain Res. 157:11, 1978.
8. Gray, A.A.: The Labyrinth of Animals, Including Mammals, Birds, Reptiles, and Amphibians. London, Churchill, 1907.
9. Groen, J.J.: Mechanical analysis of the phenomena in response to the stimulation of the semicircular canals. In van Egmond, A.A.J., Groen, J.J., and Jongkees, L.B.W. (eds.): The function of the vestibular organ. Pract. Oto-Rhino-Laryngol. (Suppl. 2) 14:42, 1952.
10. Hillman, D.E.: Observations on morphological features and mechanical properties of the peripheral vestibular receptor system in the frog. Prog. Brain Res. 37:1, 1972.
11. Hillman, D.E.: Cupula structure and its receptor relationship. Brain Behav. Evol. 10:52, 1974.
12. Howland, H.C. and Masci, J.: The phylogenetic allometry of the semicircular canals of small fishes. Z. Morphol. Tiere 75:283, 1973.
13. Howland, H.C. and Masci, J.: The functional allometry of semicircular canals, fins, and body dimensions in the juvenile centrarchid fish, *Lepomis gibbosus*. J. Embryol. Exp. Morphol. 29(3):721, 1973.
14. Jones, G.M. and Spells, K.E.: A theoretical and comparative study of the functional dependence of the semicircular canal upon its physical dimensions. Proc. R. Soc. Med. B 157:403, 1963.

15. ten Kate, J.H.: The Oculo-Vestibular Reflex of the Growing Pike: A Biophysical Study. Ph.D. Thesis, Rijksuniversiteit te Groningen, 1969.

16. ten Kate, J.H. and Kuiper, J.W.: The viscosity of the pike's endolymph. J. Exp. Biol 53:495, 1970.

17. Landolt, J.P., Correia, M.J., Young, E.R., Cardin, R.P.S., and Sweet, R.C.: A scanning electron microscopic study of the morphology and geometry of neural surfaces and structures associated with the vestibular apparatus of the pigeon. J. Comp. Neurol. 159:257, 1974.

18. Lawrence, M.: Dynamic range of the cochlea transducer. Cold Spring Harbor Symp. Quant. Biol. 30:159, 1968.

19. Llinas, R. and Hillman, D.E.: Comments on structure-function relationships in the peripheral vestibular system. In Iberall, A.S. and Guyton, A.C. (eds.): Regulation and Control in Physiological Systems. Pittsburgh, IFAC Publications, 1973, pp. 180–184.

20. Mayne, R.: The Constants of the Semicircular Canal Differential Equation. Goodyear Aerospace Report (GERA-1083). Litchfield Park, Arizona, Goodyear Aerospace Corporation, 1965.

21. McLaren, J.W.: The Configuration of Movement of the Semicircular Canal Cupula. Ph.D. Thesis, The University of Iowa, 1977.

22. McLaren, J.W. and Hillman, D.E.: Configuration of the cupula during endolymph pressure changes. Society for Neuroscience Meeting Abstracts, Sixth Annual Meeting, Toronto, Canada, 1976, p. 1060.

23. Money, K.E., Bonen, L., Beatty, J.D., Kuehn, L.A., Sokoloff, M., and Weaver, R.S.: Physical properties of fluids and structures of vestibular apparatus of the pigeon. Am. J. Physiol. 220:140, 1971.

24. Oman, C.M., Frishkopf, L.S., and Goldstein, M.H.: An upper limit on the physiological range of cupula motion in the semicircular canal of the skate. Society for Neuroscience Meeting Abstracts, Sixth Annual Meeting, Toronto, Canada, 1976, p. 1053.

25. Oman, C.M., Frishkopf, L.S., and Goldstein, M.H.: Cupula motion in the semicircular canal of the skate, *Raja erinacea:* An experimental investigation. Acta Otolaryngol. 87:528, 1979.

26. Oman, C.M. and Young, L.R.: Physiological range of pressure difference and cupula deflections in the human semicircular canal: Theoretical considerations. Acta Otolaryngol. 74:324, 1972.

27. Steer, R.: The Influence of Angular and Linear Acceleration and Thermal Stimulation on the Human Semicircular Canal. Sc.D. Thesis, Massachusetts Institute of Technology, Cambridge, Mass., 1967.

28. Steinhausen, W.: Uber Sichbarmachung und Funktionsprufung der Cupula Terminalis in den Bogengangsampullen des Labyrinthes. Pflügers Arch. Ges. Physiol. 217:747, 1927.

29. Steinhausen, W.: Uber die Beobachtung der Cupula in den Bogengangsampullen des Labyrinths des lebenden Hechts. Pflugers Arch. Ges. Physiol. 232:500, 1933.

30. Trincker, D.: The transformation of mechanical stimulus into nervous excitation by the labyrinthine receptors. Symp. Soc. Exp. Biol. 16:289, 1962.

31. Van Buskirk, W.C.: The effect of the utricle on fluid flow in the semicircular canals. Ann. Biomed. Engin. 5:1, 1977.

32. Van Buskirk, W.C. and Morse, E.W.: Fluid flow in the semicircular canals: A non-Newtonian model. Recent Adv. Engin. Sci. 7:273, 1977.

33. Vilstrup, J.: Studies on the completed structure and mechanism of the cupula. Ann. Otol. 59:46, 1950.

34. Vries, H.L. de: Physical aspects of the sense organ. Progr. Biophys. Chem. 6:207, 1956.

35. Zalin, A.: On the function of the kinocilia and stereocilia with special reference to the phenomenon of directional preponderance. J. Laryngol. Otol. 81:119, 1967.

Discussion

HOWLAND: I think we're indebted to you for pointing out the problems with elliptic cross sections of semicircular canals, and I'm sure I may even have been guilty of approximating an ellipse with a circle. It sounds like the sort of thing that one would do. But when thinking about what sections of semicircular canals are elliptic, I wanted to make a comment and ask you about a problem. In many of the fishes the semicircular canals are really very round in cross section. But the place where they are elliptic is in the pars communis, between the anterior and posterior semicircular canals. And it seems to me that the pars communis might even play a more important role in some of these terms than the utricle. I'm saying that with rough intuition because it's a narrower tube. But I would also like to point out that in the pars communis you have a rather more complicated dynamic problem, because when the animal pitches forward you have flow in two directions in the pars communis. If the canals were separate the flow would be seen to go one way in the anterior canal and the other way in the posterior canal. That seems to me a dynamic problem worthy of the prowess of MIT mathematicians to think about and perhaps more critical for the total equation than the problem of the utricle.

OMAN: You're right: it's a difficult problem. Let me comment on the effect of a relatively localized constriction in the duct, such as an elliptic section in the pars communis. An elliptic duct shape may be particularly significant, at least in the human, because the elliptic cross section is continued for a long distance around the duct. In terms of the quasi-steady flow solution, the inertial pressure difference which drives a given flow velocity is proportional to the length of the duct segment. To the extent that the constricted section of the pars communis is a relatively small segment of the semicircular canal, the effect of the constriction on the flow will be less important.

ANDERSON: Lest we think that all the imagination that Dr. Oman has presented this morning is at an end, I'm sure that there are more possibilities for further analysis. One of the most interesting investigations that I think is available is the possibility of looking into the fractional S-terms and the possibility of looking at the ampulla in more detail mechanically. It's an ideal distributed parameter system for looking at the aspects of the dynamics.

OMAN: Absolutely. One thing I should emphasize with respect to frequency response is that a very important simplification implicit in this model is that entrance effects (pressure gradients across the ampulla produced by rapid changes in ampulla cross section) have been left out, and in fact may be quite significant in certain instances, particularly at high frequencies. Bill Van Buskirk of Tulane has looked at this and shown that the quasi-steady approximation of the type I have used here probably applies for normal head movements. But for high frequency movements, it's a different story. How can this be approached? Well, one way to proceed is to develop a fluid scale model. One would try to write the scaling laws that apply for both the fluid mechanics and the cupula mechanics, although the latter will be quite complicated, and data will be hard to obtain. With respect to the cupula mechanics, the solutions for deformation of thick beams and diaphragms are extremely difficult to handle. I've talked to a few people about this who are experts in the aeroelastic field, and who consider fluid interactions with deformable membranes. They suggest a scale model simulation, or possibly an approach using finite element techniques.

O'LEARY: I'm wondering how much credibility you have put on the Navier-Stokes equation, given that this is a highly viscoelastic fluid and not a newtonian fluid. Those of us who have attempted to suck endolymph up into a pipette are aware that it's extremely viscoelastic.

OMAN: I agree with you that the newtonian assumption is important. As some of you may know, the question of whether endolymph is a newtonian fluid is a subject of some debate. When Bob Steer looked at this question experimentally, using large samples of human endolymph, his conclusion was that it is fundamentally newtonian. But others aren't so sure. Perhaps it depends on what portion of the labyrinth you take your sample from. The fundamental question is whether fluid in the ampulla behaves differently. I think we have to regard this as an open question. And until it's settled, we should be careful not to use an analytic model like this in the same manner that a drunk uses a lamppost—to support our theories rather than to shed illumination on the subject.

GUALTIEROTTI: I have two questions. The first one is, the frogs that we've been examining live in a widely changing temperature range. You have pointed out the influence of temperature on laminar flow. This introduces a complication for the animal, that has to depend on information varying with the environmental temperature. Temperature ranges from, say, 10 to 40°C in the swamp. The second question is, there will be obviously a distribution of velocity due to the laminar flow within the ampulla. And that means that central parts of the crista will be subjected to greater deformation as central velocity is higher, whereas the velocity is nearly nil near the wall of canal. So, we will also have a deformation of the cupula due to the velocity ranges, I suppose, for your calculation?

OMAN: The analysis I presented today suggests that endolymph viscosity changes with temperature physically may be a significant effect. With respect to the characteristics of the flow in the ampulla, the local flow solution depends intrinsically on the behavior of the cupula in the fluid lamina that's associated with the point in the ampulla that you're talking about. But that, unfortunately, is dependent on strains which are transmitted up and down the face of the cupula. It's a very complicated modeling problem. I think that it might not be easily subjected to analytic approaches. It perhaps ought to be looked at in a physical simulation. But in order to do analog experiments to set up a fluid mechanical model of the canal and study this phenomenon, we'd have to know more about the mechanical characteristics of the cupula locally, for example, the Young's modulus of the material high up in the cupula and also in the subcupular space.

LIM: I have been looking at the fresh freeze fracture of the cupula in pigeons, frogs, and mammals. It seems to me that what your model showed is pretty sound. One, I think cupulas do have skirts, that go all the way down to the periphery, with respect to which I think Dean Hillman also agrees with me, and which are also seen in...the fixed tissue which is very difficult to show; but a fresh tissue does. And then secondly, when you do the fresh freeze fracture, when you look at the midportion, you do have a gap just as if you see the fixed tissue. I think Igarashi, and many others, showed that. I think perhaps, if I recollect Hillman's experiment, in which he put a pipette in the center of the cupula, the motion of the center of the cupula might be slightly different from the peripheral part of the cupula where the endings are fixed through the base.

OMAN: Hillman's basic concept is that the cupula is fixed to the vault of the ampulla...in the normal physiology. And so right at the boundary you would presumably have no local motion. But you would have shear deformation of the material. Or there would be a deformation of the cupula the higher you went up in the material. With respect to your comment about the drapes described by Hillman, I guess it's fair to say that the discrepancies between Hillman's description, the description by Dohlman, and the comments by Van Buskirk about the possible role of a gelatinous matrix within the ampulla alongside of the cupula (and its possible contribution to the cupula stiffness), are reasons for my caution in interpreting the results of our experiments in the skate in terms of whether they confirm or deny the Hillman results in the frog.

15

Mechanical Properties of Sensory Hairs in the Semicircular Canal Crista

ÅKE FLOCK

The work outlined herein was designed to investigate the micromechanics of sensory hair bundles and to identify the responsible structural components at the ultrastructural and molecular level. Our reason for interest stems from the fact that very little is known about the first step in the excitation of sensory hair cells in the vestibular system: that of the mechanical properties and action of the sensory hairs and their substructures.

This investigation, which is in progress, involves controlled micromanipulation of sensory hairs under visual observation combined with electron microscopy, biochemical work including gel electrophoresis, identification of proteins by immunofluorescence and the study of membrane components by specific labeling techniques. Since this work is not completed it is premature to discuss conclusions and functional implications here. Rather, one of the difficulties involved will be pointed out.

Our finding that the stereocilia of the sensory hair bundle and the cuticular plate contain actin (4), a protein involved in contractile mechanisms in many nonmuscle cells (8), demands particular attention. Several possible functions for a contractile mechanism in hair cells immediately come to mind, one being that the mechanical properties of the vestibular system and the cochlea are partly under physiologic control (9). One would therefore like to know whether a change in interrelations among the actin filaments, or between them and other components of the cell, can be induced by physiologic stimulation. Such changes can be quite subtle, as they are in the acrosomal process of *Limulus* sperm where a change of twist of 0.2°/subunit of the actin α-helix causes motion by changing the actin cable from a coiled to a stiff, extended state (1). We would like to be sure, in the ear, that our description of, for instance, the packing of actin filaments in the stereocilia (3) applies to a well-defined mechanical state. In this respect one of our findings is disastrous: fixation with glutaraldehyde changes the stiffness properties of the sensory hairs, as determined by micromanipulation, from one of stiff tilting at the base to a flexible bending (7). Obviously, ultrastructural observations of such material do not reflect the natural state of the system.

The mere preservation of actin in nonmuscle cells is not an easy task, as witnessed by several authors (10). The preservation of physiologic interrelationships will demand even another step of resolution. A high priority task in the present investigation is therefore to develop fixation methods which give the best possible preservation of proteins and which preserve the stiffness of the stereocilia. In fact, most histology textbooks, and also contemporary scanning electron micrographs (2), depict stereocilia as rather flexible structures, probably an artifact caused by fixation. One has to return to the work by Schultze (12) and Retzius (11) on unfixed material to encounter the first descriptions of sensory hair stiffness properties.

Until problems such as those outlined above have been solved reference is given to the reports cited above, to a preliminary paper on pathologic actin in the waltzing guinea pig (6), and to an article now in press (5).

ACKNOWLEDGMENT

The present work was supported by a grant from the Swedish Medical Research Council (04X-02461), by funds of the Karolinska Institutet, Bergwalls Stiftelse, Swärds Stiftelse, and Söderbergs Stiftelse.

References

1. De Rosier, D., Tilney, L., and Flicker, P.: A change in twist of the actin-containing filaments in *Limulus* sperm produces motion without myosin. J. Cell Biol. 79:266, 1978.
2. Engström, H. and Engström, B.: Structure of the hairs on cochlear sensory cells. Hearing Res. 1:49, 1978.
3. Flock, Å.: Physiological properties of sensory hairs in the ear. In: Psychophysics and Physiology of Hearing, edited by Evans, E.F., and Wilson, J.P., New York, Academic Press, 1977.
4. Flock, Å. and Cheung, H.: Actin filaments in sensory hairs of the inner ear receptor cells. J. Cell Biol. 75:339, 1977.
5. Flock, Å., Cheung, H., Flock B., and Utter, G.: Three sets of actin filaments in sensory cells of the inner ear. Identification and functional orientation deter-

mined by gel-electrophoresis, immunofluorescence and electron-microscopy. J. Neurocytol. In press.

6. Flock, Å., Cheung, H., and Wersäll, J.: Pathological actin in vestibular hair cells of the waltzing guinea pig. Adv. Otorhinolaryngol. 25:12, 1979.
7. Flock, Å., Flock, B., and Murray, E.: Studies on the sensory hairs of receptor cells in the inner ear. Acta Otolaryngol. 83:85, 1977.
8. Hitchcock, S.E.: Regulation of motility in nonmuscle cells. J. Cell Biol. 74:1, 1977.
9. Kemp, D.T.: Stimulated acoustic emissions from within the human auditory system. J. Acoust. Soc. Am. 64:1386, 1978.
10. Maupin-Szamier, P. and Pollard, T.D.: Actin filament destruction by osmium tetroxide. J. Cell Biol. 77:837, 1978.
11. Retzius, G.: Das Gehörorgan der Wirbeltiere. I. Fische und Amphibien. Stockholm, Sanson and Wallin, 1884.
12. Schultze, M.: Ueber die Endigungsweise der Hörnerven im Labyrinth. Arch. Anat. Physiol. Wiss. Med. 343, 1858.

Discussion

OMAN: Åke, I wonder if you could tell us what your thoughts are on the role of F-actin. In the mammalian skeletal muscle, as I recall, the actin filaments are really a complex protein involving G-actin globules, and the contraction is actually triggered by conformational changes in tropomyosin and troponin molecules which are interlaced along the G-actin chain. I wonder, in this particular preparation, do you think that we're dealing with a fairly simple F-actin filament composed of only G-actin globules or whether the tropomyosin/troponin calcium trigger machinery is present? Also, is there any evidence in your electrophoresis results that in fact there's a significant amount of myosin present which would presumably form the other half of this contractile system, as in both striated or nonstriated muscles? If so, could you comment on whether in the natural state bonding might occur between the two components corresponding to contractile bonding in muscle.

FLOCK: In the muscle, the actin filaments are made up of globular proteins, which attach to one another like a chain, and seven of these are covered by tropomyosin. In the relaxed state, tropomyosin covers binding sites for the myosin heads. The third control protein, troponin, holds the tropomyosin in place, and is also attached to the actin globules. Upon contraction, calcium causes the troponin to move so that the tropomyosin is swung away. It swings down so that seven actin globules get their binding sites exposed to the myosin heads. It has been shown in recent years that several proteins can interact with actin. Actin seems to be a conservative protein which develops early in the animal kingdom. There are several binding sites, for the troponin and for the myosin. Now, in the sperm acrosomal process, you have other mechanisms of binding. There you have parallel strands of actin filaments, imagine a second one here and a third and so on, and they are attached laterally by links. These links are composed of still another protein, not yet identified. It has a molecular weight of 55,000. But the interesting thing is that this protein binds close to the binding site for myosin. So maybe we will find in the end that actin is a conservative protein, and several proteins can interact with it. What we now have to do is search for all these different control proteins, the ones we know of and such we don't know of. The course we are taking is to do that immunologically. We have to isolate and purify proteins and use them to make antibodies, and then locate the antibody targets through fluorescence or electron microscopy. So far we have done immunohistochemical identification of actin. We find in gel electrophoresis a band for α-actinin, which in the muscle is a glue which holds actin together, and we have a band so far also for tropomyosin.

WIEDERHOLD: With the stiffness and the linking together at the tops of the cilia, is this picture now consistent with Hillman's "plunger" model of bending over and pushing down the base of the kinocilium? Do you have any information on how the cuticular plate might move when you bend the stiff hairs?

FLOCK: The cuticular plate does not move in the intact crista. It is a stationary structure and the stereocilia have their hinges on top of it. But if you remove the membrane, with triton X-100, then you can see the cuticular plate rock together with the stereocilia. So there is a link between the two.

WIEDERHOLD: Do you see any pulling in and out of the base of the kinocilium?

FLOCK: I have not been able to watch that, you see, because I am looking at it from behind and not in the lateral view, really. So I can't answer that question. Also, in the organ of Corti there is no kinocilium at all, of course.

LOWENSTEIN: I must congratulate you on your elegant experiment. I'm very impressed. My first question is, I don't think you discussed the kinocilium if I recall. Did you discuss or describe that it's really motile? And, I think, Wiederhold had studies where cilia . . . at least you told me that it was motile. And if I recall a study with the octopus where the entire sensory cilia are the kinocilia, would you postulate whether the kinocilia system may evolve into stereocilia, and maybe that is the advantage of having a stereocilia system.

FLOCK: We have observed spontaneous beating of the kinocilium in three cases. Larry Frishhopf at MIT has induced beating of the kinocilium in the crista of the skate by applying a certain fraction of white India ink, which causes beating of the cilia for some reason. However, there is a syndrome in man which has several components. One is the misplacement of the heart to the right side instead of the left. Also, there are respiratory tract and sinus problems and infertility. These patients have a loss of dynein, which is the equivalent of the myosin in the muscle cells. Dynein is missing because of some genetic defect. Therefore, during embryologic development the ciliary motion is not there to cause asymmetry, pushing the heart over to the left. Immobility of the cilia also explains infertility and respiratory tract disease. These patients have a normal caloric response in the vestibular system. This does not say that the kinocilium is not important in the transduction process. It might be that the base of the kinocilium has a specific membrane which is mechanically sensitive, and that in development this membrane spreads by induction from the centriole over to encompass also at least the base of the stereocilia. So, we shouldn't discount the kinocilium yet.

HONRUBIA: Would you care to speculate on how this biochemical machinery works in the transduction process and how it absorbs mechanical energy? And how then do you envision that receptor potentials are being produced?

FLOCK: Well, that's a hard one. I'd like to limit myself to the mechanics. I'd like to start by having a simple view that the actin filaments within the stereocilia only provide a simple mechanical transmission system; they are stiff and transmit force to the base. I would like to look at the cuticular plate as a separate mechanical element, and I would not be surprised to find that a contractile mechanism is present within the cuticular plate.

HONRUBIA: You are, then, dismissing the thought that has been held for many years that the mechanical deformation of mucopolysaccharides in the gelatinous tectorial membrane, or the cupula, may be important in the transduction process. You now interpret the role of these structures, the gelatin structures, as pure transformers.

FLOCK: Yes.

ROSS: Some time ago in a study of both scanning and transmission electron microscopy of a tectorial membrane, we were fortunate enough to show a very regular arrangement of filaments from the tectorial membrane to the glycocalyx material of the stereocilia which you refer to but don't describe. So I immediately thought that perhaps there was something going on between the tectorial membrane and that glycocalyx material. It's a very regular arrangement of filaments. I don't know if you recall the picture or not.

FLOCK: This refers to the organ of Corti, doesn't it?

ROSS: Yes.

FLOCK: I'd still like to look upon those as being mechanical links between the tectorial membrane and the stereocilia. Hans Engström also has described fibers in the tectorial membrane attaching to stereocilia.

LOWENSTEIN: Have you any information as to how far the stereocilia penetrate into the cupula mass?

FLOCK: No. Not really. We've discussed this, Larry Frishhopf and I, and I think there is somewhat of a difference between species. In the skate it appears that the sensory hairs extend into the gel proper. In other systems, like the frog, according to Dean Hillman, there is some distance of subcupular space, and only at the very tip do the stereocilia enter into the base of the cupula.

16

A Species Comparison of Linear and Nonlinear Transfer Characteristics of Primary Afferents Innervating the Semicircular Canal

M.J. CORREIA, J.P. LANDOLT, M.-D. NI,
A.R. EDEN, AND J.L. RAE

Vestibular primary afferents which subserve the ampullae of the semicircular canals carry digitally encoded information about angular head movements to the central nervous system. The action potentials which travel along ampullary afferents are the result of a mechanoelectric transduction process composed of several cascaded systems. Each system contributes its static and dynamic response characteristics to the transduction process which has been detailed elsewhere (8,20,26), but can be briefly summarized.

Angular acceleration of the head causes movement of the cupula-endolymph (system I) within the semicircular duct.[1] The stereocilia and kinocilium of each hair cell are coupled to the cupula, so that as the cupula moves the cilia bend relative to the hair cell body which is anchored between supporting cells in the crista ampullaris. Hyperpolarizing or depolarizing receptor potentials are set up in the hair cells (system II), which make chemical and electrical synaptic contact with multibranching axonal fibers of the primary afferents. These unmyelinated fibers (system III), which form a complicated neural plexus within the vestibular neuroepithelium, carry

[1] In accordance with the nomenclature presented in *Nomina Anatomica* (1), we will use the term semicircular duct to describe the membranous part of the labyrinth sometimes referred to as the semicircular canal.

graded postsynaptic potentials that produce a generator potential on each unmyelinated axon until it exits the basement membrane of the neuroepithelium and becomes myelinated. At this point the generator potential produces an action potential which travels along the myelinated axon (system IV) to its cell body in the vestibular (Scarpa's) ganglion. From there the neural pathway continues by way of a myelinated postganglionic fiber to cells within the vestibular nuclei in the brain stem.

Although some progress has been made recently (23,37), technical difficulties have limited the amount of information about each of the individual systems identified above. Specifically, direct visualization of the motion of the cupula-endolymph system has been limited, electrophysiological recording of receptor potentials within hair cells has been limited, and electrophysiological recordings of discrete postsynaptic potentials (components of the generator potential) from the unmyelinated portion of primary afferent axons have been sparse.

However, during the past decade, numerous investigators (5,6,9,10,13,18,19,28,30–32,41,42,45–48) have recorded action potentials from ampullary afferent fibers or their cell bodies in the vestibular ganglion in several species. From this work certain inferences can be made about events that occur in the systems involved in the transduction of head angular acceleration information into neural messages and which precede the myelinated portion of the primary afferent.

In the following pages, we will summarize these data within the framework of the dominant model posited to describe the response dynamics of the cupula-endolymph system (system I). This model, the "torsion-pendulum" model, is derived from knowledge that each semicircular duct and the utricle can be likened to a closed circular tube filled with fluid, the endolymph. As the head rotates, this fluid moves relative to the wall of the tube because of inertial forces. At the same time, the fluid movement is opposed by viscous wall resistance and the distortable cupula which acts as a fluid-tight flap across the tube (in the plane of the crista). The cupula is thought to contribute slight inertial effects as well as elastic properties, and possibly viscous resistance. These conditions suggest an analogy to the torsion pendulum from classical physics, which can be expressed (49) by the second order differential equation,

$$\Theta\ddot{\xi}(t) + \Pi\dot{\xi}(t) + \Delta\xi(t) = \Theta K\ddot{\theta}(t) \qquad [16\text{-}1]$$

where $\xi(t)$, $\dot{\xi}(t)$, and $\ddot{\xi}(t)$ are cupula-endolymph system angular displacement, velocity, and acceleration, respectively; $\Theta \equiv$ cupula-endolymph system moment of inertia; $\Pi \equiv$ cupula-endolymph system rotational resistance (viscous damping); $\Delta \equiv$ cupula-endolymph system rotational stiffness; $\ddot{\theta}(t) \equiv$ head angular acceleration; and $K \equiv$ constant of proportionality between endolymph displacement and cupular displacement. From Eq. 16-1 the frequency response function (transfer function estimate; ref. 3) relating head angular acceleration, $\ddot{\theta}(s)$, to cupula deflection, $\xi(s)$, is

$$\frac{\xi(s)}{\ddot{\theta}(s)} = \frac{K}{s^2 + \Pi/\Theta s + \Delta/\Theta} \quad [16\text{-}2]$$

where $s = j(2\pi f)$ with $j = (-1)^{1/2}$

Since it is believed that the cupula-endolymph system behaves as a heavily damped torsion pendulum (10 times critically damped), the denominator of Eq. 16-2 has real roots which can be determined by solving the quadratic equation, i.e., $r_{1,2} = \{-\Pi/\Theta \pm (\Pi^2/\Theta^2 - 4\Delta/\Theta)^{1/2}\}/2$. Furthermore, since Π appears to be large relative to Δ and Θ, the square root term in $r_{1,2}$ can be approximated by the first two terms of its binomial expansion, i.e., $(\Pi/\Theta - 2\Delta/\Pi)$. Hence $r_1 \cong -\Delta/\Pi$ and $r_2 \cong -\Pi/\Theta$. Setting $\tau_L = -1/r_1 \cong \Pi/\Delta$ and $\tau_S = -1/r_2 \cong \Theta/\Pi$, Eq. 16-2 can be rewritten as

$$\frac{\xi(s)}{\ddot{\theta}(s)} = \frac{K\tau_L\tau_S}{(\tau_L s + 1)(\tau_S s + 1)} \quad [16\text{-}3]$$

or, in terms of head velocity, $\dot{\theta}(s) = \ddot{\theta}(s)/s$, as

$$\frac{\xi(s)}{\dot{\theta}(s)} = \frac{Ks\tau_L\tau_S}{(\tau_L s + 1)(\tau_S s + 1)} \quad [16\text{-}4]$$

or, in terms of head position, $\theta(s) = \ddot{\theta}(s)/s^2$, as

$$\frac{\xi(s)}{\theta(s)} = \frac{Ks^2\tau_L\tau_S}{(\tau_L s + 1)(\tau_S s + 1)} \quad [16\text{-}5]$$

Values for Π, Θ, Δ, and K, and hence the response dynamics of the cupula-endolymph system, can be estimated from the physical dimensions of the semicircular ducts and the physical properties of the endolymph and/or the cupula. For example, the viscous drag coefficient, Π, for a thin circular ring is $\Pi = 16\eta\pi^2R^3$. However, since the utricle occupies one-quarter of the ring and probably introduces minimal drag, it has been suggested (38) that $\Pi = 12\eta\pi^2R^3$, where η is the viscosity of the endolymph, $\pi = 3.14\ldots$, and R is the effective radius of curvature of the semicircular duct. Oman and Young (43) have suggested that the moment of inertia is $\Theta = \rho R^2 (V_d + V_a + V_u)$, where ρ is the density of endolymph, and V_d, V_a, and V_u are the volumes of endolymph in the semicircular duct, the ampulla, and the utricle, respectively. The elastic restoring couple, Δ, and the constant of proportionality between cupular displacement and endolymph displacement, K, have been defined (25) as $\Delta = \mu\pi r^2 R$ and $K = \pi^2 r^2 R/V$, respectively, where μ = coefficient of cupular stiffness and V = volume of the ampulla. A mean value, K = 0.5, was determined by Jones and Spells (25) using measurements taken from 44 species. Several investigators (11,21,25,38) have calculated values of Π and Θ in different species. For example, Curthoys et al. (11), using the equations cited above, determined values for Π and Θ and suggested that $\tau_S \cong \Theta/\Pi = 0.015$ seconds (sec), $\tau_S = 0.007$ sec, and $\tau_S = 0.004$ sec for the lateral semicircular canals in man, cat, and guinea pig, respectively. Based on the work of other investigators, Landolt and Correia (28) and Fernandez and Goldberg (13) used values $\tau_S = 0.002$ sec and

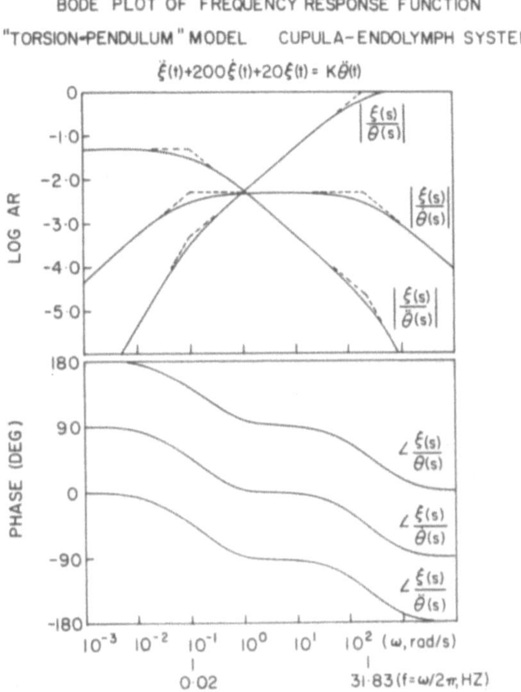

Fig. 16-1. Bode plots showing predicted curves of log AR and phase (in degrees) relating cupula-endolymph system displacement, $\xi(s)$, to head acceleration, $\ddot{\theta}(s)$, head velocity, $\dot{\theta}(s)$, and head position, $\theta(s)$. The curves are derived from the differential equation with the coefficients indicated at the top of the figure. For these coefficients, the upper corner frequency, $\omega_u = 200$ rad/sec (corresponding to $f = 31.83$ Hz), and the lower corner frequency, $\omega_L = \Delta/\pi = 10^{-1}$ rad/sec (corresponding to $f = 0.02$ Hz). After Belanger, F. and Mayne, R.: The Dynamics of the Semicircular Canals. Goodyear Aerospace Corp., 1965.

$\tau_S = 0.003$ sec for the anterior semicircular canal of the pigeon and the semicircular canals of the squirrel monkey, respectively. Since the coefficient of cupular stiffness, μ, has not yet been measured, no direct calculation of Δ has been possible. However, estimates of $\tau_L \cong \Pi/\Delta$ have been obtained using rotational stimuli and a variety of responses which have been reviewed elsewhere (22). Bode plots of three frequency response functions for one set of values of τ_L and τ_S are shown in Fig. 16-1. These curves are plots of the logarithmic amplitude ratio (log AR) and phase (ϕ) versus frequency, and represent evaluation of Eqs. 16-3–16-5 with assumed values of $\tau_L = 10.0$ sec, $\tau_S = 0.005$ sec, and $K = 1.0$ (2).

As Mayne (36) has indicated, the regions in which log AR does not change with frequency determine the mode of behavior of the semicircular canals (Fig. 16-1). Thus, in the frequency range $0 \leq f < f_L$, the canals act as angular accelerometers; in the frequency range $f_L \leq f < f_U$, as integrating accelerometers (i.e., as velocity meters); and as position meters for $f_U \leq f < \infty$. Assuming values of $\tau_L = 10.0$ sec and $\tau_S = 0.005$ sec, $f_L = 1/2\pi\tau_L \cong 0.02$ Hz and $f_U = 1/2\ \pi\tau_S \cong 32.0$ Hz. Since most of man's normal head movements are in the frequency range $f = 0.02–32.0$ Hz, it seems reasonable to assume that output from the semicircular canal is normally interpreted by the central nervous system (CNS) as being a measure of head velocity.

Does neural discharge of single ampullary afferents confirm the predictions made in Fig. 16-1 or do the systems which follow the cupula-endolymph system in the transduction process introduce deviations?

Before we attempt to answer this question, we must first determine whether ampullary afferent responses behave as the output of cascaded linear systems.

NONLINEAR RESPONSE CHARACTERISTICS OF AMPULLARY AFFERENTS

Analyses of single ampullary afferent responses of different species to sinusoidal, random, and transient angular accelerations of the head have revealed, in general, that some afferents demonstrate linear response properties, whereas others, which are exposed to the same stimuli, behave in a nonlinear manner. In general, nonlinear afferent responses (9,13,21,28,31,46,48) can be roughly categorized as (1) asymmetric response amplitudes about the resting (spontaneous) level; (2) rectified responses; (3) harmonic distortion of the response to sine wave stimuli; and (4) range compression of responses as indicated by the intensity (input-output) function.

Without exception, ampullary afferents in all species seem to respond to rotation in accordance with the morphological-functional relationship noted by Lowenstein and Wersäll (33); i.e., an afferent's action potential firing rate increases relative to the resting rate when the direction of rotation causes the stereocilia of the hair cell (or hair cells), which the afferent innervates, to deflect toward the hair cell's kinocilium. Conversely, rotation in the opposite direction causing the stereocilia to deflect away from the kinocilium is signaled by a decrease in firing rate relative to the resting level. However, asymmetric responses (neural activity above the resting level relative to that below) have been noted in most species, particularly for pulse angular accelerations of opposite polarity. In the squirrel monkey (18), cat (5), pigeon (28), and frog (46), peak excitatory responses (increase in firing rate above resting level) to pulse angular accelerations are usually greater than peak "inhibitory responses"[2] (decrease in firing rate below resting level). The tendency for asymmetric responses to occur during sine wave rotations has

[2] The usual caveat is issued. A decrease in firing rate probably results from a graded diminution of excitation (viz. a decrease in the depolarization of hair cells). For correctness, the word inhibition should be replaced by the word disfacilitation. To be consistent with other investigators, however, we have used the term inhibition throughout this paper.

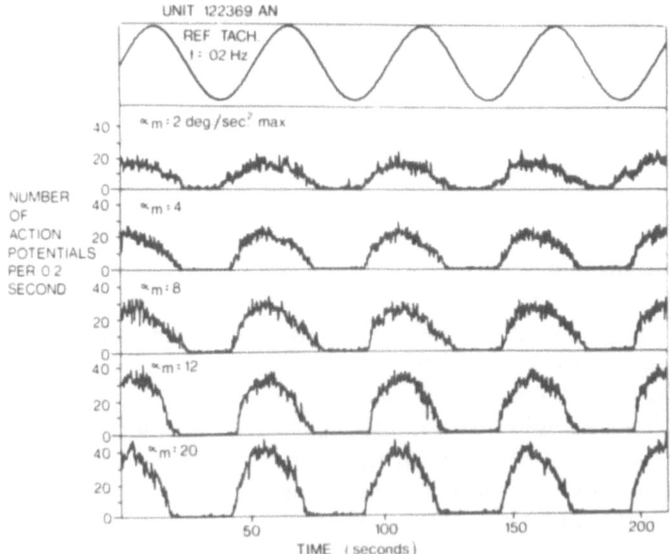

Fig. 16-2. A rectified neural response from a pigeon's anterior ampullary afferent. The top panel shows the frequency of the sinusoidal stimulus ($f = 0.02$ Hz). The remaining panels illustrate the neural response (units: $I \cdot sec^{-1}/0.2$ sec) to increasing levels of maximum (peak) angular acceleration (α_m). From Correia, M.J. and Landolt, J.P.: Adv. Otorhinolaryngol. 19:134, 1973.

been less clearly defined. In the squirrel monkey (13), the average discharge rate during stimulation is higher than the resting level. This finding, however, has not been observed following similar analysis of ampullary afferent data from the gerbil (47).

Rectification, as illustrated by a pigeon's neural response in Fig. 16-2, does appear to be a ubiquitous nonlinear response of some ampullary afferents in all species. According to Groen et al. (21), this amplitude asymmetry appears to be a reflection of a unidirectional dynamic range restriction imposed on the afferent's response by the nearness of its resting level to zero. For example, Blanks and Precht (6) noted that 11 of 13 (85%) frog ampullary afferents showed rectification at $f = 0.25$ Hz to a peak angular acceleration of 98.7°/sec² ($deg \cdot sec^{-2}$). These units were part of a sample of 31 afferents which exhibited a mean resting rate of 8.1 impulses/sec (I/sec) (range: 1 to 25 I/sec). However, it is equally clear that the tendency for an ampullary afferent response to rectify during angular head acceleration is a function of the afferent's gain (expressed as $I \cdot sec^{-1}/deg \cdot sec^{-2}$) at a particular frequency of oscillation or the afferent's response to different magnitudes of long duration pulse angular accelerations, i.e., its sensitivity, $\bar{S}$ (expressed as $I \cdot sec^{-1}/deg \cdot sec^{-2}$). Warm-blooded animals such as the squirrel monkey and pigeon with an average sensitivity of 2.24 (18) and a gain at $f = 0.2$ Hz of 1.72 I·sec^{-1}/deg·sec^{-2} (Fig. 16-5), respectively, and relatively high average resting discharges of approximately 90 I/sec (9,10,18,30), also produce rectified responses when the level of inhibitory stimulation (given the afferent's gain or sensitivity factor) is sufficient to silence the response.

Experiments using the frog (50) and the guinea pig (52, 53) suggest that as with ampullary afferent action potential responses, ampullary generator potentials are also asymmetric for excitatory and inhibitory stimuli. Peak depolarizing potentials in response to excitatory stimuli are greater than peak hyperpolarizing potentials to inhibitory stimuli. However, rectification appears to be a property associated solely with action potentials carried on the myelinated portion of the primary afferent system since Taglietti et al. (50) were unable to demonstrate, in the frog, rectified generator potentials to 0.1 Hz sinusoidal rotations with peak angular accelerations as large as 119.2 deg·sec^{-2}. This stimulus level is above that which produces a rectified action potential response at 0.25 Hz in the frog (6). However, this stimulus level does produce other nonlinear distortions in the generator potential response during the peak inhibitory portion of the stimulus cycle.

Random noise analysis (14,35) provides further evidence for the presence of nonlinearities in the pigeon's ampullary afferent response. For example, Fig. 16-3A illustrates the analysis of a single anterior ampullary afferent's response to bandlimited (DC-10 Hz) Gaussian noise rotation of the head. Figure 16-3A shows a trace of a portion of the stimulus, $x(t)$ (rotator tachometer voltage); entrained action potentials (A.P. train); and a trace corresponding to a voltage which is proportional to the instantaneous firing frequency, $y(t)$. Figure 16-3B is a plot of the first order Wiener kernel, $h_1(\tau)$, calculated by the cross-correlation method of Lee and Schetzen (29). The kernel, $h_1(\tau)$, relates $x(t)$ to $y(t)$ and indicates the afferent's response to a unit impulse of head velocity. Figure 16-3C shows, in relief, a plot of the second order Wiener kernel, $h_2(\tau_1,\tau_2)$. The contributions of the higher order kernels to a system response, $y(t)$, for a given input, $x(t)$, can be determined by the following equation.

$$
\begin{aligned}
y(t) = h_0 &+ \int_0^\infty h_1(\tau)x(t-\tau)d\tau \\
&+ \int\int_0^\infty h_2(\tau_1,\tau_2)x(t-\tau_1)x(t-\tau_2)d\tau_1 d\tau_2 \qquad [16\text{-}6] \\
&- P\int_0^\infty h_2(\tau,\tau)d\tau \\
&+ \ldots
\end{aligned}
$$

where P is the variance of the input signal, $x(t)$.

When second and higher order kernels are zero, Eq. 16-6 reduces to the familiar convolution integral for a linear system with $h_1(\tau)$ as its unit impulse response. Higher order kernels also represent the response to impulses. For example, the second order kernel represents that part of the system output

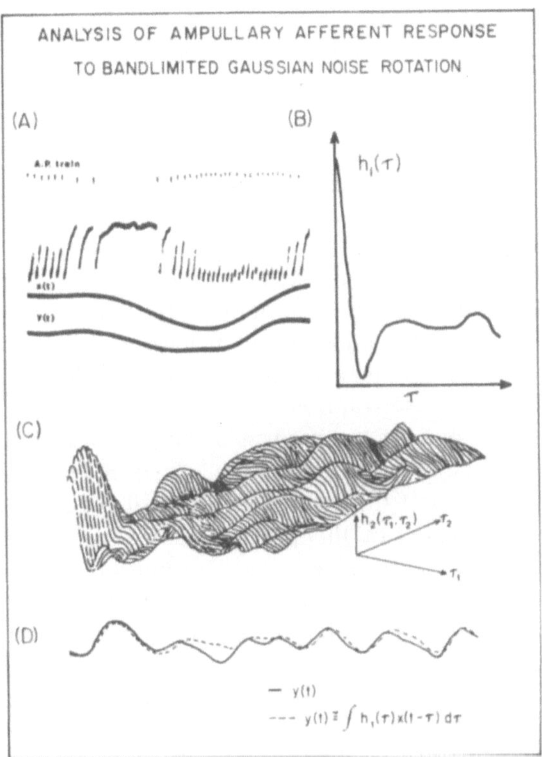

Fig. 16-3. Analyses of a pigeon's anterior ampullary afferent response to bandlimited Gaussian noise rotation. **(A)** The response waveform, $y(t)$, represents a voltage that is proportional to the instantaneous frequency of the action potential (A.P.) train. The stimulus waveform, $x(t)$, represents a voltage that is proportional to the velocity of the pigeon's head. **(B)** The first order kernel, $h_I(\tau)$, calculated by the method of Lee and Schetzen (29), relates $x(t)$ and $y(t)$. This kernel (the best estimate of the linear system's unit impulse response) is shown for values of τ to 0.5 sec. **(C)** The second order kernel, $h_2(\tau_1,\tau_2)$, relates $y(t)$ to $x(t)$ for values of τ_1 and τ_2 to 0.05 sec. **(D)** A comparison of output, $y(t)$ (solid line), with that modeled by using only the first order kernel, $h_1(\tau)$ (dashed line).

which is due to two input unit impulses which are separated in time by a difference of $\tau_1 - \tau_2$. The first and second order kernels presented in Figs. 16-3B and 16-3C, respectively, resemble those presented by French and Wong (15) in their nonlinear analysis of the tactile sensilla of the trochanteral hair plate (a mechanoreceptor) of the cockroach. The second order kernel plotted in Fig. 16-3C is important in that it is *nonzero* and spread over the entire surface of the plot as opposed to being zero everywhere but along the main diagonal (15). These characteristics are indicative of a linear filter with fast but finite dynamics followed by a squaring nonlinear element (35). Since it is known that rectification is a characteristic nonlinearity of some pigeon ampullary afferent responses (Fig. 16-2), it must be assumed that the squaring element approximates the nonlinearities that may be introduced by a rectifier. French and Wong (15) have argued that the addition of two curves, $y = h_1(\tau)x$ and $y = h_2(\tau_1,\tau_2)x^2$, approximates the intensity function of a rectifier; $y = 0(x < 0)$, $y = x(x \geq 0)$ over the range $x = \pm h_1(\tau)/h_2(\tau_1,\tau_2)$.

Figure 16-3D is a plot of the actual neural response $y(t)$ (solid line) and a plot of a modeled response $y(t)$ (dashed line) using only the first order kernel as approximated by the equation shown at the bottom of the figure. It has been suggested (35) that the best way to determine the contribution of a kernel to an output is to model the output using the kernel. A comparison of the two curves in Fig. 16-3D indicates that, with certain exceptions, the first order kernel (linear element) accounts for most of this ampullary afferent's response to random noise rotations.

Other types of nonlinear responses have been reported for sinusoidal stimulation. For example, one type is illustrated in Fig. 16-4, where it is apparent that a skewness of the sinusoidal waveform exists about the maxima and minima firing rates. For this particular pigeon's anterior ampullary afferent response, the increase in the action potential firing rate has a less steep slope than the decrease in the firing rate. The data which are illustrated in Fig. 16-2 for a rectifying afferent response show an opposite skewness about the maxima firing rates. The skewness of nonrectifying sinusoidal response waveforms has been quantified in several ways. For the frog, Blanks

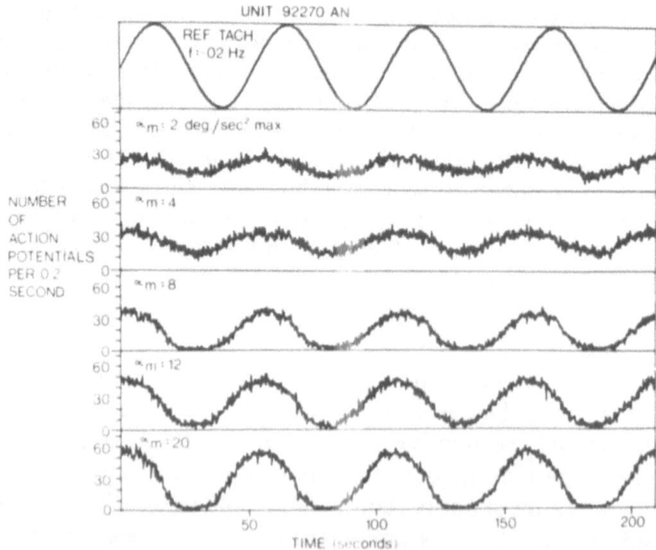

Fig. 16-4. A skewed neural response from a pigeon's anterior ampullary afferent. Details are the same as those for Fig. 16-2. From Correia, M.J. and Landolt, J.P.: Adv. Otorhinolaryngol. 19:134, 1973.

and Precht (6) measured the angular distance between the peak response amplitude and the midpoint of the excitatory portion of the response waveform, a cycle histogram. They obtained values ranging from $+10°$ (lead) to $-15.2°$ (lag) with an average of $-2.96 \pm 7.89°$ (mean$\pm$SD, $N = 13$). Louie and Kimm (31) analyzed rhesus monkey lateral ampullary afferent responses using percent harmonic distortion [defined as the square root of the ratio of the sum of the second through the tenth harmonic to the first harmonic of a single cycle of averaged firing rate multiplied by 100 (average based on 10 cycles)]. They reported mean $\pm$ SD harmonic distortion of $30.0 \pm 3.0\%$ (23 ampullary afferents) which appears to be constant over the frequency range, $f = 0.10$–4.00 Hz.

Landolt and Correia (28) used the same statistical method to calculate harmonic distortion of a continuous series of sinusoidal responses from the pigeon's ampullary afferents and observed considerable variability between responses of different afferents even in the same bird. They found a mean$\pm$SD harmonic distortion of $20 \pm 13\%$ at $f = 0.02$ Hz for a peak angular acceleration of 20 deg$\cdot$sec^{-2}. Fernandez and Goldberg (13) quote a mean harmonic distortion of 7.3% at an oscillation frequency of 0.025 Hz and a peak angular acceleration of 40 deg$\cdot$sec^{-2} for squirrel monkey ampullary afferent responses.

Several investigators (5,9,18,28,46) have examined the intensity functions (stimulus-response relationships) for ampullary afferents in different species using a variety of waveforms. Figure 16-5 illustrates intensity functions derived from ampullary afferent neural responses in the frog, bird, cat, and monkey.

A comparison of these curves prompts several general comments. First, fewer and fewer ampullary afferents appear to have nonlinear intensity functions as one proceeds upward in evolutionary development. For example, Precht et al. (46) reported that 61% (30/49) of the frog ampullary afferents they tested, using angular acceleration pulses, showed nonlinear relationships between firing rate above the spontaneous level and magnitude of angular acceleration. Four units illustrating both linear and nonlinear relationships are presented in Fig. 16-5A. Figure 16-5B shows mean ($\pm$SD) values relating the pigeon's ampullary afferent firing rate versus peak angular acceleration. Two frequencies of oscillation are represented. It is clear that at the low frequency of oscillation ($f = 0.02$ Hz), the mean response is nonlinear. Two of 15 (13%) of the ampullary afferent responses showed significant reduction ($P < 0.05$) in regression sum of squares when fit by a second degree polynomial using curvilinear regression. At a higher frequency, $f = 0.2$ Hz, only 1 of 15 (7%) of the pigeon ampullary afferent responses was significantly nonlinear ($P < 0.05$). Blanks et al. (5) and Goldberg and Fernandez (18) reported that most of the ampullary afferents they tested in the cat and squirrel monkey, respectively, produced linear intensity functions in response to different magnitudes of angular acceleration pulses. The intensity functions for representative afferents, showing different degrees of adaptation, are presented in Fig. 16-5C for the cat, and in Fig. 16-5D for the squirrel monkey.

Secondly, of the four species whose intensity functions are presented in Fig. 16-5, only the frog appears to have a noticeable threshold to rotation. Extrapolation of the frog's intensity function to the zero firing rate above the spontaneous level results in threshold values, illustrated in Fig. 16-5A, and reported by Precht et al. (46) to range from 0.3 to 2.5 deg·sec^{-2}.

Thirdly, average afferent sensitivity, $\bar{S}$, derived from intensity functions using pulse angular accelerations, appears to be quite different for the frog when compared with mammals. The mean slope of the three approximately linear curves in Fig. 16-5A, with data represented by closed triangles, circles, and squares, was calculated to be 4.98 I·sec^{-1}/deg·sec^{-2}. This value falls within the range of values reported by Precht et al. (46), i.e., $4.0 \leq S \leq 6.8$, and is about 2.5 times larger than mean sensitivity values reported for mammals such as the cat ($\bar{S} = 1.8$) and the squirrel monkey ($\bar{S} = 2.24$).

Amplitude ratios (gains) have been determined for different species using different levels of peak angular acceleration for sinusoidal oscillations at different frequencies. Blanks and Precht (6) reported an average ($\pm$SD) gain of 1.58 $\pm$0.69 I·sec^{-1}/deg·sec^{-2} for 13 frog lateral ampullary afferents at $f = 0.25$ Hz. They noted that the mean gain at this frequency did not change with higher levels of peak angular acceleration. We calculated the "best-fit" linear regression line for each of 10 of the 15 pigeon ampullary afferents which produced the intensity function shown for $f = 0.20$ Hz in Fig. 16-5B. We had data at each acceleration level for each of only 10 units. The mean regression line we obtained was $y = 1.72x + 1.23$. Average gain at the angular acceleration levels of 2, 4, 8, 12, and 20 deg·sec^{-2} was 2.91, 2.12, 1.84, 1.92, and 1.75 I·sec^{-1}/deg·sec^{-2}, respectively. A repeated measurement analysis of variance revealed a statistically significant difference ($F = 7.21$; d.f. = 4, 36; $P < 0.01$) between mean gain values for all five levels. However, when all the unusually high gain values corresponding to the lowest level of stimulation (2 deg·sec^{-2}) were deleted from the sample, differences between the mean gain levels corresponding to the four remaining intensity levels were not statistically significant ($F = 2.09$; d.f. = 3, 22; $P > 0.05$).

Schneider (47) reported that, in general, gerbil primary afferent response gain values, tested at 0.5 Hz, decreased linearly with increasing levels of stimulus intensity. He presented data for three units: one showed no change over a 3.5-fold increase in stimulus intensity; the other two showed decreasing values of gain for increasing levels of stimulation. Specifically, the two units' gains changed 33% and 47% over an 8-fold change in intensity level (i.e., from 101 to 804 deg·sec^{-2}). Such nonlinear behavior is not surprising considering the high levels of the stimuli used. Fernandez and Goldberg (13) presented a "representative" squirrel monkey's ampullary afferent response to a 16-fold increase in stimulus intensity (i.e., from 5.0 to 80.6 deg·sec^{-2}) at

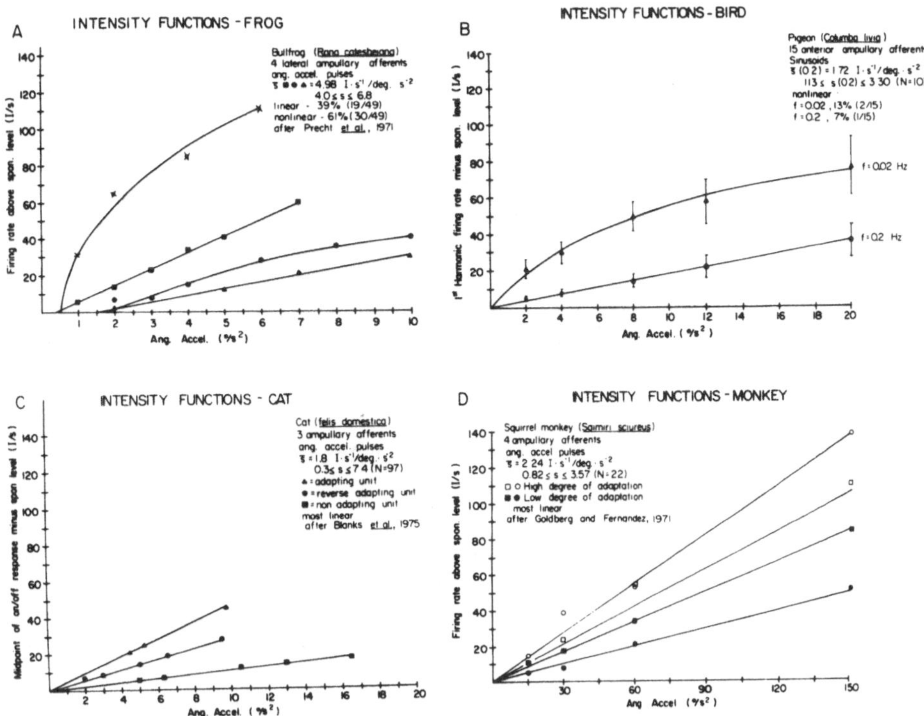

Fig. 16-5. A composite figure illustrating ampullary afferent responses of four species. Representative intensity functions are shown for individual ampullary afferents for the frog **(A)**, bird **(B)**, cat **(C)**, and squirrel monkey **(D)**. Functions in **(A)**, **(C)**, and **(D)** were obtained using angular velocity trapezoidal (angular acceleration pulse) stimuli. Mean intensity functions for the pigeon **(B)** were obtained using sinusoidal rotations. These functions are presented for two frequencies, $f = 0.02$ and $f = 0.2$ Hz. For each species, estimates of mean sensitivity, $\bar{S}$, and dispersion about the mean are presented. **(A)** After Precht, W., et al.: Exp. Brain Res. 13:378, 1971. **(C)** After Blanks, R.H.I., et al.: J. Neurophysiol. 38:1250, 1975. **(D)** After Goldberg, J.M. and Fernandez, C.: J. Neurophysiol. 34:635, 1971.

a frequency of oscillation of 0.05 Hz. They noted that the unit's gain varied less than 10% and that the phase angle varied no more than 3°.

In summary, it appears that ampullary afferent responses, regardless of species, demonstrate a rectification nonlinearity when the product of the sensitivity factor (or gain) and the stimulus magnitude is greater than the resting discharge and the response is silenced during inhibitory stimuli. Also, it appears there is a tendency for the excitatory response to a pulse angular acceleration to be larger in peak amplitude than an inhibitory response to an equivalent stimulus of opposite polarity. Furthermore, some species show a sine wave amplitude asymmetry (13) whereas others do not (47), and some species (28,31) show more harmonic distortion of the sinusoidal waveform response than do others (13). Moreover, it seems that extreme stimuli, either high or low [e.g., in the pigeon, angular accelerations < 2 deg $\cdot$ sec^{-2} (28); and in the gerbil, angular accelerations > 201 deg $\cdot$ sec^{-2} (47)], cause nonlinearities in ampullary afferent responses. Finally, it appears that the lower a species is in evolutionary development, the greater is the proportion of its ampullary afferents having nonlinear intensity functions (Fig. 16-5A–D).

However, as with all sensory systems and regardless of species, the stimulus range can either be restricted or ampullary afferents selected so that an afferent's linearized response can be subjected to a linear systems analysis and the afferent's transfer function estimated by a frequency response function (3).

LINEAR RESPONSE CHARACTERISTICS OF AMPULLARY AFFERENTS

Determination of the Best Form of the Frequency Response Function

Groen et al. (21) determined the coefficients of Eq. 16-1 for a number of lateral ampullary afferents in the ray (*Raja clavata*). These investigators used sinusoidal oscillations and pulse angular accelerations and suggested that the average undriven differential equation for the ray's cupula-endolymph system was $\ddot{\xi}(t) + 35\dot{\xi}(t) + \xi(t) = 0$. Using the algebraic manipulations we described earlier, values of $\tau_L = 35$ sec and $\tau_S = 0.03$ can be obtained from this equation. Groen et al. (21) measured the inner diameter of the lateral semicircular duct in one ray. Using this value ($d = 2r = 0.68$ mm), and values of $\eta = 0.01$ poise and $\rho = 1.0$ g $\cdot$ cm^{-3} for endolymph viscosity and density, respectively, and the equation $\Pi/\Theta = 4\eta/\rho r^2$, they determined a value of 35 sec^{-1} (identical to Π/Θ in the equation derived from average ampullary responses).

Fernandez and Goldberg (13) noted some important differences between the average frequency response function they obtained for the anesthetized squirrel monkey's ampullary afferents and the one predicted by the torsion-pendulum model of the cupula-endolymph system. Typical Bode plots of the latter are shown in Fig. 16-1. A Bode plot of the frequency response function of their data is shown in Fig. 16-6.

In this and subsequent plots, *mean* data from investigators other than ourselves were obtained from equations, tables, and graphs contained in their publications. Amplitude ratio and phase values were entered into a PDP 11/20 computer program (7) and best-fit transfer functions were iteratively obtained. In some cases we evaluated equations suggested by the investigators themselves and "forced" curves of their models through their published values of amplitude ratio and phase. In all cases, aside from visualization of how well the predicted curves fit the data, a "goodness-of-fit" statistic, MSE, is presented. The mean square error, MSE = LSE/(2L − S), where the least square error, LSE, is formed from a matrix of residuals which is

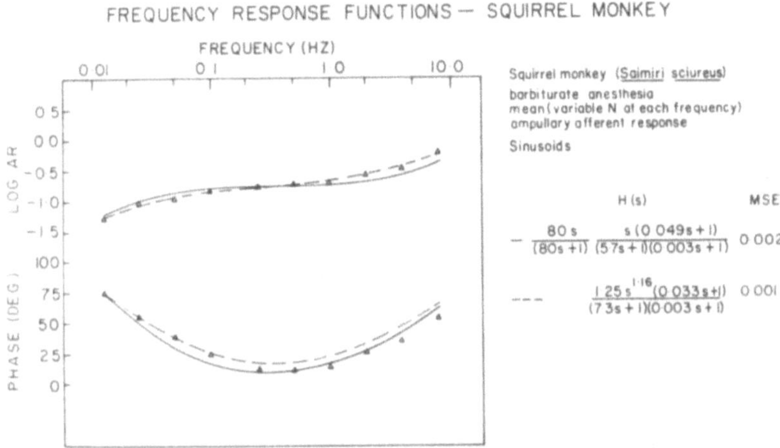

Fig. 16-6. A Bode plot of squirrel monkey mean ampullary afferent data. The plots represent neural response re head velocity. Frequency response functions (estimated transfer functions) were either forced through or fit to the data. Goodness-of-fit (in the least squares sense) is indicated by the mean square error (MSE). The equation corresponding to the solid line is taken from Fernandez and Goldberg (13). The second equation contains best-fit parameters in an equation in which the operator $\tau_A s/(\tau_A s + 1)$ was replaced by the operator s^k. Each fit was constrained since the value, $\tau_s = 0.003$ sec, was fixed. The operators and their associated time constants are defined in the text.

made up of error terms that express the difference between experimental data and best-fit or "forced-model" data. The degrees of freedom for the MSE statistic consist of L ≡ number of frequency points and S ≡ number of unknown parameters in the frequency response function.

Thus, as indicated, Fig. 16-6 presents a Bode plot of the mean data from the publication by Fernandez and Goldberg (13), together with the equation which they suggested as a best-fit to their data (plotted as solid line). The amplitude ratio (gain) and phase values for each afferent's response were determined by relating the amplitude of the first harmonic of the average response of several cycles to peak head velocity. The two major discrepancies between the frequency response function of the torsion-pendulum model and the squirrel monkey's mean ampullary afferent frequency response function are that the latter shows an unpredicted phase lead at low frequencies and an unpredicted high frequency gain enhancement with progressive phase lead (compare Figs. 16-1 and 16-6). Fernandez and Goldberg (13), after showing that different squirrel monkey ampullary afferent responses demonstrate different degrees of adaptation to pulse angular accelerations, suggested that the low frequency phase advance discrepancy could be accounted for by an adaptation operator (34,54), $\tau_A /(\tau_A s + 1)$. Moreover, they suggested that the simplest operator which could account for the high

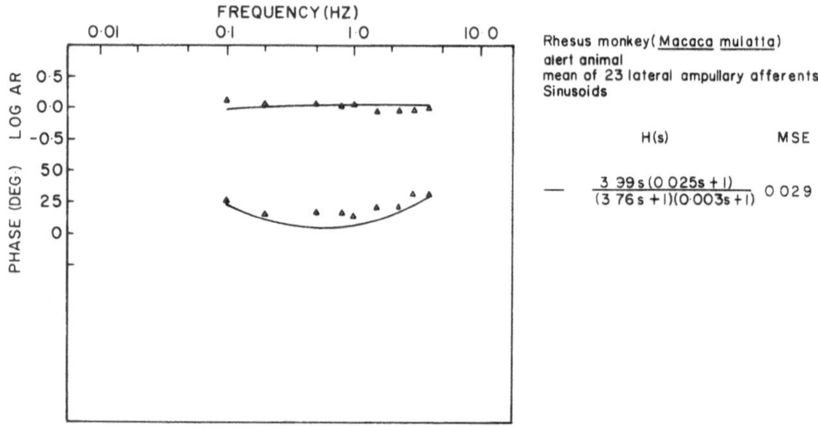

Fig. 16-7. A Bode plot of rhesus monkey mean lateral ampullary afferent data. The plots represent neural response re head velocity. The equation corresponding to the solid line represents the simplest frequency response function which described the data. Best-fit unconstrained values of $\tau_L = 3.76$ sec and $\tau_V = 0.025$ sec were determined. From Louie, A.W. and Kimm, J.: Exp. Brain Res. 24:447, 1976.

frequency gain enhancement and progressive phase advance was a zero term, $(\tau_V s + 1)$, which would imply a system sensitive to both cupular displacement and cupular velocity. The time constant, τ_V, would determine the frequency at which the system would shift from one mode of response to the other (viz., displacement to velocity). Fernandez and Goldberg (13) incorporated these operators into a general transfer function of the form,

$$\frac{R(s)}{\dot{\theta}(s)} = \frac{\tau_A s}{(\tau_A s + 1)} \frac{s(\tau_V s + 1)}{(\tau_L s + 1)(\tau_S s + 1)} \qquad [16\text{-}7]$$

Louie and Kimm (31) investigated the frequency response of 23 lateral ampullary afferents (re peak angular acceleration of the head) in the alert rhesus monkey and obtained data over a mid to high frequency range (0.1–4.0 Hz). They suggest that their gain (log AR) decreases with increasing frequencies at a rate of 1.09 log units/decade, approximating the slope of a first order lag system (or a second order lag system over a frequency range well below the high frequency cutoff). Moreover, these investigators noted a progressive phase increase at higher frequencies suggesting an operator including a time constant, τ_V, as for the responses from the squirrel monkey. Figure 16-7 presents a frequency response function of the mean of their data, except that in this figure the neural response is referenced to the peak head velocity. Over the frequency range, $f = 0.1$–4.0 Hz, the gain is essentially flat (0.194 log unit change), but a progressive phase increase is noted between 1.0 and 4.0 Hz. The constrained (τ_S set to 0.003 sec;

based on biophysical calculations) best-fit frequency response function of the simplest form, which accounts for these response characteristics, is shown in the figure. Time constants of $\tau_V = 0.025$ sec and $\tau_L = 3.760$ sec were obtained. Since data were not available at low frequencies ($f < 0.1$ Hz), it was not necessary to include the adaptation operator, $\tau_A s/(\tau_A s + 1)$. Moreover, the value of τ_L *is suspect*, since it also is most clearly defined in the low frequency range (0.005–0.1 Hz).

Thus, from data available, it appears that the primate's ampullary afferent response reflects the cupula-endolymph system (torsion-pendulum model), modified by a lead operator (high-pass filter) and a frequency-dependent differentiator. It has been suggested (13) that the high-pass filter represents adaptation, and the frequency-dependent differentiator represents a shift in response mode from that of sensing cupular displacement to that of sensing cupular velocity.

Schneider and Anderson (48) examined the frequency response of single lateral ampullary afferents of a small mammal, the gerbil. Following a suggestion by Goldberg and Fernandez (19), Schneider and Anderson (48) divided units into two groups; those with "regular" and those with "irregular" resting discharge patterns. They presented mean frequency response data for units whose resting discharge produced a coefficient of variation, CV, which was greater than 0.1 (triangles in Fig. 16-8), and mean frequency response data for units with a resting discharge having a $CV \leq 0.1$ (squares

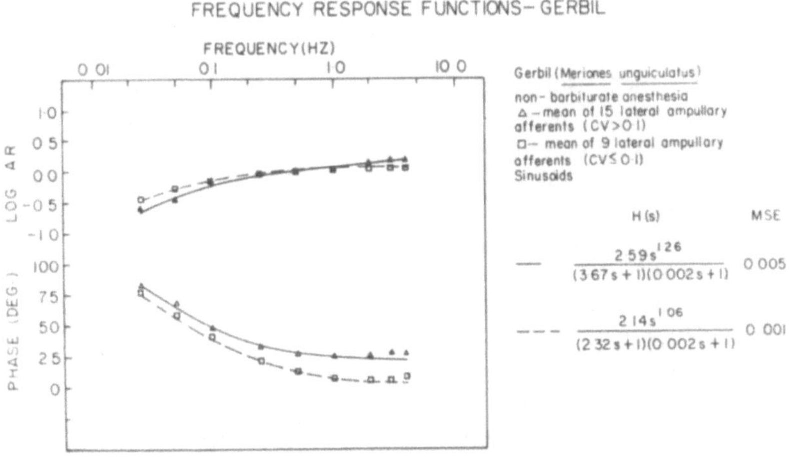

Fig. 16-8. A Bode plot of gerbil mean lateral ampullary afferent data. The plots represent neural response re head velocity. Squares represent mean data from units with a regular resting discharge, i.e., from units having $CV \leq 0.1$; triangles represent mean data from irregular units ($CV > 0.1$). Best-fit frequency response functions combining the s^k operator with the torsion-pendulum model are also presented (τ_S is constrained to 0.002 sec). After Schneider, L.W. and Anderson, D.J.: Brain Res. 112:61, 1976.

in Fig. 16-8). As indicated by the Bode plots in Fig. 16-8, the frequency response functions for mean first harmonic neural responses (re head velocity) are different in the two cases. Those units with a $CV > 0.1$ show a greater slope at lower (0.02–0.10 Hz) and higher frequencies (2.0–4.0 Hz) and a greater phase advance (re peak head velocity) at all frequencies. Schneider and Anderson (48) noted that, in contrast to the primate, their data did not show gain enhancement and progressive phase advance at high frequencies. They did observe, however, that all their units showed a constant positive gain slope for frequencies above 0.1 Hz, and a constant large phase lead above 0.5 Hz. Because of these characteristics, they proposed a frequency response function with the general form

$$\frac{R(s)}{\ddot{\theta}(s)} = \frac{Cs^k}{(\tau_L s + 1)(\tau_S s + 1)}, \; 0 \leq k \leq 1. \qquad [16\text{-}8]$$

This function relates neuron firing rate, $R(s)$, to angular acceleration of the head, $\ddot{\theta}(s)$. The constant, $C = K\Gamma(1 - k)$, where $\Gamma(1 - k)$ is the gamma function with parameter $(1 - k)$. The fractional differentiator, s^k, has a greater or lesser effect on the frequency response function as k varies from one to zero.

Although they did not fit parameters, Schneider and Anderson (48) presented values for $\tau_L = 2.5$ sec, $\tau_S = 0.002$ sec, and suggested a general frequency response function, relating neuron firing rate, $R(s)$, to angular head velocity, $\dot{\theta}(s)$, to be

$$\frac{R(s)}{\dot{\theta}(s)} = \frac{K\Gamma(1 - k)s^{k+1}}{(\tau_L s + 1)(\tau_S s + 1)}. \qquad [16\text{-}9]$$

Furthermore, they suggested that for irregular units, k had an average value of 0.29, whereas for regular units, k equaled zero. We determined parameters for equations of the form of Eq. 16-9 which provided best-fit frequency response functions to the data of Schneider and Anderson (48). These parameters, which are included in the equations presented in Fig. 16-8, are slightly different from those presented by Schneider and Anderson (48); notably, we found a larger torsion-pendulum long-time constant, $\tau_L = 3.67$ sec, for afferents with a $CV > 0.1$ and a nonzero value of $k = 0.06$ for units with a $CV < 0.1$.

Over a similar frequency range, ampullary afferent frequency response data from the pigeon, unlike primates but like the gerbil, are inadequately described by a frequency response function of the form of Eq. 16-7. A Bode plot of a typical anesthetized pigeon's anterior ampullary afferent response to sinusoidal oscillations (with a peak angular acceleration of 12 deg $\cdot$ sec^{-2}) is presented in Fig. 16-9. Clearly, the curve which is the poorest fit is the one generated by evaluation of the equation form (Eq. 16-7) corresponding to the torsion-pendulum model cascaded with an adaptation operator (MSE = 1.93). The data are not fit much better by an equation which corresponds to the torsion-pendulum model only (MSE = 1.61). However, a

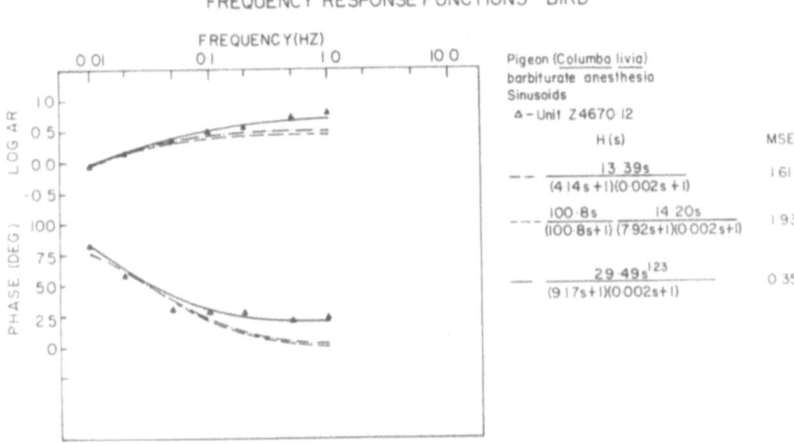

Fig. 16-9. A Bode plot of a pigeon's single anterior ampullary afferent response to sinusoidal head velocity. Best-fit curves of three forms of frequency response functions are also shown. One curve (dot-dash line) represents the torsion-pendulum model. A second curve (dashed line) represents the torsion-pendulum model plus an adaptation operator. A third curve (solid line) represents the torsion-pendulum model with the fractional differentiator (adaptation operator), s^k.

5-fold improvement in the MSE occurs when the equation (Eq. 16-9) corresponding to the torsion-pendulum model with a fractional differentiator, s^k, is fit to the data (MSE = 0.35). The resulting curve accommodates two characteristics of the frequency response data; namely, increasing gain over the frequency range $f = 0.1-1.0$ Hz, and a phase lead at low frequencies which levels off to constant phase advance over the frequency range from approximately 0.1–1.0 Hz.

Landolt and Correia (28) noted that each pigeon anterior ampullary afferent they tested demonstrated a frequency response which produced a unique value for k and τ_L. Thus, a range of values was produced, and in a sample of 14 ampullary afferents, $0.02 \leq k \leq 0.66$ and $4.45 \leq \tau_L \leq 22.17$. Moreover, Landolt and Correia (28) showed that a linear regression of CV on k for 28 ampullary afferents had a significant positive slope ($P < 0.05$). They concluded that each ampullary afferent had its own signature response with a unique value of τ_L and k. The degree of adaptation, as indicated by the value of k (high adaptation, fractional values of k approaching one; low adaptation, fractional values of k approaching zero) is correlated with the regularity of the resting discharge. Figure 16-10 illustrates the responses of two pigeon anterior ampullary afferents (re head angular acceleration). These responses are presented as typical of units with different degrees of adaptation. Unit Z4670.12 whose Bode plot is also presented in Fig. 16-9 (except re head angular velocity) is presented again as an example of a unit having an

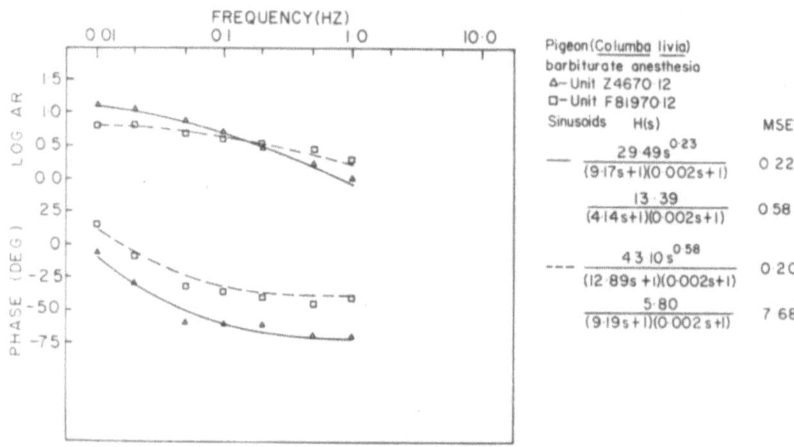

Fig. 16-10. A Bode plot of two pigeon anterior ampullary afferent responses to sinusoidal head acceleration. These data typify the frequency responses of two units which have medium ($k = 0.23$) and high ($k = 0.58$) exponent values for the s^k operator. The value of k, thought to reflect degree of adaptation, is correlated with the *CV* of the unit's resting discharge (28). Frequency response equations and associated MSEs for best-fit, torsion-pendulum models are presented for comparison.

"average" degree of adaptation. A best-fit curve and the corresponding equation for this unit are indicated by the solid line in Fig. 16-10. For comparison, the best frequency response function corresponding to the torsion-pendulum model is presented along with the corresponding MSE. Also presented in Fig. 16-10 is the Bode plot of Unit F81970.12, which has a high degree of adaptation ($k = 0.58$); best-fit frequency response functions of forms corresponding to those fit to data from Unit Z4670.12 are shown.

One important observation emerges from a comparison of the parameters in the equations in Fig. 16-10, namely, that when the fractional differentiator (adaptation operator) is not included in the frequency response function for units showing adaptation, *serious underestimations* of the value of τ_L can occur (compare $\tau_L = 9.17$ sec with $\tau_L = 4.41$ sec for Unit Z4670.12 in Fig. 16-10).

Landolt and Correia (28) determined an average frequency response function for all the anterior ampullary afferents that they tested. The mean best-fit parameters ($\pm$ SEM) determined (using an equation of the form of Eq. 16-8) were $C = 26.96$ (± 7.45) $\mathrm{I \cdot sec^{-1}/deg \cdot sec^{-2}}$, $k = 0.24$ (± 0.04), and $\tau_L = 10.24$ ($\pm$ 1.20) sec. The value of τ_S was not determined, but it was assumed to be 0.002 sec [based on biophysical calculations of Money et al. (38)]. Data used to determine the mean best-fit parameters were obtained from barbiturate anesthetized pigeons using sine wave stimuli over a limited frequency range (typically 0.01–2.0 Hz).

It has been argued (16) that barbiturate anesthesia may inhibit vestibular efferent activity and thereby open a feedback loop which might influence ampullary afferent activity.

In Fig. 16-11 we present mean (± SD) log AR and phase values (neural response re head angular velocity) from ampullary afferents ($N = 9$) of unanesthetized *encéphale isolé* preparations and from ampullary afferents ($N = 2$) of animals anesthetized with a nonbarbiturate anesthetic (Ketalar). Using bandlimited Gaussian random noise rotations and cross-spectral digital Fourier techniques, we have extended our bandwidth of analysis to $f = 16.0$ Hz. A predicted curve (solid line) which is plotted is based on the average frequency response function of Landolt and Correia (28). It can be seen that when we normalized the mean log AR data from the nonbarbiturate anesthetized preparations to the value of the predicted curve at 0.5 Hz, the remaining mean log AR values fell along the predicted curve over the range $f = 0.5–16.0$ Hz. Like the gerbil (Fig. 16-8, $2.0 \leq f \leq 4.0$ Hz), the pigeon phase data show a tendency for a progressive phase advance beginning at 5.0 Hz. Mean coherence function values corresponding to selected frequency values between 1.0 and 16.0 Hz are also presented in Fig. 16-11.

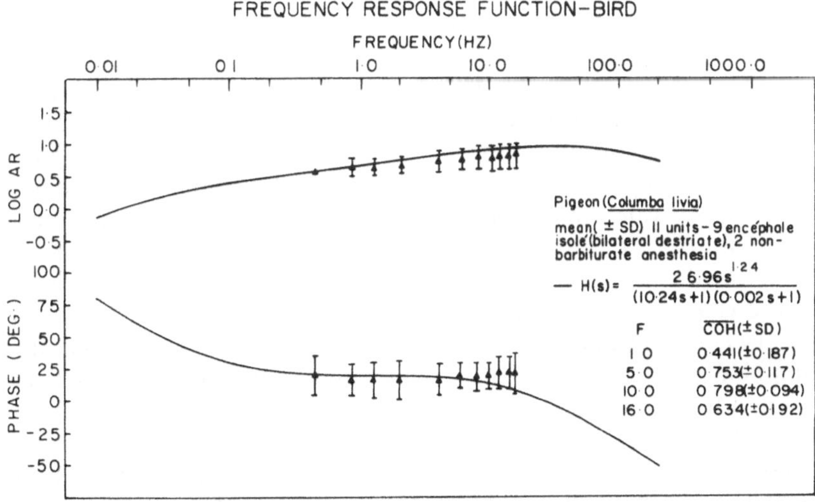

Fig. 16-11. A Bode plot of pigeon mean ampullary afferent data. The plots relate neural response to angular head velocity. Mean (±SD) log AR, phase, and coherence values are presented for the frequency range $f = 0.5 - 16.0$ Hz. These data were obtained from unanesthetized (*encéphale isolé*) and nonbarbiturate (Ketalar) anesthetized preparations which were exposed to Gaussian bandlimited noise rotations. The mean AR value at 0.5 Hz was normalized to the curve (solid line) which represents the frequency response function derived from the equation presented in the figure, which is the mean best-fit equation obtained for barbiturate anesthetized preparations using sinusoidal rotations over the range of frequencies, $f = 0.01 - 2.0$ Hz (28).

As indicated elsewhere (39), coherence values, which vary between zero and one, provide confidence estimates about the reliability of experimental data and the validity of using a linear systems approach at specific frequencies or over a band of frequencies. A coherence value, $\gamma^2_{xy}(f) = 1$, for the ideal case of a time invariant linear system with a high signal-to-noise ratio.

Landolt and Correia (28) also showed that the fractional differentiator, s^k, could be expressed as the product of poles and zeros as follows:

$$s^k = G \prod_{i=1}^{M} \left[\frac{D_i s(\zeta_{i-1} s + 1)}{\tau_i s + 1} \right] \qquad [16\text{-}10]$$

where G and D_i are constants and $\zeta_{i-1}(\zeta_o \equiv 0)$ are real and separate time constants.

The special case of a single adaptation time constant is interesting, and can be written as

$$s^k = \frac{GD_1}{\tau_1} \left[\frac{\tau_1 s}{\tau_1 s + 1} \right]. \qquad [16\text{-}11]$$

Eq. 16-11 is clearly recognizable as the adaptation operator (34,54) which was used by Fernandez and Goldberg (13) to fit their squirrel monkey frequency response functions (Fig. 16-6; Eq. 16-7).

We wished to determine if the s^k operator, which fits gerbil and pigeon data, would also fit squirrel monkey data. Figures 16-12 and 16-13 present data from Fernandez and Goldberg (13) relating neural response of irregular and regular units, respectively, to head angular velocity. Also presented in the figures are their best-fit equations and corresponding plots (solid lines). In the case of irregular units (Fig. 16-12), substitution of the s^k operator for the $\tau_A s/(\tau_A s + 1)$ operator results in a curve which is closer to the data and results in an equivalent MSE but with one less degree of freedom. For regular units (Fig. 16-13), substitution of the s^k operator results in a 5-fold reduction in the MSE. Moreover, substitution of the s^k operator leads to a reduction in the value of τ_V for irregular units and its elimination for regular units.

O'Leary and Honrubia (41) determined the guitarfish's ampullary afferent unit impulse response, $h(\tau)$, to pseudorandom binary sequence rotations. They noted four classes of unit impulse responses. The function, $h(\tau)$, with mean best-fit parameters, which they obtained for the class of afferents showing adaptation, is presented in Fig. 16-14. We scaled their amplitude coefficients, as they did in some cases (ref. 41, Fig. 1), by dividing by 92 (their bin width in msec). The log AR and phase values (relating neural response to head angular acceleration), which are presented in Fig. 16-14, were obtained from the Laplace transform of $h(\tau)$. It is clear from plots of the best-fit frequency response functions to the log AR and phase data in Fig. 16-14 that a frequency response function including τ_L and s^k (solid line, Fig. 16-14) does not provide a good fit. However, a frequency response function with adaptation time constants $\tau_1 = 3.05$ sec, $\tau_2 = 0.10$ sec, $\zeta_1 = 0.37$ sec, and $\tau_L = 1.79$ sec (τ_S not considered) provides an excellent fit to the data.

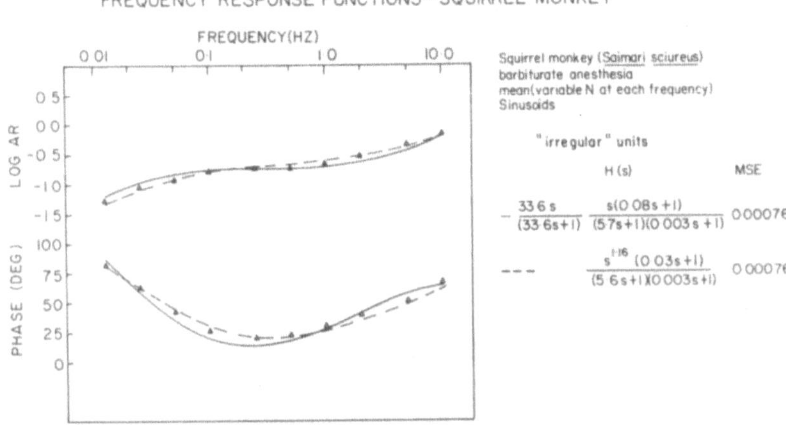

Fig. 16-12. A Bode plot of squirrel monkey mean ampullary afferent data from irregular units. The plots relate neural response to head angular velocity. Curves, equations, and MSE values are presented which correspond to the best-fit equation from Fernandez and Goldberg (13) and an equation which replaces the $\tau_A s/(\tau_A s + 1)$ operator with a fractional differentiator, s^k. After Goldberg, J.M. and Fernandez, C.: J. Neurophysiol. 34:635, 1971.

Recently, Lowenstein and Compton (32) analyzed the frequency response characteristics of ampullary afferents of the ray (*Raja clavata*) and two species of dogfish (*Scyliorhinus canicula* and *S. stellaris*). Although these investigators did not present *AR* values, they did present phase angles for six ampullary afferents over the frequency range $f = 0.01-1.0$ Hz. These values

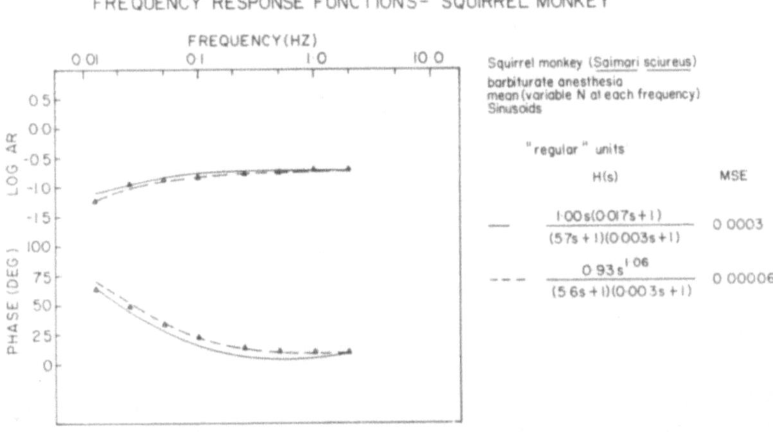

Fig. 16-13. A Bode plot of squirrel monkey mean ampullary afferent data from regular units. Details are the same as those presented in Fig. 16-12. After Goldberg, J.M. and Fernandez, C.: J. Neurophysiol. 34:635, 1971.

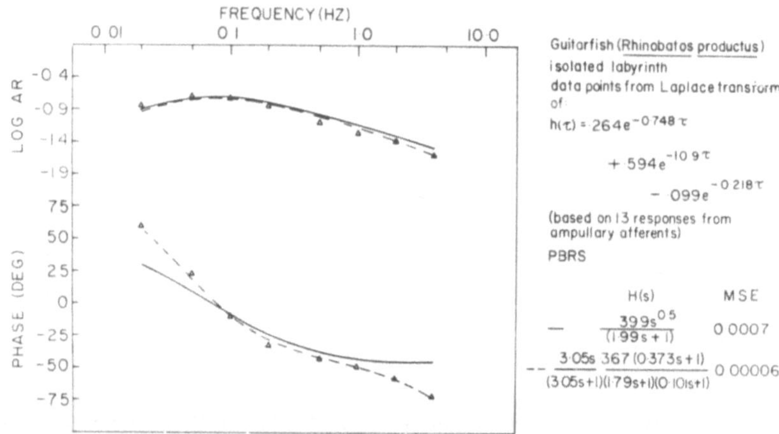

Fig. 16-14. A Bode plot of guitarfish mean ampullary afferent data. The plots relate neural response to angular acceleration of the head for one class of afferents. The curves, equations, and MSE values indicate that while an equation including the s^k operator and the torsion-pendulum model (τ_L only) does not provide a good fit, an equivalent expression (see text) with two adaptation time constants provides an excellent fit. After O'Leary, D.P. and Honrubia V.: J. Neurophysiol. 39:645, 1976.

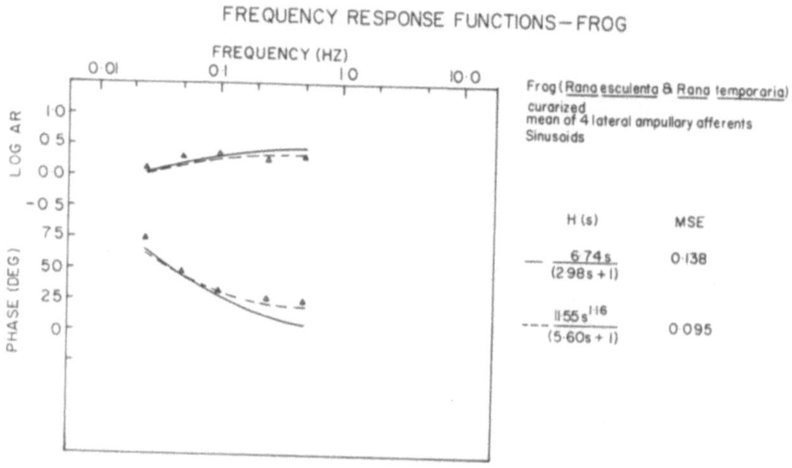

Fig. 16-15. A Bode plot of frog mean ampullary afferent data. The plots relate neural response to head angular velocity. Best-fit curves based on equations representing the torsion-pendulum model (τ_S ignored) and the torsion-pendulum model with a fractional differentiator are presented. After Blanks, R.H.I. and Precht, W.: Exp. Brain Res. 25:369, 1976.

were superimposed on and noted to fall around two curves (one for adapting units, one for nonadapting units) obtained by evaluation of Eq. 16-7 using the following parameters: $\tau_A = 47.8$ sec (for adapting units), $\tau_A = 500$ sec (for nonadapting units), $\tau_L = 9.5$ sec, $\tau_S = 0.005$ sec, and $\tau_V = 0.015$ sec.

Finally, we examined the frequency response function of the frog. We averaged AR and phase values of five lateral ampullary afferents tested over the frequency range, $f = 0.02–0.50$ Hz, by Blanks and Precht (6). These average values (re head angular velocity) are displayed in Fig. 16-15. Over this range, and for these afferents' responses, a nonzero value of k ($k = 0.16$) reduced the mean square error by 31% when incorporated in an equation of the form of Eq. 16-9. The time constant, τ_S, was not considered since its effect would be negligible over the frequency range $f = 0.02–0.5$ Hz. Although the equation incorporating a nonzero value of the exponent of the fractional differentiator, s^k, did provide a better fit to the data, it increased the value of τ_L to 5.6 sec. This value is nearly twice that ($\tau_L = 2.98$ sec) obtained by a fit of the data to the torsion-pendulum model (Eq. 16-9; $k = 0$), and the value of $\tau_L = 3.0$ sec obtained by Precht et al. (46) who tested frog ampullary afferent responses using angular acceleration pulses. There are three possibilities for this discrepancy. First, the units whose responses were available to us for analysis were nonadapting (Eq. 16-9; $k = 0$). Secondly, the five units which we analyzed actually had a mean value of $\tau_L = 5.6$ sec [16% of those obtained by Precht et al. (46) were in the 4- to 6-sec range]. Thirdly, the absence of data for low (below 0.02 Hz) and high (above 0.5 Hz) frequencies as well as the small number of afferents analyzed did not produce a fit which is representative. Further experimentation with this species is necessary to evaluate each of these possibilities.

In summary, for the species we have considered, it appears that ampullary afferent responses can be assumed to represent the output of a series of cascaded linear systems when either afferents which show clear nonlinear behavior are not considered, or when their responses are linearized by restricting the range of stimulus amplitudes. From a review of published data, it seems clear that the percentage of ampullary afferents which show nonlinear behavior decreases as one proceeds upward in evolutionary development. Moreover, it is equally clear that the two-pole operators for the torsion-pendulum model of the cupula-endolymph system are necessary but not sufficient to describe completely the response dynamics of all ampullary afferents. Across species, it appears that an operator must be added to account for the phenomenon of neural adaptation. As indicated herein and elsewhere (28), a fractional differentiator of the form s^k is the most general and best descriptor of neural adaptation since (1) it appears applicable to other sensory systems (51); (2) it can be decomposed into a product of cascaded adaptation operators, the first of which is the $\tau_A s/(\tau_A s + 1)$ operator which has been shown to describe adaptative phenomena of a variety of vestibular responses (13,34,54); (3) its Laplace transform (28) predicts phenomena which have been noted for ampullary afferent responses to pulse angular accelerations; namely, per-rotatory response decline during constant angular acceleration of the head and undershoot (or overshoot) of the response following termination of the stimulus; and (4) there are physiological processes within the mechanoelectric transduction of angular head motion to encoded action potentials on the primary ampullary afferents which provide a basis for the s^k model.

Physiologic Basis for the s^k Operator in the Vestibular Neuroepithelium

Thorson and Biederman-Thorson (51) have reviewed the pertinent neurophysiological literature and point out that in certain mechanoreceptors (e.g., cockroach leg mechanoreceptor), baroreceptors, joint receptors, and stretch receptors (slowly adapting), the decay of nerve impulse-firing frequency, $R(t)$, due to adaptation, following a step stimulus input at $t = 0$, is

$$R(t) \equiv y(t) = \beta t^{-k},\ t > 0\ , \qquad [16\text{-}12]$$

where β is an amplitude scale constant. Making use of the definition of the gamma function, $\Gamma(x)$, one can write

$$t^{-k} = \frac{1}{\Gamma(k)} \int_0^\infty a^{k-1} e^{-at} da\ . \qquad [16\text{-}13]$$

Equation 16-13 indicates that a t^{-k} response might arise from first order kinetics of a summation of different exponential decays weighted by each decay's rate constant, a, raised to the power $(k-1)$. Exponential decays are a common feature of distributed relaxation processes. The frequency domain equivalent of Eq. 16-12 in Laplace transform notation (3) is

$$L\{y(t)\} \equiv Y(s) = H(s) \cdot X(s)$$
$$= \beta\Gamma(1-k)s^{k-1}\ ,$$

and, since $X(s) = 1/s$ for a unit step increase in stimulus at $t = 0$,

$$H(s) = \beta\Gamma(1-k)s^{k},\ 0 \le k \le 1. \qquad [16\text{-}14]$$

There is morphological and physiological evidence for distributed relaxation processes in the cascaded systems which precede action potential generation on vestibular afferents. Taglietti et al. (50), using the frog, were able to abolish components of adaptation in the generator potential response to angular acceleration pulses of various durations. When they replaced all Na^+ by Li^+ in the fluids which bathed the ampullae, they eliminated response decline during constant head angular acceleration and undershoot following termination of head angular acceleration. Because of these results and those they obtained by bathing the ampullae with other solutions, Taglietti et al. (50) suggested that a Na^+/K^+ ion pump, located in each unmyelinated nerve terminal of ampullary primary afferents, was responsible for neural adaptation. Moreover, Taglietti et al. (50) reported that receptor potentials did not show the adaptive phenomena which were present in the generator potential.

These results suggest that, at least for the frog, the phenomena associated with adaptation are not produced by mechanoelectric processes in the hair cells (reflected by receptor potential changes), but by electrogenic pumps located in the unmyelinated terminals of the ampullary afferents.

Correia and Landolt (10) have deduced from a stochastic point-process analysis (27) of resting discharge on single ampullary afferents that the generator potential which initiates action potentials on an afferent is the result of spatial and/or temporal summation of excitatory postsynaptic potentials (EPSPs) and inhibitory postsynaptic potentials (IPSPs), which may be found on branches of unmyelinated afferent and efferent fibers, respectively, in the vestibular neuroepithelium. Figure 16-16 (4) shows the profuse branching of unmyelinated afferent and efferent (stippled) fibers as they innervate one or more type I and/or type II (adjacent and nonadjacent) hair cells. If these terminals carry postsynaptic potentials from hair cells which are widely distrib-

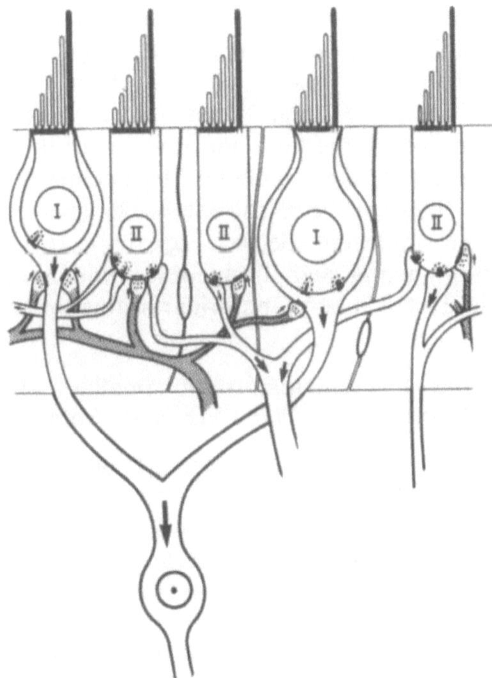

Fig. 16-16. A drawing indicating the complex neural plexus within the vestibular neuroepithelium of the mammal. Unmyelinated afferent and efferent (shaded) fibers branch and innervate multiple hair cells (both type I, amphora shaped; and type II, cylindric shaped). Arrows indicate direction of neural information flow. Horizontal line at bottom of figure (through which nerve fibers pass) represents the basement membrane, the region where nerve fibers lose their myelin sheaths (thickened part of fiber) before innervating the neuroepithelium. After Bergström, B. and Engström, H.: Equilib. Res. 3:27, 1973.

uted over the crista, if the hair cells are not stimulated homogeneously (12), and if the postsynaptic potentials exhibit properties of a relaxation process following hair cell stimulation, then the composite generator potential for a single afferent could exhibit properties of a distributed relaxation process.

Generality of the ($\tau_V s + 1$) Zero Operator

As more data become available, it may become evident that another operator must be added to those representing adaptation and the torsion-pendulum model in the frequency response function relating head motion and ampullary afferent response. Irregular ampullary afferents in the squirrel monkey show a frequency response function characterized by a high frequency gain enhancement and a progressive phase increase (Fig. 16-12). A similar characteristic has been noted for the frequency response function of ampullary afferents in the rhesus monkey (Fig. 16-7), gerbil (irregular units Fig. 16-8), pigeon (Fig. 16-11), and in isolated afferents of the frog (6), ray and dogfish (32). The origin of this phenomenon is unknown. However, Fernandez and Goldberg (13) have suggested a high frequency zero operator ($\tau_V s + 1$) to account for these *AR* and phase characteristics. They speculated that this operator might be necessary because, above certain frequencies (determined by the time constant, τ_V, of the operator), afferent discharge might represent rate of cupular displacement rather than cupular displacement itself. It appears from the data we have presented that the time constant, τ_V, varies for different species (Figs. 16-7, 16-11, and 16-12). Further experimentation, using high frequency stimulation for a variety of species, should clarify the nature and origin of this operator.

Comparison of Values of $\hat{\tau}_L$ Obtained from Ampullary Afferent Electrophysiological Studies with Those Predicted from Biophysical Studies

It is obvious from our analyses of different frequency response function forms (Figs. 16-9 and 16-10), that if adaptation is not taken into account, serious underestimation of the true value of $\hat{\tau}_L$ results. An interesting question follows. If adaptation is accounted for, how close are mean values of $\hat{\tau}_L$ derived from ampullary afferent electrophysiology to those derived from a biophysical analysis of the properties of the endolymph and cupula and the dimensions of the semicircular ducts? To estimate values of $\hat{\tau}_L$ for different species based on the dimensions of their semicircular ducts, we used Eqs. 16-1–16-5 to determine $\hat{\tau}_L \simeq \Pi/\Delta = 12\eta\pi^2R^3/\mu\pi r^2R \sim R^2/r^2$. We have assumed that across species, the following quantities are constant: the viscosity, η, of endolymph and the coefficient of cupula stiffness, μ. Since adaptation was accounted for in the determination of $\hat{\tau}_L$ from ampullary afferent responses of the squirrel monkey, we used values of $R = 1.80$ mm, $r = 0.12$ mm (24), and $\hat{\tau}_L = 5.73$ sec (13), and calculated values of $\hat{\tau}_L$ for other species by solving the proportionality $R^2/r^2/225 = \hat{\tau}_L/5.73$. The resulting values of $\hat{\tau}_L$ and values of R and r for different species are presented in Table 16-1. For

Table 16-1.

COMPARISON OF ESTIMATES OF τ_L. BIOPHYSICAL ESTIMATES, $\hat{\tau}_L$, ARE COMPARED WITH MEAN ESTIMATES, $\bar{\tau}_L$, DERIVED FROM ANALYSIS OF AMPULLARY AFFERENT RESPONSES IN THE FREQUENCY (A) AND (B) AND TIME (C) DOMAINS.

Species	Type of Ampullary Afferent	Parameter (unit)					
		R (mm)	r (mm)	τ_L (sec)	$\bar{\tau}_l$ (sec) (mean ± 1 SEM) (A)	$\bar{\tau}_L$ (sec) (mean ± 1 SEM) (B)	$\bar{\tau}_L$(sec) (mean ± 1 SEM) (C)
Rhesus monkey (green monkey)[b]	Lateral	2.11 (25)	0.110 (25)	9.37	3.76[c]	—	—
Squirrel monkey	Pooled (lateral, anterior, posterior)	1.80 (24)	0.120 (24)	—	5.73 ± 0.23 (13), 5.6[a]	—	5.41 ± 0.38 (18)
Cat	Lateral	1.64 (11)	0.117 (11)	5.00	4.95 ± 1.87 (42)	—	3.88 ± 1.54 (5)
Gerbil (jerboa/mouse)[b]	"Regular" lateral	0.94 (25)	0.080 (25)	3.51	2.50 (48), 2.32[a]	—	—
	"Irregular" lateral	0.94 (25)	0.080 (25)	3.51	2.50 (48), 3.67[a]	—	—
White king pigeon (domestic pigeon)[b]	Anterior	2.40 (25)	0.120 (25)	10.19	10.24 ± 1.20 (28)	3.33 ± 0.38 (9)	—
Frog (giant toad)[b]	Lateral	1.40 (25)	0.140 (25)	2.55	2.98[a] [k = 0] 5.60[a] [k = 0.16]	1.90 (6)	3.00 (46)
Guitarfish	Pooled	7.00 (40)	0.500 (40)	4.99	2.66 ± 1.22 (42); 1.79[a]	—	—
Ray and dogfish (spiny dogfish)[b]	Pooled	4.70 (25)	0.240 (25)	9.75	9.5 [range, 3–27 s] (32)	—	—

Numbers in parentheses are references. Values of $\bar{\tau}_L$ are are given as mean ± 1 SEM where possible.

[a]Value we obtained from "best-fit" transfer function of the form $H(s) = Ks^{k+1}/(\tau_L s + 1)(\tau_S s + 1)$.

[b]Animal in parentheses was used to obtain values of R and r.

[c]As pointed out in the text, this value is suspect since it is based on a paucity of low frequency data.

(A) Values obtained from frequency domain analysis when adaptation was taken into account.

(B) Values obtained when transfer function of the form $H(s) = K/(\tau_L s + 1)(\tau_s s + 1)$ was assumed and the expression $\tau_L = \omega^{-1} \tan \phi$ was used for low frequency phase angles, ϕ. Adaptation was not taken into account.

comparison, values of $\tilde{\tau}_L$ derived from primary ampullary afferent electrophysiological studies, with and without accounting for adaptation, are also presented in Table 16-1. Most values of $\tilde{\tau}_L$ in Table 16-1 were obtained from data derived from frequency domain analyses of ampullary afferent responses. However, where available, values of $\tilde{\tau}_L$ from time domain analyses are also included. A comparison of the columns headed $\hat{\tau}_L$ and $\tilde{\tau}_L$ in Table 16-1 indicates that although there is a range of values of τ_L for different species, the values determined from frequency domain analysis of ampullary afferent responses (after accounting for adaptation) are quite close to those determined from the proportionality equation presented above. In fact, when the SEM was available, the biophysical estimate $\hat{\tau}_L$ generally fell within one SEM of the mean estimate $\tilde{\tau}_L$ as determined from frequency domain analysis of ampullary afferent responses.

A Comparison of Mean Amplitude Ratios (Gains) and Phase Angles (re Head Angular Velocity) for Different Species

Table 16-2 presents estimates of the mean AR and phase angles of ampullary afferent responses (re head angular velocity) for different species at decade increments from $f = 0.025$–2.5 Hz. We calculated the mean AR values by two methods: (1) For investigators who normalized their data, we used the mean unnormalized values they published (or provided us) at a specific frequency and scaled the best-fit frequency response function, which they presented or which we determined, to that AR value. We then applied the same scale constant to AR values at frequencies of $f = 0.025$, 0.25, and 2.5 Hz (2). For those investigators who presented best-fit frequency response functions which included a constant gain term, we simply evaluated their equations at $f = 0.025$, 0.25, and 2.5 Hz.

The data presented in Table 16-2 provide the following observations. First, for the squirrel monkey and the gerbil, as previously reported (19,47), irregular units have a higher gain over the bandwidth 0.025–2.5 Hz than do regular units. Secondly, the pigeon and frog have a 3- to 5-fold higher mean gain over the bandwidth 0.025–2.5 Hz than do mammals. Thirdly, for the rhesus monkey, squirrel monkey, cat, gerbil, pigeon, and frog there exists a clear phase advance of the ampullary afferent response (re head angular velocity) in the region where the log AR curve is relatively flat (Figs. 16-6–16-9 and 16-15). The average phase advance for the species listed in Table 16-2 is around 25° at $f = 0.25$ Hz.

In summary, the general conclusion which emerges from the data in Table 16-2 is that ampullary afferent output does not produce a frequency response function which has a "flat" AR region with zero phase (Fig. 16-1), but probably because of adaptation, a phase advance exists over the entire bandwidth we have considered (Fig. 16-11). This phase advance must be equalized by processing within central nervous system pathways to produce zero phase between head velocity (head position) and eye velocity (eye position). Moreover, animals such as the pigeon and the frog which have a clear

Table 16-2.
SPECIES COMPARISON OF MEAN AMPLITUDE RATIOS (GAINS) AND PHASE ANGLES (RE HEAD VELOCITY) DERIVED FROM FREQUENCY DOMAIN ANALYSIS OF AMPULLARY AFFERENT RESPONSES.

Species	Type of Ampullary Afferent	Sample Size (N)	Frequency: 0.025 Hz		0.25 Hz		2.5 Hz	
			AR (I · sec^{-1}/deg · sec^{-1})	Phase (deg)	AR (I · sec^{-1}/deg · sec^{-1})	Phase (deg)	AR (I · sec^{-1}/deg · sec^{-1})	Phase (deg)
Rhesus monkey (31)	Lateral	23	0.53	59.63	1.05	11.59	1.14	19.71
Squirrel monkey (13)	Pooled (lateral, anterior, posterior)	52						
	"Regular"		0.22 [0.25][a]	54.04 [48.29]	0.38 [0.38]	11.62 [7.63]	0.44 [0.40]	3.35 [12.89]
	"Irregular"		0.28 [0.41]	63.31 [59.58]	0.62 [0.62]	23.31 [14.35]	0.99 [0.99]	37.59 [49.54]
Cat (45)	Lateral	3	0.21	65.42	0.64	20.78	1.12	10.34
Gerbil (48)	Lateral							
	"Regular"	8	0.12	75.36	0.40	20.57	0.48	5.17
	"Irregular"	15	0.16	83.42	0.56	33.06	1.04	22.59
Pigeon (28)	Anterior	14	1.43 [1.07][b]	53.45 [52.03]	2.93 [2.10]	24.98 [23.23]	5.09 [3.49]	20.16 [18.36]
Frog (6)	Lateral	5	1.01	63.06	2.20	20.89	3.20	15.05
Guitarfish (41)	Pooled	13	0.02	141.14	0.22	52.71	0.65	25.77
	Eq. 16-10 (41)							

Numbers in parentheses are references. All phase angles in the above table are positive and represent a phase lead at each of the frequencies which are presented.

[a] All *AR* and phase values for the squirrel monkey which are in brackets were obtained from evaluation of Fernandez and Goldberg (13) equations using an unnormalized *AR* at 0.25 Hz of 0.38 for "regular" and 0.62 for "irregular" units (17).

[b] *AR* and phase values for the pigeon which are in brackets were predicted from *median* parameters of frequency response function (28).

vestibulocollic response (head movement resulting from vestibular stimulation) have a much higher gain (and gain slope) than do mammals which rely primarily on the vestibuloocular response to stabilize images on the retina. As indicated elsewhere (44), a vestibulocollic response in itself affects cupula-endolymph movement within the semicircular ducts. Thus, it seems likely, that for vestibulocollic reflexes, the semicircular canals appear as feedback transducers, rather than as open-loop components as in the vestibuloocular system. It is not inconceivable that such a system requires the high gain levels observed for the frog and the pigeon.

ACKNOWLEDGMENTS

The authors wish to express their appreciation to Mr. Johnnie Moses and Ms. Pat Groves for their photographic and secretarial assistance, respectively. Some of the results reported herein were made possible by NASA Contract NAS9-14641 and ONR Contract NR 201-185 with M.J. Correia. This work is dedicated to W. Carroll Hixson, a superbly insightful bioengineer.

References

1. Anonymous: Nomina Anatomica, 4th ed. Amsterdam, Excerpta Medica Foundation, 1977.
2. Belanger, F. and Mayne, R.: The Dynamics of the Semicircular Canals. GERA-1085, Litchfield Park, Arizona, Goodyear Aerospace Corp., 1965.
3. Bendat, J.S. and Piersol, A.G.: Measurement and Analysis of Random Data. New York, Wiley, 1966.
4. Bergström, B. and Engström, H.: The vestibular sensory cells and their innervation. Equilib. Res. 3:27, 1973.
5. Blanks, R.H.I., Estes, M.S., and Markham, C.H.: Physiologic characteristics of vestibular first-order canal neurons in the cat. II. Response to constant angular acceleration. J. Neurophysiol. 38:1250, 1975.
6. Blanks, R.H.I. and Precht, W.: Functional characterization of primary vestibular afferents in the frog. Exp. Brain Res. 25:369, 1976.
7. Brassard, J.R., Correia, M.J., and Landolt, J.P.: A computer program for graphical and iterative fitting of low and high order transfer functions to biological data. Computer Prog. Biomed. 4:1, 1974.
8. Correia, M.J. and Guedry, F.E., Jr.: The vestibular system: basic biophysical and physiological mechanisms. In Masterton, R.B. (ed.): Handbook of Behavioral Neurobiology, Vol. 1. Sensory Integration. New York, Plenum Press, 1978, pp. 311–351.
9. Correia, M.J. and Landolt, J.P.: Spontaneous and driven responses from primary neurons of the anterior semicircular canal of the pigeon. Adv. Otorhinolaryngol. 19:134, 1973.
10. Correia, M.J. and Landolt, J.P.: A point process analysis of the spontaneous ac-

tivity of anterior semicircular canal units in the anesthetized pigeon. Biol. Cybern. 27:199, 1977.

11. Curthoys, I.S., Markham, C.H., and Curthoys, E.J.: Semicircular duct and ampulla dimensions in cat, guinea pig, and man. J. Morphol. 151:17, 1977.

12. Dohlman, G.F.: Experiments on the mechanism of Ménière attacks. J. Otolaryngol. 6:135, 1977.

13. Fernandez, C. and Goldberg, J.M.: Physiology of peripheral neurons innervating semicircular canals of the squirrel monkey. II. Response to sinusoidal stimulation and dynamics of peripheral vestibular system. J. Neurophysiol. 34:661, 1971.

14. French, A.S.: Practical nonlinear system analysis by Wiener kernel estimation in the frequency domain. Biol. Cybern. 24:111, 1976.

15. French, A.S. and Wong, R.K.S.: Nonlinear analysis of sensory transduction in an insect mechanoreceptor. Biol. Cybern. 26:231, 1977.

16. Gleisner, L. and Henriksson, N.G.: Efferent and afferent activity pattern in the vestibular nerve of the frog. Acta Otolaryngol. (Suppl.) 192:90, 1963.

17. Goldberg, J.M.: Personal communication.

18. Goldberg, J.M. and Fernandez, C.: Physiology of peripheral neurons innervating semicircular canals of the squirrel monkey. I. Resting discharge and response to constant angular accelerations. J. Neurophysiol. 34:635, 1971.

19. Goldberg, J.M. and Fernandez, C.: Physiology of peripheral neurons innervating semicircular canals of the squirrel monkey. III. Variations among units in their discharge properties. J. Neurophysiol. 34:676, 1971.

20. Goldberg, J.M. and Fernandez, C.: Vestibular mechanisms. Ann. Rev. Physiol. 37:129, 1975.

21. Groen, J.J., Lowenstein, O., and Vendrik, A.J.H.: The mechanical analysis of the response from the end-organs of the horizontal semicircular canal in the isolated elasmobranch labyrinth. J. Physiol. 117:329, 1952.

22. Guedry, F.E., Jr.: Psychophysics of vestibular sensation. In Kornhuber, H.H. (ed.): Handbook of Sensory Physiology, Vol. VI/2. Vestibular System, Part 2, Psychophysics, Applied Aspects, and General Interpretations. New York, Springer-Verlag, 1974, pp. 4–154.

23. Hudspeth, A.J. and Corey, D.P.: Sensitivity, polarity, and conductance change in the response of vertebrate hair cells to controlled mechanical stimuli. Proc. Natl. Acad. Sci. U.S.A. 74:2407, 1977.

24. Igarashi, M.: Dimensional study of the vestibular apparatus. Laryngoscope 77:1806, 1967.

25. Jones, G.M. and Spells, K.E..: A theoretical and comparative study of the functional dependence of the semicircular canal upon its physical dimensions. Proc. R. Soc. B. 157:403, 1963.

26. Klinke, R. and Galley, N.: Efferent innervation of vestibular and auditory receptors. Physiol. Rev. 54:316, 1974.

27. Landolt, J.P. and Correia, M.J.: Neuromathematical concepts of point process theory. I.E.E.E. Trans. Biomed. Engin. BME-25:1, 1978.

28. Landolt, J.P. and Correia, M.J.: Neurodynamic response analysis of anterior semicircular canal afferents in the pigeon. J. Neurophysiol. 43:1746, 1980.

29. Lee, Y.W. and Schetzen, M.: Measurement of the Wiener kernels of a nonlinear system by cross correlation. Int. J. Control 3:237, 1965.

30. Lifschitz, W.S.: Responses from the first order neurons of the horizontal semicircular canal in the pigeon. Brain Res. 63:43, 1973.

31. Louie, A.W. and Kimm, J.: The response of 8th nerve fibers to horizontal sinusoidal oscillation in the alert monkey. Exp. Brain Res. 24:447, 1976.

32. Lowenstein, O. and Compton, G.J.: A comparative study of the responses of isolated first order semicircular canal afferents to angular and linear acceleration, analyzed in the time and frequency domains. Proc. R. Soc. B. 202:313, 1978.

33. Lowenstein, O. and Wersäll, J.: A functional interpretation of the electron microscopic structure of the sensory hairs in the cristae of the elasmobranch *Raja clavata* in terms of directional sensitivity. Nature 184:1807, 1959.

34. Malcolm, R. and Melvill Jones, G.: A quantitative study of vestibular adaptation in humans. Acta Otolaryngol. 70:126, 1970.

35. Marmarelis, P.Z. and Marmarelis, V.Z.: The White-noise Approach. Analysis of Physiological Systems. New York, Plenum Press, 1978.

36. Mayne, R.: The dynamic characteristics of the semicircular canals. J. Comp. Physiol. Psychol. 43:309, 1950.

37. McLaren, J.W.: The configuration of movement of the semicircular canal cupula. Ph.D. Thesis. Iowa City, University of Iowa, 1977.

38. Money, K.E., Bonen, L., Beatty, J.D., Kuehn, L.A., Sokoloff, M., and Weaver, R.S.: Physical properties of fluids and structures of vestibular apparatus of the pigeon. Am. J. Physiol. 220:140, 1971.

39. Ni, M.-D., Correia, M.J., Rae, J.L., and Koblasz, A.J.: A real-time, mini-computer program for calculation of frequency response function and coherence function by digital Fourier methods. MIMI 77, Proc. of the International Symposium on Mini and Microcomputers. I.E.E.E. Cat. #77CH1347-4C. 1978, pp. 205–212.

40. O'Leary, D.P.: Personal communication.

41. O'Leary, D.P. and Honrubia, V.: Analysis of afferent responses from isolated semicircular canal of the guitarfish using rotational acceleration white-noise inputs. II. Estimation of linear system parameters and gain and phase spectra. J. Neurophysiol. 39:645, 1976.

42. O'Leary, D.P., Tomko, D.L., Black, F.O., and Peterka, R.J.: Comparative analysis of cat and guitarfish semicircular canal afferent responses. In Hood, J.D. (ed.): Vestibular Mechanisms in Health and Disease. New York, Academic Press, 1978, pp. 28–34.

43. Oman, C.M. and Young, L.R.: The physiological range of pressure difference and cupula deflection in the human semicircular canal. Acta Otolaryngol. 74:324, 1972.

44. Outerbridge, J.S.: Experimental and theoretical investigation of vestibularly driven head and eye movement. Ph.D. Thesis. Montreal, McGill University, 1969.

45. Peterka, R.J., O'Leary, D.P., and Tomko, D.L.: Linear system techniques for the evaluation of semicircular canal afferent responses using white noise rotational stimuli. In Hood, J.D. (ed.): Vestibular Mechanisms in Health and Disease. New York, Academic Press, 1978, pp. 10–17.

46. Precht, W., Llinas, R., and Clarke, M.: Physiological responses of frog vestibular fibers to horizontal angular rotation. Exp. Brain Res. 13:378, 1971.

47. Schneider, L.: Responses of first and second order vestibular neurons to thermal and rotational stimuli. Ph.D. Dissertation. Ann Arbor, Mich., University of Michigan, 1973.

48. Schneider, L.W. and Anderson, D.J.: Transfer characteristics of first and second order lateral canal vestibular neurons in gerbil. Brain Res. 112:61, 1976.

49. Steinhausen, W.: Uber die Beobachtung der Cupula in den Bogengangsampullen des Labyrinths des Lebenden Hechts. Pfluegers Arch. Ges. Physiol. 232:500, 1933.

50. Taglietti, V., Rossi, M.L., and Casella, C.: Adaptive distortions in the generator potential of semicircular canal sensory afferents. Brain Res. 123:41, 1977.

51. Thorson, J. and Biederman-Thorson, M.: Distributed relaxation processes in sensory adaptation. Science 183:161, 1974.

52. Trincker, D.: Resting potentials in the system of semicircular canals of the guinea pig and their changes caused by experimental cupula displacements. Pfluegers Arch. 264:351, 1957.

53. Trincker, D.: The transformation of mechanical stimulus into nervous excitation by the labyrinthine receptors. Symp. Soc. Exp. Biol. 16:289, 1962.

54. Young, L.R. and Oman, C.M.: Model for vestibular adaptation to horizontal rotation. Aerospace Med. 40:1076, 1969.

Discussion

GOLDBERG: Your excellent review has shown that there are substantial differences in the afferent discharge characteristics of different species. Several years ago, Mayne (ref. 36) and later Jones and Spells (ref. 25) speculated that the dynamics of the semicircular canals were matched to the kinds of head movements that different animals make. Do the newer data support this or any other unifying concept concerning the variations among species in their canal dynamics? Let me cite a particular example. Canal afferents in the frog have a low resting discharge, and yet are characterized by high sensitivity to excitatory angular accelerations. On this face, this would imply that the response in the frog would be essentially unidirectional. At the other extreme, the resting discharge in the monkey is quite high so that it can respond almost as well to inhibitory and excitatory accelerations. I, for one, cannot explain the differences between the frog and monkey in terms of their behavior. Could you enlighten me as to the functional meaning of this or the other species differences summarized in your paper?

CORREIA: As I tried to indicate in our presentation, we were struck not so much by the differences, but by the remarkable similarities between the transfer characteristics of ampullary primary afferents of different species. For example, I suggest that we have shown that over a restricted range of frequencies (0.01–5.0 Hz), one general form of frequency response function appropriately describes the linear transfer characteristics of average ampullary afferent data from a variety of species ranging in evolutionary development from the guitarfish to the squirrel monkey. In this frequency response function, we need only include terms or groups of terms (operators) which appear to represent movement of the cupula-endolymph system and adaptation phenomena. I hasten to add, however, that the values of the time constants associated with each of the operators of the general frequency response function are different for different species. I suggest that these time constant values have changed during evolution by modification of the morphology of one or more of the cascaded systems which we identified in our presentation as being involved in the mechanoelectric transduction of angular head motion into neuroelectric signals on the ampullary primary afferents. Moreover, it seems clear to me that these evolutionary changes were guided by teleology. I find it understandable, for example, that the dimensions of the semicircular ducts might change during evolution to produce response dynamics which match the milieu in which a species must respond to survive. Consider for example, the environmental demands placed on a bird of prey such as a hawk. His survival depends on his ability, during flight, to locate his prey on the ground and maintain visual fixation on his prey as he swoops to capture it. During flight, because of wind drafts and turbulence, the hawk may be exposed to large and small head motions containing *both* very low and very high frequencies. Let us suppose the hawk's prey was a rat. The rat, a terrestrial animal, probably would not be exposed to low frequency head motions as he forages for food or runs to escape potential captors. However, he too must make appropriate vestibuloocular responses to clearly view the terrain as he runs to survive. From an analysis of the above situation, we might expect the hawk and the rat to have evolved with semicircular canals whose response bandwidth matched their environments. In our presentation, we considered two animals which are similar to the hawk and the rat. We considered an aerial/terrestrial animal, the pigeon, and a small terrestrial animal, the gerbil. We determined the time constant which specified the lower corner frequency of the pigeon's ampullary afferent response gain curve. From the dimensions of the semicircular ducts, we speculated that the lower corner frequency would be around 0.02 Hz and the upper corner frequency would be around 80 Hz. We have performed a frequency analysis of the pigeon's ampullary afferent response to rotation and we have confirmed the lower corner frequency prediction. We have not yet been able to confirm the upper corner frequency prediction by analysis of ampullary afferent responses to rotation, but we have evidence that it is above 20 Hz (Fig. 16-11). The dimensions of the gerbil's semicircular ducts suggest that its ampullary afferent response gain curve should have a low corner frequency near 0.05 Hz and a high corner frequency near 80 Hz. Electrophysiological studies (ref. 48) of the frequency response of the gerbil's lateral semicircular canal afferents have confirmed the predicted value of the lower corner frequency, which is nearly three times higher than that of the pigeon. If future electrophysiological studies show that the gerbil and the pigeon have the same upper corner frequency, then the gerbil will indeed have a narrower bandwidth of response than the pigeon, at least regarding information coming from the labyrinthine periphery.

As you point out, there are exceptions to the concept that evolution, guided by teleology, has found the most efficient way for the semicircular canals to signal head motions. One example is the restriction of dynamic range in the frog's ampullary afferent output because of a high sensitivity factor and a low resting discharge rate. However, we should mention that this condition has apparently been corrected in warm-blooded animals, since it has been clearly demonstrated that in these species the ampullary afferent resting discharge is higher and sensitivity lower. Moreover, it may be that this apparent inefficient transduction process works well for the frog when you consider that both it and the pigeon have a strong vestibulocollic response. Both animals have higher gain and sensitivity factors than do mammals which have smaller vestibulocollic responses. It may be that high gain is necessary to offset attenuation of cupula-endolymph movement relative to the semicircular duct which may occur because compensatory head movements are made in response to vestibular stimulation. Intuitively, this suggestion seems reasonable since compensatory head movements would be in the same direction as endolymph flow, and therefore movement of endolymph relative to the duct (head) would be decreased.

YOUNG: First, my congratulations on pulling all that data into an extremely useful paper. I have a comment about the adaptation operator and a question about extrapolating from it. The original adaptation operator work was done at McGill and MIT ten years ago, at a time when we did not know the actual canal afferent time constants. We had the biophysical models to deal with, but it was before the extensive afferent recordings were available. And at the time, we thought that the cupula return time constant was longer than those of postrotatory nystagmus or sensation recordings. Consequently, the notion of the adaptation operator then was to account for nystagmus and sensation slope differences in cupulometry. In that context alone the adaptation operator notion really is outdated in terms of current knowledge of the cupula time constant. Now my question: we know that the afferent time constant in squirrel monkeys is short, on the order of 5 seconds. Can you extrapolate as to what you would expect the dominant time constant to be in man?

CORREIA: No. [The following note was added in proof.] If we utilize values provided by Curthoys et al. (ref. 11) for the radius of curvature, R, and the internal radius, r, of the human lateral, anterior, and posterior semicircular ducts, and if we use the proportionality relationship which provided estimates of τ_L for other species (Table 16-1), we obtain the following estimates: $\tau_L = 7.5s$ for the lateral semicircular canal; $\tau_L = 15.0s$ for the anterior semicircular canal; and $\tau_L = 12.6s$ for the posterior semicircular canal.

YOUNG: Two other easy ones: One sacrifices a certain sense of physical reality in going from an integer exponent on s to a fractional operator. And I'm not sure that just having a better fit is worth that loss of any sense of being able to pin down what in the physics could account for it. Obviously, there are ways of accounting for fractional operators, but they're not obvious. And I wondered if you can comment on that, and the second, well, let's ask you about the first.

CORREIA: You are quite right. It is unwise to substitute one mathematical operator for another in a frequency response function just to obtain a better fit to amplitude ratio and phase data. The replacement operator should be more general and flexible, and should have a clear anatomic and/or physiologic basis for its existence. I feel that the s^k operator has all these features. We have shown in this presentation and elsewhere (ref. 28) that when the s^k operator is substituted for the "adaptation" operator, $\tau_A s/(\tau_A s + 1)$, in a general frequency response function (Eq. 16-7), an equivalent or possibly better fit of frequency domain *and* time domain ampullary afferent data results. We have also shown in our presentation that the s^k operator, or an equivalent form, fits mean frequency domain ampullary afferent data from a variety of species, including the squirrel monkey, gerbil, pigeon, and guitarfish. Moreover, we (ref. 28) have shown that the "adaptation" operator, $\tau_A s/(\tau_A s + 1)$, is a special case of the s^k operator and therefore the latter is more general. Furthermore, the s^k operator is more flexible since it can be expressed as a product of poles and zeros and thereby account for processes with more than one adaptation time constant. Finally, Thorson and Biederman-Thorson (ref. 51) have shown that the s^k frequency domain operator or its equivalent time domain operator, t^{-k}, describe adaptive responses of a variety of receptors including mechanoreceptors, baroreceptors, and joint receptors. Moreover, they point out that with certain assumptions, the s^k operator describes the first order kinetics of a distributed series of relaxation processes. These authors find anatomical and physiological correlates to a distributed series of relaxation processes in the Limulus eye. In our presentation, we have suggested anatomical and physiological correlates of a distributed series of relaxation processes in the vestibular neuroepithelium. For the above reasons, I feel that replacement of the $\tau_A s/(\tau_A s + 1)$ operator by the s^k operator is a sensible strategy.

GUALTIEROTTI: I'm addressing myself to Dr. Goldberg's observation about unidirectionality of the receptor owing to the lower basic frequency. I think there is a misconception here. If we take the bullfrog, for instance, we can find a discharge at rest of the vestibular receptor of 2 to 8 impulses/sec at a temperature of 58°F. The increase in the firing rate/degree Fahrenheit, as an average, is nearly one pulse/second. Suppose that the bullfrog lives in an environment like Louisiana, where the temperature is up to 80°. The firing rate will increase accordingly. We tested this effect: a rate of firing of 8/sec at 58°F increased to 20/sec at 75°F and up to 30/sec at 80°F. At that particular rate, of course, the bidirectionality of response would be quite feasible. As the coefficient of variation decreases proportionally, the firing becomes regular and the difference between frog and cat vestibular units nearly disappears. And, in fact, we are now trying to get some feeling about how the dynamic properties of the vestibular receptors, for instance, the semicircular canal, change at temperatures at which the basic firing rate is so high that it's quite feasible that a negative response appears.

17

The Origin and Functional Significance of the Resting Activity and Peripheral Adaptation in the Vestibular System

O. LOWENSTEIN

The resting activity in afferent sensory channels may be defined as activity in the absence of an overt adequate stimulus. Alternative terms encountered in the literature are spontaneous, basic, or background activity. The term "resting discharge" was first used by Adrian and Zotterman (1) in their pioneer work on the mode of information transmission in sensory nerves. They were rather surprised to encounter impulse activity in sensory nerves in skin and muscle preparations in the absence of stimulation.

So far as the acousticolateralis system is concerned, the first serious attempts to investigate the origin of such "spontaneous" activity were made by Hoagland (9) in his work on the lateral line organs of the catfish and the trout. He came to the conclusion that the "spontaneously" discharging lateral line receptors were end organs which essentially show no adaptation and that their activity was maintained by metabolic processes resulting in an excitatory state, such that nerve impulses are continuously being initiated at a frequency determined by the quantity of the product of the metabolic process. Hoagland found this hypothetical process to be dependent on an intact blood circulation. He also convinced himself that the "spontaneous" receptor discharge was not identical with the injury discharge observable in "broken" nerve endings. However, Hoagland failed to establish a possible functional role in the utilization by the animal of the information derived from the lateral line receptors.

Work on the lateral line of elasmobranch fishes furnished a further step in the recognition of the resting discharge as an integral part of receptor behavior. In 1937 Sand (21) published the results of an intensive study of the responses of the lateralis organ of *Raja clavata* in which he perfused the hyomandibular canal system with sea water with a cannula allowing reversal of the fluid current in the canal. These experiments yielded the first evidence of a bidirectional modulation of the resting activity in response to oppositely directed water currents. The biologic significance of this receptor behavior was, however, obscured by the fact that the relation between the direction of the perfusion current and increase or decrease in the resting activity could be found to be reversed in closely adjacent units in the same nerve strand. The explanation of this phenomenon had to await the ultrastructural study of the hair cell and its characteristic deployment within the lateralis end organ.

The ultimate clarification of the question of the functional significance of the resting activity became possible when it was found to be characteristically present in the afferent nerves from the semicircular canals in the elasmobranch labyrinth (15). Experiments based on the operative elimination of parts of the vestibular system had yielded conflicting results with respect to the question of functional uni- or bidirectionality of semicircular canals. McNally and Tait (4,18,19), in their exhaustive study of the effects of isolated elimination of the various vestibular end organs of the frog labyrinth, had come to the conclusion that a given semicircular canal responds to one direction of angular acceleration only, namely, the vertical canals to ampulla-leading, and the horizontal canal to ampulla-trailing, rotary acceleration in the plane of the canal. In contrast to this finding, elimination experiments in bony fishes (minnow and pike) had pointed strongly to a bidirectional eye reflex control by the horizontal semicircular canal, i.e., to the capability of a single intact canal to elicit compensatory eye movements to angular acceleration in opposite directions (11,12).

An electrophysiologic study of this problem appeared, therefore, most desirable, and when this was carried out (16,17) it produced evidence for the by now well-established response behavior of semicircular canals, and at the same time demonstrated the role of the resting activity in providing a "zero" level of impulse discharge which is increased by angular acceleration in one and decreased by angular acceleration in the opposite direction, thus furnishing a substrate for receptive bidirectionality.

When the total of the continued outflow of impulse activity from all end organs of the vestibular system was assessed, it became clear that this must be considered as *a chief source of vestibular tonus.* Its smooth modulation by dynamic and static stimuli, moreover, makes the peripheral organs practically thresholdless.

THE ORIGIN OF RESTING ACTIVITY

After this short historical note we may now step into the present to consider the most generally accepted views on the origin and role of the resting activi-

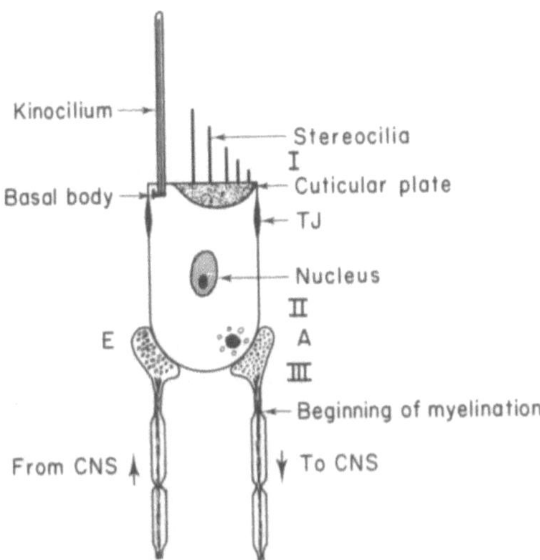

Fig. 17-1. Schema of hair cell and its innervation. *A*, afferent synapse; *E*, efferent synapse; *TJ*, tight junction with neighboring supporting cell; *I*, *II*, *III*, possible regions of origin of adaptive process, as discussed in the text.

ty in stimulus transduction and information processing in the vestibular system.

Figure 17-1 shows the vestibular hair cell [type II of Wersäll (24)] with its afferent and efferent innervation. The type II cell has been chosen here because of its ubiquitous presence in all vertebrates. This picture is by now so well known as not to warrant any detailed explanation. Attention should, however, be drawn to the tight junctions at the top of the hair cell between it and the supporting cells. These junctions form the so-called reticular membrane and are thought to represent an ion-tight barrier between the chemical environments of the apical and basal parts of the hair cell, i.e., between the obviously mechanosensitive structures on the one hand and the synaptic structures on the other hand. Further, it is noteworthy that the myelination of the peripheral processes of the first order afferent and the efferent neurons begins (ends) fairly close to the hair cell basis below the limiting membrane of the sensory epithelium.

After the ultrastructural studies of the vestibular hair cells had revealed the presence of chemotransmitting synapses both in the afferent and the efferent peripheral pathways, it appeared reasonable to consider the origin of the resting activity to be an ongoing transmitter leakage across the synapse between the hair cell and the first order afferent neuron, resulting in a lasting depolarization of its peripheral naked axoplasmic filament. This would produce the generator potential for the generation of all-or-nothing impulses at the first node of Ranvier of the afferent fiber.

Resting activities vary considerably in different types of vertebrates so far as their average frequency and regularity are concerned. In the amphibians the resting activity is of relatively low frequency and irregular (3), whereas in fishes such as the elasmobranchs resting frequencies are moderately high and in some instances highly regular (8,17). Very high resting frequencies of varying regularity are reported for the mammalian and avian vestibular organ (4,6,7).

This background of activity is thought to be smoothly modulated by an increase or decrease of neurotransmitter producing peak frequencies of impulse discharge on maximum displacement of apical hair cell processes in one, and abolition of discharge on their maximum displacement in the opposite direction. A number of hypotheses have been proposed as to the nature of transduction of the mechanical disturbance to the changes in the release of neurotransmitter (14), but it is fair to say that no clear picture has emerged so far. Furthermore, the chemical nature of afferent transmission is as yet unknown, although it is fairly certain not to be cholinergic. It is clear that the "leaky synapse" hypothesis puts the onus for the generation of the resting activity squarely on the hair cell and its afferent synaptic apparatus. Without the hair cell—no resting discharge!

The resting activity can be modulated by DC polarization (10,13), and this modulation can be made to mimic the effects of adequate mechanical stimulation (i.e., angular accelerations in opposite directions in the case of semicircular canal units). The two modes of stimulation, if applied simultaneously, can be seen to summate both in strength and in direction.

In the light of the above hypothesis, this may be interpreted in terms of a direct effect of the polarizing electric current on the hair cell membrane, and more specifically on the presynaptic membrane of the afferent synapse.

An alternative locus of origin of the resting activity might be the peripheral axoplasmic terminal of the first order afferent neuron itself. If its membrane were permanently moderately depolarized, the myelinated axon of the first order afferent neuron would propagate a continuous sequence of action potentials whose frequency would be a function of the persistent indigenous state of depolarization of the terminal. The advent at this terminal of neurotransmitter from the mechanically stimulated hair cell increases the state of depolarization and causes an increase of activity commensurate with the intensity of mechanical stimulation.

However, the question arises how the cessation of impulse activity in the afferent channel in response to an oppositely directed mechanical stimulus is brought about. The fact that a complete abolition even of the high frequency resting activity in mammalian preparations does occur in response to such a stimulus, poses a problem.

It has recently been found that it is possible to uncouple the hair cell functionally from its first order neuron. The relevant experiments were carried out on the lateral line organs of the mudpuppy by Sand (22). The author subjected the cupula, and with it the apical aspects of the hair cells, to a change in chemical environment by the application of a saline solution containing the chelating agent EGTA, thus removing all Ca^{++} from the cupula, the hair processes, and the apical hair cell surface. A preparation so treated no longer responds to mechanical sinusoidal oscillation of the cupula by a synchronous modulation of the resting discharge. The basic level of the resting discharge, however, remains largely unaffected by the treatment. Moreover, the preparation can still be made to respond to sinusoidal variation of a polarization current applied extracellularly to the basal region of the hair cell, and so to mimic the response picture no longer obtainable by mechanical stimulation.

Sand postulates that the neurotransmitter release from the hair cell in response to adequate mechanical stimulation is based on a Ca^{++} ionic mechanism, whereas the resting activity is Ca^{++} insensitive. It should be kept in mind, in this context, that the chemical environments of the apical and basal (synaptic) portions of the hair cell are separated by ion-tight junctions forming the reticular membrane (Fig. 17-1). The ionic mechanism underlying the generation of the resting activity may, therefore, be quite different (e.g., based on Na^+ and K^+).

On the basis of results obtained by mechanical and electric stimulation of the mudpuppy lateral line organs (22), Sand argues that electric stimulation by sinusoidally oscillating polarization currents causes synchronous modulation of the resting discharge by modulating the neurotransmitter discharge from the hair cell and not by a direct electric effect on the terminals of the first order afferent neuron. The uncoupling of the Ca^{++} sensitive mechanical stage in the transduction process responsible for transmitter release takes place within the hair cell, where the mechanical generation of a receptor potential is believed to be abolished by Ca^{++} removal. Electric stimulation is, however, capable of replacing it in its role of causing modulation of transmitter release.

Earlier interpretations of the functioning of the hair cell system as a direct mechanoelectric transducer (20) were formulated before the discovery of the synaptic structures in the hair cell.

So far as the Ca^{++} sensitivity of the initial mechanosensitive stage in the transduction process is concerned, it is clear that its possible implications for vestibular physiology would have to be assessed taking the chemical environment of the apical hair cell structures in both cases into account. The mudpuppy is a freshwater amphibian and the composition of the fluid surrounding the apical surface of its hair cells is bound to be very different from that of the endolymph bathing the corresponding structures in the semicircular canal cupulae and otolith membranes in the vestibular end organs. Here we find, generally in the vertebrates, but especially in mammals, a situation characterized by high K^+ and low Na^+ concentrations approaching those characteristic for intracellular levels, whereas the basal portions of the hair cells and their synaptic regions are surrounded by interstitial fluid resembling in its ionic composition that of blood or cerebrospinal fluid.

It appears, therefore, quite likely that the modulation of neurotransmitter release is in both cases a two-stage process, based on two different ionic mechanisms. However, a final acceptance of this for the vestibular organ will have to await relevant experiments, including measurements of intracellular changes brought about by experimental changes in ionic content of the endolymph, such as manipulation of its Ca^{++} content.

Meanwhile, it appears reasonable to retain the assumption that the resting activity as such originates in response to constant depolarization of the postsynaptic membrane by leakage of neurotransmitter, and that the impulse activity in the myelinated axon of the first order afferent neuron is modulated by changes in transmitter release. A complete cessation of transmitter release would cause the complete abolition of the resting activity associated with hair process displacement in the so-called "inhibitory" direction. Similar silencing of the resting activity could also be brought about by hyperpolarization of the postsynaptic membrane under the influence of inhibitory transmitter released at the neighboring efferent synaptic sites. In contrast to the hair cell membrane, the postsynaptic membrane would thus have to be described as electrically unexcitable electrogenic membrane.

It must, however, be kept in mind that the membrane of the naked axoplasmic filament of the peripheral terminal of the first order afferent neuron may have mixed functional properties; i.e. it may represent a mosaic or regionally separated portions of electrically unexcitable and excitable constituents. If that were the case, the possibility of a contribution to the transduction process by direct mechanoelectric transduction could not be ruled out. The responsibility for the maintenance of the resting activity could in such a case be shared by the hair cell and the peripheral terminals of the afferent neuron.

PERIPHERAL ADAPTATION

At the beginning of this discussion it was pointed out that, whatever its origin, the existence of a resting activity in a considerable proportion of afferent sensory units of the acousticolateralis system warrants the assumption that these units must, at least potentially, be slowly or nonadapting to lasting external or internal excitatory agents. Peripheral adaptation and its time course have been extensively studied throughout the relatively recent history of vestibular research.

In some of these studies the adaptive properties of different types of vestibular unit have been correlated with other fundamental parameters, especially with average resting frequency and regularity expressed as coefficient of variation (2,7). For the semicircular canal two extreme functional types of unit may be distinguished even in the absence of a morphologically overt difference in the associated hair cells and their innervation, e.g., type I and II of Wersäll (24). There now exists a consensus that hair cells located at or near the top of the crista ampullaris are innervated by afferent fibers at the large-diameter end of the continuous spectrum of fiber diameter, are often "spontaneously" silent, have a relatively high threshold, and tend to fire irregularly and to adapt relatively rapidly; whereas units controlled by hair cells located on the slopes of the crista have opposite functional properties, i.e., they are innervated by relatively thin afferent fibers, are "spontaneously" active (in mammals at astonishingly high frequencies), fire with high regularity, and are therefore practically thresholdless. They are also slowly or nonadapting. These two contrasting types may be characterized as phasic and tonic, respectively. Obviously, peripheral adaptation is best studied in first order afferent neurons in the open-loop condition, i.e., with interrupted efferent outflow. In the whole organism this peripheral adaptation is grafted upon the adaptive processes occurring in central nervous transmission and also upon the various associated feedbacks channeled from the effectors to the sense organ itself via the various peripheral efferent pathways (5).

The question of the origin and mechanism of peripheral adaptation is as complex as that concerning the origin of the resting activity, and is, as pointed out above, related to it. An interpretation of the adaptation of canal units to constant angular acceleration as due to mechanical accommodation to the lasting displacement by viscous rearrangement of internal cupula constituents or by slippage between cupula and hair processes under the influence of elastic restoring couples (2) is put in question by the experiments in which the effects of constant acceleration are mimicked by electric polarization, including adaptation, despite the fact that here no mechanical displacement is involved (13).

Again, as in the case of the resting discharge, alternative sites of origin of the adaptive change in activity have to be considered. We confine ourselves here to the excitatory part of the bidirectional hair cell response in canal controlled units. It is assumed that this is due to an increase in neurotransmitter release and consequent increased depolarization of the postsynaptic peripheral terminals of the first order afferent neuron. This change in transmitter release is assumed to be governed by a stimulus-analog change in a receptor potential. This, in accordance with the hypothesis described earlier, is caused by an inward Ca^{++} current, which in turn is a function of the mechanical displacement of the hair cell processes. A possible locus *I* of adaptation is, therefore, the apical part of the hair cell membrane involved in the generation of this Ca^{++} current. Only intracellular measurements of such a generator potential and the time course of its changing amplitude could reveal the occurrence at locus *I* of adaptation to a constant mechanical displacement.

It was suggested above that the second stage in the transduction process, also located within the hair cell, is the depolarization of the presynaptic membrane by the receptor potential, and the consequent change in neurotransmitter release. This would furnish a possible locus *II* for the origin of adaptive change. Finally, the postsynaptic membrane itself may represent a third possible point of origin (locus *III*).

In contrast to some of the "giant" hair cells of the amphibian lateral line system, the vestibular hair cell has so far not been reliably accessible to prolonged measurements of intracellular potentials or to the manipulation of its ionic contents. We must, therefore, rely on circuitous reasoning. Locus *I* is excluded from qualifying as the "necessary" site of adaptation by the observed effects of DC polarization. Locus *II* is, according to Sand (22), directly affected by electric stimulation, and could therefore be assumed to be the origin of the adaptive process. The findings of Sand et al. (23) appear to speak against the involvement of locus *III* in this instance, as the postsynaptic membrane was found to be insensitive to electric stimulation. However, this site may, at least in the rapidly adapting "spontaneously" inactive types of unit, adapt to the presence of neurotransmitter, and must therefore be included among the possible sites of origin of peripheral adaptation.

So far only the semicircular canals have been considered. In the otolith organs we encounter the difficulty concerning the concept of a spontaneous or resting discharge, that the topographic orientation of the hair cells within the macula presents a complex map, and that the macula is not confined to one spatial plane. This makes the relationship between the sensory cells and the overlying otolithic material difficult to assess for any given population of hair cells. The zero point of hair process alignment associated with a so-called resting discharge is bound to differ from cell to cell within such a map.

We encounter a range of adaptability in otolith-controlled units which is similar to that found in the canal cells. A resting discharge proper would be confined to the static type of otolith-controlled unit, which shows little or no adaptation in any given spatial position of the labyrinth. The continuous presence of the specifically heavy otolithic mass in the gravitational field makes an "unstimulated" state difficult to conceive. We should, therefore, be well advised to use the term "background activity" in describing the impulse discharge picked up from an otolith-controlled unit observed in the so-called normal orientation of the labyrinth. If this activity is observed to be unchanging in the stationary labyrinth, it had best be described as the adapted state of activity of a unit related to a certain estimated vector of gravitational stimulation.

Having suggested that the adaptation to a constant stimulus is chiefly a neurologic process (the hair cell having full neuronal status), we may assume that the time constants of adaptation are an invariant characteristic of a given hair cell type (including its nerve supply). If it were possible to observe the behavior of a selected unit for a prolonged period of time, its adaptive behavior should change only within narrow limits. This means that if a unit is characterized initially as a slowly or rapidly adapting unit, its behavior in that respect should remain unchanged throughout the period of observation. In view of the correlation between background discharge and adaptation behavior, we should also expect the average background activity of such a unit to remain materially unchanged outside the limits of "normal" fluctuation.

RESTING ACTIVITY AND ADAPTATION IN ORBITAL FLIGHT

It is, however, of considerable interest that Bracchi et al. (3), in their report on multiday observations of the activity of specific single otolith-controlled units in the labyrinth of the bullfrog (*Rana catesbeiana*) in orbital flight, draw attention to radical changes both in resting activity and in adaptive behavior of such units during the period of observation (7 days). Units characterized by a low frequency irregular discharge at rest showed a fluctuation in resting activity of over 20 times that observed on the ground. This significantly increased occurrence of short-period bursts in resting activity was observed in one unit. Moreover, two of the units unexpectedly showed a reversal in adaptive behavior when stimulated by linear acceleration steps. A change from phasic to tonic behavior or vice versa was observed in these units, which had been characterized on the ground as rapidly and slowly adapting, respectively. The authors attribute these changes to an "unloading" of the macula in the condition of free fall, depriving it of the "tensioning" it normally (on the ground) receives by the action of the weight of the otolith mass. The conjecture is that the mechanical condition of the sensory epithelium might be a contributing factor to the characteristic behavior of single receptor cells.

Of course, the efferent innervation to the hair cells might be an alternative source of behavior modification. Yet, the authors appear to have had no record of comparable changes in behavior characteristics during their extensive studies of unit behavior on the ground, despite the fact that such units were observed under "closed-loop" conditions, i.e., fully under efferent influence. Unless the nature of the efferent outflow from the CNS were to change radically in free fall, the observed changes in unit behavior might have to be attributed to a change in the mechanical state of the macula.

One can only look forward to an extension of the study of the behavior characteristics of individual otolith-controlled units in future experiments with this material in orbital flight.

References

1. Adrian, E.D. and Zotterman, V.: The impulses produced in sensory nerve endings. III. Impulses set up by touch and pressure. J Physiol. (Lond.) 61:465, 1926.
2. Blanks, R.H.I., Estes, M.S., and Markham, C.H.: Physiologic characteristics of vestibular first-order canal units in the cat. II. Response to constant angular acceleration. J. Neurophysiol. 38:1250, 1975.
3. Bracchi, F., Gualtierotti, T., Morabito, A., and Rocca, E.: Multiday recordings from primary neurons of the statoreceptors of the labyrinth of the bullfrog. Acta Otolaryngol. (Suppl.) 334:1, 1975.
4. Correia, M.J. and Landolt, J.P.: A point process analysis of the spontaneous activity of anterior semicircular canal units in the anesthetized pigeon. Biol. Cybern. 27:199, 1977.
5. Gacek, R.R.: Morphological aspects of the efferent vestibular system. Handb. Sensory Physiol. 6/1:213, 1974.
6. Goldberg, J.M. and Fernandez, C.: Physiology of peripheral neurons innervat-

ing semicircular canals of the squirrel monkey. I. Resting discharge and response to constant angular accelerations. J. Neurophysiol. 34:635, 1971.

7. Goldberg, J.M. and Fernandez, C.: Physiology of peripheral neurons innervating semicircular canals of the squirrel monkey. III. Variations among units in their discharge properties. J. Neurophysiol. 34:676, 1971.
8. Groen, J.J., Lowenstein, O., and Vendrik, A.J.H.: The mechanical analysis of the responses from the end-organs of the horizontal semicircular canal in the isolated elasmobranch labyrinth. J. Physiol. (Lond.) 117:329, 1952.
9. Hoagland, H.: Electrical responses from the lateral line nerves in fishes. IV. The repetitive discharge. J. Gen. Physiol. 17:195, 1933.
10. Lifschitz, W.S.: Responses from first order neurons of the horizontal semicircular canal in the pigeon. Brain Res. 63:43, 1973.
11. Lowenstein, O.: Experimentelle Untersuchungen Uber den Gleichgewichtssinn der Elritze (*Phoxinus laevis*) Vergl. Physiol. 17:806, 1932.
12. Lowenstein, O.: The tonic function of the horizontal semicircular canals in fishes. J. Exp. Biol. 14:473, 1937.
13. Lowenstein, O.: The effect of galvanic polarization on the impulse discharge from sense endings in the isolated labyrinth of the thornback ray (*Raja clavata*). J. Physiol. (Lond.) 127:104, 1955.
14. Lowenstein, O.: Functional aspects of vestibular structure. In: Ciba Symposium on Myotatic, Kinesthetic, and Vestibular Mechanisms. London, Churchill, 1967, pp. 121–137.
15. Lowenstein, O. and Sand, A.: The activity of the horizontal semicircular canal of the dog fish (*Scyllium canicula*). J. Exp. Biol. 13:416, 1936.
16. Lowenstein, O. and Sand, A.: The individual and integrated activity of the semicircular canals of the elasmobranch labyrinth. J. Physiol. (Lond.) 99:89, 1940.
17. Lowenstein, O. and Sand, A.: The mechanism of the semicircular canal. A study of the responses of single-fibre preparations to angular accelerations and to rotations at constant speed. Proc. R. Soc. Lond. B. 129:256, 1940.
18. McNally, W.J. and Tait, J.: Ablation experiments on the labyrinth of the frog. Am. J. Physiol. 75:155, 1925.
19. McNally, W.J. and Tait, J.: Some results of section of particular nerve branches to the ampullae of the four vertical semicircular canals of the frog. Q. J. Exp. Physiol. 23:147, 1933.
20. Murray, R.W.: The response of the lateralis organs of *Xenopus laevis* to electrical stimulation by direct current. J. Physiol. (Lond.) 134:408, 1956.
21. Sand, A.: The mechanism of the lateral sense organs of fishes. Proc. R. Soc. Lond. B. 123:472, 1937.
22. Sand, O.: Effects of different ionic environments on the mechanosensitivity of lateral line organs in the mudpuppy. J. Comp. Physiol. 102:27, 1975.
23. Sand, O., Ozawa, S., and Hagiwara, S.: Electrical and mechanical stimulation of hair cells in the mudpuppy. J. Comp. Physiol. 102:13, 1975.
24. Wersäll, J.: Studies on the structure and innervation of the sensory epithelium of the crista ampullaris in the guinea pig. Acta Otolaryngol. (Suppl.) 126:1, 1956.

Discussion

OMAN: I wonder whether you feel it's necessary to assume that there's just one form, one locus of peripheral adaptation, in hair cell systems in general. In any given system a number of these different sites could perhaps be involved. Observation with respect to the possibility of synaptic adaptation, e.g., when Frishhopf and I have applied burst-type mechanical stimuli to the free-standing lateral line organ in the mudpuppy and observed receptor potentials intracellularly using averaging techniques, did not show an adaptation in the receptor potential, but rather an adaptive phenomenon in the response of the peripheral neuron. A second point is that Taglietti, of course, has been looking at the question of the origin of adaptation in the peripheral neuron in the frog semicircular canal crista. His evidence implicates a postsynaptic mechanism.

WIEDERHOLD: Actually, I want to follow up on both comments that Chuck made. Furukawa's work and Weiss's work in the alligator lizard show adaptation in the nerve firing and not in the receptor cell potential. And another bit of caution on Sand's implication of calcium. So many things change when calcium is lowered; for one thing, the mechanics and electron microscopy to be looked at in lowered calcium. There are two recent papers showing extremely low calcium concentrations in endolymph.

LOWENSTEIN: I perfectly agree with this comment since I myself, as you may have noticed, am not 100% happy about the calcium story. But we have got to account for the facts of this uncoupling, and this is a first step in our interpretation.

GOLDBERG: Three questions or comments. Perhaps we should take them one at a time. Concerning the interpretation of your galvanic polarization experiments, weren't the polarization currents applied through the nerve?

LOWENSTEIN: Yes, that's right. We polarized through the afferent channel.

GOLDBERG: Your results showed that anodal currents were excitatory, cathodal currents inhibitory. We have found similar rules in the squirrel monkey. The only way to explain the results is in terms of the currents acting on the afferent terminal. Were the predominant effect on the hair cell, then the directional rules would have to be reversed. Given this, the adaptation you saw with the currents might be ascribed to mechanisms residing in the afferent terminal, rather than in the hair cell. This in no way excludes the possibility that some of the afferent adaptation might reflect transmitter depletion. But a contribution from the afferent terminal is definitely implicated.

LOWENSTEIN: That's why I felt that the afferent terminal might have mixed functional properties. Dr. Murray analyzed this and came to the conclusion that possibly the afferent postsynaptic membrane is the site of electrical excitability.

GOLDBERG: You suggested that the lateral line organs might be exceptional among acousticolateralis systems in that the tops of the hair cells would not be in a high K^+ environment. Recently, Russell and Sellick (1976) have shown that there is a high K^+ concentration, presumably in the space between the hair cells and the cupula. They also found a positive DC potential in the same space, which was attributed to an electrogenic K^+ pump. These are important observations since they imply that transduction in lateral lines, as in other end organs, can be interpreted in terms of Davis' (1965) mechanoelectric theory. The last point concerns the existence of a resting (or zero-force) discharge in otolith afferents. Our results in the squirrel monkey can only be explained by assuming the existence of such a discharge. While confirmatory experiments in the zero-*g* environment of space flight would be welcome, the existence of a zero-force activity is in my view already established.

LOWENSTEIN: It's a question of quantity. How much and how variable? This is one of the factors why we look forward so very much to the experiment in space. Because that question is an open one.

HONRUBIA: I'd like to make a comment to your presentation. I think the issue of the origin of the spontaneous activity is very capital to the problem of excitation of the system, and my experience has been different from what you have reported. What I have done is tried to perform experiments to obtain data from the semicircular canals. I did these experiments with Dr. O'Leary, and the cochlea and lateral line with Dr. Strelioff. I found that DC polarization of these systems has really no effect on the resting activity. Now I did the experiments a little differently from yours. I put the electrode in the endolymphatic spaces. No, I did not polarize the whole structure as you did, but I fail to see any change in activity on the systems. On the other hand, in the lateral line organ Strelioff and I saw that we can modulate the transduction mechanism with alternating currents, and the effect is not algebraic, it's actually a multiplicative factor. We interpret that experiment to support a notion that the physiologic stimulus actually is a mechanical modulation of current flowing through the hair cells. But then the experience of the failure of increased spontaneous activity when the potential gradient is increased across the hair cells is not clearly understandable. Perhaps it is not solely the effect of the electric currents which is important. Because my observations are confusing I am asking your interpretation. But I want to make it clear that experience is contradictory to your observations.

LOWENSTEIN: So far as our different experience with regard to the effect of DC polarization on resting activity is concerned, I feel it would be difficult to resolve this problem here.

FREE PAPERS

18

Variability of the Spontaneous Firing Rate of Vestibular Receptors in a Stable, Controlled Bullfrog Preparation

F. BRACCHI, T. GUALTIEROTTI, AND A. MORABITO

The time invariance of the spike train data conveying sensory information is a basic assumption on which both the logic of the sensory function and the existing statistical method of analysis are based (6,13,17). How stationary a sensory spike volley in the ordinary acute experimental condition really is remains to be seen. Irregular discharges of action potentials, considered the finer form of sensory transmission of information within the nervous system (16), are known to show "wandering" even at zero input (1,8,12,14,15–17). Neurons showing a regular firing rate seem more stable at least for a limited period of time (9). It could be argued that the instability of the irregular firing rate might depend on the unsteady condition of the acute biologic preparation resulting from destructive surgery, anesthetic drugs, curare and curare-like agents, and so on, and from environmental variables not properly controlled. The emitter-follower-microelectrode system in current use and the microelectrode-nerve time junction are also not stable. This factor is a less important agent of variability since the recording time is usually short, seldom more than a few hours. But, on the other hand, the extremely limited period of observation of the unit activity might not provide enough samples, and large enough samples, for proper tests of stationarity (13).

In order to evaluate the true physiologic variability of the resting or adaptive discharge of a sensory unit, a physiologically stabilized and environmentally fully controlled bullfrog preparation has been developed and the spike train data recorded from first order vestibular afferents during several days or weeks. As known, the frog vestibular receptors discharge irregularly at a relatively low rate (1), and can be considered an acceptable biologic model of an irregular sensory activity.

STABILIZED BULLFROG PREPARATION

Laboratory adapted healthy bullfrogs fed with an artificial diet (Table 18-1) (amino acids, maltose, and vitamins in diluted frog Ringer's solution with lipovitamins in oil, injected) (5) were monitored in a controlled aquarium at

Table 18-1.
BULLFROG ARTIFICIAL DIET (300 gm/frog/day)

1 mg each of		
L-isoleucine		
L-leucine		
L-lysine		
L-valine		
L-methionine		
L-threonine		
L-tryptophan		
5 mg of		
L-alanine		
0.48 g of		
Maltose in 6.5 ml distilled water		
2 ml Ringer's		
(*Ringer's Formula*)	*gm/liter*	
NaCl	6.5	
KCl	0.14	
$NaHCO_3$	0.20	
NaH_2PO_4	0.91	
Vitamin B^1	−0.02 mg	
Vitamin B^2	−0.03 mg	
Vitamin B^6	−0.03 mg	
Vitamin B^{12}	−0.2 mg	
Pantothenic	−0.03 mg	
Nicotinic	−0.01 mg	
Calcium Chloride	−2.2 mg	
Magnesium Sulfate	−1.1 mg	
Vitamin C	−0.8 mg	
Tetracycline	−5 mg	
Vitamin D^2	0.075 mg	Given monthly
Vitamin K	0.10 mg	Given monthly
Vitamin A	0.06 mg	Given monthly

constant temperature, constant water flow, and constant light/dark cycle. The water flow assured catabolite elimination and periodic lowering of water pH to 4 maintained bacterial growth within safe limits. Skin debris was filtered out. Nondestructive surgery gave access to the VIIIth nerve, keeping the nervous centers, the nervous connections, and blood supply intact. Full recovery was permitted and vestibular function tested for normal response by observation of vestibular reflexes, swimming, and jumping pattern. In some test animals microelectrodes were implanted chronically under light anesthesia at this stage (2–4) and the recovered animals tested as above. Upon recovery from anesthesia, the frogs behaved normally. In the experimental animal the thoracic plexus branches and part of the lumbar ones were cut for immobilization and the head fixed by means of a specially designed head holder which didn't produce any skin lesions. Spike recording was performed using the neutral buoyancy microelectrode technique (3), but the isolation of the microelectrode was made with parylene to assure a practically constant impedance (10).

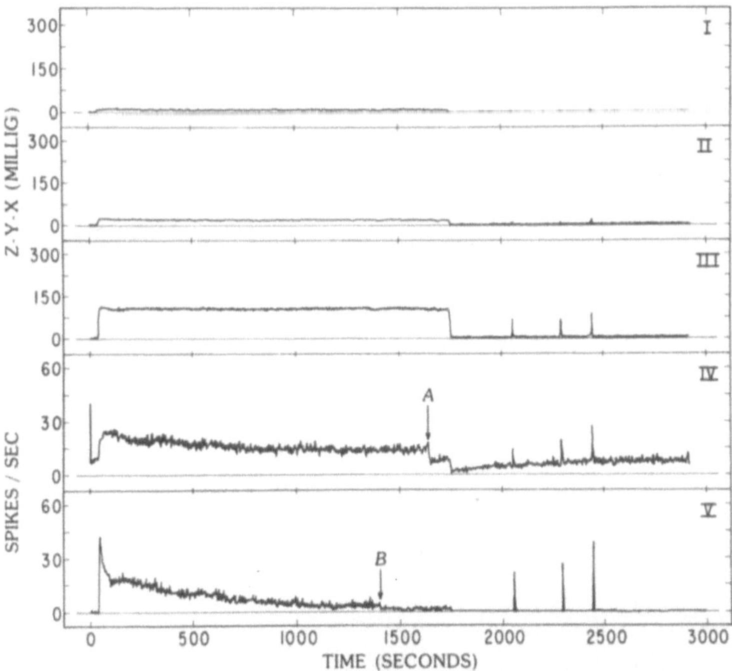

Fig. 18-1. Sudden and independent change of gain of two statoreceptors during a sustained stimulus of 60 Hz vibration, recorded by *x-y-z* accelerometer and integrated (*I–III*). **A** Slowly adapting statoreceptor (record *IV*). **B** Fast-adopting statoreceptor (record *V*). For details, see text.

SPIKE IDENTIFICATION AND CONTROL

Spike recognition was performed via double time and amplitude gates (7). Amplitude and time differences were determined between each minimum and the following maximum, and the gates set at the parameters of the required signal, including a variability factor of ±10%. Through the minimum interval technique the existence of one spike only was continuously controlled.

RESULTS AND CONCLUSIONS

The spike train data of units belonging to different parts of the vestibule, semicircular canals, and phasic and tonic statoreceptors were recorded for 15 to 23 days, when the bullfrog preparations appeared stationary, being several days away both from the microelectrode implant and the terminal stage. The units were not shielded from environmentally generated vibratory accelerations consisting of 60-cycle vibrations with a maximum intensity

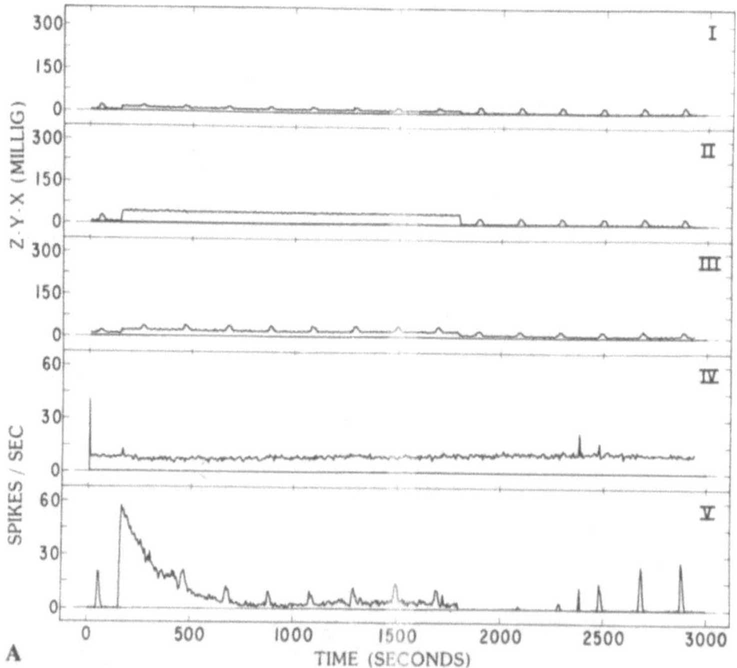

Fig. 18-2. **A** Time history of the activity of an anterior semicircular canal unit (record *IV*) and a phasic statoreceptor (record *V*) during environmental acceleratory inputs: the prolonged vibration induced by the turbulent motion of the environmental water, and the periodic short bursts of added acceleratory stimuli. A relatively rapid adaptation is shown during the prolonged stimulus. After the end of the stimulus a

of ± 0.1 g. In fact such environmentally generated perturbations were used to study short-term stability (12). All statoreceptors, phasic and tonic, responded to such stimuli (Fig. 18-1), but not the semicircular canals (Fig. 18-2). The resting (semicircular canal unit, fast-accommodating phasic statoreceptors) and adaptive (tonic statoreceptors) unitary activities were recorded twice or three times in 24 h for periods of 2 h each time and analyzed in samples of 500 s. Even in these conditions sudden shift of firing rate was observed with no apparent cause (Fig. 18-3). During a steady stimulus changes of gain were recorded (Fig. 18-1). These variations appeared independently in the different units, and could be sudden (Fig. 18-1) or slowly developing (Fig. 18-3). The coefficient of variation nearly always changed inversely with respect to the firing rate (Fig. 18-3). Some environmental factors, such as variations of temperature, sudden increase of pressure, and changing the environment from water to air and vice versa at constant temperature modified the firing rate (Fig. 18-4); this occurred simultaneously for all units observed. The mechanism of the changes of the resting or adaptive firing rate and of gain cannot be studied with the present data. However,

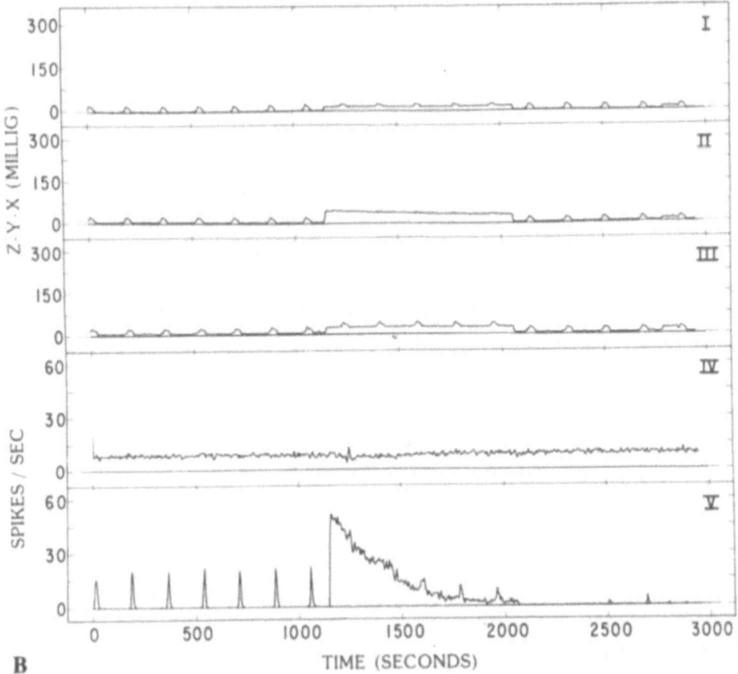

long period of inhibition follows during which the unit responses to the periodic short accelerations are absent or reduced. Recovery takes up to 10 to 25 min. The semicircular canal unit is not stimulated at all by the environmental perturbation, not even indirectly through the firing rate changes of the otolith cells. **B** As shown, 24 h later the response characteristics of the two units remain the same.

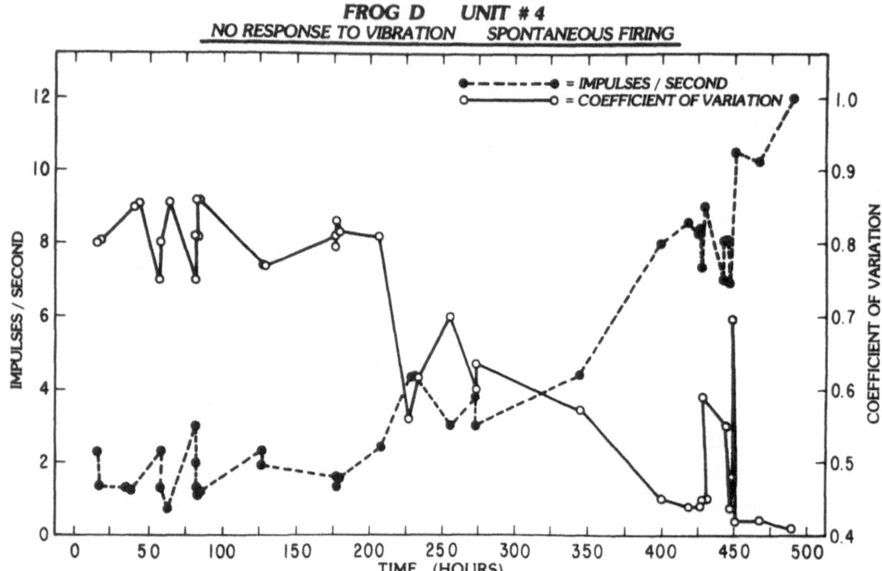

Fig. 18-3. Semicircular canal unit of Fig. 18-2. Mean frequency and coefficient of variation (CV) during 500 h of recording of the activity at rest. The mean frequency and CV remain within constant variability limits for approximately 230 h; then a progressive increase of frequency of discharge is observed, reaching a peak of five times the original value after 400 h. Note the reciprocal changes of the CV and the mean frequency.

cellular factors of different kinds must be involved since (1) the variations happen for each cell independently; and (2) these changes involve both the resting activity and the responses to stimulation, either completely adapted or in phase of adaptation. Different origins and sites of action are postulated for these functions (11). To solve this problem, endocellular recording in chronic preparations is needed; however, this technique is at present not available.

ACKNOWLEDGMENTS

This work was supported by grant NSG-2197 and contract NAS2-9871 from the National Aeronautics and Space Administration. We are indebted to Dr. Bingham for suggesting the method for bacterial control and to Drs. Peterks and Tomko for the design of the frog headholder.

References

1. Bracchi, F., Gualtierotti, T., Morabito, A. and Rocca, E.: Multiday recordings from the primary neurons of the statoreceptors of the labyrinth of the bull frog. Acta Otolaryngol. Suppl. 334:1, 1975.

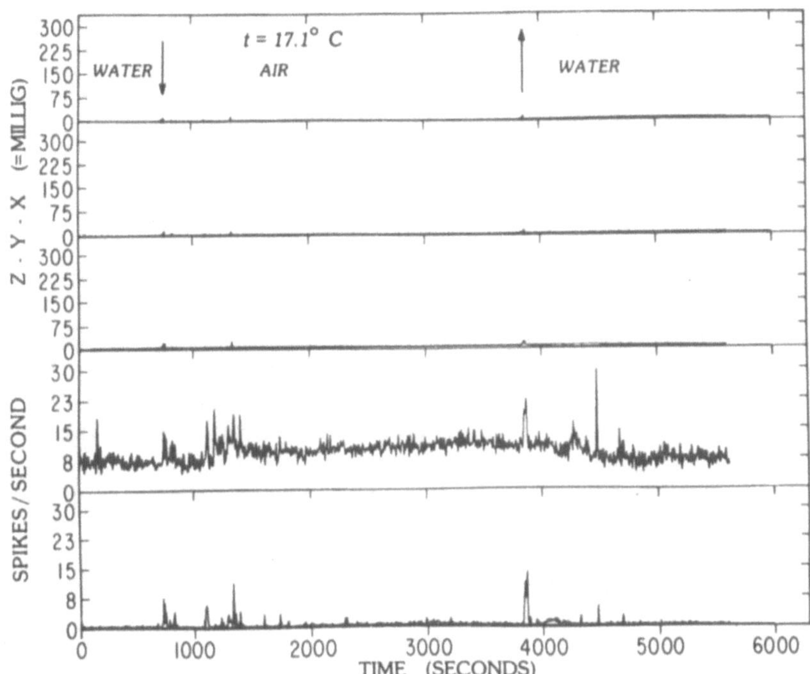

Fig. 18-4. Effect of the change from a water to air environment in two units. As shown, a significant increase in the mean frequency is observed in both units during the period in air. Temperature is constant in both situations at 17.1°C.

2. Geisler, C.D. and Goldberg, J.M.: A stochastic model of the repetitive activity of neurons. J. Biophys. 6:53-69, 1966.
3. Gualtierotti, T. and Alltucker, D.S.: Prolonged recording from single vestibular units in the frog during plane and space flight, its significance and technique. Aerospace Med. 38:513, 1967.
4. Gualtierotti, T. and Bailey, P.: A neutral bouyancy microelectrode for prolonged recording from single nerve units. Electroencephalog. Clin. Neurophysiol. 25:77, 1968.
5. Gualtierotti, T. and Gerathewohl, S.J.: Spontaneous firing and responses to linear acceleration of single otolith units of the frog during short periods of weightlessness during parabolic flight. In: The Role of Vestibular Organs in the Exploration of Space. NASA SP-77, 1965, pp. 221–232.
6. Johnson, D.H. and Kiang, N.Y.S.: Analysis of discharges recorded simultaneously from pairs of auditory nerve fibers. J. Biophys. 16:719, 1976.
7. Kojima, G.K. and Bracchi, F.: Microprocessor-based instrument forineutral pulse wave analysis. I.E.E.A. Trans. Biomed. Engin. BME-27(9):515, 1980.
8. Levine, M.W. and Shefner, J.M.: A model for the variability of interspike intervals during sustained firing of a retinal neuron. J. Biophys. 19:241, 1977.
9. Loe, P.R., Tomko, D.L., and Werner, G.: The neural signal of angular head position in primary afferent vestibular nerve axons. J. Physiol. 230:29, 1973.
10. Loeb, G.E., Bak, M.J., Salcman, M., and Schmidt, E.M.: Parylene as a

The Vestibular System: Function and Morphology. New York, Springer-Verlag, 1981.

11. Lowenstein, O.: The origin and functional significance of the resting activity and peripheral adaptation in the vestibular system. In: Gualtierotti, T. (ed.): The Vestibular System: Function and Morphology. New York, Springer-Verlag, 1981.

12. O'Leary, D.P., Segundo, J.P., and Vidal, J.J.: Perturbation effects on stability of gravity receptors. Biol. Cybern. 17:99, 1975.

13. Perkel, D.H., Gerstein, G.L., and Moore, G.P.: Neuronal spike trains and stochastic point processes. J. Biophys. 7:391, 1967.

14. Stein, R.B.: A theoretical analysis of neuronal variability. J. Biophys. 5:173, 1965.

15. Vidal, J., Jeannerod, M., Lifschitz, W., Levitan, H., Rosenberg, J., and Segundo, J.P.: Static and dynamic properties of gravity and sensitive receptors in the cat vestibular system. Kybernetik 9:205, 1971.

16. Viernstein, L.J. and Grossman, R.G.: Neural discharge patterns in the transmission of sensory information. In: 4th London Symposium on Information Theory. London, Butterworths, 1961, pp. 252–269.

17. Werner, G. and Mountcastle, V.B.: The variability of central neural activity in a sensory system, and its implications for the central reflection of sensory events. J. Neurophysiol. 26:958, 1963.

19

The Comparative Sensitivity of Selected Receptor Systems

SHYAM KHANNA AND CARL SHERRICK

In the myriads of adaptive modes conferred on living organisms by Nature, there have appeared receptor systems that detect nearly every form of energy found in the universe. Electromagnetic energy in the form of ultraviolet, visible, and infra-red light, sound energy, electric fields, chemical solutions, angular acceleration, and mechanical pressure are all detected by one or another specialized receptor organ, such as the eye, ear, vestibular system, thermal receptors, chemoreceptors, electroreceptors, or mechanoreceptors.

During the process of specialized adaptation, the receptor systems have developed higher and higher degrees of sensitivity as well as of selectivity. In designing physical devices and methods for measuring the adequate stimuli for receptive systems, physicists, physiologists, and others have tended to express receptor sensitivity in the stimulus dimensions that it seems to follow, or that are most readily measured. Thus, whereas the sensitivity of the eye can be expressed as the number of photons required for excitation (11), the common measure of stimulus intensity is in candles /m^2; the threshold of hearing is expressed in relative or absolute units of sound pressure; electroreceptor sensitivity is expressed as field strength in V/cm; cutaneous mechanoreceptor sensitivity is expressed as displacement or velocity of motion.

Because the purpose of the present paper is to compare the absolute sensitivities of various receptor systems, it will be necessary to convert the conventionally employed dimensions of the stimulus to a common unit, viz., energy or power. If the stimulus is a transient signal that waxes and wanes quickly, the "threshold" of excitation may be given in the dimensions of work or energy. For a continually applied signal the threshold is better defined as power flux.

In the great majority of cases, liminal energy or power requirements have not been calculated or measured directly, and it will be necessary to compute these values from available data that provide the stimulus dimensions commonly employed, along with data that specify certain dynamic characteristics of the systems. For example, given the voltage across a simple circuit as well as its impedance, one may calculate the power absorbed by the circuit elements.

The receptor systems dealt with in this report divide readily into two groups along both structural and, as will be seen, functional dimensions. The first group consists of encapsulated receptors found in the skin, and includes the mechanoreceptors and thermal receptors. Such receptor systems have been characterized as having primary sensory neurons (7), and have a receptor-nerve interface with no intervening structure, in which the stimulus acts directly on a specialized receptor membrane of the nerve ending. The stimulus-produced generator current triggers a nerve spike when it is sufficiently strong (14,32). The second group comprises receptors with hair cells, described as having secondary sensory neurons (7), in which the hair cell synapses with the primary afferent neuron. In these receptors the stimulus acts on the hair cells, which probably produce chemical transmitters that initiate nerve activity by synaptic action. Among this group are the electroreceptors, the vestibular receptors, and the auditory receptors.

THRESHOLD POWER FOR MECHANORECEPTORS

Mechanical Vibration of the Skin

The power dissipated in a mechanical system that is vibrated sinusoidally is given by the equation:

$$P = F \times V \times \cos \theta$$

where P is power in erg/s, F is the force in dynes, V is the velocity amplitude in cm/s, and θ is the phase angle between force and velocity.

Values of the force, velocity amplitude, and amount of power required for threshold sensations of vibration for human observers over the frequency range of best response for the Pacinian corpuscle are shown in Table 19-1. The area stimulated was the fingertip, which is considered to be one of the most sensitive of the body areas (5). The measurements shown were made with an impedance head (Bruel & Kjaer, Model 8001), with which force, ve-

Table 19-1.
VALUES OF THRESHOLD FORCE AND VELOCITY, AND COMPUTED VALUES OF POWER, FOR MECHANICAL VIBRATION OF THE FREE FINGERTIP.

Frequency (Hz)	Threshold Force (dyn)	Threshold Velocity (cm/s)	cos θ	Power (W × 10^{-7})	Displacement (dB re 10^{-4} cm rms)	Power (dB re 10^{-6} W)
80	63	0.02	0.96	1.2	−8	−9.2
110	47	0.02	0.91	0.9	−14	−10.5
150	47	0.009	0.89	0.4	−20	−14.0
175	32	0.009	0.79	0.2	−22	−17.0
200	32	0.01	0.81	0.3	−21	−15.2
225	47	0.02	0.66	0.6	−18.4	−12.2
250	47	0.02	0.81	0.8	−17.7	−11.0
300	32	0.03	0.85	0.8	−17	−1.0
350	32	0.03	0.97	0.9	−17.7	−10.5
400	32	0.03	1.0	0.9	−17.7	−10.5
500	32	0.04	0.95	1.2	−17	−9.2

From Sherrick, C.: Unpublished data.

locity, and phase angle measurements could be made at the values of the observer's threshold. The area of contact was 0.64 cm², and a static force of 15 *g* was exerted against the finger (Sherrick, unpublished data).

From the values obtained, the power required for stimulation at threshold was calculated by the equation given above, and the values are listed in Table 19-1. It is apparent that the range of powers required for excitation is ± 5 dB around a value of 0.1 μW (Fig. 19-1).

We assume that the power flux is dissipated over an area of 0.64 cm², and that it is radiated coherently, i.e., not spherically but as a ray bundle. We assume as well that the Pacinian corpuscle is the receptor excited, since it has

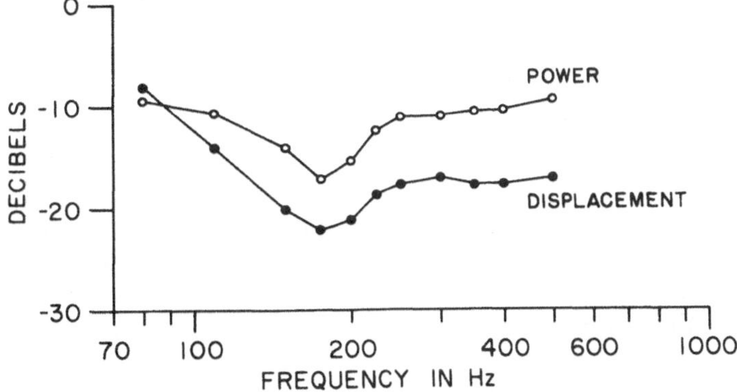

Fig. 19-1. Threshold sensitivity at the fingertip in terms of displacement and power. Displacement shown in dB below 10^{-4} cm rms. Power expressed in dB below 10^{-6} W.

been shown to be present in human digital skin (21) and responsive to frequencies above 60 Hz (26). The area of the corpuscle is about 0.5 mm^2 (15), and the power reaching a single Pacinian corpuscle at the best frequency (175 Hz) is (at most):

$$P = \frac{0.2 \times 10^{-7} \cdot 0.5 \times 10^{-2}}{0.64} = 1.6 \times 10^{-10}$$

Considering the fact that there may be no more than 100 Pacinian receptors in the finger (21) whose responses may or may not summate, and that the power radiated from the mechanical contactor is absorbed by inert tissue as well as attenuated according to the inverse square law, the estimate may be an order of magnitude too large.

From measurements of threshold for bursts of vibrotactile stimuli, Verrillo (27) has determined that durations as short as 300 ms at 250 Hz show no significant increase in threshold amplitude over that for very long bursts. The energy in such a burst can be computed readily as follows:

$$\frac{1.6 \times 10^{-10}\text{W} \times 10^{7}\ \text{erg/s/W}}{300 \times 10^{-3}\ \text{s}} = 0.005\ \text{erg}$$

Wolf (33) measured the minimum perceptible collision of the skin with a delicate ballistic pendulum arrangement, and calculated the threshold energy for a contact area of 0.005 cm^2 on the thumb as 0.026 erg. Von Frey (29) computed the energy required to elicit a report of a pressure sensation when a hair on the arm was deflected by a tiny weight, consisting of a piece of tinsel. By measuring the amplitude of deflection of the hair and the weight used, von Frey calculated a minimal value of 0.014 erg. It has been shown by a number of authors (6) that the hair follicle region harbors sensitive receptor complexes, and the data given above suggest that the energy required to excite a single receptor is comparable to that calculated from Sherrick's values, given variations in method, bodily locus, and duration of impact.

Thermal Receptors in the Skin

According to Hartline (10), the power flux density of a radiant heat source required for electrophysiologic or behavioral threshold responses in the pit receptors of Boid snakes is 8×10^{-5} W. The receptors found in the pits are unmyelinated nerve endings with terminal fields of 30 to 50 μm diameter. Multiplying these areas by the power flux density yields liminal power at the receptor field of between 6×10^{-10} and 2×10^{-9} W. These values agree closely with those computed for liminal mechanoreceptor power requirements.

Electroreceptors

The ampullae of Lorenzini are a much studied receptor system found in the ray, to which a variety of functions were assigned before it was concluded that they act as detectors of weak electric field gradients in the environment of the fish (1). From a pore-like opening on the surface of the skin of the ray there leads a jelly-filled canal that ends on a cluster of small swellings containing the sensory tissues. The latter are the ampullae of Lorenzini, for which the canals, which may be as long as 16 cm in a medium-sized ray, act as conducting leads (attenuating voltage by 4 dB/m) from the body surface to the receptor. The special character of the tissues surrounding the canals prevents conduction of electricity from inside to outside the canal walls (18). From the original tabulation by Kalmijn (12) there have been extracted representative values of threshold voltages and currents. Calculations of the power required for threshold levels of excitation can be made under the following assumptions:

1) Let the openings of a pair of canals be Y cm apart on the surface of the ray. If the voltage gradient along the line joining the openings is E V/cm, then the potential difference between the openings is $E \times Y$V. Under this assumption, the farther apart the openings, the higher the potential difference and hence the possible sensitivity of the system.

2) The threshold current flowing through the canal is given by the current density times the average cross-sectional area of the canal. If we assume that the conductive system contains well-matched and lossless elements, the calculated liminal power will be somewhat larger than for natural conditions.

For a medium-sized ray, having a distance between canal openings of 20 cm and a canal diameter of 0.2 cm, the liminal voltage gradient given by Kalmijn (12) is 0.01 μV/cm. Liminal current density is 0.5 nA/cm^2. Threshold power flux is then given by:

$$\begin{aligned} &0.01 \times 10^{-6}\text{V/cm} \times 20 \text{ cm} \times 0.5 \\ &\quad \times 10^{-9}\text{A/cm}^2 \times \pi(0.1)^2 \text{ cm}^2 \\ &\quad = 3.14 \times 10^{-18} \text{ W}. \end{aligned}$$

Because there may be as many as 200 sensory cells per ampulla in these species (25), a uniform distribution of power among them would yield an estimate of about 1.6×10^{-20} W per receptor.

It is clear that the electroreceptor system is infinitely more sensitive than that of the cutaneous mechanoreceptors. It is also apparent from several studies (18,25) that electroreceptors are structurally more like the hair cells of the auditory, vestibular, and lateral line systems than like cutaneous receptors.

Vestibular Receptor Systems

Data available from Oman and Young (20) indicate that, at threshold, an angular acceleration of 0.01 degree/s^2 is required, from which the pressure acting on the cupula was calculated to be 1.25×10^{-4} dyn/cm^2. Under the same conditions, the deflection of the cupula from its midpoint was computed as

0.01 μm. The force on the cupula, which has a cross-sectional area of $\pi \times (0.06)^2$ cm^2, is

$$1.25 \times 10^{-4}\ \text{dyn/cm}^2 \times \pi \times (0.06)^2\ \text{cm}^2 = 1.4 \times 10^{-6}\ \text{dyn}.$$

The work done on the cupula is

$$1.4 \times 10^{-6}\ \text{dyn} \times 0.01 \times 10^{-4}\ \text{cm} = 1.4 \times 10^{-12}\ \text{erg}.$$

Oman (personal communication) has reported that the threshold is reached in about 10 s of stimulation. The power delivered to the system, therefore, is

$$\frac{1.4 \times 10^{-12}\text{erg} \times 10^{-7}\text{W/erg/s}}{10\ \text{s}} = 1.4 \times 10^{-20}\text{W}.$$

The cupular macula contains about 7600 hair cells, but it is difficult to know how many of these are involved in the liminal response, especially because the vestibular system is well known to be one in which the cells fire spontaneously at a regular rate, and increase or decrease their rate depending on the direction of rotation. It is possible to make an arbitrary assumption that probably only a small fraction of the total population of sensory cells (perhaps 10 to 100) are excited at threshold accelerations, in which cases the liminal power requirements for the vestibular system are computed as 10^{-21} to 10^{-22} W.

Auditory Receptor System

The acoustic power flux in the cochlea can be calculated from the impedance of the scalae and the pressure in the scalae at threshold. From Nedzelnitsky's (19) data on cats, we can obtain the ratio of the sound pressure at the tympanic membrane to that in scala vestibuli. Because these data were obtained with the bulla open, the correction for the closed bulla from the work of Guinan and Peake (8) must be applied. An additional correction from the work of Wiener et al. (31) permits the computation of the ratio at the external ear to that in scala vestibuli. To provide the necessary threshold sound pressure values for cats, the behavioral data of Miller et al. (17) were employed. By appropriate combination of these sources of data, the pressure in the scala vestibuli at threshold for the cat can be computed as a function of frequency, as Fig. 19-2 shows. It can be seen that the sound pressure level is *about 25 dB* (re 0.0002 dyn/cm^2) $\pm$ 10 dB from 100 to 5000 Hz, which is a surprisingly flat function, given the usual 50-dB range of values found for minimum audible field pressures (17).

The measurements of Lynch et al. (16) on the impedance of scala vestibuli show it to be resistive above 100 Hz, ranging around a value of 1.9×10^6 acoustic Ω between 100 and 6000 Hz. Acoustic power is calculated from the ratio of the square of the pressure to the resistance, and for liminal power,

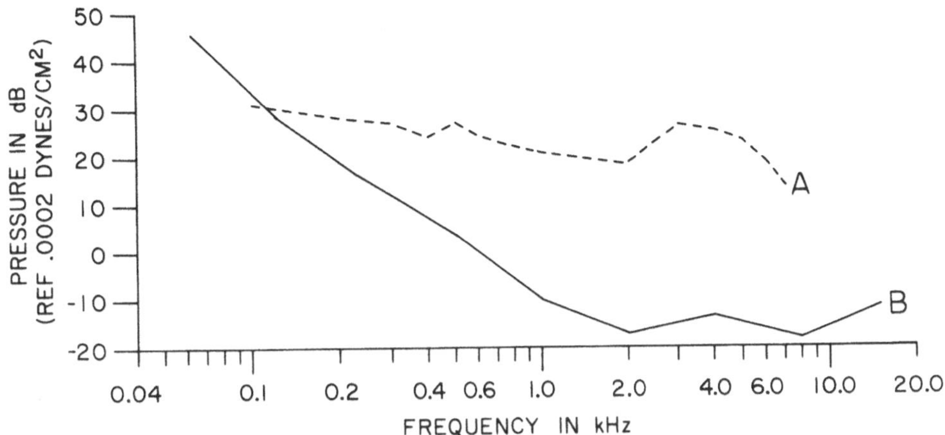

Fig. 19-2. Hearing threshold of cat. **A** Pressure in scala vestibuli. **B** Minimum audible field (17).

given the present values, it amounts to

$$\frac{(3.38 \times 10^{-3}\ \text{dyn/cm}^2)^2}{1.9 \times 10^6} = 6 \times 10^{-19}\ \text{W}.$$

To calculate the power reaching the hair cells, it is necessary to show how the power is spatially distributed along the basilar membrane. According to the findings of Russell and Sellick (23), who recorded intracellularly from hair cells, the tuning curves thus obtained are identical with the neural tuning curves in the first order afferents. This suggests that the frequency selectivity commonly seen in such units is a reflection of the mechanical tuning afforded by the cochlea. The data of Kiang et al. (13) and Evans (4) for primary auditory fibers can be used to determine the half-power values of bandwidth for particular frequencies. This effective bandwidth can be converted to a spatial function on the basilar membrane, given a set of data that specify the point of maximal displacement on the membrane as a function of frequency of sound. By estimating the half-power points on the spatial envelope, we have delineated the area of the basilar membrane over which half the sound power is effective.

Schuknecht's (24) data for the cochlea of the cat have been adapted to provide the figures in Table 19-2, to give the bandwidth in terms of frequency and membrane length as a function of three frequencies of stimulation. In addition, the number of hair cells in the spatial bandwidth is given. The power reaching each hair cell is approximately the half power divided by the number of hair cells in the effective region, and this is given in Table 19-3 for three frequencies. Note that the calculated values agree with those obtained for the vestibular receptors and the electroreceptors, as shown in

Table 19-2.
EFFECTIVE REGIONS FOR DELIVERY OF POWER FLUX ON THE BASILAR MEMBRANE, CALCULATED FROM BANDWIDTHS AT SELECTED FREQUENCIES.

Best Frequency (Hz)	3 dB Bandwidth (Hz)	Spatial Bandwidth (mm)	No. of Hair Cells in the Spatial Region of Bandwidth: Outer (415/mm)	Inner (120/mm)
500	75	0.51	211	61
2000	150	0.3	124	36
6000	200	0.14	58	17

Table 19-4, which summarizes the sensitivity values for all the systems discussed.

Three points can be emphasized from the values of Table 19-4:

1) The threshold power requirements for receptors without hair cells (Pacinian, sensory hairs, and pit receptors) are very similar, ranging from 0.3 to 2.3×10^{-9} W.
2) The threshold power requirements for receptors with hair cells (ampulla of Lorenzini, vestibular, and auditory receptors) are also very similar, ranging from 1.6×10^{-20} to 10^{-22} W.
3) The difference between the two receptor classes is very large, of the order of 10^{11} to 10^{12} times.

The remarkable gain in sensitivity made possible by the interposition of the hair cell in the receptor-nerve nexus implies an amplification function for the hair cell. If so, and if it behaves like man-made amplifiers, its limiting sensitivity is determined by its resistance noise and the amplifier noise factor. Given a perfect amplifier the resistance noise must still be a limiting factor, and its value is given by:

$$P = 4 \times K \times T \times \Delta F$$

where P is the noise power in W; K is Boltzmann's constant; T is temperature, °K; and ΔF is the bandwidth of the system in Hz (30).

Table 19-3.
POWER DELIVERED TO SINGLE HAIR CELLS AT SELECTED SOUND FREQUENCIES.[a]

Frequency (Hz)	Power/Hair Cell (W)
500	2.2×10^{-21}
2000	3.7×10^{-21}
6000	8.0×10^{-21}

[a] See Table 19-2.

Table 19-4.
THRESHOLD SENSITIVITY OF SINGLE RECEPTORS.

Receptors without Hair Cells	Threshold Energy (erg)	Threshold Power (W)
Mechanoreceptors		
Skin (Pacinian) (33)	0.026	1.6×10^{-10}
Skin (Pacinian)	0.005	
Skin (sensory hair) (29)	0.014	
Thermal receptors		
Pit receptors (free nerve ending) (10)		2.3×10^{-9} to 5.6×10^{-10}
Receptors with Hair Cells		
Electroreceptors		
Ampulla of Lorenzini (*Raja clavata*) (12)		1.6×10^{-20}
Vestibular system		
Human semicircular canal (20)		10^{-21} to 10^{-22}
Auditory system		
Cat (various sources; see text)		2.2×10^{-21} to 8×10^{-21}

Numbers in parentheses are reference citations.

From Russell and Sellick's (23) data for the hair cells, $\Delta F = 230$ Hz; and the temperature of the hair cells is 37°C = 310°K. Hence,

$$P_{noise} = 1.38 \times 10^{-23}\ \text{J/°K} \times 310\text{°K} \times 230\ \text{Hz} \times 4 = 3.9 \times 10^{-18}\ \text{W}.$$

Thus, signal detection would seem to be occurring at a signal/noise ratio of −10 to −30 dB.

Additional evidence that a considerable power gain is available within the hair cell has been reported by Russell and Sellick (23), who measured its intracellular summation potential (in the guinea pig) as 2 mV. In addition, the average value of resistance within the hair cell was calculated by these authors as $5 \times 10^7\ \Omega$. The average power dissipation inside the hair cell is, therefore,

$$\frac{(2 \times 10^{-3})^2\ \text{V}}{5 \times 10^7\ \Omega} = 8 \times 10^{-14}\ \text{W}.$$

Given that the estimate of mechanical energy per hair cell was of the order of 10^{-21} W, the power gain for the hair cell would seem to be at least 10^7. Early work by von Békésy (28) demonstrated that the electric energy in the DC cochlear potentials was more than that supplied by the acoustic stimulus, but could provide no quantitative estimate of the power gain in the hair cells.

Harris (9) and de Vries (3) considered the effects of Brownian motion of air molecules on the level of mechanically produced noise in the hair cells.

As Dallos (2) has pointed out, this source of random noise is probably well below the values of energy required for detection, assuming some degree of coupling among individual hair cells. It would therefore appear that the limiting value of noise is attributable to that within the hair cell itself, as suggested by the just cited Nyquist equation relating noise power and temperature.

The precise mechanism of hair cell transduction is not yet clear, as Dallos (2) has shown in his excellent treatise on the subject. Until the details of the production of the hair cell generator potential and neurochemical distributive function are known, the exact mechanism of the amplification afforded by the hair cell, and its limiting noise level, cannot be discussed in any but the broadest terms.

ACKNOWLEDGMENTS

The preparation of this paper was supported in part by grants from the DHEW, USPHS, No. NS 03654 and 1 K04 NS 00292, to Columbia University College of Physicians and Surgeons, and in part by a grant from DHEW, USPHS, No. NS 04755 to Princeton University.

References

1. Bullock, T.H.: An essay on the discovery of sensory receptors and the assignment of their functions together with an introduction to electroreceptors. In Fessard, A. (ed.): Handbook of Sensory Physiology, Vol. III/3. Electroreceptors and other Specialized Receptors in Lower Vertebrates. New York, Springer-Verlag, 1974, pp. 1–12.
2. Dallos, P.: The Auditory Periphery: Biophysics and Physiology. New York, Academic Press, 1973.
3. de Vries, H.: Brownian motion and transmission of energy in the cochlea. J. Acoust. Soc. Am. 24:527, 1952.
4. Evans, E.F.: Cochlear nerve and cochlear nucleus. In Keidel, W.D. and Neff, W.D. (eds.): Handbook of Sensory Physiology, Vol. V/2. New York, Springer-Verlag, 1975, pp. 2–96.
5. Geldard, F.A.: The Human Senses, 2nd ed. New York, Wiley, 1972.
6. Geldard, F.A.: Vibratory reception in hairy skin. In Flores d'Arcais, G.B. (ed.): Studies in Perception. Milan, Martello-Giunti, 1974, pp. 301–311.
7. Grundfest, H.: The general electrophysiology of input membrane in electrogenic excitable cells. In Loewenstein, W.R. (ed.): Handbook of Sensory Physiology, Vol. I. New York, Springer-Verlag, 1971, pp. 135–165.
8. Guinan, J.J. and Peake, W.T.: Middle ear characteristics of anesthetized cats. J. Acoust. Soc. Am. 41:1237, 1967.
9. Harris, G.G.: Brownian motion in the cochlear partition. J. Acoust. Soc. Am. 44:176, 1968.
10. Hartline, P.H.: Thermoreception in snakes. In Fessard, A. (ed.): Handbook of

Sensory Physiology, Vol. III/3: Electroreceptors and Other Specialized Receptors in Lower Vertebrates. New York, Springer-Verlag, 1974, pp. 297–312.

11. Hecht, S., Shlaer, S., and Pirenne, M.H.: Energy, quanta, and vision. J. Gen. Physiol. 25:819, 1942.
12. Kalmijn, A.J.: The detection of electric fields from inanimate and animate sources other than electric organs. In Fessard, A. (ed.): Handbook of Sensory Physiology, Vol. III/3: Electroreceptors and Other Specialized Receptors in Lower Vertebrates. New York, Springer-Verlag, 1974, pp. 147–200.
13. Kiang, N.Y.S., Watanabe, T., Thomas, E.C., and Clark, L.F.: Discharge patterns of single fibers in the cat's auditory nerve. Research Monograph No. 35. Cambridge, Mass., MIT Press, 1965.
14. Loewenstein, W.R.: Facets of a transducer process. Cold Spring Harbor Symp. Quant. Biol. 30:29, 1965.
15. Loewenstein, W.R.: Mechano-electric transduction in the Pacinian corpuscle. Initiation of sensory impulses in mechanoreceptors. In Loewenstein, W.R. (ed.): Handbook of Sensory Physiology, Vol. I. New York, Springer-Verlag, 1971, pp. 269–290.
16. Lynch, T.J., Nedzelnitsky, V., and Peake, W.T.: Measurements of acoustic input impedance of cochlea. J. Acoust. Soc. Am. (Suppl.) 59:1, 1976.
17. Miller, J.D., Watson, C.S., and Covell, W.P.: Deafening effect of noise on the cat. Acta Otolaryngol. (Suppl.) 176:1, 1963.
18. Murray, R.W.: The ampulla of Lorenzini. In Fessard, A. (ed.): Handbook of Sensory Physiology, Vol. III/3: Electroreceptors and Other Specialized Receptors in Lower Vertebrates. New York, Springer-Verlag, 1974, pp. 125–146.
19. Nedzelnitsky, V.: Measurements of sound pressure in the cochleae of anesthetized cats. Sc.D. Thesis. Cambridge, Mass., Massachusetts Institute of Technology, 1974.
20. Oman, C.M. and Young, L.R.: The physiological range of pressure difference and cupula deflections in the human semicircular canal. Acta Otolaryngol. 74:324, 1972.
21. Quilliam, T.A.: The structure of fingerprint skin. In Gordon, G. (ed.): Active Touch. New York, Pergamon, 1978, pp. 1–27.
22. Rosenhall, U.: Mapping of the cristae ampullares in man. Ann. Otol. 81:882, 1972.
23. Russell, I.J. and Sellick, P.M.: Intracellular studies of hair cells in the mammalian cochlea. J. Physiol. (Lond.) 284:261, 1978.
24. Schuknecht, H.F.: Neuroanatomical correlates of auditory sensitivity and pitch discrimination in the cat. In Rasmussen, G.L. and Windle, W.F. (eds.): Neural Mechanisms of the Auditory and Vestibular Systems. Springfield, Ill., Thomas, 1960, pp. 76–90.
25. Szabo, T.: Anatomy of the specialized lateral line organs of electroreception. In Fessard, A. (ed.): Handbook of Sensory Physiology, Vol. III/3: Electroreceptors and Other Specialized Receptors in Lower Vertebrates. New York, Springer-Verlag, 1974, pp. 13–58.
26. Talbot, W.H., Darian-Smith, I., Kornhuber, H.H., and Mountcastle, V.B.: The sense of flutter-vibration: Comparison of the human capacity with response pat-

terns of mechanoreceptive afferents from the monkey hand. J. Neurophysiol. 31:301, 1968.

27. Verrillo, R.T.: Temporal summation in vibrotactile sensitivity. J. Acoust. Soc. Am. 37:843, 1965.

28. von Békésy, G.: Experiments in Hearing. New York, McGraw-Hill, 1960.

29. von Frey, M.: Über die zur ebenmerklichen Erregung des Drucksinns erforderlichen Energiemengen. Z. Biol. 70:333, 1920.

30. Westman, H.P.: Reference Data for Radio Engineers. International Telephone & Telegraph Corporation, 1956.

31. Wiener, F.M., Pfeiffer, R.R., and Bakus, A.S.N.: On the sound pressure transformation by the head and auditory meatus of the cat. Acta Otolaryngol. 61:255, 1965.

32. Wolbarsht, M.L.: Electrical characteristics of insect mechanoreceptors. J. Gen. Physiol. 44:105, 1960.

33. Wolf, H.: Exakte Messungen über die zur Erregung des Drucksinnes erforderlichen Reizgrössen. Inaugural dissertation, Jena. In Skramlik, E.: Psychophysiol. Ergänz. 4:xi, 935, 1937.

Discussion

LOWENSTEIN: Have you calculated the threshold power for the visual and olfactory systems?

KHANNA: Yes.

LOWENSTEIN: And your conclusion?

KHANNA: We did not include the optical system because the eye has yet a third type of sensory neurite. The sensitivity of the visual system is very similar to that of the ear. We have not done any calculation of the olfactory system, because we were unable to calculate the chemical energy involved.

20

Horizontal Canal Afferent Dynamics Measured Using White Noise and Cross Spectral Analysis

DENNIS P. O'LEARY, DAVID L. TOMKO, AND ROBERT J. PETERKA

Research in vestibular receptor physiology has expanded significantly during the past decade, particularly in areas concerning aspects of information processing and control (3,6,12). The rapid development of system theory and control techniques for engineering applications has provided vestibular researchers with new ways of analyzing, describing, and classifying results from dynamic studies of this biologic control system. The use of computers for accurate stimulus control and data analysis has enabled use of experimental designs that would have been impractical only a decade ago. Our research in the cat semicircular canal afferent system was aided both theoretically and technically by use of certain control applications, programed via a laboratory computer system, as an integral part of the experimental design.

This report presents an overview of the operations that were employed for a dynamic analysis of horizontal canal afferent responses. It provides the technical framework for a companion paper (Chap. 21) describing tests of certain hypotheses concerning the processing and representation of afferent information from this receptor.

OVERVIEW OF EXPERIMENTAL DESIGN

In order to avoid problems associated with a single, unwieldy computer program to control the experiments, program modules were written for specific tasks using the technique of structured programming. In this approach, a number of smaller programs were written and tested independently, and then linked together to communicate in a cooperative way (8,13). The final result was a structure with a goal supported by subgoals, each corresponding to individual program modules as shown schematically in Fig. 20-1. The determination of vestibular system dynamics, represented as the major goal at the top, can be subdivided into five sequential operational procedures described in the second row: (1) the choice of an appropriate stimulus waveform; (2) a method for filtering (or "preprocessing") the response data; (3) the choice and implementation of specific mathematical transform operators; (4) the computation and display of specific system descriptors; and (5) the correlation of these descriptors with other biologic properties of the system. This structure is an accurate representation of the specific operations employed during this investigation, as stated in the third row of Fig. 20-1. The purpose of this report is to describe the first four procedures in the above sequence. Chapter 21 will describe the results of applying procedure 4 to our experimental results.

METHODS

Stimulus Waveform

The stimulation device was a servocontrolled rate-of-turn table interfaced via a digital controller with a PDP-11 computer. Computer control of the turntable enabled particular stimulus waveforms to be programed, tested independently, and selected as programed stimulus modules in the structure of

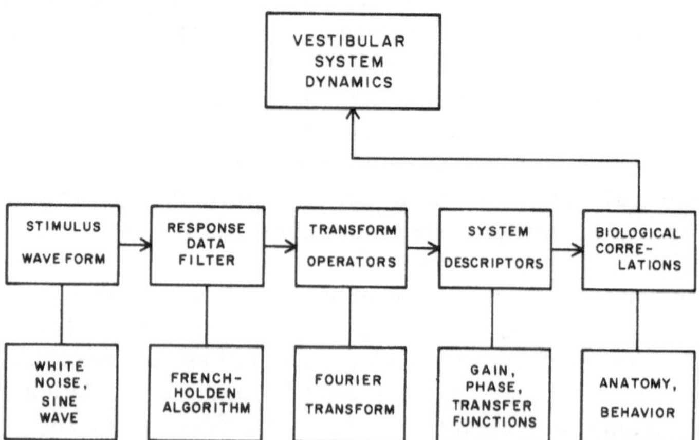

Fig. 20-1. Description of the operations used in an investigation of peripheral vestibular system afferent responses. The general operations shown in the second row were implemented using the specific stimulus, analysis, or interpretive procedures described in the third row.

Fig. 20-1. Rotational acceleration stimuli were employed consisting of either sinusoidal or pseudorandom binary sequence (PRBS) waveforms. The former have been used in other vestibular dynamic response studies (3,5,12) and the latter were adapted recently for specific applications in vestibular research (6,9). The PRBS waveforms were generated digitally and commanded the direction of rotational acceleration according to a specific pattern that repeated cyclically. As described previously, this stimulus can be considered bandlimited white noise for the vestibular system, with the mathematical properties of a superposition of individual sine waves of discrete frequencies over a selectable bandwidth (2,7). The stimulus bandwidths used in this study extended from 0.0075 to 5.0 Hz, approximately. Single frequency sinusoidal stimuli were used also in order to compare the results obtained from the same physiologic response data. The actual stimulus waveforms delivered at the head were calibrated with an accelerometer, and correction terms were included to compensate for the nonideal frequency response of the rotational table.

Response Data Filter

Extracellular afferent spike train responses were recorded from the VIIIth nerve of barbiturate anesthetized cats using micropipettes filled with 2 M sodium chloride. A hydraulically sealed chamber, attached to the skull with acrylic cement, both provided recording stability and served as a rigid attachment point. Spike times were recorded digitally with a resolution of 10 μs, and stored as disk files. They were cross correlated with the stimulus waveform to provide an on-line display of the dynamic response to either PRBS or sinusoidal stimuli.

As an alternative to spike train binning procedures used in other studies, the responses were first processed with a digital filter known as the French-Holden algorithm (FHA) (4). As a mathematical operator, this filter is a convolution of a $(\sin x)/x$ function with each spike, considered as a delta function, with the results extending both forward and backward in time. This operation merely replaces each spike with a $\sin x/x$ function, centered at the time of spike occurrence. By summing the results of this procedure from each spike, and sampling the summed waveform at a sufficiently rapid rate to include the entire system bandwidth, the result is an accurate representation of the response (10). Furthermore, this operation has the dual advantages of resulting in low-pass filtering for reduction of high frequency noise and an equally timed series of discrete sample points which facilitates further calculations.

Transform Operators

Output data from the FHA were a representation of the manner in which each cell's firing rate varied in time in response to the stimulus. In the subsequent analysis, these data were transformed into the frequency domain by use of the discrete Fourier transform (DFT). The use of these transform techniques has been described both theoretically and in engineering applications (11). In practice, each period of the response to PRBS stimuli was transformed into real and imaginary Fourier components via the DFT, and the results from six to ten periods were then averaged in order to reduce the variance of the frequency components. This operation resulted in more than 100 discrete frequency components and those at higher frequencies were averaged together to reduce the number of points to less than 12.

A major advantage of the use of frequency domain data is that the various substages of the system could be treated independently. For example, Fourier transform techniques were applied to the PRBS input, including correction terms for the response characteristics of the rotating table, and the results were combined with those from the spike train output in order to derive useful system descriptors.

RESULTS

System Descriptors

Appropriate formulae were used to compute system gain, phase, and transfer functions from both the PRBS and sine wave data (9). In addition, formulae were used also to compute coherence functions at each frequency point, which were used for two purposes. As derived in the literature, coherence values close to one imply that a system is both linear and free from corrupting noise, whereas values close to zero imply that the system is nonlinear and/or was perturbed by extraneous noise sources (1). Moreover, the use of appropriate formulae provide confidence intervals bracketing the gain and phase system descriptors. An example of the use of these descriptors is shown in Fig. 20-2 for one afferent unit.

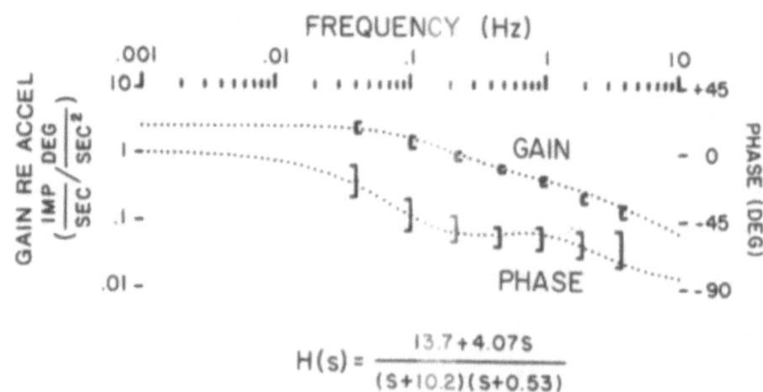

Fig. 20-2. The gain, phase, and transfer function of an afferent unit response, as determined from the use of pseudorandom sequence white noise rotational acceleration, and application of the operations shown in Fig. 20-1. The brackets represent two standard deviation intervals about the experimental gain and phase points.

A final system descriptor was the determination of a transfer function by fitting the gain and phase points with an appropriate nonlinear regression procedure. The dotted lines in Fig. 20-2 are the theoretical fit to the transfer function shown expressed in Laplace transform notation. In effect, the fitted transfer function describes the system underlying this afferent response in terms of only four parameters, which can be used for quantitative comparisons among responses from different nerve fibers. Certain of these comparisons, in addition to the relative agreement of results from sinusoidal and PRBS responses, are discussed in Chapter 21.

CONCLUSIONS

The modern era of digital signal processing has provided new technical advantages for experimental vestibular research. It proved useful in these experiments to follow certain principals of structured programing in order to obtain a well-controlled stimulus and accurate analysis. The incorporation of a specific spike train filtering algorithm resulted in a data format that was amenable to signal processing techniques based on the discrete Fourier transform, and therefore similar to the signal processing employed in other scientific areas such as industrial process control, seismology, and communications engineering. The resulting system descriptors provided the basis for quantitative transfer functions that can be related to the underlying mechanisms of information processing in the vestibular system.

SUMMARY

We have characterized information processing by vestibular system neurons using an input-output analysis based on analytic digital signal processing techniques. The goal of this approach was to determine accurate system descriptors of semicircular canal afferent neurons. The behavior of such neurons is sufficiently linear to enable calculation of a transfer function providing a useful and precise specification of a small number of parameters which characterize its dynamic responses under a wide variety of stimulus conditions. These parameters are of general use to physiologists studying reflex systems involving eye movements or posture, for example, where a precise description of the information carrying capacity of the afferent limb of the reflex is essential for the understanding of its normal operation.

Our experiments were performed using two powerful experimental tools. We used pseudorandom binary sequences of rotational acceleration to approximate an efficient "white noise" stimulus; and we used spectral analysis to measure dynamic response characteristics. Our results of utilizing this approach to study horizontal semicircular canal afferents in the cat will be used to illustrate some of the practical aspects, advantages, and limitations of the technique. In particular, we have studied the effects of spike train spectral properties on the analysis, the proper choice of stimulus bandwidth, the necessity of accurately describing the stimulus delivered, the estimation of goodness of fit of the resulting model parameters, and the use of on-line analysis for a quick look at the data during the experiment. Chapter 21 presents additional results of these studies.

ACKNOWLEDGMENT

This work was supported by NIH grants # NS 12308, NS 12494, and GM 01455.

References

1. Bendat, J.S. and Piersol, A.G.: Random Data; Analysis and Measurement Procedures. New York, Wiley Inter Science, 1971.
2. Davies, W.D.T.: System Identification for Self-Adaptive Control. New York. Wiley Inter Science, 1970.
3. Fernandez, C. and Goldberg, J.M.: Physiology of peripheral neurons innervating semicircular canals of the squirrel monkey. II. Response to sinusoidal stimulation and dynamics of peripheral vestibular system. J. Neurophysiol. 34:661, 1971.
4. French, A.S. and Holden, A.V.: Alias-free sampling of neuronal spike trains. Kybernetik 8:165, 1971.
5. Groen, J.J., Lowenstein, O., and Vendrik, A.J.H.: The mechanical analysis of the responses from the end-organs of the horizontal semicircular canal in the isolated elasmobranch labyrinth. J. Physiol. 117:329, 1952.
6. O'Leary, D.P., Dunn, R.F., and Honrubia, V.: Analysis of afferent responses from the isolated semicircular canal of the guitarfish using rotational acceleration white noise imputs. I. Correlation of response dynamics with receptor innervation. J. Neurophysiol. 39:631, 1976.
7. O'Leary, D.P. and Honvubia, V.: On-line identification of sensory systems using pseudorandom binary noise perturbations. Biophys. J. 15:505, 1975.
8. Orr, K.T.: Structured System Development. New York, Yourdon Press, 1977.
9. Peterka, R., O'Leary, D.P., and Tomko, D.L.: Linear system techniques for the evaluation of semicircular canal afferent responses using white noise rotational stimuli. In Hood, J.D. (ed.). Vestibular Mechanisms in Health and Disease, London, Academic Press, 1978, pp. 10–17.
10. Peterka, R.J., Sanderson, A.C., and O'Leary, D.P.: Practical considerations in the implementation of the French-Holden algorithm for sampling of neuronal spike trains. I.E.E.E. Trans. Biomed. Engin. BME-25:192, 1978.
11. Rabiner, L.R., Schafer, R.W., and Rader, C.M.: The chirp z-transform algorithm. I.E.E.E. Trans. Audio Electroacoust. AU-17:89, 1969.
12. Schneider, L.W. and Anderson, D.J.: Transfer characteristics of first and second order lateral canal vestibular neurons in gerbil. Brain Res. 112:61, 1976.
13. Warnier, J.D.: Logical Construction of Programs. New York, Van Nostrand Reinhold, 1976.

Discussion

OMAN: Dennis, could you contrast the practical and computational advantages of the pseudorandom binary sequence as opposed to just simply taking a sum of sines when generating a random input?

O'LEARY: The pseudorandom sequence has a computational advantage. It is easily generated through a simple algorithm; and it provides up to, in this case, 127 discrete frequency points. It would be impractical to sum many individual sines. However, the results should be equivalent if Fourier analysis is used to filter the data at each frequency.

21

Analysis of Horizontal Canal Afferent Dynamics Using White Noise and Sinusoids in the Barbiturate-Anesthetized Cat

DAVID L. TOMKO, DENNIS P. O'LEARY,
AND ROBERT J. PETERKA

Recent evidence has indicated that primary afferent neurons innervating a single semicircular canal can be subdivided into more than a single population on the basis of their dynamic response properties (6,11,14). Such variations in dynamic properties have been found to be correlated with neuronal variability (6,14), and thus indirectly with conduction velocity and axon diameter (7) or with the presumed anatomic locus in the crista of the receptors which the afferents innervate (2,11). It has been hypothesized that variations in first order afferent dynamic properties may be associated with their innervation patterns (5), or with local mechanical characteristics of the cupula near the innervated receptor cells (10).

The present study addresses five basic questions about the response properties of VIIIth nerve horizontal canal afferents. (1) Are measurements of dynamic response properties (i.e., gain and phase) invariant over time, and therefore statistically stationary? (2) Are such characteristics the same when determined with different experimental stimuli [i.e., sinusoidal rotations of various frequencies or pseudorandom binary sequences (PRBSs) of rotational acceleration]? (3) How different are individual units from one another in their dynamic properties? (4) Are differences between individual cells correlated with the resting discharge characteristics of the cells? (5) Should the response properties of a sample of these neurons be averaged to produce an average transfer function of an entire canal?

METHODS

We have recorded extracellular action potentials from 102 VIIIth nerve neurons using glass micropipettes in barbiturate-anesthetized cats. Following venous catheterization and tracheal intubation, anesthesia was maintained using supplementary doses of sodium pentobarbital. After the head was mounted in a stereotactic instrument, the right vestibulocochlear nerve was exposed at its exit from the internal auditory meatus by removing the bony cover of the cerebellar fossa, and rapidly aspirating the flocculus and adjacent portions of the right cerebellar hemisphere. To improve access to the nerve for microelectrode placement, the top of the internal auditory canal was removed with dental burrs to expose approximately 2.5 mm^2 of the dorsal surface area of the VIIth and VIIIth cranial nerves lateral to the internal auditory meatus (4).

While the head was still in the stereotactic device, a chamber was sealed to the skull with dental acrylic. The ear bars were removed and the animal's head was then suspended rigidly by clamping the chamber to a supporting framework tilted back 30° with respect to the Horsley-Clarke horizontal plane. This device enabled penetration of the nerve without removal of the bony tentorium, and ensured that the head could be accurately positioned on the turntable. The head and its supporting structure were tilted forward 30° to approximately align the horizontal canals in the plane of rotation (3). In addition, the head was placed so that the turntable axis of rotation passed through the midpoint of the intraaural axis (i.e., the point 0, 0, −10 in the Horsley-Clarke coordinate system). The body was firmly supported to minimize movement during stimulation and rectal temperature was maintained at 35 to 38°C. Recording details have been described fully elsewhere (9).

Cats were mounted on a computer-controlled rate table in a gimbal device which permitted the head to be tilted for typing canals. When a cell's sensitivity was to ipsilateral yaw accelerations in the plane of the horizontal canal, the head was again positioned so that the horizontal canals were in the plane of the rotational accelerations. Theoretical and practical details of the generation and delivery of the PRBS stimulus have been considered elsewhere (12); sinusoids for driving the rate table were either synthesized by the computer or obtained from a waveform generator. The transfer function of the table was taken into account in determining the actual stimulus delivered to the animal. The analysis procedures for deriving gain and phase measurements were considered in an earlier paper (13). The rationale for our approach was considered in Chap. 20.

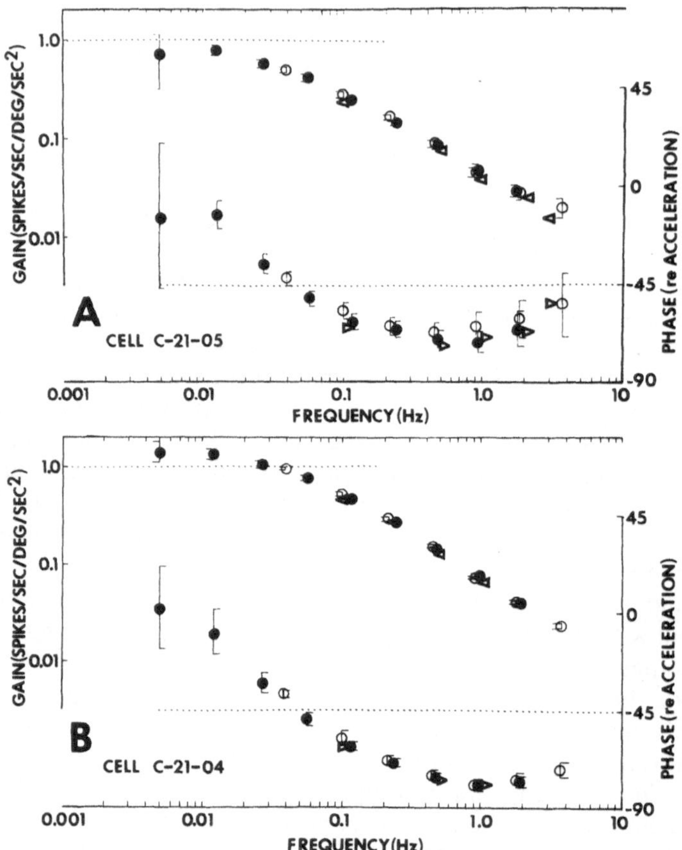

Fig. 21-1. Transfer function characteristics (gain and phase) for horizontal canal afferent neurons C21-05 (**A**) and C21-04 (**B**). Open and closed dots in both plots are gain (*upper*) and phase (*lower*) derived from the application of two different PRBSs with overlapping bandwidths. The brackets indicate 95% confidence intervals as calculated from the coherence function (1). Left-pointing triangles are gain points derived from application of the sine stimuli. Right-pointing triangles are phase points from sines.

RESULTS AND DISCUSSION

Stationarity of PRBS Measurements

Figure 21-1 shows the transfer function gain and phase estimates of two neurons over the bandwidth 0.005 to 3.3 Hz. Transfer function points for each of the two units were derived from the application of two different PRBSs with overlapping bandwidths (represented as open and closed dots). The transfer function points derived for unit C21-05 (Fig. 21-1A) from the two PRBS measurements appear to belong to a single family, as do those of the triangular points which were derived for the same unit from the application of sinusoids. All measurements for this cell required about 90 min. The transfer function points for a second cell, C21-04 (Fig. 21-1B), also appear to belong to a single family, but a different one than those of C21-05. The close replicability of gain and phase measurements for these two units over long periods of time was typical of this population of afferents. Because of this stability over time, we attribute small differences between individual units to underlying physiologic differences between them rather than to effects of variability of measurements over time.

Agreement of Gain and Phase Measured from PRBS and Sines

For 60 of 94 horizontal canal afferents identified with the PRBS, we calculated a total of 164 gain and phase points using sinusoids of various frequencies between 0.01 and 3.33 Hz. For example, unit C21-05 (Fig. 21-1A) yielded five gain and five phase points from sine stimulation (0.1, 0.5, 1.0, 2.0, and 3.3 Hz). Gain and phase values at these same frequencies were estimated from the PRBS data by interpolation.

The transfer function points measured from the two techniques were compared by calculating the difference between the gains and the difference between the phases, and are shown in Fig. 21-2. There was no statistically significant difference between gains measured with the two techniques at any

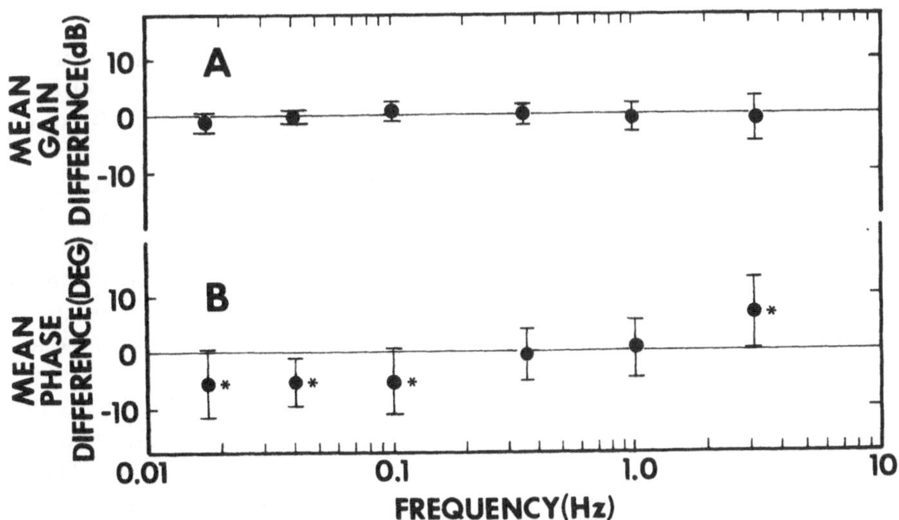

Fig. 21-2. Mean difference (±1.0 SD) between gains (**A**) and mean difference between phase (**B**) of 60 horizontal canal afferents measured using sinusoids and PRBSs of rotational acceleration over the bandwidth tested. Statistically significant differences are indicated by asterisks. The N for each point from lowest to highest frequency: 8, 22, 39, 42, 40, and 13.

frequency in the bandwidth tested (paired t test, $p<0.01$). Phase measurements showed small, but consistent, differences between the two techniques. At the lowest frequencies tested, phase lags obtained from sine stimuli were 5° greater than those from PRBS, whereas they were 5° smaller at higher frequencies. The four frequencies at which phase measurements showed statistically significant differences are indicated in Fig. 21-2 by asterisks.

Differences Among Units

There were marked differences among the transfer functions of individual neurons, as can be seen in Fig. 21-1. The gain of unit C21-04 (Fig.21-1B) was approximately double that of C21-05 (Fig. 21-1A) over the entire bandwidth tested. Phase points for the two units matched well, up to about 0.2 Hz. Above this frequency, the phase lag of unit C21-05 began to decrease, whereas that of C21-04 continued to increase up to 1.0 Hz. At 3.3 Hz, there was about a 20° relative phase difference between the two units. Such phase differences between individual units were as large as 60 at the highest frequencies tested (3.69 Hz). Such high frequency differences among units imply that different parameters are required for quantitative fits to the transfer function points. The implications of such differences for the capacity of each neuron to encode head movements are profound, and will be considered in detail in a later report.

Correlation of Resting Discharge Variability with Dynamic Properties

Previous investigators have found that differences in the dynamic response properties of canal neurons were correlated with morphologic and physiologic properties of the cells (6,14,15). We examined the correlation between transfer function properties and neuronal discharge variability at rest.

We arbitrarily subdivided the population of cells on the basis of their resting discharge variability (i.e., in the absence of an experimental stimulus) by calculating the coefficient of variation (CV). Low variability (LV) cells were characterized by regularly spaced action potentials (i.e., CV < 10%); high variability (HV) cells had more irregularly spaced action potentials under the same conditions (CV > 30%); the remaining cells were of intermediate variability (IV).

LV neurons had on average the largest phase lags of all cells, across the entire bandwidth tested (solid line histogram in Fig. 21-3C and G); whereas HV cells had the smallest phase lags (dashed line histogram in Fig. 21-3C and G). The phase lags of these two types of cells paralleled each other across the entire badwidth tested. In contrast, IV units (shaded histogram in Fig. 21-3C and G) had phase lags at high frequencies which were intermediate between those of the LV and HV cells; however, at lower frequencies the phase lags of the IV cells were more similar to those of the LV cells. Simi-

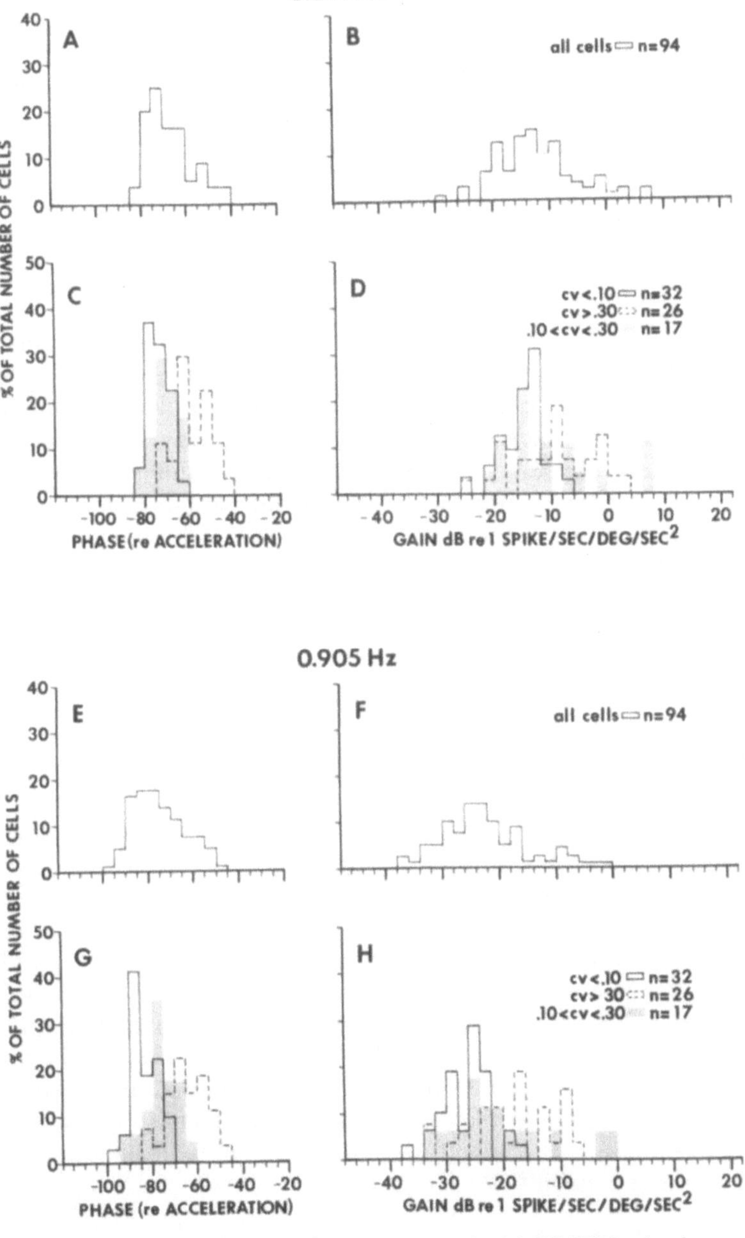

Fig. 21-3. Histograms of gains and phases measured with PRBS for horizontal canal afferents at 0.218 Hz (**A–D**) and 0.905 Hz (**E–H**). **A–B** and **E–F** show the distributions of these parameters for populations ($N = 94$) at these two frequencies, whereas **C–D** and **G–H** show distributions of these same parameters in subpopulations ($N = 75$) characterized by different resting coefficients of variation.

lar trends were observed for gain, but these were not as striking. To determine whether these subpopulations differed from one another statistically, a nonparametric distribution-free multiple comparisons test based on the Kruskal-Wallis rank sums test was used (8). The three subpopulations of phases differed from each other statistically ($p < 0.05$) at each of the two frequencies shown in Fig. 21-3. At both frequencies gains of cells with CVs greater than 30% differ significantly from those of cells with CVs less than 10%.

Averaging Gain and Phase Estimates

Our data (Fig. 21-3A,B,E, and F) show a broad distribution of transfer function properties. Averaging the responses of all members of such a population of afferents in either the time or frequency domain will always minimize the importance of differences between them, differences which we regard as having potential physiologic significance. These differences should be expected in light of the variety of factors influencing the sensory transduction process (i.e., local differences in the type of hair cell, its connection to the cupula, the structure of the cupula itself, and the synaptic connection of the hair cells to the afferents). Because of all these factors, we conclude that averaging responses from this population of neurons should be applied judiciously, if at all, until we understand which cells have common operating characteristics. Final resolution of the question of whether these differences in dynamics are important to nervous system function can only be achieved by studying their central projections.

CONCLUSIONS

Gain and phase values for 60 horizontal canal primary afferents were similar when measured using either sinusoids or PRBSs of rotational acceleration (± 1 dB, $\pm 10°$ respectively, over the bandwidth 0.01 to 3.5 Hz). Gain and phase results were stationary but varied widely from cell to cell at the same test frequency. Transfer function properties were correlated with resting discharge variability. We hypothesize the existence of a continuum of dynamic response types because of the broad distribution of transfer function properties. We conclude that averaging gain and phase characteristics of these cells could obscure important differences in their individual information processing capabilities.

SUMMARY

We have used both pseudorandom binary sequences (PRBSs) of rotational acceleration and sinusoidal oscillations to characterize the dynamic response properties of each of 60 horizontal canal afferents. Our *first goal* was to determine accurate input-output transfer function characteristics (gain and phase) for these neurons. This information is important to the physiologist studying the various oculomotor and postural reflexes which have vestibular inputs. Our *second goal* was to determine the relative agreement of gain and phase measured with PRBS with gain and phase measured using sinusoidal inputs for the same neuron. We felt it particularly important to test in a physiologic preparation the correspondence of transfer function results from the relatively new PRBS technique with those from the already accepted sinusoidal rotations.

Our results may be summarized as follows:

1) The input-output transfer function gain and phase values varied widely from cell to cell at the same test frequency.
2) These values were closely replicable for each individual cell regardless of the measurement technique used for the frequency bandwidth 0.01 to 3.5 Hz.
3) They also varied systematically with the intrinsic variability of the neuron's discharge properties.

Because transfer function properties (gain and phase) for individual cells were closely replicable, and varied widely from cell to cell, we concluded that an accurate description of canal afferent dynamics must not be based on an "averaged" transfer function, but must be considered as a population of neurons, each having a unique input-output relationship.

ACKNOWLEDGMENTS

We gratefully acknowledge our collaboration with Dr. Robert Schor in the early phases of this study, the expert technical assistance of Mr. Frank N. Ali, and the support of NIH grants NS 12308, NS 12494, and GM 01455. We thank Dr. Schor and Dr. Karen Mudry for criticizing an earlier version of this paper.

References

1. Bendat, J.S. and Piersol, A.G.: Random Data: Analysis and Measurement Procedures. New York, Wiley Inter Science, 1971.
2. Dunn, R.F.: Horizontal ampullary nerve fiber projections to the crista angustarum in the crista ampullaris. J. Comp. Neurol. 183:779, 1979.
3. Estes, M.S., Blanks, R.H.I., and Markham, C.H.: Physiologic characteristics of vestibular first order canal neurons in the cat. I. Response plane determination and resting discharge properties. J. Neurophysiol. 38:1232, 1975.
4. Gacek, R.: The course and central terminations of first order neurons supplying the vestibular end organs in the cat. Acta Otolaryngol. (Suppl.) 254:1, 1969.
5. Goldberg, J.M.: Discussion. In Naunton, R.F. (ed.): The Vestibular System. New York, Academic Press, 1975, p. 183.

6. Goldberg, J.M. and Fernandez, C.: Physiology of peripheral neurons innervating semicircular canals of the squirrel monkey. III. Variations among units in their discharge properties. J. Neurophysiol. 34:676, 1971.

7. Goldberg, J.M. and Fernandez, C.: Conduction times and background discharge of vestibular afferents. Brain Res. 122:545, 1977.

8. Hollander, M. and Wolfe, D.A.: Nonparametric Statistical Methods. New York, Wiley, 1973.

9. Loe, P.R., Tomko, D.L., and Werner, G.: The neural signal of angular head position in primary afferent vestibular nerve axons. J. Physiol. (Lond.) 230:29, 1973.

10. McLaren, J.W. and Hillman, D.E.: Differential response of the crista to sinusoidal rotation. Neurosci. Abstr. 4:612, 1978.

11. O'Leary, D.P., Dunn, R.F., and Honrubia, V.: Analysis of afferent responses from isolated semicircular canal of the guitarfish using rotational acceleration white-noise inputs. I. Correlation of response dynamics with receptor innervation. J. Neurophysiol. 39:631, 1976.

12. O'Leary, D.P. and Honrubia, V.: On-line identification of sensory systems using pseudorandom binary noise perturbations. Biophys. J. 15:505, 1975.

13. Peterka, R.J., O'Leary, D.P., and Tomko, D.L.: Linear system techniques for the evaluation of semicircular canal afferent responses using white noise rotational stimuli. In Hood, J.D. (ed.): Vestibular Mechanisms in Health and Disease. New York, Academic Press, 1978, pp. 10–17.

14. Schneider, L.W. and Anderson, D.J.: Transfer characteristics of first and second order lateral canal vestibular neurons in gerbil. Brain Res. 112:61, 1976.

15. Yagi, T., Simpson, N.E., and Markham, C.H.: The relationship of conduction velocity to other physiological properties of the cat's horizontal canal neurons. Exp. Brain Res. 30:587, 1977.

Discussion

ROBINSON: Could you briefly summarize the physiologic results and say why you think central structures just don't average out differences between units?

TOMKO: I didn't say central structures don't average out differences between cells. Peripheral structures encode them the way they encode them for a reason. Obviously, the transfer function phase differences in Figure 21-1 are predominantly high frequency differences and primarily affect the system's short time constants. During rapid head movements, cells with different short time constants will behave differently. This point will be elaborated in a later paper.

YOUNG: The use of the white noise analysis technique has its real power in the things you can do beyond the first-order describing function, beyond the transfer function, in investigating nonlinearities in a systematic way or investigating multiple inputs, and I wonder whether you feel that you are just at a first stage in its usage or that you are satisfied with limiting yourself to the linear transfer function or describing function approach, since you have in your hands the power to get a much more general description.

O'LEARY: We feel that because coherence is as close to one as it is, that the first-order description is an excellent description. As a first approximation, that description leads to understanding the system without going to a higher-order, more convoluted analysis. However, understanding receptor dynamics may involve some subtle nonlinearities which we are also investigating by combining pseudorandom sequences with other types of inputs.

ANDERSON: Since you have covered details around the analysis, and due to the fact that you currently have some kind of small discrepancy, that you still have a question about it in terms of phase, I wonder what is the length of the pseudorandom sequence you used, and if you used an integral number of lengths of the sequence.

PETERKA: We used primarily two PRBS sequences. One was a 255 state sequence and the other one was a 1023 state sequence.

ANDERSON: If it has 255 states, then you would presumably have recorded 255 points per period. Now, the discrepancy in phase I was concerned with is the possibility of using an exceedingly long sequence and then having to window the response versus having one end matching up against the succeeding one. If it was not periodic, then you could run into the problem of having to induce a window to get rid of some discrepancy and some leakage in the Fourier analysis.

PETERKA: We always used an integral number of either the 255 or 1023 state sequences and therefore did not require this windowing. Also, we always rejected the first period, assuming that it was a transient response. The French-Holden algorithm, since it's a two-sided filtering operation, requires that you have data before each PRBS period and data after the last complete period. That's another consideration.

PRECHT: A brief comment. This seems to be the year of the cat canal analysis. There are three studies now: yours, the one in Tokyo, and ours, happening in the same year. And they all agree quite well in the results except for some minor differences, it seems, but they really are in the minor range. What was the time constant that you came up with? The mean time constant or, if I may ask, after your critical comments about a mean time constant, one we can compare to the rest. And then a practical question: Is it much more difficult to hold cells in a non-sealed chamber when you do your white noise analysis? Just comparing it to the sinusoid stimulation where you don't need a closed sealed chamber, at least when you do the horizontal rotation.

TOMKO: The time constant was 4.95 ± 1.87. I've never recorded with an unsealed chamber. Ours are hydraulically sealed. I have no idea whether you could record with a micropipette that way.

SCHOR: I also participated in some of these experiments. I'm just going to remind David that his chamber is not only sealed, but filled with agar. You would occasionally get a small air bubble inside, not have a complete seal and not be able to hold anything.

PRECHT: During white noise?

SCHOR: During white noise, yes. And during sinusoid. I guess maybe our motions are bigger. I'm not an expert on sealed chamber technology.

GOLDBERG: I don't care if it's sealed or unsealed. Let me make a theoretical point. I never divided populations into twos or threes. That was always clearly stated as a rhetorical device. It is dependent on the time you have to write a paper or how much time you have to give a presentation. Theoretically, as I stated in my talk yesterday, there is no physiological evidence for saying there are hard and fast boundaries between afferents or between their innervation patterns or between hair cells. I think Dr. Engström made that point also. So, sometimes you don't have enough data and you plot the data as a scatter diagram, and other times you have enough data, so you group them into three groups usually because you want to get rid of the middle, and look at the difference between the two ends.

22

The Response of Primary Semicircular Canal Neurons to Angular Accelerations of Varying Magnitude

IAN S. CURTHOYS

Goldberg and Fernández (6) and Blanks et al. (3) both found that a given constant angular acceleration causes a larger increase in firing in irregular primary canal afferents than in regular afferents. In the present study this difference between regular and irregular cells was studied over a range of acceleration values. The main result was observed in a study of guinea pig and rat primary afferents and was confirmed by reanalysis of previously published data on cat afferents.

Standard procedures were used for acute single neuron recording in the VIIIth nerve of the three species. The cats were maintained under pentobarbital for the whole experiment. The guinea pigs and rats were *encéphale isolé* preparations. The lateral cerebellar structures were removed to expose the VIIIth nerve and recording was by means of glass micropipettes filled with 2 M NaCl. The horizontal canal was placed in the plane of rotation of the turntable. Each cell encountered was classified as being regular or irregular at its resting rate, using the criterion of regular if the coefficient of variation of interspike intervals was less than 0.0579 or equivalent. Each cell was subjected to various numbers of long duration constant angular accelerations, from two to eight values ranging from 2 to 33°/s^2. At each acceleration the maximum increase in firing during acceleration was measured and a logarithmic regression line was fitted to these increases (15). For most cells a logarithmic function provided an adequate fit (Fig. 22-1).

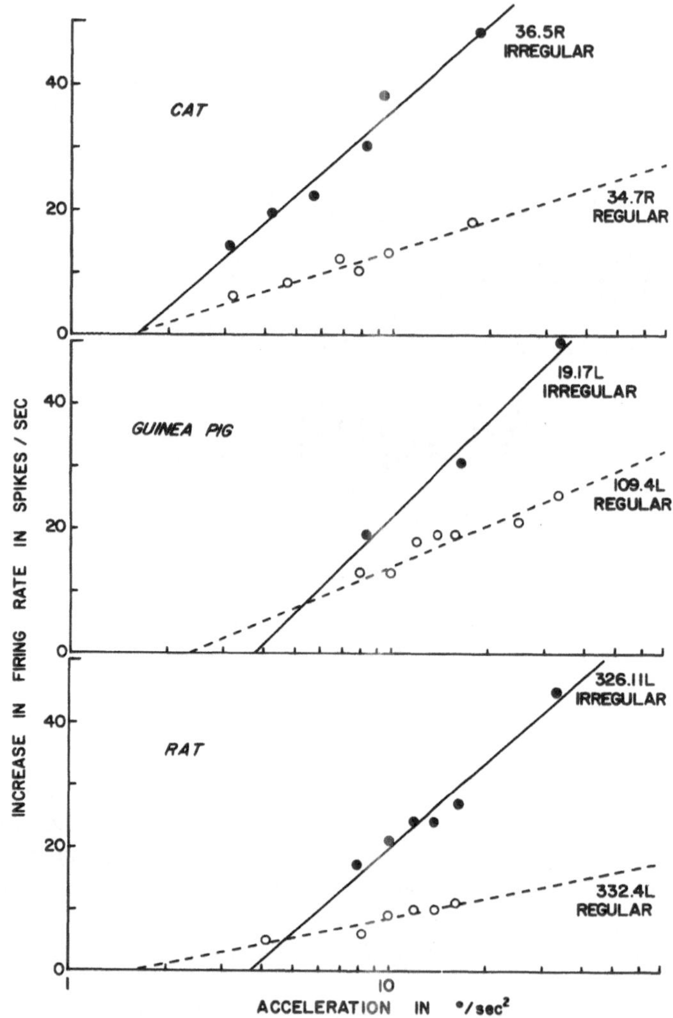

Fig. 22-1. The relation between angular acceleration magnitude and increase in firing rate for regular (open circles) and irregular (closed circles) individual horizontal canal neurons of the three species.

Figure 22-1 shows examples of typical acceleration-firing rate functions for individual regular and irregular cells in each species. These acceleration-firing rate functions are essentially intensity or input-output functions.

Each line yields two measures of the cell's response to the acceleration stimuli: (1) the slope of the acceleration-firing rate function; and (2) an estimate, by extrapolation, of the acceleration yielding zero increase in firing (the threshold for the cell). The slopes and thresholds were averaged for regular and irregular neurons separately. Figure 22-2 shows the regression lines

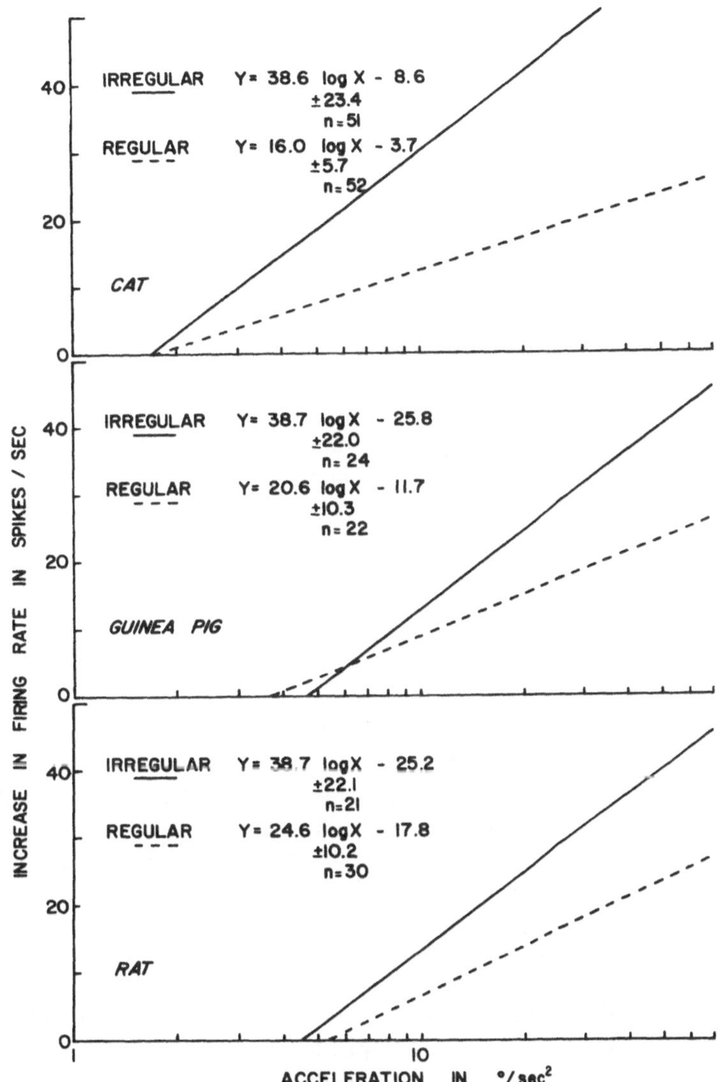

Fig. 22-2. Average g regression lines in relating angular acceleration magnitude and increase in firing rate for regular (dashed lines) and irregular (solid lines) neurons of the three species. The lines were obtained by averaging the coefficients for all the appropriate cells. The standard deviation of the slope and the number of neurons are given under each regression equation.

corresponding to the average values in each species, together with the respective equation and the standard deviation of the slope.

The average result is clear in each species. Irregular cells have significantly steeper slopes than do regular cells, but the (extrapolated) thresholds are not significantly different. These results are also found if linear rather than logarithmic fits are used. The different slopes mean that regular cells increase their firing to increasing accelerations at a slower rate than do irregular cells. This confirms the results of Goldberg and Fernández (6) and Blanks et al. (3); i.e., that at any given acceleration above threshold irregular cells increase their firing more than do regular cells.

One possible explanation of these results is that they are solely due to regional differences in the response of hair cells to cupula deformation. In the crista, thick neurons which contact predominantly type I receptor cells are probably irregular (7,18). Thin neurons with many branches which contact predominantly type II receptor cells are probably regular. Wersäll (17) has reported that type I receptors tend to be found in the central region of the crista and type II receptors are localized in the periphery. On the other hand, Lindeman (11) maintains there is a constant proportion of type I to type II receptors throughout the crista (60%:40%). Both agree that the thickest neurons (i.e., probably irregular) supply the central region of the crista.

Hillman (9) has observed that during stimulation the central region of the cupula bends more than the lateral parts near the ampullary wall. If that observation applies in the present situation, then the hair cells in the central region of the crista should be bent more than hair cells in the lateral parts. So one might expect a larger increase in firing in afferents from this central region. Thick irregular afferents supply the central region so one might expect a larger increase in firing in irregular cells than in regular cells, and indeed this is the result obtained in the present study.

However, at very small accelerations only hair cells in the central region would be bent enough to activate their afferents, so one might expect these same irregular cells to have lower thresholds than regular cells. This is not obtained. In the three species there is no evidence of any consistent difference between the thresholds of regular and irregular cells (Fig. 22-3). It seems that regional factors alone are not sufficient to explain the present results.

It may be that the different slopes are the result of the restricted branching of thick irregular afferents and the profuse branching of thin regular afferents.

In 1961 van Bergeijk (2) conducted neural modeling studies of afferent networks with varying numbers of branches. He showed that the input-output functions of networks with many branches tended to have shallower slopes than those of networks with few branches. He assumed that the cause of this slope reduction as branching increased was that each branch was capable of spike initiation, and an action potential initiated in one branch could block action potentials in other branches by means of the refractory period at the junction between branches. In this way, the probability of blocking or interference increases with the number of branches and the firing rate. Murray and Capranica (13) and Murray (12) have shown that antidromic invasion from one spike-initiating branch into others would cause a

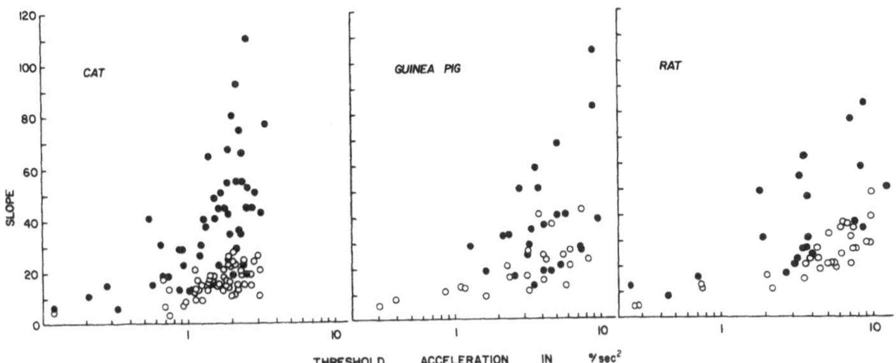

Fig. 22-3. Relation between threshold and slope of the acceleration-firing rate function for regular (open circles) and irregular (closed circles) neurons of the three species. The ordinate, slope, is the value of *b* in the regression equation and the threshold is the acceleration yielding zero increase in firing as derived by extrapolation from the regression equation.

similar slope reduction with increasing branching. Possibly other mechanisms may produce the same result (8). These mechanisms act to reduce the increase in firing with increasing stimulus intensity; in other words, to reduce the slope of the input-output functions of multiple branching networks. Multiple sites of spike initiation and antidromic invasion have been demonstrated in a variety of neurons, including lateral line afferents (1,4,10,13,14).

Applying these neural modeling results to the data from vestibular primary afferents suggests that the acceleration-firing rate functions of regular canal neurons should have shallow slopes as a result of their profuse branching, whereas irregular neurons should have steep slopes as a result of their restricted branching. These are the results obtained. The lack of difference in average threshold seems to be consistent with Lindeman's observation (11) that there is a constant proportion of type I to type II hair cells throughout the crista and that thin regular afferents are also present in the central zone.

There is one other piece of indirect support for the hypothesis linking regularity with branching density. In another study I have shown that in the newborn rat regular primary afferents are not encountered until about day 4. All the afferents are irregular at birth. The first regular cells to appear have a lower resting rate than regular afferents in adults. With increasing age the proportion of regular cells gradually increases and their average resting rate also increases (5). Anatomically, during development of the mouse labyrinth, afferent fibers grow to contact more and more receptor cells (16), exactly as one would expect if regularity is determined by branching density.

Other physiologic characteristics of regular and irregular cells may be the result of different branching densities, as Murray (12) has suggested.

References

1. Alnaes, E.: Peripheral inhibition of eel lateral line receptors as caused by antidromic sensory terminal invasion, Acta Physiol. Scand. 88:35, 1973.
2. Bergeijk, W.A. van: Studies with artificial neurons. II. Analog of the external spiral innervation of the cochlea. Kybernetik 1:102, 1961.
3. Blanks, R.H.I., Estes, M.S., and Markham, C.H.: Physiologic characteristics of vestibular first order canal neurons in the cat. II. Response to constant angular acceleration. J. Neurophysiol. 38:1250, 1975.
4. Calabrese, R.L. and Kennedy, D.: Multiple sites of spike initiation in a single dendritic system. Brain Res. 82:316, 1974.
5. Curthoys, I.S.: The development of function of horizontal semicircular canal primary neurons in the rat. Brain Res. 167:41, 1979.
6. Goldberg, J.M. and Fernández, C.: Physiology of peripheral neurons innervating semicircular canals of the squirrel monkey. I. Resting discharge and response to constant angular accelerations. J. Neurophysiol. 34:635, 1971.
7. Goldberg, J.M. and Fernández, C.: Conduction times and background discharge of vestibular afferents. Brain Res. 122:545, 1977.
8. Highstein, S.M. and Politoff, A.L.: Relation of interspike baseline activity to the spontaneous discharges of primary afferents from the labyrinth of the toadfish, *Opsanus tau*. Brain Res. 150:182, 1978.
9. Hillman, D.E.: Cupular structure and its receptor relationship. Brain Behav. Evol. 10:52, 1974.
10. Lindblom, U. and Tapper, D.N.: Integration of impulse activity in a peripheral sensory unit. Exp. Neurol. 15:63, 1966.
11. Lindeman, H.H.: Studies on the morphology of the vestibular apparatus. Ergebn. Anat. Entwickl.-Gesch. 42:1, 1969.
12. Murray, M.J.: Systems of mutually triggering event generators: basic properties and functions in information transmission and rhythm generation. J. Comp. Physiol. 117:63, 1977.
13. Murray, J.J. and Capranica, R.R.: Spike generation in the lateral line afferents of *Xenopus laevis*: evidence favoring multiple sites of initiation. J. Comp. Physiol. 87:1, 1973.
14. Pabst, A.: Number and location of the sites of impulse generation in the lateral line afferents of *Xenopus laevis*. J. Comp. Physiol. 114:51, 1977.
15. Shimazu, H. and Precht, W.: Tonic and kinetic responses of cat's vestibular neurons to horizontal angular acceleration. J. Neurophysiol. 28:991, 1965.
16. van de Water, T.R., Anniko, M., Nordemar, H. and Wersäll, J.: Embryonic development of the sensory cells in macula utriculi of mouse, in M. Portmann and J.-M. Aran (eds.) Inner Ear Biology, Paris, INSERM. 1977, p. 25–35.
17. Wersäll, J.: Studies on the structure and innervation of the sensory epithelium of the cristae ampullares in the guinea pig. Acta Otolaryngol. (Suppl.) 126:1, 1956.
18. Yagi, T., Simpson, N.E., and Markham, C.H.: The relationship of conduction velocity to other physiological properties of the cat's horizontal canal neurons. Exp. Brain Res. 30:587, 1977.

Discussion

GOLDBERG: If I understand your model, you postulate that there are several independent spike generators in the terminal arborizations of thin, regularly discharging afferents. Offhand, I would have thought that such a spike generating mechanism could only give rise to irregular discharge patterns. Am I wrong?

CURTHOYS: I think that this means studies of branching on regularity by Murray last year. He conducted a large number of simulation studies on exactly this question, looking at the effect of branching in a series of systems. He found an increase in regularity with an increase in branching. I think antidromic invasion is probably the more correct view of it rather than the refractory period explanation. But, as I pointed out, both mechanisms work in the same direction. I think the idea of spike initiation in each branch would tend to do exactly what I said; it would decrease the slope and increase the resting rate, tending to minimize outliers. That's what happens when you go from irregular to regular: outlying interspike intervals are getting eliminated. And that's the kind of thing Murray has said in his paper. I don't know whether that answers your question.

23

Dynamic Properties from Utricular Afferents

RUBEN BUDELLI AND OMAR MACADAR

Otolithic organs have been classically considered as accelerometers with a practically flat gain-frequency curve (26). However, a closer look into the responses recorded from the otolithic afferent nerves reveals a more complex input-output relationship (9,10,22,23). Otolithic organs do not respond to changes in the acceleration vector in a linear way: they are sensitive to high frequency vibrations in a nonlinear fashion (21). Linear accelerations of the same amplitude but opposite sense elicit different responses from otolithic organs (9). Because of adaptation (23), otolithic organs can respond phasically to a sustained mechanical stimulus, and as a result their gain-frequency curves have a positive slope. Furthermore, all the afferents innervating a given organ do not respond identically to the same stimulus (23). This complexity makes it difficult to consider the otolithic organs as simple accelerometers, and makes a characterization of the different afferents necessary.

This paper identifies the distinctive physiology of the various afferent populations that innervate an otolithic organ. To accomplish this, we characterized the dynamic behavior of the different utricular afferents according to their responses to steps of the acceleration vector and to sinusoidal changes of the acceleration. We also explored the cellular mechanisms responsible for the differential characteristics of the adaptative responses to steps.

PREPARATION AND METHODS

We chose a preparation of the isolated utriculus from elasmobranchs similar to the preparation described by Lowenstein and Roberts (20). This preparation permits visual selection of utricular afferents; in other preparations, the determination of the origin of the afferent recorded is based on the characteristics of the response. The preparation was mounted in a platform that was tilted and vibrated (Fig. 23-1).

CLASSIFICATION OF UTRICULAR AFFERENTS

To differentiate the afferent populations, we classified them according to their spontaneous activity, their responses to tilts, and their responses to vibrations (2). The resulting groups were further characterized by examining their dynamic properties.

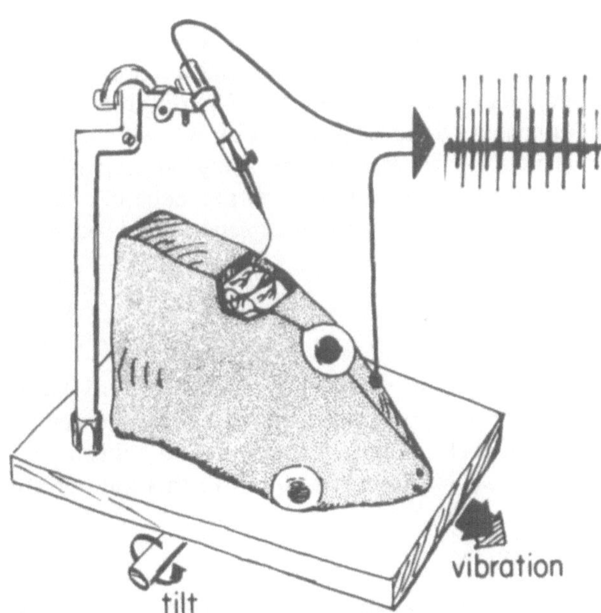

Fig. 23-1. Stimulation and recording arrangement. The head of the fish is held in a platform forming an angle of 60° with the horizontal plane (ventral side up and contralateral side down). The direction of the vibration (in the horizontal plane) is rostrocaudal. The tilts are around a horizontal axis perpendicular to the direction of the vibration. The recording is made between a forceps to which a bundle of fibers is attached and another electrode in the head of the animal. **Insert** Actual recording from a bundle showing spikes originating in two different fibers. From Budelli, R. and Macadar, O.: Stato-acoustic properties of utricular afferents. J. Neurophysiol. 42:1479, 1979.

We found three different types of utricular afferents:

Type I

When the head of the animal was in a fixed position, type I neurons fired regularly, as shown in the time interval histogram (Fig. 23-2). When the head was tilted, the afferent response consisted of a phasic-tonic change in firing rate. When the head was stimulated with a sinusoidal vibration, type I neurons responded by following the frequency of the applied stimulus only when the difference between the frequency of the stimulus and the spontaneous rate was less than 50%. In Fig. 23-2, the type I afferent spontaneous rate was 41 spikes/s. When a 58-Hz vibration was applied, the firing rate increased to 58 spikes/s. When the stimulus frequency was 34 Hz, the firing rate decreased to 34 spikes/s.

Type II

The spontaneous firing rate and the responses to tilts of type II afferents were similar to the rate and responses of type I afferents. Lowenstein and Roberts (20) grouped these two types together according to their responses to tilts and called them gravity receptors. The differences between types I and II became apparent when the preparation was subjected to a vibrational stimulus: type II neurons increased their firing rate regardless of the stimulus frequency. Another difference was the presence of adaptation to vibrational stimuli in type II neurons and the absence of this property in type I afferents.

Type III

This group of afferents did not show spontaneous activity but fired only during the transition from one position to another, regardless of the direction of the movement. Type III afferents responded to vibrational stimuli with maintained firing and showed no adaptation to this kind of stimulus.

VIBRATION SENSITIVITY

The dependence of afferent sensitivity on the frequency of a sinusoidal stimulus and the presence of adaptation indicate that the utricle transmits information to the central nervous system not only about the position of the head but also about the rate of change in the acceleration vector applied to the animal's head. Analysis of this dynamic aspect of the utricular function requires measurements of the afferent sensitivity at different frequencies of a vibrational stimulus. The accomplishment of this task in type III neurons is straightforward because these afferents do not have spontaneous activity. Therefore, the determination of the threshold or its inverse (sensitivity) can

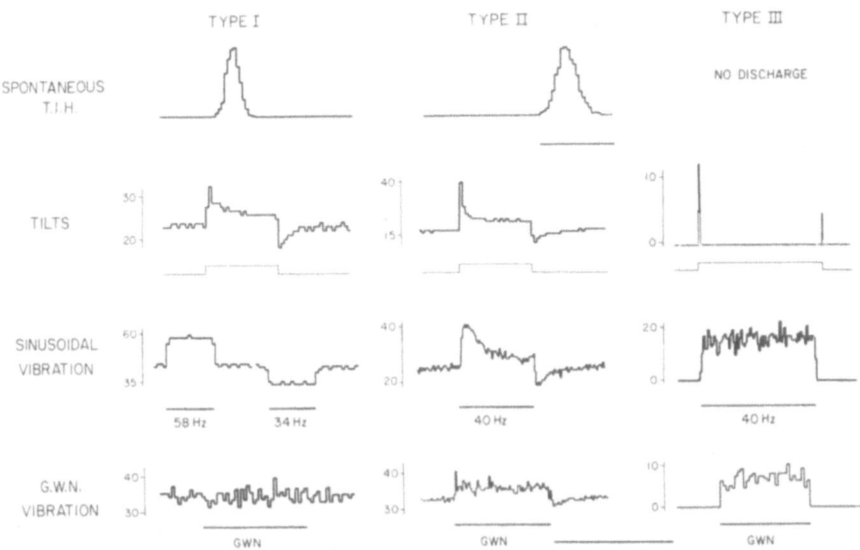

Fig. 23-2. Types of afferents. Each column contains the time interval histogram (T.I.H.) obtained during absence of stimulation and the spike frequency as a function of time during periods of application of different kinds of stimuli: Tilts, sinusoidal vibrations, and Gaussian White Noise (GWN) vibrations. The time calibration for the time interval histogram (*top*) is 25 ms and the one for the spike frequencies (*bottom*) is 30 s. The units in the ordinates are spikes/s. From Budelli, R. and Macadar, O.: Stato-acoustic properties of utricular afferents. J. Neurophysiol. 42:1479, 1979.

be easily done by observing the minimum stimulus that produces spikes. In type I and II afferents, however, this task is complicated since these afferents show a resting activity that makes determination of their threshold nearly impossible without a statistical approach. Figure 23-3 summarizes the problems involved in determining the sensitivity for these afferents. Figure 23-3A shows the records obtained when a vibration is applied to a hypothetical neuron. Visual inspection of this figure is not sufficient to detect the stimulus-response relationship. Visual detection improves if the records are represented as a cycle histogram (Fig. 23-3B), where the response may consist of a localization of the firing around a specific phase of the stimulus. However, to prove statistically that there is a response, we must test the null hypothesis of nondependence. One way to solve this problem is to consider the phase at which a spike fires during a sinusoidal vibration as a random variable (24). This variable is represented in Fig. 23-3C by a unit vector whose angle (α) with respect to a fixed direction (r) indicates the phase relation between the spike and the maximum acceleration of the stimulus. If there is no relation between the stimulus and the afferent's electric activity, the spikes are expected to appear at any point in the stimulation cycle and

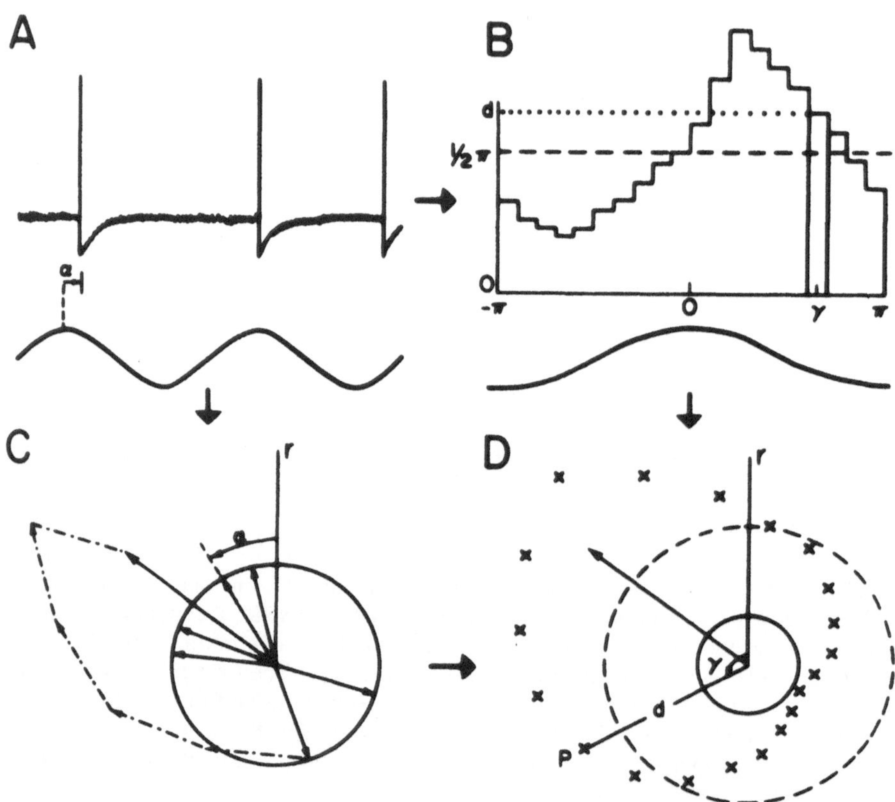

Fig. 23-3. Statistical process. **A** Recordings of afferent spikes (*top*) and the accelerometer output (*bottom*). **B** Cycle histogram tracing using the data from neuron in **A**. **C** Circular distribution where the spikes from **A** are represented as unit vectors and the average vector is calculated. **D** Polar representation of the cycle histogram, the average vector, and the confidence region. Modified from Budelli, R. and Macadar, O.: Stato-acoustic properties of utricular afferents. J. Neurophysiol. 42:1479, 1979.

the unit vectors are expected to be equally distributed in the circle. In this case, the average of all these vectors would be small, falling within a circular region (confidence region) whose size can be determined. If the vector falls outside this confidence region, we assume that the vibration affects neuronal firing. Further details about this method can be found elsewhere (2,11,24).

In Fig. 23-3D, each bar in the cycle histogram is represented as a point (*P*) in the plane whose distance (*d*) to a fixed point is equivalent to the height of the bar in Fig. 23-3B, and the angle γ formed with a given direction (*r*) represents the phase of the middle point at the base of the bar. To analyze the response characteristics of type I and II afferents, we used the polar representation of the cycle histogram, the mean vector, and its confidence region (represented as a solid line), as shown in Fig. 23-3D. Finally, we defined the gain vector as the quotient of the average vector divided by the stimulus amplitude measured in gravity units. This can be converted into spikes/s·*g* by multiplying the gain times twice the frequency of firing.

Type I Afferents

The gain curve of type I afferents reaches its maximum at a frequency close to the spontaneous rate of the neuron (Fig. 23-4). All type I neurons recorded presented this characteristic. The maximum gain varies with changes in the firing frequency that result from positional changes or spontaneous rate variation.

Since the average vector for frequencies 20 and 100 falls within the confidence region, we cannot assume that it is a good estimation of the central phase and, consequently, of the gain. However, neurons in a region close to the threshold would presumably behave more linearly. Therefore, we used several stimulus intensities for most of the cells studied, as shown for the type II neuron. We plotted the gain and phase curves by using the responses of each frequency to the smallest effective stimulus.

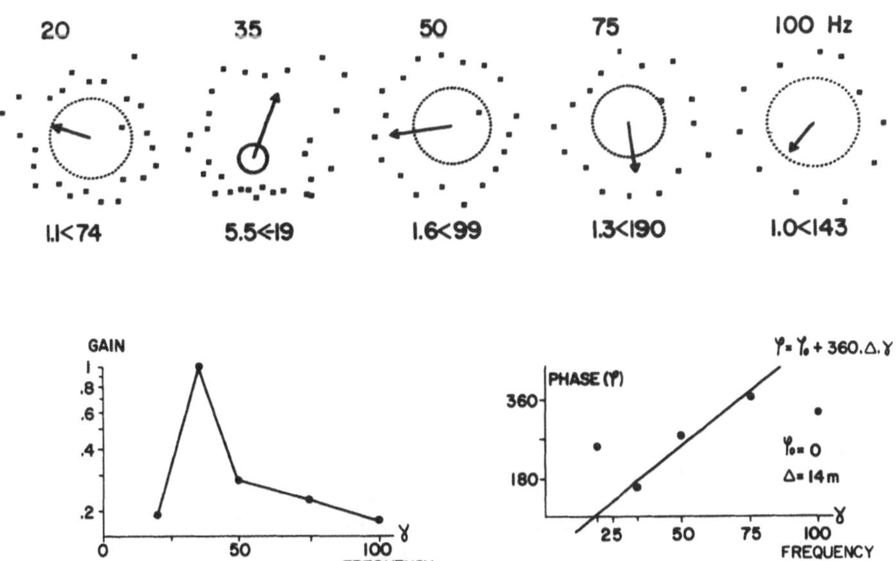

Fig. 23-4. Dynamic response of a type I neuron. Cycle histogram in polar coordinates, average vector, and confidence region for different frequencies of the stimulus. The numbers below each histogram represent the gain amplitude and its phase. The plots at the bottom are the gain amplitude and phase as a function of the stimulus frequency.

Type II Afferents

The dependence of the gain on the stimulus amplitude is less pronounced in type II than in type I neurons; the maximum gain of type II corresponds to frequencies higher than their spontaneous rate; and the corresponding peak of type II is less pronounced than that of type I cells (Fig. 23-5).

Type III Afferents

Study of silent afferents easily yields determination of the vibration sensitivity. The acceleration threshold and its inverse, the sensitivity, were determined for different frequencies. Sensitivity curves were obtained by plotting the sensitivity as a function of the frequency. These curves divided type III neurons into the three main subtypes shown in Fig. 23-6. Curves of subtype IIIA are bimodal with an ascending region preceding two peaks and a de-

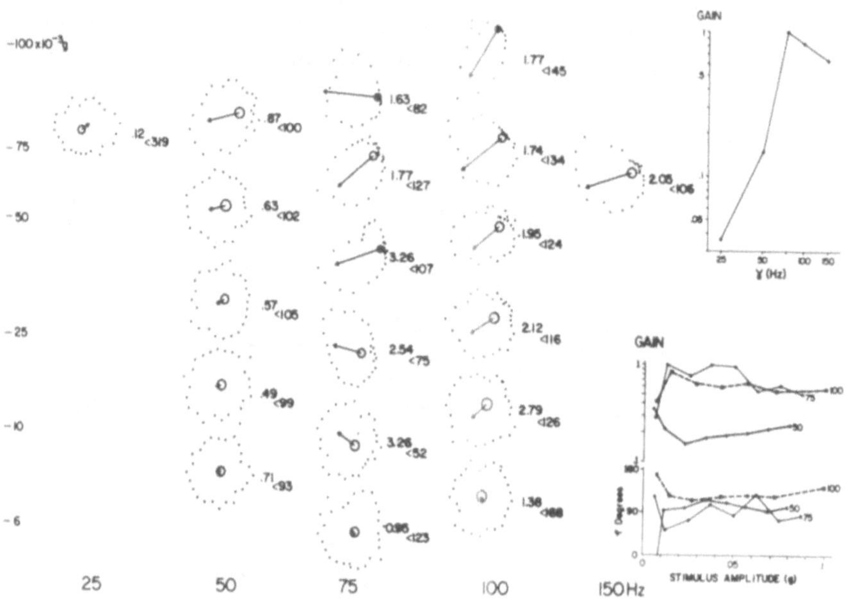

Fig. 23-5. Dynamic response of a type II neuron. Each column includes the response to sinusoidal stimuli of the same frequency and different amplitudes represented by the cycle histogram in polar coordinates, the average vector, and the confidence region. The ordinates indicate the acceleration amplitude peak-to-peak of the stimulus, measured using 10^{-3} g as a unit. The numbers at the right of each histogram represent the gain amplitude and phase. The plots at the right are the gain of the smallest stimulus producing a detectable response for each frequency as a function of its frequency (*top*), and the gain amplitude and phase as a function of the stimulus amplitude for different frequencies (*bottom*). From Budelli, R. and Macadar, O.: Stato-acoustic properties of utricular afferents. J. Neurophysiol. 42:1479, 1979.

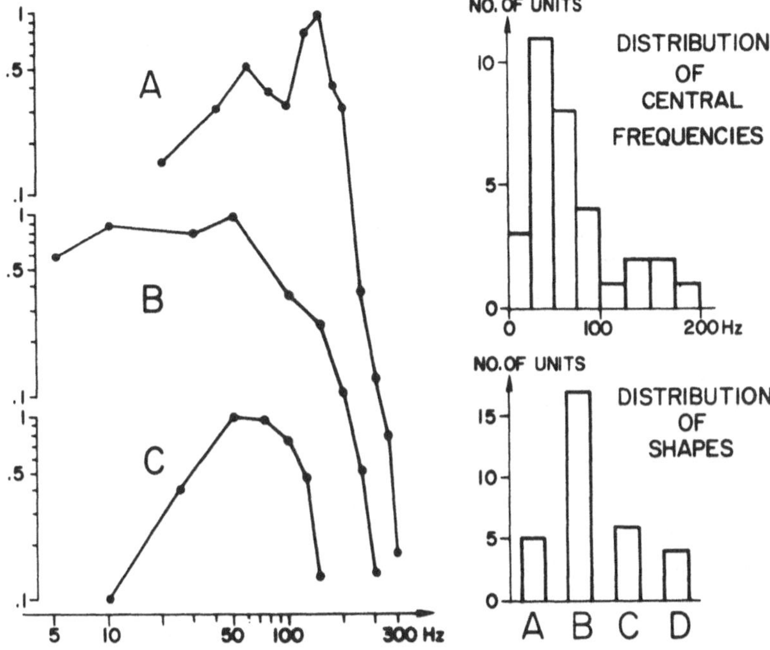

Fig. 23-6. Sensitivity curves for type III neurons. **Left** The ratio between the sensitivity at a particular frequency and the maximum sensitivity was plotted as a function of the frequency of the stimulus. The data were obtained from three type III neurons. **Right top** Histogram of central frequencies (maximum sensitivity). **Right bottom** Histogram of frequencies of subtypes recorded from the population of type III neurons. *D* represents fibers with sensitivity curves that do not fit the description of A, B, or C. From Budelli, R. and Macadar, O.: Stato-acoustic properties of utricular afferents. J. Neurophysiol. 42:1479, 1979.

scending region at high frequencies. The sensitivity curves of subtype IIIB resemble those of an accelerometer, with a flat region at low frequencies and a monotonically descending region at high frequencies, sometimes preceded by a discrete peak. The sensitivity curves of subtype IIIC neurons resemble those of acoustic afferents, with a sharp peak. These curves support the hypothesis that some utricular afferents have a role in hearing (6).

GENERALIZATION OF THE FINDINGS

The results obtained using elasmobranchs are not exclusive to these animals. There are reports in the literature suggesting the versatility of otolithic afferents in several animal species. The sensitivity to high frequency stimulation has been described in the saccular afferents of the frog (3) and has been inferred from the otolith's resonant frequency in fishes (8). The presence of afferent nerves with different properties in otolithic organs has been reported in the monkey (9,10) and in the frog (4). In the frog, type I and III fibers were recorded from the sacculus (3,15). The absence of type III fibers in the mammalian otolithic organs has led to the conclusion that all otolithic afferents have a resting discharge (12). This conclusion may be premature since the lack of silent fibers (type III) may be a result of the sample technique used in those experiments (9,10,17). On the other hand, it is possible that silent fibers are part of the otolithic apparatus only in the species lacking a cochlea, in which type III fibers may carry acoustic information. Another general property of otolithic afferents is adaptation, which occurs not only in elasmobranchs (20,23) but also in amphibians (15) and mammals (9).

ROLE OF AFFERENT DIVERSITY IN INNER EAR FUNCTION

Equilibrium

The utricle may be considered an organ in which different types of afferents carry different information to the brain (12).

The high frequency responses of the utricular afferents allow the animal to detect a rapid head movement. If such a movement is applied to the animal's head, the utricular afferents respond in different and complementary ways. The most immediate change occurs in type III neurons, which code the positional change in a sudden burst of action potentials. Since these afferents are sensitive to relatively high frequencies of stimulation, a movement will be coded quickly. However, the information transmitted by type II neurons does not allow the central nervous system to discriminate between a positional change and a vibratory stimulus. To achieve this discrimination, type I and II neurons change their firing rate during a change of position. The existence of adaptation implies that a short time after a movement, the frequency change will be larger than that predicted by the projection of the polarization vector (26), thus allowing faster coding of the movement amplitude.

Hearing

In discussing acoustic perception, we should consider first the relationship between the natural acoustic stimulus for this species and the vibrational stimulus used in these experiments. Underwater sound waves have two different components: pressure and displacement. The relationship between them depends on the distance from the sound source. The displacement wave decays quickly with distance and is effective only within a few meters from the source (near-field). The pressure wave can reach longer distances (far-field) and affect animals with swim bladder, but cannot affect animals without it (5). A 100-Hz sound with an underwater wave length of approximately 15 m produces a nearly uniform change of pressure around the animal. Fish without swim bladder are not affected by this pressure. In fish with swim bladder or any other compressible structure, the pressure wave will produce movement of different structures, including the vestibular system. This movement can be detected by the hair cells. Thus, the swim bladder interferes with the pressure wave to generate a secondary source, which has both a far-field pressure wave and a near-field vibration wave. Fish without a compressible structure (such as elasmobranchs) are not sensitive to the pressure wave; therefore, they do not hear in the far field.

Since the density of the elasmobranch body is similar to the density of the surrounding water, the vibration wave will move the whole animal. Otoconia, which are solid particles heavier than water and the surrounding tissue, offer greater resistance to the vibration, thus generating a relative displacement that bends the ciliary apparatus of the hair cells. This bend has been considered a step in the mechanoelectric transduction mechanism operating in hair cell systems. Since vibrations of the whole body of elasmobranchs appear to be the natural acoustic stimulus, a vibration of the utricle rather than an actual sound simulates the acoustic stimulus.

Experimental behavioral responses have yielded evidence that the vibration wave is the actual stimulus in some sharks (13). The vibrational threshold for the horn-shark approximates thresholds for type III fibers and also those of type I and II in the guitarfish. This proximity indicates that the utricle cannot be dismissed as a possible acoustic receptor. We found that type III afferents in general and subtype IIIC afferents in particular share properties with the cochlear afferents described in mammals (14). They respond to tone bursts, exhibit a tuning curve, and do not fire spontaneously (Figs. 23-2 and 23-6). Type I and II neurons also possess tuning curves. Type I neurons generate tuning curves with a high sensitivity for frequencies around their firing rate (Fig. 23-4). Type II neurons extend the sensitivity to higher frequencies, complementing type I neurons (Fig. 23-5). These two afferent types do not respond with a simple code to a vibrational stimulus, as do type III cells. If these were the only afferents carrying acoustic information, the detection of a sound would be a complicated task for the central nervous system.

ADAPTATION

Another way to characterize the dynamic behavior of utricular afferents is to determine their responses to a pulse or a step of the acceleration vector. These responses are equivalent to the gain-frequency curves for the characterization of a linear system, and complementary for a nonlinear system, such as the utricle (10).

The responses of the utricular afferents to steps of the acceleration vector, produced by tilting the head of the animal, are characterized by the presence of adaptation. This is defined as the decline in the frequency of impulse firing that occurs in first order sensory nerves stimulated with a constant intensity (25). This phenomenon occurs in most receptors (7), including the utricle (1,10,20,23).

Adaptation may occur at any level involved in the transformation of mechanical signals into changes in the firing of the afferent fibers. Nakajima and Onodera (25) divided this transformation into two main steps: the transformation of the stimulus into a generator potential in the afferent terminals, and the spike generation mechanism. Consequently, they differentiate adaptation into generator adaptation and spike adaptation.

If the electric properties of the membrane of the nerve terminals where the generation of the spikes occurs were responsible for the adaptation, one would expect that electric polarization applied to the nerve terminals would also produce such an adaptation (16,19). Since polarization does not produce the same adaptation as that produced by tilting the head of the animal (Fig. 23-7), the difference might be due to some previous process. According to Nakajima and Onodera (25), this previous process is generator adaptation, which also occurs in the muscle spindle (16) and in the Pacinian corpuscle (18).

Fibers with different patterns of adaptation to tilts usually have similar

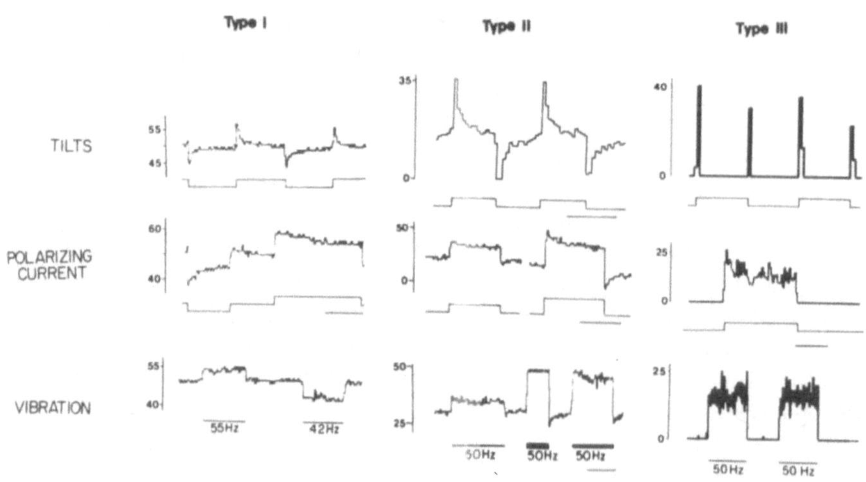

Fig. 23-7. Patterns of response adaptation in the three types of utricular afferents. For each type, the responses of the same unit to the three modalities of stimulation are represented. The upper trace in each recording represents firing rate in impulses/s as a function of the ongoing time. The lower trace represents the stimulus. Tilts were 30°. Polarization currents were hyperpolarizations when the stimulus trace showed a downward deflection and depolarizations when it was upward. The frequency of vibration is indicated in each case, and the intensities are represented by the trace thickness. **Type I** Tilts and polarizations elicit adapting responses, whereas vibrations are able to increase or decrease the rate, without adaptation. **Type II** Response to tilts has larger adaptation than do responses to polarizations and vibration. **Type III** Fast and complete adaptation to tilts, uncertain adaptation to polarization, and no adaptation to vibrations. Time bars: 30 s. From Macadar, O. and Budelli, R.: In preparation.

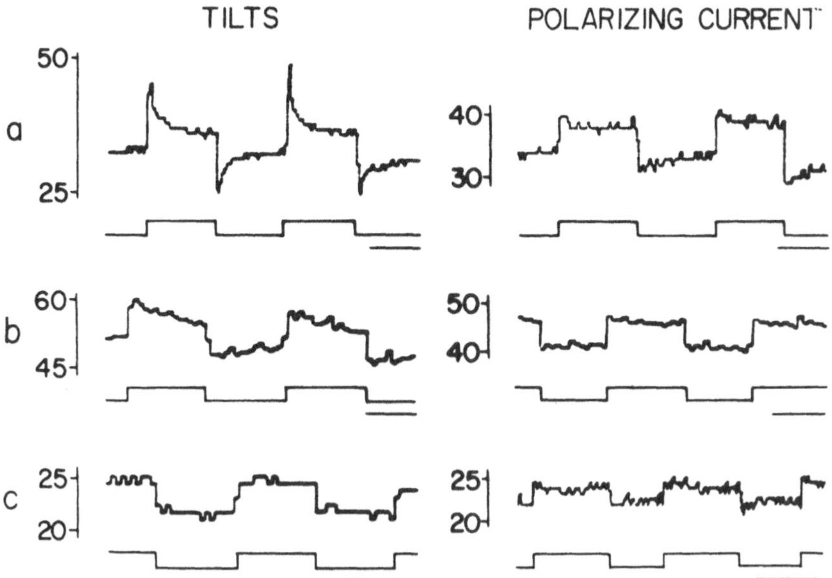

Fig. 23-8. Adaptation to tilts and polarization. The responses of three afferents to tilts and currents are represented. The amplitude of adaptation in the responses to tilts decreases from the afferent *a* (dominantly phasic) to the afferent *c* (almost tonic). A low intensity polarizing current (20 nA) elicited almost purely tonic responses, similar in the three units. Time bar: 60 s. From Macadar, O. and Budelli, R.: In preparation.

adaptive responses to polarization currents (Fig. 23-8). Thus, we conclude that the differences among afferents are generated by generator adaptation rather than by spike adaptation. The opposite—differences in adaptation produced at the spike generator mechanism—occurs in the stretch receptor of the crayfish (25).

We conclude that both generator adaptation and spike adaptation contribute to the global adaptation of the utricle, with generator adaptation responsible for the different degrees of adaptation among afferents.

CONCLUSION

The different properties displayed by the utricular afferents (Table 23-1) are a consequence of the specialized information each type carries. The properties of type I and II fibers are suited to carry information about movements and position of the head and also to transmit acoustic information. Type III fibers are more suited to provide information about acoustic stimuli, but can also convey information about head movements. At the evolutionary level of the elasmobranch, then, the differentiation between acoustic stimuli and constant linear accelerations is not well developed at the receptor level.

Table 23-1.
SUMMARY OF RESULTS.

Characteristic	Type I	Type II	Type III
Spontaneous firing			
Mean rate	32	37	0
Variation coefficient	<0.20	<0.17	—
Response to tilts			
Adaptation	Yes	Yes	Yes
Rectification	No	No	Yes
Response to sinusoidal vibration			
Change of rate	Depending on frequency	Increase	Increase
Adaptation	No	Yes	No
Relationship between firing rate (FR) and frequency of maximum gain (FMG)	FR = FMG	FR < FMG	0 = FR < FMG
Response to white noise vibration			
Change of rate	No	Increase	Increase
Adaptation	No	Yes	No
Response to polarizing current			
Adaptation smaller than adaptation to tilt	Yes	Yes	Yes

Some of the dynamic properties of the utricular afferents, such as adaptation and, consequently, the positive slope in the gain-frequency curve, are generated in a mechanism previous to the spike generation.

ACKNOWLEDGMENTS

We thank Diane Foster, Sheilla Odnert, Cynthia Smith, and Ibrahim Hernandez for help in the preparation of the manuscript; Janet Lewis, Carl Muller, and Manuel Don for critically reading the manuscript; and Humberto Bracho for comments and suggestions about the organization of the material. The presented results were obtained with the support of grants from NSF, NIH, and United Cerebral Palsy to Jose P. Segundo.

References

1. Blanks, R.H.I. and Precht, W.: Functional characterization of primary vestibular afferents in the frog. Exp. Brain Res. 25:369, 1976.
2. Budelli, R. and Macadar, O.: Stato-acoustic properties of utricular afferents. J. Neurophysiol. 42:1479, 1979.
3. Cazin, L. and Lannou, J.: Response du saccule a' la stimulation vibratoire directe de la macule, chez la grenouille. C. R. Soc. Biol. (Paris) 169:1067, 1975.
4. Cazin, L. and Lannou, J.: Two populations of afferent fibers in the saccular nerve of the frog (*Rana esculenta*). Brain Res. 114:501, 1976.
5. Chapman, C.J. and Sand, O.: Field studies of hearing in two species of flatfish: *Pleuronects platessa* (L) and *Limanda limanda* (L) (Family *Pleunectidae*). Comp. Biochem. Physiol. 47A:371, 1974.
6. Colnaghi, G.L.: Saccular potentials and their relationship to hearing in the goldfish (*Carassius auratus*). Comp. Biochem. Physiol. 50A:605, 1973.
7. Davis, H.: Some principles of sensory receptor action. Physiol. Rev. 41:391, 1961.
8. de Vries, H.: The mechanics of the labyrinth otoliths. Acta Otolaryngol. 38:262, 1950.
9. Fernandez, C. and Goldberg, J.: Physiology of peripherel neurons innervating otolith organs of the squirrel monkey. I. Response to static tilts and to long duration centrifugal force. J. Neurophysiol. 39:970, 1976.
10. Fernandez, C. and Goldberg, J.: Physiology of peripherel neurons innervating otolith organs of the squirrel monkey. III. Response dynamics. J. Neurophysiol. 39:996, 1976.
11. Goldberg, J. and Brown, P.B.: Response of binaural neurons of dog superior olivary complex to dichotic tonal stimuli: some physiological mechanisms of sound localization. J. Neurophysiol. 32:613, 1969.
12. Goldberg, J. and Fernandez, C.: Vestibular mechanism. Ann. Rev. Physiol. 37:129, 1975.
13. Kelly, J.C. and Nelson, D.R.: Hearing thresholds of the horn shark (*Herodontus francisci*). J. Acoust. Soc. Am. 58:905, 1975.

14. Kiang, N.Y.: Discharge Patterns of Single Fibers in the Cat's Auditory Nerve. Cambridge, Mass., M.I.T. Press, 1965.

15. Lannou, J. and Cazin, L.: Response to tilting of the fibers of the frog's saccular nerve. Pfluegers Arch. 366:143, 1976.

16. Lippold, O.C.J., Nicholls, J.G., and Redfearn, J.W.T.: Electrical and mechanical factors in the adaptation of a mammalian muscle spindle. J. Physiol. (Lond.) 153:209, 1960.

17. Loe, P.R., Tomko, D.L., and Werner, G.: The neural signal of angular head position in primary afferent vestibular nerve axons. J. Physiol. (Lond.) 230:29, 1973.

18. Loewenstein, W.R. and Mendelsohn, M.: Components of receptor adaptation in a pacinian corpuscle. J. Physiol. (Lond.) 177:377, 1965.

19. Lowenstein, O.: The effect of galvanic polarization on the impulse discharge from sense endings in the isolated labyrinth in the thornback ray (*Raja clavata*). J. Physiol. (Lond.) 127:104, 1955.

20. Lowenstein, O. and Roberts, T.D.M.: The equilibrium function of the otolith organs of the thornback ray (*Raja clavata*). J. Physiol. 110:392, 1949.

21. Lowenstein, O. and Roberts, T.D.M.: The localization and analysis of the response to vibration from the isolated elasmobranch labyrinth. A contribution to the problem of the evolution of hearing in vertebrates. J. Physiol. (Lond.) 114:471, 1951.

22. Macadar, O., Wolfe, G.E., Budelli, R., and Segundo, J.P.: Multivalued stimulus-response relation in isolated elasmobranch utricles. Biol. Cybern. (in preparation).

23. Macadar, O., Wolfe, G.E., O'Leary, D.P., and Segundo, J.P.: Response of the elasmobranch utricle to maintained spatial orientation, transitions, and jitter. Exp. Brain Res. 22:1, 1975.

24. Mardia, K.V.: Statistics of Directional Data. New York, Academic Press, 1972.

25. Nakajima, S. and Onodera, K.: Membrane properties of the stretch receptor neurones of crayfish with particular reference to mechanisms of sensory adaptation. J. Physiol. (Lond.) 200:161, 1969.

26. Young, L.R.: Role of the vestibular system in posture and movement. In Mountcastle, V.B. (ed.): Medical Physiology. St. Louis, Mosby, 1974, pp. 704–721.

Discussion

GOLDBERG: Could you tell from which part of the utricular macula your recordings were derived?

BUDELLI: We recorded from all the bundles innervating the utriculus. For each unit recorded we registered the bundle from which it was isolated. We could not establish a correlation with clear bounds between types of units and regions of the utricle: usually, in a single bundle, we simultaneously recorded two units of a different type.

GOLDBERG: The reason I ask is that Lowenstein and Roberts (1951) thought that the lacinia, which is not covered by the otolithic membrane, was particularly sensitive to vibrations. Perhaps Professor Lowenstein could comment on this.

LOWENSTEIN: Well, in all animals the lacinia of the utriculus is highly vibration sensitive, but we came to the conclusion that it isn't necessary to assume that it serves in the vibration detection because we find that the other two sense endings in the sacculus and the crista quarta or macular fenestra of Scarpa neglect were, in fact, lying opposite a window . . . and so we felt these were the preferred pathways to the vibrations. Our final conclusion was that the sense endings in the otolith organs of the elasmobranch are potentially vibration receptors. But it depends on the exposure of these sense endings to the sound pathway to determine which one is used by the animal. And we didn't assume that the utriculus was in fact utilized with this purpose.

BUDELLI: In relation to the hearing capabilities of the utricle, we added to the classic results of Lowenstein and Roberts the measurement of the movement of the head. This enabled us to compare the threshold of the utricular afferents with the behavioral acoustic threshold determined in similar fish. Both thresholds are of the same order of magnitude. Therefore, the possibility that utricular afferents transmit acoustic information cannot be discarded.

PLATT: Following up again on the utricle story, I wonder whether you could say anything about the directional sensitivity of the responses that you saw. Referring back to Dr. Lowenstein's earlier work on the elasmobranch ray, and supported by my observations on some tissue of the guitarfish, the utricle in elasmobranchs appears to have a rather patchy distribution of orientation rather than the consistent pars interna and pars externa that we see in the other vertebrates.

BUDELLI: For each fiber recorded, we established the direction of maximum sensitivity and the bundle from which it was isolated. Nevertheless, we could not establish a distinct correlation between direction of maximum sensitivity and regions of the utricle. I think this result matches your "patchy distribution".

Part 4

Visual-Vestibular Interactions

INVITED LECTURE 24

Visual and Vestibular Influences in Human Self-Motion Perception

LAURENCE R. YOUNG

Visual-vestibular interaction is interpreted as one of a more general case of interaction of multiple sensory inputs in the process of estimation of state. Our general approach to this problem, as illustrated in the diagram of Fig. 24-1, has been to consider the human as an optimal state estimator (30,31). He takes information coming from the various indicated sensors and combines them with an "expected state" signal which is generated from his internal model of what he believes his control mechanisms to be, as reflected in active movements. The estimator produces both an estimate of his current state (rotation, translation, and orientation with respect to the vertical) and an estimate of the conflict in these cues. If the conflicts exceed some threshold, disorientation, motion sickness, and vertigo may result. The development of this conflict is also seen as a necessary one for the process of habituating to an altered environment. The altered environment may either be associated with external changes such as the well-known prism-wearing experiments or the altered environment associated with motion in a rotating vehicle or in a weightless environment, or it may be an internally altered environment, for example, by the removal of one labyrinth.

This paper will begin with some very basic questions about human spatial orientation before proceeding to detailed discussion of visual-vestibular interaction. The presumed purposes of human spatial orientation are to deter-

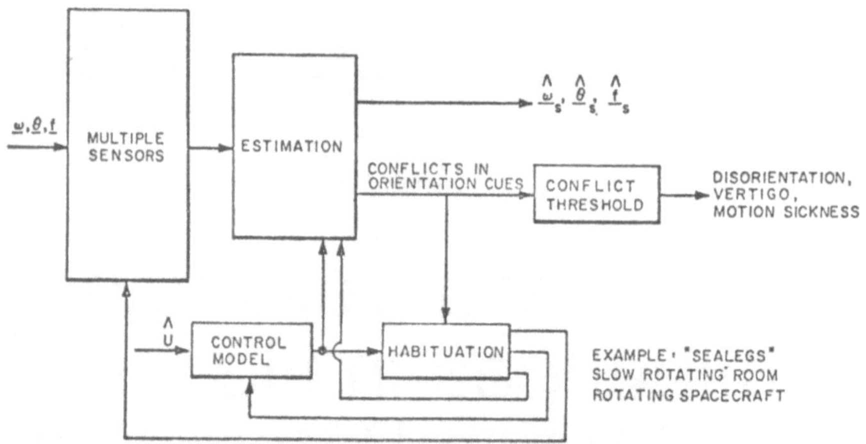

Fig. 24-1. Schematic model for spatial orientation, including active motion, all sensors, and habituation to general motion in unusual environment. From Young, L.R.: Developments in Modelling Visual-Vestibular Interactions. AMRL-TR-71-14. Wright-Patterson Air Force Base, Dayton Ohio, 1971.

mine orientation with respect to the vertical, heading, and a velocity vector. This information is necessary for ocular and head stabilization, posture control, and navigation.

The chief contributor to the early psychophysics of visual-vestibular interaction in orientation was Ernst Mach, a physicist of some renown, who applied his experimental skills with the rather simple-looking wooden device shown in Fig. 24-2 (19,24). This apparatus was used for rotating subjects,

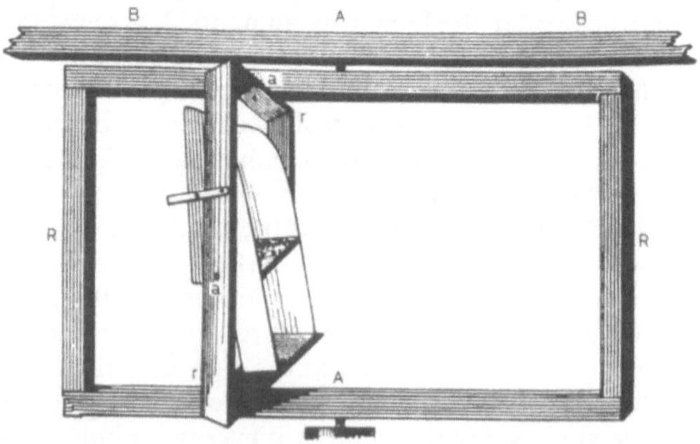

Fig. 24-2. Mach's rotating chair, movable about vertical axis through *AA* and through *a*. Subject sits in box which can be covered with light-proof paper and can be tilted about a horizontal axis α. From Mach, E.: Grundlinien de Lehre von den Bewegungsempfindungen. Leipzig, Englemann, 1875.

tilting them, producing centrifugal forces, and measuring the perception of orientation as they were subjected to unusual linear and angular accelerations. The question of conflict between visual and vestibular cues entered serious study in the domain of the psychologist in the 1930s and 1940s. In a typical situation, a subject would either be placed in or have a view of a tilted room (2,3). He would be asked to indicate the perceived vertical, which he could judge on the basis of the visual vertical (typically offset) and the veridical vertical signal detected by his gravity receptors. Typically, his indicated vertical was located somewhere between visual "vertical" and the otolith-indicated vector. There was a great deal of interest in associating the individual variations with personality differences, as to whether one was more or less field dependent, and with sex differences (29).

Real interest in the practical problems of visual-vestibular orientation came about during the early days of aviation, in which the problems of blind flying and disorientation in flight became rather severe. This question came very much to the forefront with the introduction of manned space flight in the 1960s (31).

VISUALLY INDUCED MOTION: VECTION

Several laboratories, including our own, have been systematically exploring visually induced motion and vestibularly induced motion. Figure 24-3 shows the modified flight trainer used for this purpose in our laboratory at M.I.T. We can produce controlled rotations about three orthogonal axes, while

Fig. 24-3. LINK GAT-1 trainer with visual display.

also independently controlling the movement of the visual field. In this case, the visual field is projected on the windows of the modified GAT-1 flight trainer. In other cases, in which the visual field need be more compelling, the visual field can be produced by actually rotating a drum around the subject (14). When a stationary subject sees the entire visual field begin to rotate about a vertical axis, he normally experiences a rather remarkable sensation, referred to as circularvection. At first, the perception is merely that the visual field has begun to rotate and that one is stationary, i.e., the veridical situation is sensed. After a period of typically some 5 to 30 s, however, a quite remarkable change takes place. One develops the sensation that the field itself has slowed in space and that one has begun to rotate in the opposite direction. When the circularvection is what we call "saturated," the sensation is that the field is actually stationary in space and that *all* of the relative motion is due to the subject himself. We typically use the psychophysical method of magnitude estimation: the subject is trained, using well-known motions, to indicate his subjective velocity on an arbitrary scale, i.e., the speed at which he feels he is rotating to the right or left under a yaw rotation stimulus. As seen in Fig. 24-4, when the field begins to rotate to the right, after a short delay, the subject feels that he is slowly increasing his speed of rotation to the left (32). When the field stops, he quickly returns to the null state. If he closes his eyes during the stimulation, the sensation of motion ceases, but opening his eyes will result in a build-up of the illusion, within 5 to 10 s, to the so-called steady state circularvection. Sometimes, for reasons that are still not entirely clear, there will be a rapid drop out of vection and then a reacquisition of it. Following cessation of the field motion, there is usually a brief outlasting effect, sometimes followed by a negative aftereffect, which can be a rather long-lasting feeling of motion in the opposite direction to that which was felt during the field motion.

The major characteristics of visually induced motion and visual-vestibular

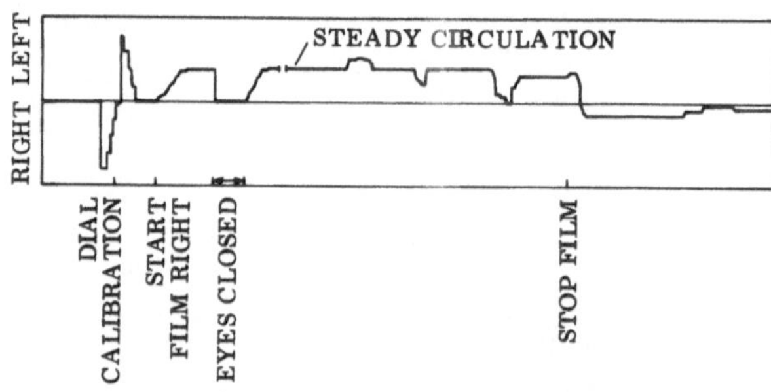

Fig. 24-4. Typical recording of magnitude estimation showing dial calibration, onset of circularvection, waxing and waning of circularvection with trainer stationary, outlasting circularvection following stopping of film, and long-lasting aftereffect in the opposite direction. Eyes are closed for 7 s. From Young, L.R., et al.: Acta Otolaryngol. 76:24, 1973.

interaction, which will be reviewed below, are the results principally of work carried out by Held and his coworkers in the Psychology Department at M.I.T.; Dichgans and Brandt, formerly of Freiburg, Germany; Berthoz in Paris; and our laboratory at M.I.T.

It has become apparent that the necessary stimulus for visually induced motion is a homogeneous motion in the peripheral field and not in the foveal field. In fact, a pure foveal stimulus will generally not induce the sensation of motion. Furthermore, the nystagmoid eye movements associated with this optokinetic stimulus have been shown not to be necessary for the generation of this visually induced motion. Even when the central part of the visual field is reversed in its direction from the peripheral field, one can get the eye movements following the central field, and yet the sensation of motion is determined by the peripheral field (17,18).

Background visual information rather than the foreground is important in generating visually induced motion. If the foreground is stationary and the background moves, one develops circularvection. If the contrary is the case, it is very difficult to develop circularvection. If one looks out through a fixed window at a moving scene, one can generate self-motion, but if one looks at a fixed background through a foreground that is moving, it is unlikely that one would develop any self-motion sensation. The spatial frequency of the field appears important only up to the saturation level; for circularvection, this appears to be in the region of 30% coverage (9), providing that there are a sufficient number of contrast borders that move. The actual makeup of the field is relatively unimportant (33). For induced linear motion, however, it does appear that the finer the structure, the stronger the sensation (23). The velocity of the visual field determines, to a great extent, the velocity of the self-motion. A vection threshold apparently corresponds to the thresholds for actual motion detection (4). There is also a vection saturation. If the field motion is greater than about 60°/s in yaw or some 10 m/s in linear motion, further increases of field velocity will not result in any further increase in self-motion. If one exceeds this saturation level by a great deal, in fact, the sensation of self-motion disappears.

Figure 24-5 shows examples of the dynamics of the onset transients associated with visually induced motion. In this case, it is linearly induced motion in the horizontal plane. When the stripes are moving back, there is a delay and then the onset of visually induced advancement. The onset does not always follow the same time course. The onset latency depends upon the acceleration of the field. If the field is accelerated rapidly, say a step in velocity or an impulse in acceleration, the latency may be of the order of 5 s or more. However, if the field is accelerating quite slowly and, in particular, for circularvection, if the acceleration of the field is of the order of fractions of a degree/s^2 (i.e., if the acceleration of the field is commensurate with the threshold of the semicircular canals), then this latency is, in fact, reduced to almost zero.

The quality of the visually induced motion sensation is, in almost all re-

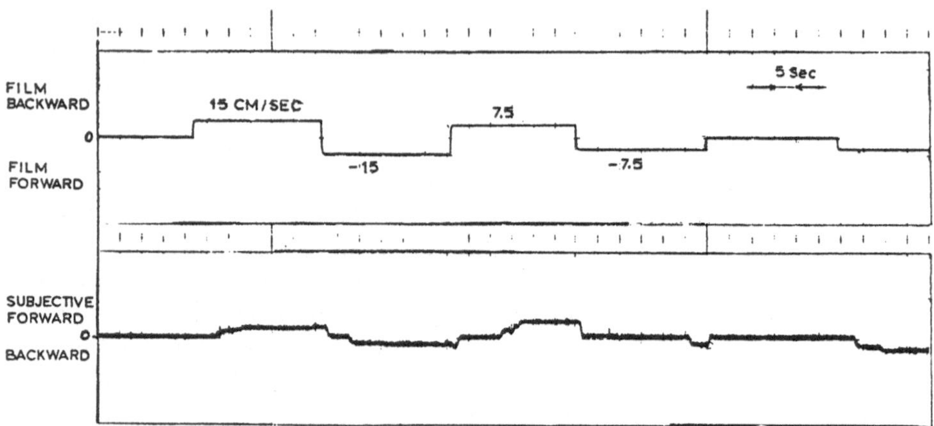

Fig. 24-5. Typical horizontal linearvection response to moving stripes with constant step velocity. From Chu, W.H.N.: S.M. Thesis, M.I.T., 1976.

spects, identical to the sensation of motion that is felt with a true vestibular stimulus. There is only one motion sensation, which is the result of multiple sensory inputs, rather than separate visual, vestibular, or proprioceptive sensations of motion. This extends even to effects such as the pseudo-Coriolis effect, whereby one makes a head movement during visually induced motion. When the subject is actually stationary, but has a sensation of yaw motion due to a visual stimulus, a nodding head movement will produce a sensation of tilt to the side just as would be the case if the subject were actually rotating about the yaw axis (15). One exception to this generalization about response quality is that changes in subjective velocity, during vection, do not necessarily entail a simultaneous perception of acceleration. Paradoxically, during vection experiments, one may suddenly notice that he has stopped or suddenly notice that he is moving without any concurrent sensation of acceleration or deceleration.

The longer a visual scene is presented, the less effective it is in producing the steady state sensation. When a subject is asked to control the linear velocity of the visual scene in order to maintain his own self-velocity constant, as shown in Fig. 24-6, he calls for ever increasing field velocities (4). A familiar example occurs when we have been driving along a highway at high speeds for many minutes or hours and then exit from the highway, only to feel that we are riding very slowly, when the speed has actually been reduced only modestly. Similarly, when we drive for a long period of time on a highway without referring to the speedometer, we find that our speed has generally crept up.

Visually induced motion sensation is predominantly a low frequency phenomenon which has its maximum effect at frequencies below 0.1 Hz. The measurements that one takes of the frequency response depend upon

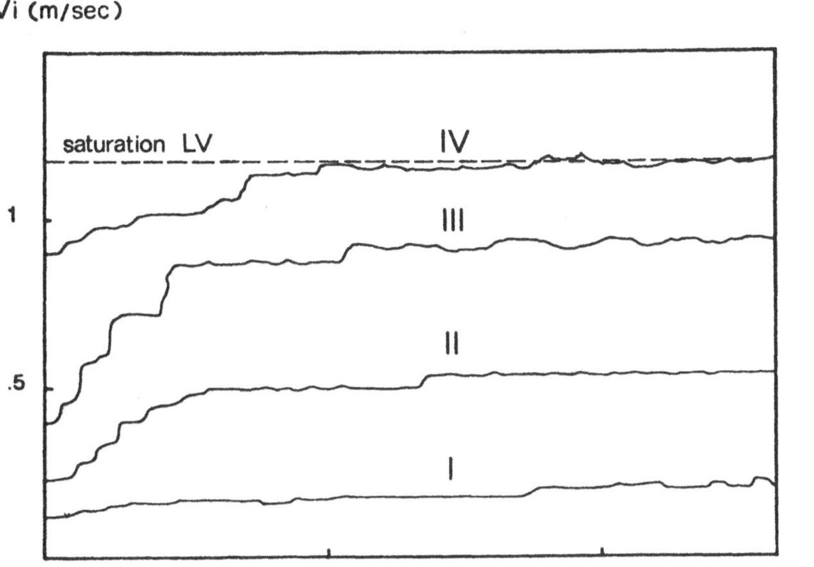

Fig. 24-6. Adaptation of linearvection during prolonged exposure to a constant velocity moving scene. In this experiment, the subject actively controls the velocity of the visual stimulus (*Vi*) so as to keep his sensation of self-motion (*LV*) constant. Response curves (*I*, *II*, *III*, *IV*) are drawn for different initial film velocities. The increase of film velocity (*Vi*) due to the progressive adaptation of linearvection is limited in the case of *LNV* by a saturation effect. From Berthoz, A., et al.: Exp. Brain Res. 23:471, 1975.

whether one uses a predictable signal, such as single sinusoids, or whether one uses a random-appearing signal made up of a sum of sines or other random signal. In the latter case, the gain fall-off is considerably more rapid. Figure 24-7 shows the subjective velocity gain and phase for vertical linearvection (12). The ratio of perceived velocity to field velocity is the gain of the system at each frequency. For each of the four subjects in this experiment, the single sinusoid gain is nearly uniform below 0.1 Hz, and then drops off above 1.6 Hz. When the same subjects are tested with a random-appearing input and a Fourier analysis performed on the gain at each frequency, one finds a very different sort of curve with a much steeper fall-off and a high gain at low frequencies. As in many other biologic systems which involve cortical function, predictability is extremely important. The subjects are able to eliminate phase lag and lock in their gain to the stimulus when the input is a predictable one. Horizontal acceleration is also illustrated for comparison (4).

Several important asymmetries appear in visually induced motion. The zero velocity state appears to be preferred. The vestibular accelerometers

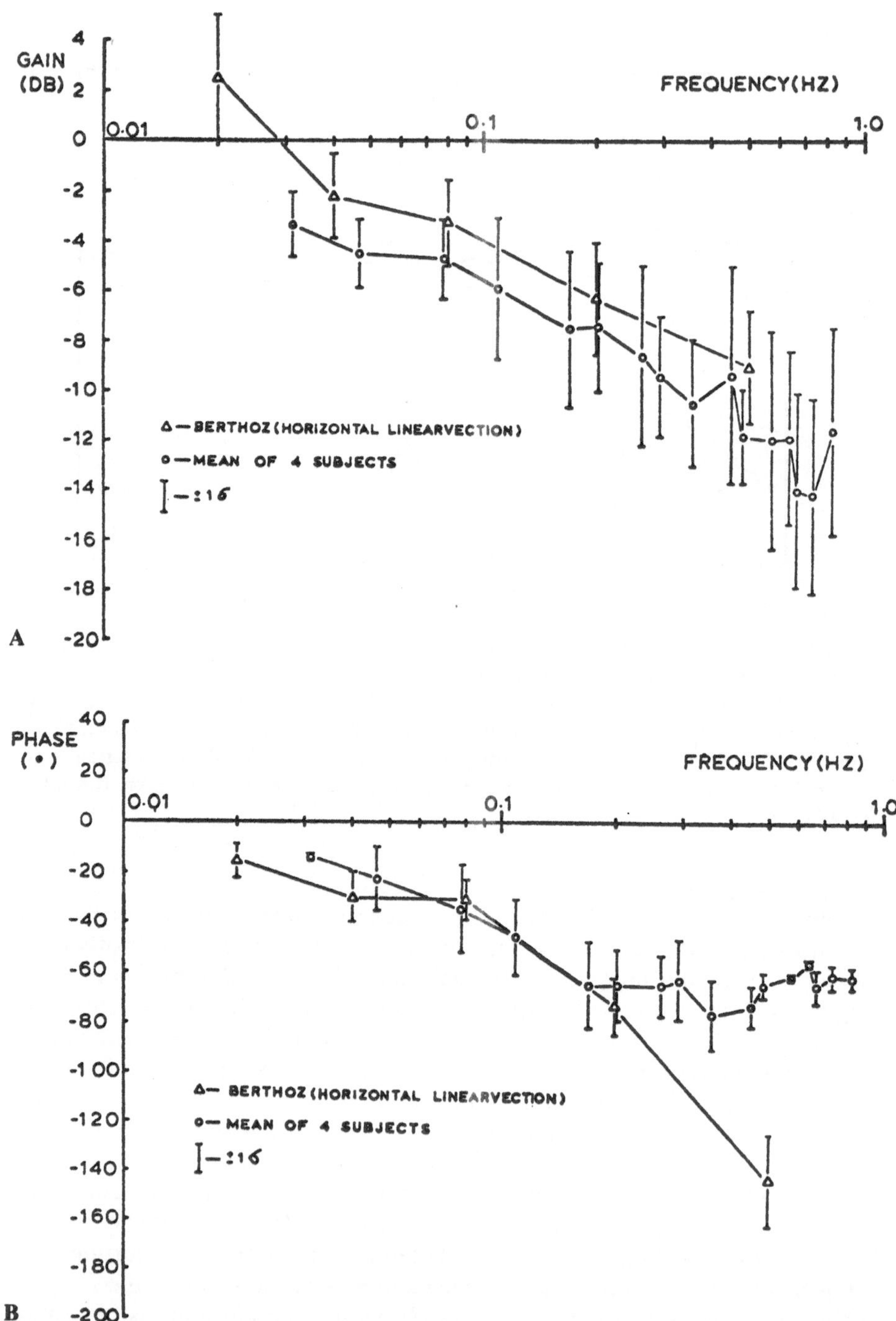

Fig. 24-7. Frequency response of linearvection. Gain (**A**) and phase (**B**) of perceived velocity relative to negative field velocity for sinusoidal vertical field motion, compared to results of Berthoz et al. (4) for horizontal fore-aft motion. Gain (**C**) and

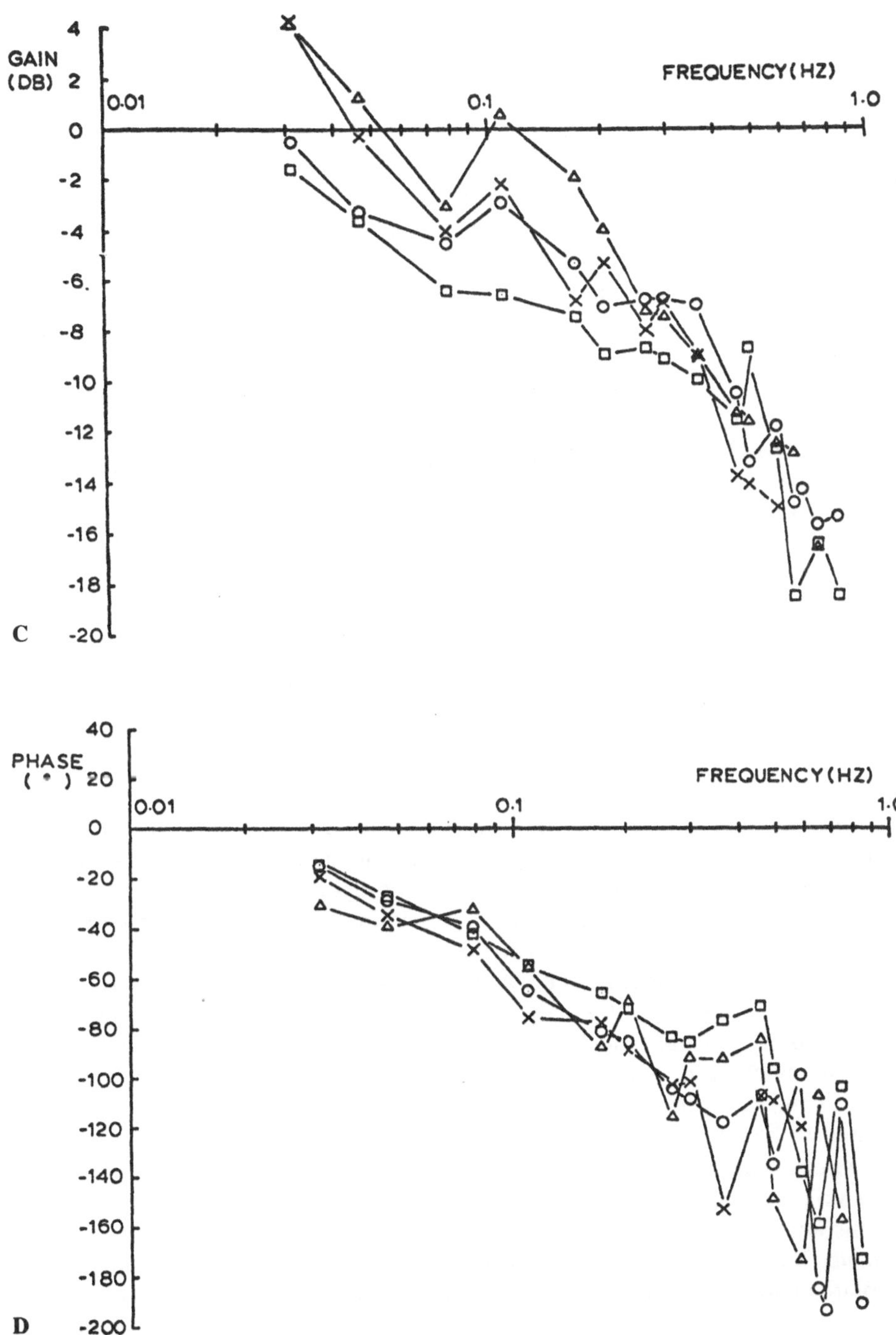

phase (**D**) for pseudorandom (sum of sines) vertical field stimulus. From Borah J., et al.: AFHRL-TR-78-83, 1979.

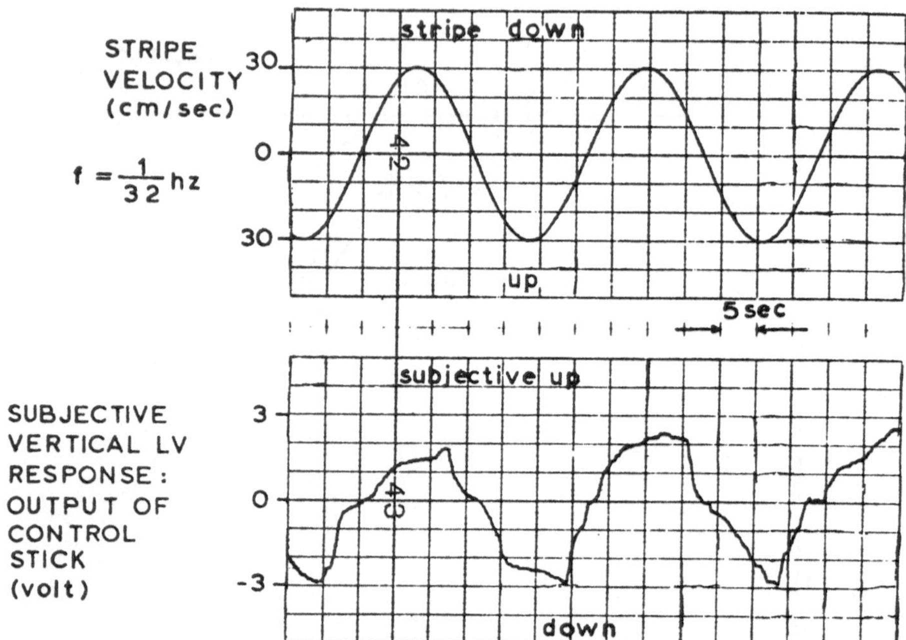

Fig. 24-8. Vertical linearvection. Response to sinusoidal vertical (*above*) field motion is a nonlinear indication of self-velocity. From Chu W.H.N.: S.M. Thesis, M.I.T., 1976.

are obviously indifferent to the steady velocity and measure only changes in that velocity. Nevertheless, response to a sinusoidal field motion is typically a slow build-up of vection and then a rapid decay, as seen in Fig. 24-8 for linearvection. There is also a fore-aft asymmetry, wherein symmetric pattern motion introduces a much stronger sensation of moving backward than it does of moving forward. The explanation of this may possibly be related to the fact that we do not have eyes in the back of our head; consequently, the absence of a visual field behind does not inhibit the sensation that we are moving backward. Alternatively, it may be related to the fact that so much of our normal motion is forward, as during locomotion, and that we are habituated to having the field move backward.

The response that one gets to a rotation of the visual field about a horizontal axis is a complex paradoxical one (17). One feels that he is rotating at a constant pitch or roll velocity, but also that he has pitched only to a given angle. The tilt sensation when one is exposed to a pitching visual field, which rotates about a lateral horizontal axis, exhibits a strong asymmetry. Strong sensations of pitching down result from a field which moves upward; however, when the field pitches down the sensation is not one of pitching up, but rather of upward linear motion (34). Interestingly enough, there is a similar asymmetry in vertical optokinetic nystagmus.

INTEGRATION OF VESTIBULAR SIGNALS WITH VECTION

We now turn to the main point of this paper: how is visually induced motion integrated with vestibular signals? There are two parts to this question: (1) How does the presence of the visually induced motion affect the way vestibular information is interpreted? (2) How does vestibular stimulation modulate the way visual information is interpreted? To start with the latter question, we begin by examining how changing static vestibular signals affect simple visually induced tilt motion. As mentioned above, a steady rotating field will produce a sensation of both rotation and a steady angle of tilt. The perceived tilt angle as a function of the angular velocity of the visual field for the head erect is shown in Fig. 24-9. When the experiment is repeated with the subject's head tilted to one side, the magnitude of the visually induced tilt is greatly increased. With the head on the side, the vestibular stimulus condition is significantly changed. The utricles are then in a relatively ineffective orientation for indicating anything about further change in body angle. The major plane of the utricular macula is then aligned more or less parallel to the vertical (27), in the position where the individual hair cells provide the least information about further changes in the orientation of the body. With the head erect, however, the utricle is in a plane in which it is reasonably sensitive to any actual head tilt.

One might ask why the utricle comes into play here, since it is not, in fact, stimulated since the head remains stationary. Our theory is that the signals from the utricle, indicating to the subject that he has not tilted, conflict with the visual signal indicating that he has rotated, and that it is only in the resolution of this conflict that the compromised angle of subjective tilt is reached (30). When one produces a situation in which the otolith cues are weighted less—by putting the otoliths in an orientation which is undesirable from the point of view of their maximum sensitivity (for the moment neglecting the influence of the sacculus in orientation)—then there is a relative increase in the magnitude of the visually induced tilt. When the head is moved from the head erect to the head on the side, the magnitude of the illusion increases. With the head inverted, it decreases slightly. Interestingly enough, the magnitude of the visually induced tilt is minimized by pitching the head 25° forward from the normal head orientation. Moving the head 25° forward places the major plane of the utricles approximately in the horizontal plane (5), and puts them in the theoretically optimal position for utricular sensation of any tilt angle. Gravitoinertial acceleration along the longitudinal axis, primarily stimulating hair cells in the sacculus, leads to uncertain orientation perceptions which may not be utilized in the tilt estimates (25,26). These data support the notion that the indications of visually induced tilt for rotations about horizontal axes are mediated by otolith signals, even though the otoliths are not, in fact, stimulated. The constraint of the otoliths dictates the maximum tilt that is achieved. For pitch, a similar effect is seen, with the additional asymmetry that pitch down is a much stronger function than pitch up.

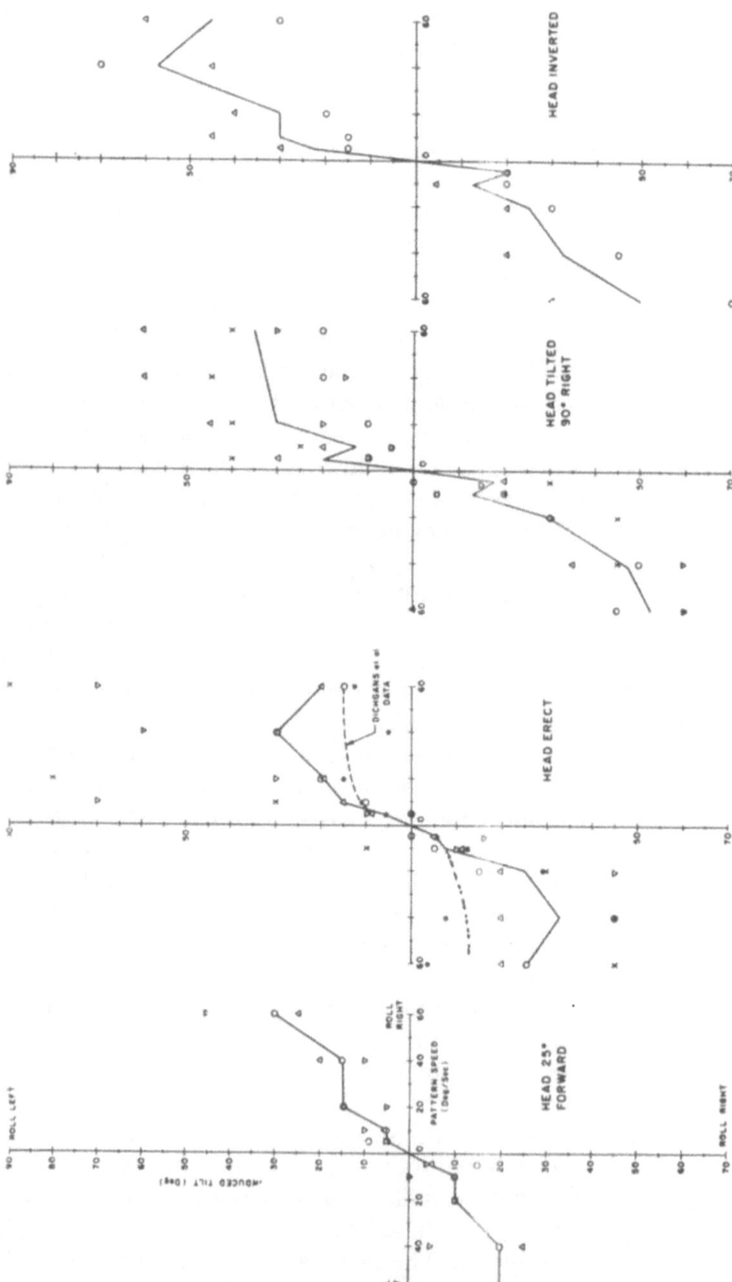

Fig. 24-9. Roll tilt angle induced by full field rolling visual stimulus as a function of pattern speed and head position. Each symbol is always associated with the same subject. Cockpit subject (x) could not participate in head forward or head inverted cases. Solid lines connect median values for data points from individual subjects. Data from Dichgans et al. (17), shown for comparison, used a 130° field, pooling tilt in both directions. From Young, L.R.., et al.: Aviat. Space Environ. Med. 24:264, 1975.

To test the influence of changing visual and vestibular cues simultaneously, it is necessary to have some way of producing independent rotations of the visual field and the vestibular stimulus. The first set of experiments to be discussed are cases in which we either rotate the visual field together with the subject (the so-called fixed condition), rotate the visual field alone (circularvection conditions), or finally rotate the subject within the laboratory-fixed field.

We have demonstrated that the presence of circularvection masks the direction of true acceleration in the opposite direction (32). The time to detect acceleration, as a function of acceleration magnitude, is increased when the direction of this acceleration is opposite to the direction of the preexisting circularvection. Circularvection also biases, to some extent, the perception of angular velocity. If, at the very instant the visual field is beginning to move to the left, for example, the subject is given a true (although brief) acceleration to the right, the latency to onset and maintenance of circularvection is reduced to a negligible level.

Finally, some important nonlinear aspects of visual-vestibular interaction emerge from our experiments on dual random inputs and visual-vestibular stimuli. In order to investigate the interactions between visual and vestibular inputs which are not in consonance, Zacharias employed our moving trainer in the manner indicated schematically in Fig. 24-10 (36). When the visual field is fixed with respect to the subject and the trainer given a step of yaw angular velocity, the subjective sensation is the well-known perrotatory sensation, in which the subject's velocity sensation is initially the true velocity, but which decays almost exponentially with a time constant of the order of 10 s. (Adaptation time and sensation reversal are omitted from this discussion.) For the circularvection case, in which the subject is stationary and the field rotated, there is a latency to onset and then an increase in sensed velocity with a time constant of the order of 5 s. For the counterrotating case, in which the visual field is fixed in space and one rotates within it, the perception is ideally initially correct and is maintained at that level thereafter. In these experiments, the trainer is given a random yaw disturbance (Fig. 24-11A) and we ask the subject to make a velocity magnitude estimation indirectly by using the nulling method. The subject merely tries to maintain the trainer velocity at what he perceives to be null by driving the trainer through a control stick. When the vertical stripes are counterrotated in space (analogous to the case of rotating in a laboratory-fixed visual field), the lower frequency components of the trainer disturbance are adequately nulled as shown in Fig. 24-11B. The residual trainer velocity contains only small high frequency deviations from zero. When the field is fixed with respect to the trainer, however, so that it provides the subject with no low frequency information concerning his true rotation in space, one finds a constant drift, in a direction which appears reasonably consistent for any test subject. The magnitude of this drift acceleration is somewhat below the semicircular canal threshold. In the case illustrated in Fig. 24-11C, it is $0.05°/s^2$. When one at-

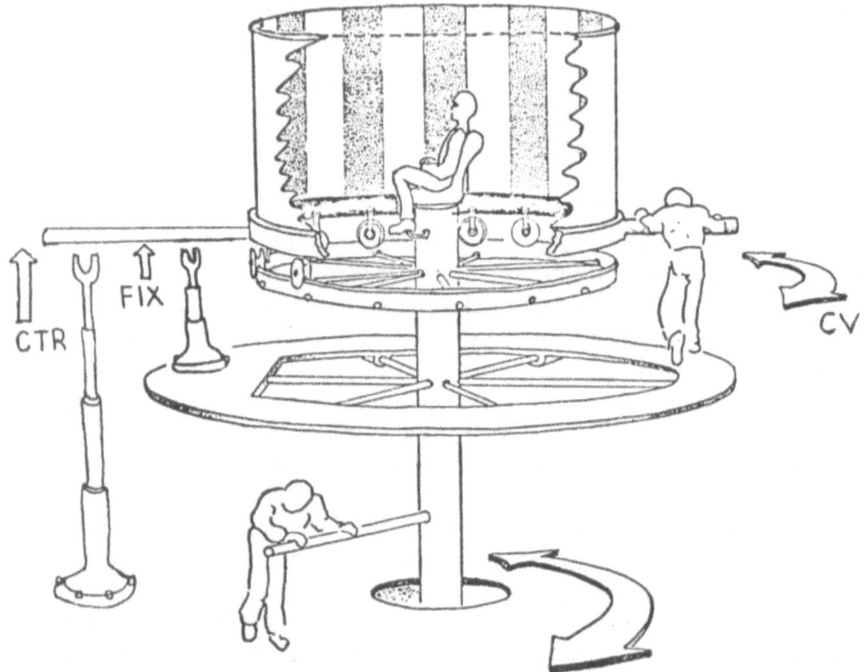

Fig. 24-10. A cartoon illustrating the stimulus modes used for independent visual and vestibular stimulation. Lower sketches indicate schematically subjective angular velocity step responses to chair motion with a subject-fixed field (*left*), rotation of the visual surround about a stationary subject (*middle*), and counterrotation of the field around the moving subject, so that the field remains fixed in space (*right*). From Zacharias, G.L.: Ph.D. Thesis, M.I.T., 1977.

tempts deliberately to induce the sensation of motion through circularvection, by adding constant visual field motion to the vestibular field stimulus, the subject superimposes a trainer acceleration in a direction to try to catch up with the stripes (Fig. 24-11D).

Although these experiments are useful in trying to deduce the nature of the visual-vestibular interaction, they are no substitute for a more regular investigation in which one can examine the interactions in terms of the dual input describing function. To do this, a somewhat more complex experiment is required, as illustrated in Fig. 24-12. We assume that the subject, whose function is indicated within the dotted lines, generates his motion sensation on the basis of the responses of both the vestibular and visual systems. He has a single perceived motion, $\hat{\omega}$. Since his task is to maintain the trainer at a null velocity, he will process this perceived motion through some control logic C(s) to produce a given deflection in the control stick or wheel. We are able to measure both the control stick deflection (λ) and the actual velocities of the trainer (ω_1) and the field (ω_2). Zacharias (35) applied independent visu-

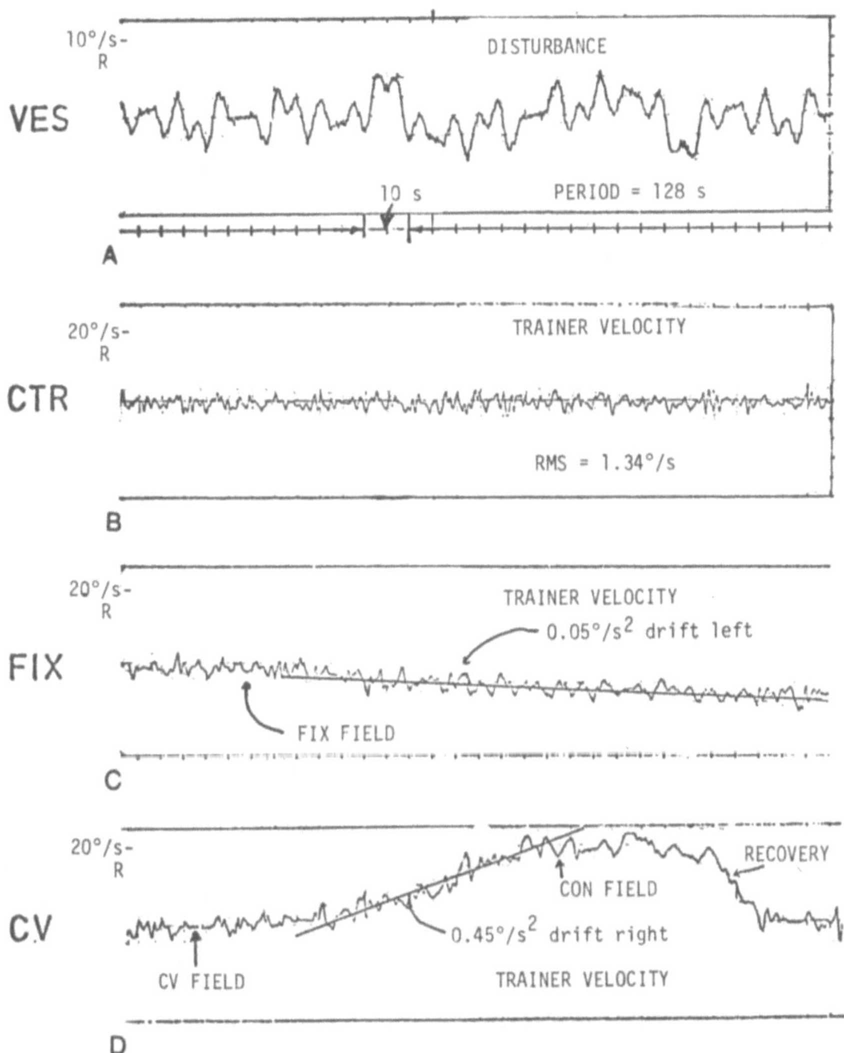

Fig. 24-11. Yaw angular velocity of the trainer under different visual field conditions. *VES*: Random disturbance applied to the trainer. *CTR:* Counterrotating stripes condition, closed-loop nulling. *FIX:* Field condition switched from *CTR* to stripes fixed relative to the subject, closed-loop nulling. *CV:* Circularvection condition, field switched to constant rate of 4°/s to the right. From Zacharias, G.L. and Young, L.R.: Exp. Brain Res. 41:159–171, 1981.

al and trainer velocity disturbances, which were each the sum of several sine waves, combined in such a way that they were random appearing. The frequencies were interwoven so that the vestibular spectrum was completely different from the visual spectrum. Any response of the subject to a vestibular disturbance (the actual motion of the trainer) would be inconsistent

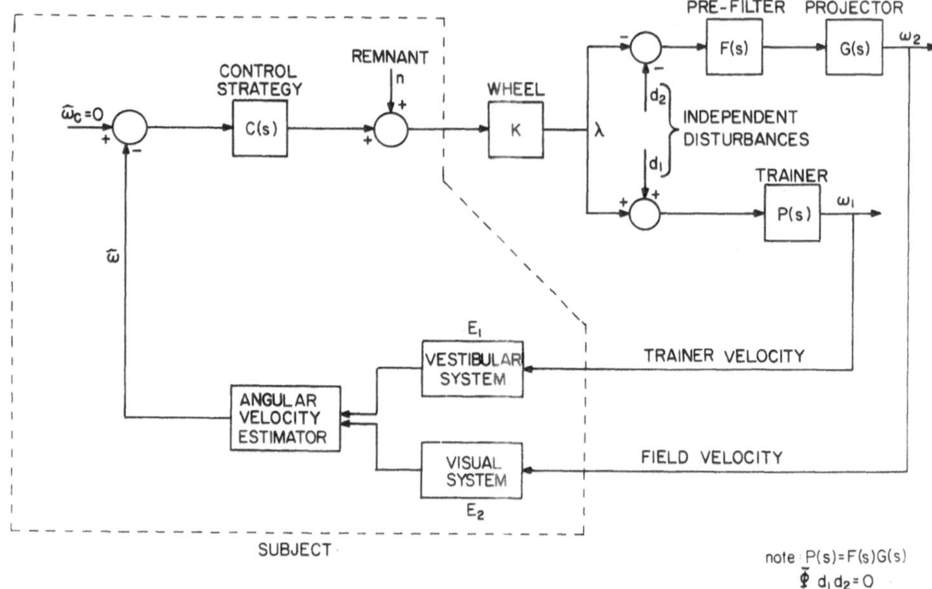

Fig. 24-12. Closed-loop velocity nulling task (dual disturbance input). From Zacharias, G.L. and Young, L.R.: Exp. Brain Res. 41:159–171, 1981.

with the response to try to null the visual stimulus. The object of this exercise was to determine the transfer functions (or more precisely, the describing functions) of the vestibular contribution and the visual contribution to velocity perception. We first, of course, had to determine the control logic, which was accomplished by a separate manual control experiment for each subject. The ratios of the Fourier coefficients of λ to d_1 measure the vestibular transfer function times the control logic. Similarly, the Fourier coefficients of λ divided by those of d_2 provide an estimate of the visual transfer function. The visual and vestibular transfer functions are shown in Fig. 24-13. The vestibular response (Fig. 24-13A) shows the expected mid-frequency constant gain (perceived velocity to actual velocity) and a drop off in gain at lower frequencies. However, this low frequency break frequency falls at about 0.15 Hz, corresponding to a time constant of only 1 s. The mechanical time constant of the semicircular canals is longer than that, and the time constant we associate with perrotatory sensation is closer to 10 s, so that quantitatively this does not at all meet our expectations of dynamic processing of vestibular rate information. When we look at the visual contribution (Fig. 24-13B), we find a steady gain at low frequencies. We would expect the visual system to operate just as well at all very low frequencies, so that this is consistent with our expectations. It was not expected, however, that the gain of the visual system would increase at frequencies above 0.1 Hz, nor is it at all clear why the static

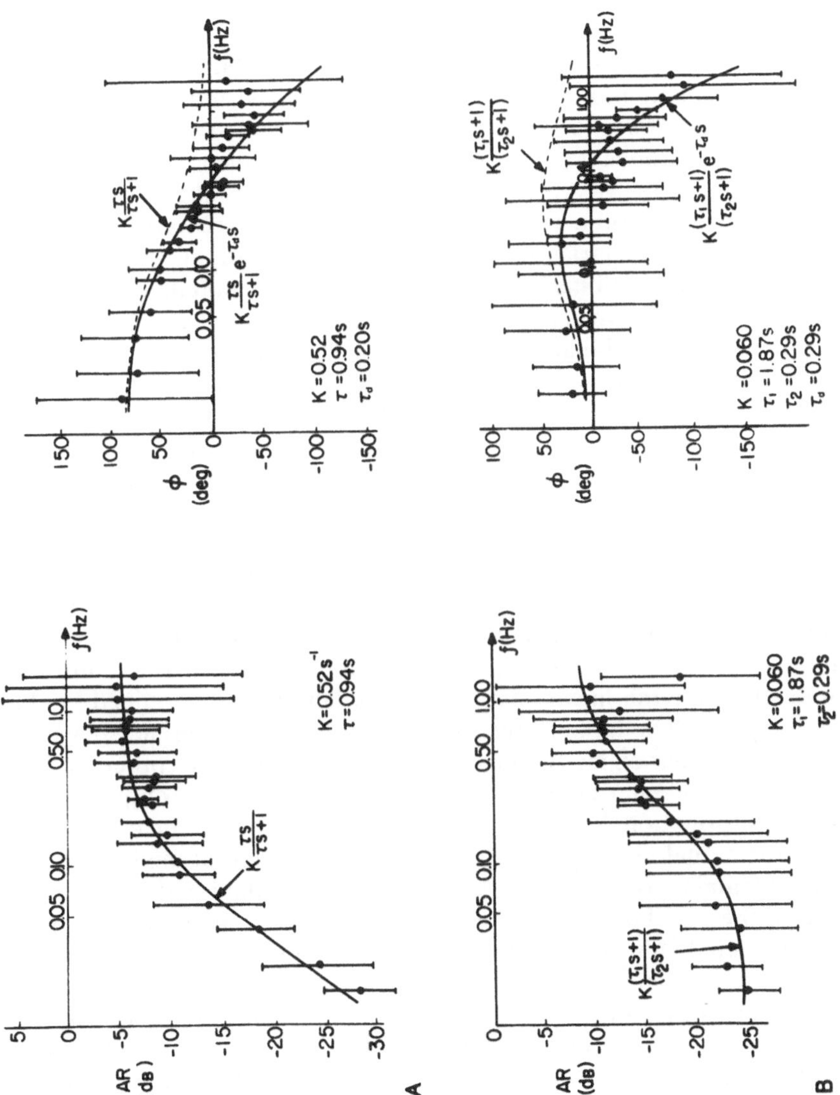

Fig. 24-13. **A** Dual input describing function: vestibular channel. **B** Dual input describing function: visual channel. From Zacharias, G.L. and Young, L.R.: Exp. Brain Res. 41:159–171, 1981.

gain of the system should be as low as −25 dB at low frequencies. To the contrary, we expected the static gain of the vestibular system to be 0 dB since, for saturated motion, one feels that one is rotating at a constant velocity just equal to that of the visual field.

CONFLICT MODEL

A complementary filter was developed in which vestibular information is processed by a simple model of the semicircular canals, and visual information is given no dynamics (36). Although this works well to match the simple step experiments, it does not in any way match the frequency response data from the random input. We were forced to reconsider the situation and see if we had been trying to force a linear model on what appears to be, evidently, a nonlinear system. Zacharias and I revised the concept of the conflict model for resolving visual-vestibular conflicts, in which we maintained the notion of a mechanism which switches between predominantly visual information and predominantly vestibular information. We do not model this as an all or nothing switch, but rather use a model which assumes that the weighting applied to the visual system depends on the conflict between the current semicircular canal signals and the current visual information, as weighted appropriately by an internal model. The soft switch conflict model, shown in Fig. 24-14, assumes that the estimates of the perceived angular velocity about the vertical axis are based upon two kinds of inputs: (1) vestibular information processed from the angular velocity of the head with respect to inertial space, which is high pass filtered by the semicircular canals; and (2) peripheral field visual motion information which is assumed to be sensed without any dynamics. The weighting that is applied to the visual information, in determining the linear sum of the two, is in turn determined by the conflict operation. The input to this conflict block is the difference between the current vestibular signal (output of the semicircular canals processed by brain stem and cerebellar circuits) and the expected vestibular output, based upon visual information. To achieve this expected output of the canals, one processes the actual visual field velocity through an internal model of these vestibular dynamics. By assuming that this internal model is a correct representation of the vestibular dynamics, we hypothesize a signal which represents the expected value of the vestibular response if the visual information (the visual field rotation) were representative of the veridical motion in space. In other words, if ω_{vis} were really self-motion, and if all the visual field motion that one saw was represented by the fact that the visual field is stationary and that one was rotating with respect to it, then this expected angular velocity, ω_{exp}, would agree with the signals coming from the vestibular system, $\hat{\omega}_{ves}$. If they are not in agreement, a conflict is produced. If the absolute value of this conflict exceeds a certain level for a sufficient time, then this conflict signal drives the visual weighting factor (K) down to some lower level. Arbitrarily, we set the magnitude of the conflict which drives K

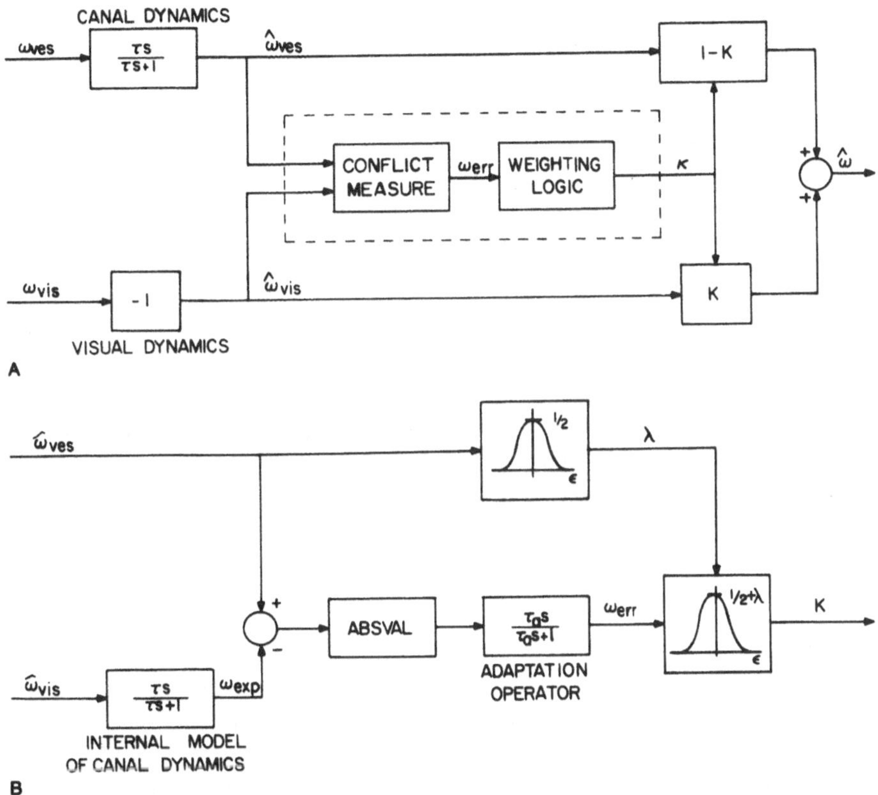

Fig. 24-14. **A** Dual input conflict model. **B** Conflict measure and weighting function. From Zacharias, G.L. and Young, L.R.: Exp. Brain Res. 41:159–171, 1981.

to zero to be equal to the threshold of the semicircular canals (approximately 2.5°/s).

In summary, when the visual information as processed by the internal model is in agreement with the vestibular system, then visual and vestibular information are both utilized. When the visual information is in disagreement for a long enough time to pass the adaptation operator, then the visual field is ignored and vestibular information is adopted. The decision rule states that when there is a conflict, in the short term, one ignores the visual input and believes the vestibular responses, even though there may be no vestibular stimulus. It is important to remember that a null signal from the vestibular system is a perfectly valid signal that indicates constant velocity. As time goes on, the presence of a null vestibular signal becomes less and less meaningful, because of the known drift within the semicircular canals, and therefore visual velocity signals begin to predominate.

The conflict model was tested for the dual input case by applying the same random inputs to the model as had been used in the experiment. The model shows the same behavior, in terms of vestibular and visual gain and phase, as was shown by the experimental results presented in Fig. 24-13. Although the model contained a time constant for the semicircular canals of 12 s, the result of the nonlinearity was to produce an effective time constant of 1 s, which was observed experimentally. It is by no means intuitively clear why the long vestibular time constant should produce this short time constant in the combined stimulus case until one considers the fact that at low frequencies in the presence of vision, the semicircular canal information is discarded. At low frequencies, the conflicts tend to be small, the visual information is given a greater weight, and consequently there is a roll-off of vestibular information at frequencies much higher than the roll-off from the semicircular canals. Similarly, the otherwise unexpected phase relationship is met by the predictions of this model. For the visual system, one finds, again, that the model calculations when treating the same dual input describing function data predict a response observed experimentally both for phase and for gain, including the reduced low frequency gain. It is important to realize that there is some semicircular canal information still being transmitted at low frequencies, and consequently the visual information is not processed with unity gain.

This encouraging result led Zacharias to look further at the predictions of the model for the simple step response experiments discussed earlier. The conflict model responses to a velocity step are shown in Fig. 24-15, first of all, for vestibular stimulation in a subject-fixed field, then for visual information only, and then for the two combined. The vestibular stimulation prediction (Fig. 24-15A) not only shows the initial rise and slow decay with about the right time constants, but interestingly enough predicts a break in this response. The response has a somewhat less steep slope for the first 5 or 6 s, and then a knee, at which point the exponential decay becomes steeper. This characteristic has been noticed in optokinetic nystagmus responses and also in vestibular nucleus unit responses.

When only visual information is present in circularvection (Fig. 24-15B), the model predicts that there will be a gradual build-up to a steady level as is seen experimentally, and it further predicts that the latency to this onset becomes shorter and shorter as the magnitude of the acceleration decreases. This results from the fact that with the small acceleration of the visual field, the conflict is small and quickly washed out, and the visual field information is more readily adopted. For combined visual and vestibular stimulation of rotation in a laboratory-fixed field (Fig. 24-15C), the model predicts that one indeed senses a rapid velocity onset and a continued steady state level. Quantitative agreement is not as good as one might like at very high angular velocities, although at low angular velocities it is reasonably good. The model further predicts that in the steady state there is an adaptation, which has been observed, and that the perception of rotation decays somewhat with time. Finally, the model predicts (Fig. 24-15D) that a simultaneous vestibular stimulus of a velocity ramp ($0.3°/s^2$) and a visual field stimulus ($4°/s$)

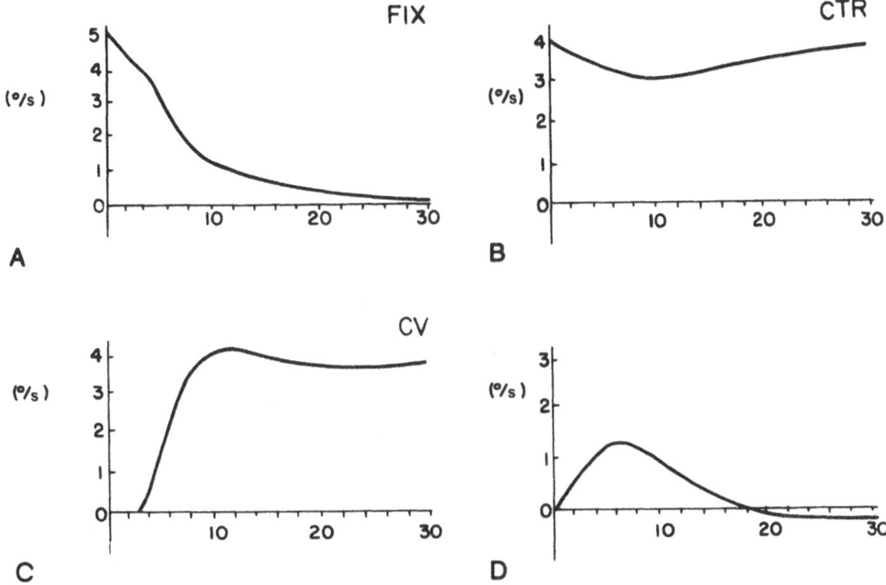

Fig. 24-15. **A** Vestibular step response. **B** Visual step response. **C** Response to visual and vestibular steps. **D** Response to visual step and vestibular ramp. From Zacharias, G.L. and Young, L.R.: Exp. Brain Res. 41:159–171, 1981.

in the same direction will lead to a steady state perception of remaining stationary as was illustrated in Fig. 24-11D.

OPTIMAL FILTERING MODEL

A newer approach to the modeling of visual-vestibular interaction that we have been pursuing is based upon the notion that the central nervous system is performing optimal filtering of the sensory information that it receives—in this case from vision, the semicircular canals, and the otoliths (6,7). If we assume that the central nervous system knows the dynamics of the sensors, that it knows the expected spectrum of the input, and that it also knows the approximate noise superimposed upon the sensory channels, then it can perform an optimal estimation. As a steady state Kalman filter, it produces estimates of angular velocity, acceleration, and orientation with respect to the vertical based upon these multiple inputs. The inputs can include semicircular canal and otolith information, visual information regarding the horizon, visual angular velocity, and proprioceptive and tactile information. The process of building this model consists of two steps. One first must perform a statistical calculation to determine the gains of the steady state Kalman filter, as indicated in Fig. 24-16. This requires assumptions about the sensory dynamics, the noise in the input, and the input spectrum to which the system will be exposed. Once the steady state Kalman filter gains are deter-

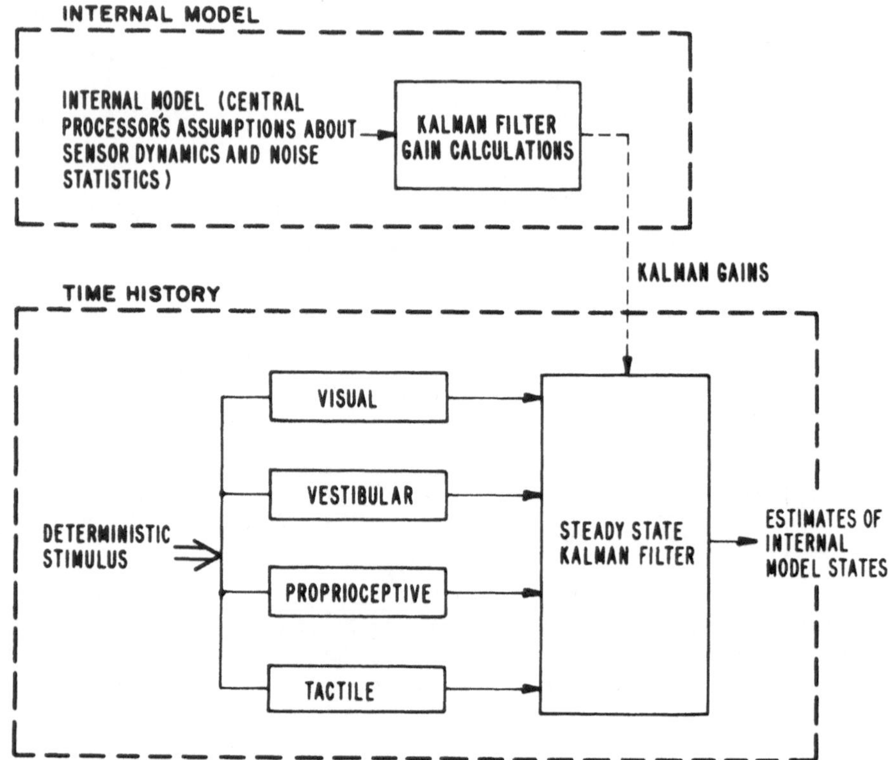

Fig. 24-16. Outline of optimal estimator motion orientation perception model. From Borah, J., et al.: Sensory Mechanism Modelling. Air Force Human Resources Laboratory Report AFHRL-TR-78-83, 1979.

mined, then the actual deterministic stimulus, the response to which is being sought, must be played through a time history program, in which one assumes simple models for the various sensors. Such nonlinearities which are important, such as thresholds or the conflict model nonlinearity, may be included.

Borah et al. (6) adopted the conventional linear semicircular canal models for rotation about a yaw axis and made certain assumptions about the spectrum of the input process that was stimulating them. We also simplified the model to provide no dynamics associated with perception of visual motion, and selected appropriate additive measurement noises for the Kalman filter gain determination. The semicircular canal model that we used for the time history did include a threshold nonlinearity. The model responses for the common experiments of visual-vestibular interaction in yaw are based on the internal model formulation shown in Fig. 24-17. These are the same kinds of body and/or field velocity step experiments which were described earlier. The only information that the model assumed was the canal dynam-

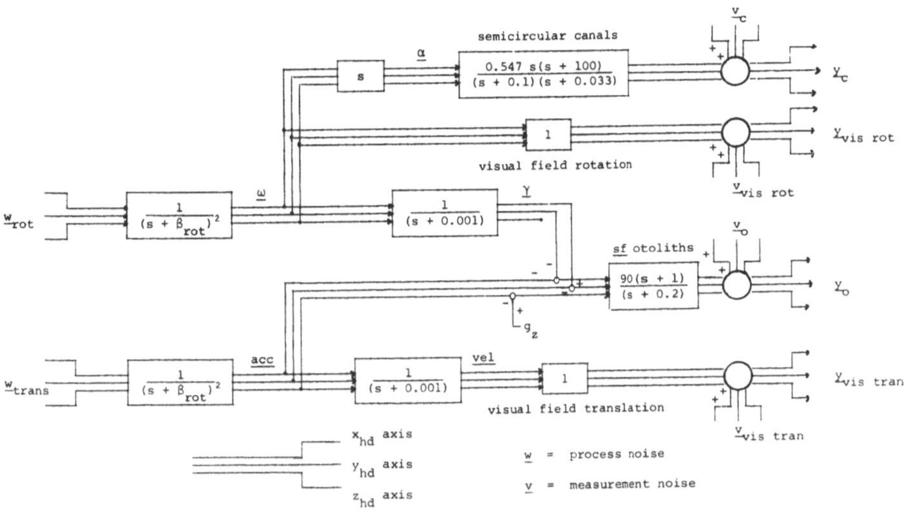

Fig. 24-17. Internal model used in the Kalman filter calculation for the optimal estimator model. Provision for effects of canal-otolith interaction as well as linear and circularvection. From Borah, J., et al.: Sensory Mechanism Modelling. Air Force Human Resources Laboratory Report AFHRL-TR-78-83, 1979.

ics and the estimates about noise on the visual and vestibular channels. The stimulus in each case was a truncated ramp representing an approximate step of just over 0.025 rad/s. First, by way of reference, Fig. 24-18 shows that the canal afferents are expected to decay fairly rapidly, crossing the zero line at about 17 s. When one rotates in the dark in the absence of visual information, the model predicts that the perception of rotation also approximately decays exponentially, with undershoot (RD). The model response to perception lags considerably behind the decay of the semicircular canal signal, although the only sensory information being received is from the canals.

A simple explanation may give an intuitive understanding of why perception lags the sensory input. With a noisy measurement estimate and an internal model of the process, the Kalman filter will not totally update based on the sensory information. Rather than rely entirely on noisy current measurements, it will process its own internal model to what it feels the updated state vector should be, weighing the canal information as it does. It is quite reasonable, therefore, to expect that this perception decay should take longer than does the presumed canal information, and this, of course, is observed in experiments. If one rotates in the light (RL), one gets a sensation of rotation that quickly rises up to nearly the correct level; the model predicts a subsequent slight decay to a steady level. If one applies a step velocity of the rotating visual field, the model response is almost an exponential rise up to some steady level (CV), again agreeing qualitatively with what

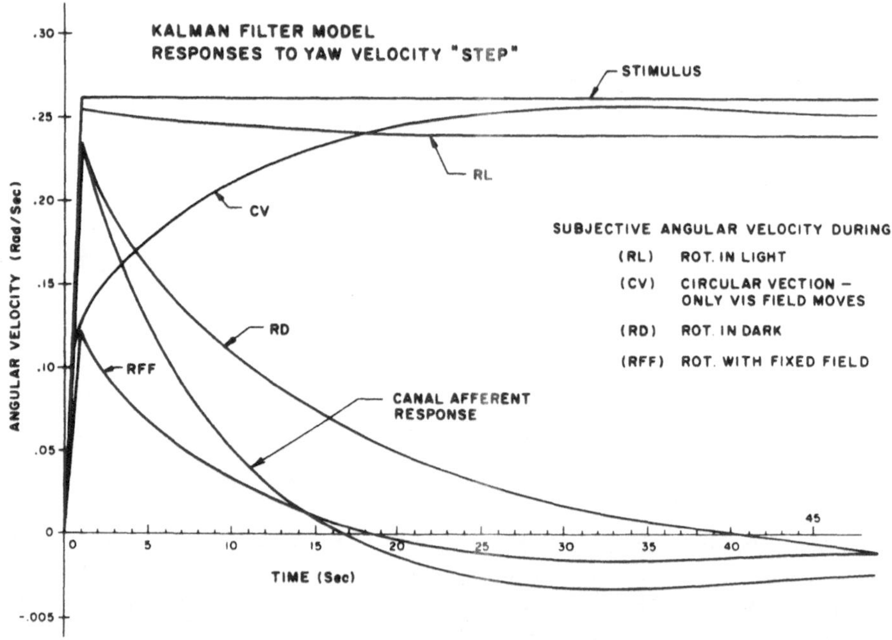

Fig. 24-18. Model response to combined platform and wide visual field yaw motion. From Borah, J., et al.: Sensory Mechanism Modelling. Air Force Human Resources Laboratory Report AFHRL-TR-78-83, 1979.

is found in practice. It does not, however, for this formulation show any initial latency. The initial latency is provided by a later modification in which the conflict model is fit into the time history generation (7). When that is done, circularvection will not appear until some time after the initiation of the stripe motion, just as was predicted by Zacharias's conflict model. Finally, for rotation with fixed field (RFF), this model predicts that for a large velocity step or acceleration pulse the perceived peak velocity would be smaller than that in the dark and decay more rapidly than in the dark. For very small accelerations, experiments do not agree with this prediction, as the oculogyral illusion heightens the perception of motion (21). For large steps, however, the linear model's prediction is borne out.

This model has also been tested for the case of otolith information, and in particular for motion in the longitudinal axis. For this case, the otoliths are modeled as simple low pass filters with a break frequency of 0.02 rad/s and a threshold with a lead processing term. When this otolith model is added to the previous semicircular canal model for inertial rotation and when one also assumes a model for perception of visual translation with no dynamics and appropriate noise on the visual channel and vestibular channels, one produces predictions for a number of commonly observed phenomena.

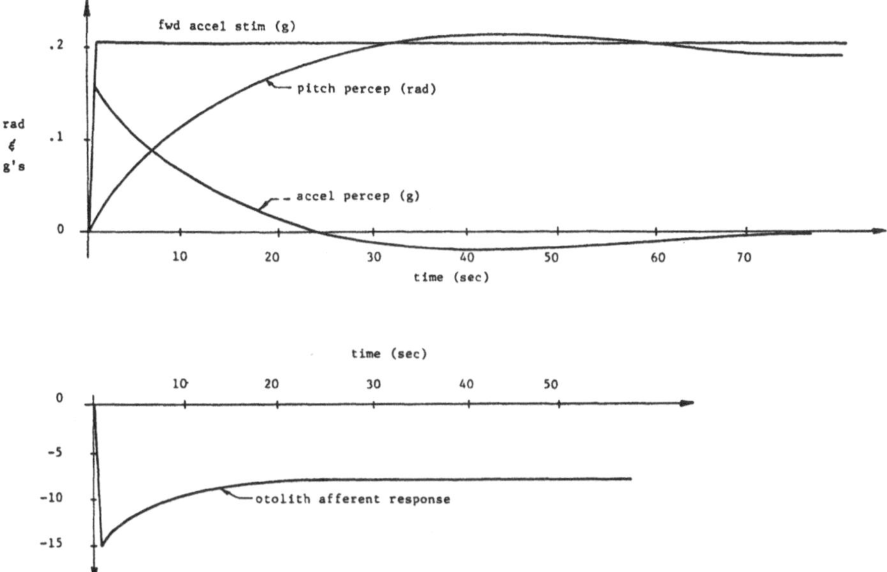

Fig. 24-19. Model response to forward acceleration "in the dark" (vestibular cues alone). From Borah, J., et al.: Sensory Mechanism Modelling. Air Force Human Resources Laboratory Report AFHRL-TR-78-83, 1979.

For example, consider the situation of constant linear acceleration. For constant forward linear acceleration in the dark, illustrated in Fig. 24-19, the actual gravitoinertial vector tips backward. The perception of motion initially is the veridical one, i.e., one initially feels forward acceleration. With time, however, the perception is that the acceleration decays and is replaced by a sensation of pitch back. If this is done in the light, so that the visual information signals the lack of rotation, then, in fact, the perception of acceleration remains accurate, and there is an indication that one should find a perception of pitch forward. (This is an interesting prediction to test experimentally, which, to my knowledge, has not yet been tried.)

CONCLUSION: TIE TO THE UNDERLYING NEUROPHYSIOLOGY

In closing, an attempt is made to tie some of the work on perceptual visual-vestibular interactions to work on visual-vestibular interactions at the neural level. Simply stated, we apply the same rotations and visual field stimuli in animal experiments as we do during the human experiments, and we measure the compensatory eye movements of the vestibular or optokinetic nystagmus in each case. If the eye movement patterns agree, then the perceptual qualities we find in the human may be related to the neural activity recorded in the animal experiments. Henn et al. (20) began rotating rhesus

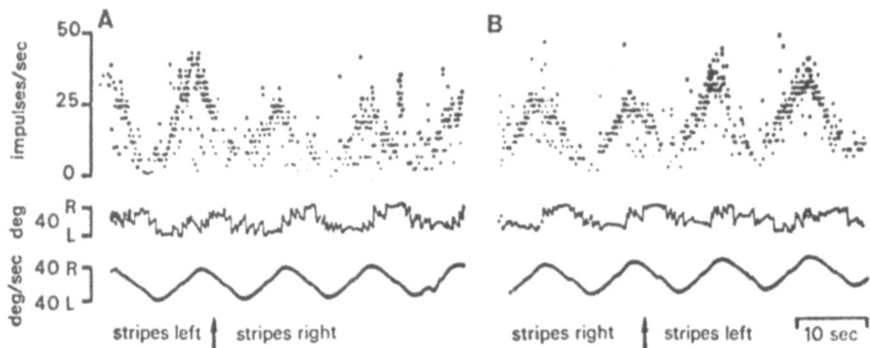

Fig. 24-20. Vestibular nucleus unit response from a monkey being turned in the light with sinusoidally modulated velocity, in the presence of a moving visual field. At the beginning of **A**, stripes in the visual field are being turned to the left; at *arrow*, direction of stripe motion is changed. The opposite applies to **B**. This unit is activated during movements to the right. When stripes are turned to the left, during monkey rotation right, unit activity is enhanced. Similarly, unit activity is depressed when stripes move to the right. From Henn, V.S., et al.: Brain Res. 71:146, 1974.

monkeys under such stimuli while recording single unit activity from the brain. Figure 24-20 shows a recording from the vestibular nucleus. During rotation of the chair to the right, the firing rate of the unit increases. This is a type II unit of the vestibular nucleus in which ipsilateral rotation decreases firing rate. At different frequencies, one produces varying phase relations of the nuclear activity to the acceleration stimulus. Changing the stripe direction changes both the average firing rate and the vestibular modulation in a complex nonlinear manner. If, for the same unit, the animal is kept stationary and one merely changes the pattern of optokinetic stimulation so that the moving stripes move to the right, there results actually a small build-up of average firing rate, along with a change in the optokinetic nystagmus, indicating that this unit activity is mediated by visual stimulation. Now, motion of the stripes to the right is consistent, in case of humans, with circularvection to the left. Actual motion to the left is the excitatory direction for this unit, so that these results are consistent. Considerably more quantitative work has since been carried out by Henn and his colleagues (10,11,28). Similar results are, in general, found in the goldfish (1,16) and cat (22).

Similar studies have been performed by Daunton et al. (27, *this volume*; ref. 13) in the cat for otolith-innervated units in the vestibular nuclei. They find similar relationships for linearvection or linearly induced motion.

It is indeed encouraging to find that for one multisensory paradigm we are able to tie together psychophysical results on motion perception, postural reactions, eye movements, and single unit activity in the periphery, brain stem, cerebellum, and cortex. Visual-vestibular interaction stands as a leading paradigm for the study of intersensory conflict and integration.

References

1. Allum, J.H.J., Graf, W., Dichgans, J., and Schmidt, C.L.: Visual-vestibular interactions in the vestibular nuclei of the goldfish. Exp. Brain Res. 26:463, 1976.
2. Asch, S.E. and Witkin, H.A.: Studies in space orientation. I. Perception of the upright with displaced visual fields. J. Exp. Psychol. 38:325, 1948.
3. Asch, S.E. and Witkin, H.A.: Studies in space orientation. II. Perception of the upright with displaced visual fields and with body tilted. J. Exp. Psychol. 38:455, 1948.
4. Berthoz, A., Pavard, B., and Young, L.R.: Perception of linear horizontal self-motion induced by peripheral vision (linearvection): basic characteristics and visual-vestibular interactions. Exp. Brain Res. 23:471, 1975.
5. Blanks, R.H.I., Curthoys, I.S., and Markham, C.H.: Planar relationships of the semicircular canals in man. Acta Otolaryngol. 80:185, 1975.
6. Borah, J., Young, L.R., and Curry, R.E.: Optimal estimator model for human orientation. IEEE Trans Systems Man-Cybernetics, 1980.
7. Borah, J., Young, L.R., and Curry, R.E.: Sensory Mechanism Modelling. Air Force Human Resources Laboratory Report AFHRL-TR-78-83, 1979.
8. Brandt, T., Dichgans, J., and Büchele, W.: Motion habituation: Inverted self-motion perception and optokinetic after-nystagmus. Exp. Brain Res. 21:337, 1974.
9. Brandt, T., Dichgans, J., and Koenig, E.: Differential effects of central versus peripheral vision on egocentric motion perception. Exp. Brain Res. 16:451, 1973.
10. Büttner, U. and Henn, V.: Thalamic unit activity in the alert monkey during natural vestibular stimulation. Brain Res. 103:127, 1976.
11. Büttner, U., Henn, V., and Oswald, H.P.: Vestibular related neuronal activity in the thalamus of the alert monkey during stimulation in the dark. Exp. Brain Res. 30:435, 1977.
12. Chu, W.H.N.: Dynamic Response of Human Linearvection. S.M. Thesis. Cambridge, Mass., Department of Aeronautics and Astronautics, M.I.T., 1976.
13. Daunton, N.D. and Thomsen, D.: Visual modulation of otolith-dependent units in cat vestibular nucleus. Exp. Brain Res. 37:173, 1979.
14. Dichgans, J. and Brandt, T.: Visual vestibular interaction: Effects on motor perception and postural control. In Jung, R., Autrun, H., Lowenstein, W.R., Mackey, D.M., and Teuber, H.L. (eds.): Handbook of Sensory Physiology, Vol. VII. Perception. Berlin, Springer-Verlag, 1978.
15. Dichgans, J. and Brandt, T.: Pseudocoriolis effects and motion sickness induced by moving visual stimuli. Acta Otolaryngol. 76:339, 1972.
16. Dichgans, J., Schmidt, C.L., and Graf, W.: Visual input improves the speedometer function of the vestibular nuclei in the goldfish. Exp. Brain Res. 18:319, 1973.
17. Dichgans, J., Held, R., Young, L.R., and Brandt, T.: Moving visual scenes influence the apparent direction of gravity. Science 178:1217, 1972.
18. Fischer, M.H. and Kornmüller, A.E.: Optokinetisch ausgeloste Bewegungswahrnehmungen und optokinetischer Nystagmus. J. Psychol. Neurol. 41:273, 1930.
19. Henn, V.S. and Young, L.R.: Ernst Mach on the vestibular organ 100 years ago. Ann. Otorhinolaryngol. 37:138, 1975.

20. Henn, V.S., Young, L.R., and Finley, C.: Vestibular nucleus units in alert monkeys are also influenced by moving visual scenes. Brain Res. 71:146, 1974.
21. Huang, J.K. and Young, L.R.: Sensation of rotation about a vertical axis with a fixed visual field in different illuminations and in the dark. Exp. Brain Res. 41:172–183, 1981.
22. Keller, E.L. and Precht, W.: Persistence of visual response in vestibular nucleus neurons in cerebellectomized cat. Exp. Brain Res. 32:591, 1978.
23. Lestienne, F., Soechtung, J., and Berthoz, A.: Postural readjustment induced by linear motion of visual scenes. Exp. Brain Res. 28:363, 1977.
24. Mach, E.: Grundlinien der Lehre von den Bewegungsempfindungen. Leipzig, Englemann, 1875.
25. Malcolm, R. and Melvill Jones, G.: Erroneous perception of vertical motion by humans seated in the upright position. Acta Otolaryngol. 77:274, 1974.
26. Melvill Jones, G. and Young, L.R.: Subjective detection of vertical accelerations: A velocity dependent response? Acta Otolaryngol. 85:45, 1978.
27. Spoendlin, H.: Ultrastructure of the vestibular sense organ. In Wolfson, R.J. (ed.): The Vestibular System and Its Diseases. Philadelphia, University of Pennsylvania Press, 1966.
28. Waespe, W. and Henn, V.: Neuronal activity in the vestibular and optokinetic stimulation. Exp. Brain Res. 27:523, 1977.
29. Witkin, H.A.: The perception of the upright. Sci. Am. 182:50, 1959.
30. Young, L.R.: Developments in Modelling Visual-Vestibular Interactions. AMRL-TR-71-14. Wright-Patterson Air Force Base, Dayton, Ohio, 1971.
31. Young, L.R.: On visual vestibular interaction. In: Fifth Symposium on the Role of the Vestibular Organs in Space Exploration. NASA SP-314, 1970.
32. Young, L.R., Dichgans, J., Murphy, R., and Brandt, T.: Interaction of optokinetic and vestibular stimuli in motion perception. Acta Otolaryngol. 76:24, 1973.
33. Young, L.R. and Oman, C.M.: Influence of head position on visually induced motion effects in three axes of rotation. Proceedings of the Tenth Annual Conference on Manual Control, Wright-Patterson Air Force Base, Dayton, Ohio, 1974.
34. Young, L.R., Oman, C.M., and Dichgans, J.M.: Influence of head orientation on visually induced pitch and roll sensation. Aviat. Space Environ. Med. 24:264, 1975.
35. Zacharias, G.L.: Motion Sensation Dependence on Visual and Vestibular Cues. Ph.D. Thesis. Department of Aeronautics and Astronautics, M.I.T., 1977.
36. Zacharias, G.L. and Young, L.R.: Influence of combined visual and vestibular cues on human perception and control of horizontal rotation, Exp. Brain Res. 41:159–171, 1981.

Discussion

GUALTIEROTTI: I'd like to make three points about this very interesting presentation. The first point is that although it has been mentioned in the latest model, it must be the muscle receptors that are one of the major contributors to the entire system. This is especially true for the neck muscles in the lower animals, from the bird down in which the nystagmus is a neck muscle nystagmus, as the entire head presents the same movement that the eye does in mammals. Even if you fix the head, you still have contraction of the muscles during rotation and during visual stimulation, and tendon receptors will respond, contributing to the sensation. Electromyographic data will help in considering this factor. My second observation is that with all these experiments on humans, we have to consider that training has to be emphasized, namely, what happens when the man is first presented with this kind of set-up, and then gets accustomed to it. In any physiologic experimentation, prepatterning is very important, as for instance in experiments on, say, respiratory physiology. If the subject is trained to run on a treadmill you can prepattern his response. We see responses preceding the onset of work. So in this case there is a great difference if the man is accustomed to a situation or if he is not accustomed to a situation. Third: all these hypotheses of conflicting sensory responses do not take into account that there might be some basic change in the cell's behavior or physical situation and function of the organ. I am especially referring to the vestibule. We have some evidence that there is some basic change in the vestibular activity, and you can't assume a priori that you have a normal responding cell's vestibular response.

YOUNG: Concerning the muscular contribution, there's no question that the sensory influence from the muscle spindle is important, particularly in its influence on the central system. Active versus passive motion has a very important effect on the perception of motion. The higher cortical processes do seem to distinguish between the sensory signals stemming from active and passive movements. With regard to training, this is both an advantage and a disadvantage in working with humans. If they know too much about the experiment, you may be measuring an unrelated event. In this regard, one of the points of the otolith-spinal test is to contrast predictable responses to behavior with the unexpected release. Finally, let me turn to the question of accounting for basic changes in the vestibular sensor. We're very much aware of the possibility of end organ changes, and in fact one of the goals of the test we have in Spacelab is to try to distinguish as best we can, within the limits of working on a human, between peripheral and central changes. The lowest level response we can measure is ocular counterrolling. Ocular counterrolling is as directly related to otolith stimulation as we can get. We measure higher level functions, fixation and perception. We would be very interested to find a different course of adaptation of ocular counterrolling than of perceptual environmental adaptation.

LOWENSTEIN: This is just a question of theoretical importance. Concerning your switch mechanism—remember the slide where you gave the switches? Did you observe any periodicity in the switching, and if so, what was the time in this periodicity? I am thinking of the reversal phenomenon. If you look at a reversing picture, concave, convex, you have about a 10-s switch. Do you find something similar here?

YOUNG: We didn't look for periodicity in the switching. The switching that we did look at was all deterministic, and keyed to the stimulus. What would be analogous to the reversal case would be the phenomenon I showed in Fig. 24-4 of state changes, sudden drop-outs of vection, and then almost unexplained returns. I just don't have any numbers in mind about the periods between them.

LAU: In manual control, the human operator transfer function is very much dependent on vehicle dynamics. In other words, in the cross-over model, the combined human operator and vehicle dynamic response seems to be approximated by $1/s$. There are three questions. Number one, what is the type of vehicle dynamics that you use in the control experiments? My second question is, do you think that manual control with motion perception as input can also use the cross-over model? Third, what is your confidence in the estimate of the transfer function of the vestibular and visual systems because of their adaptations from higher processing?

YOUNG: Thank you. You are quite correct. Vehicle dynamics are well known to influence the manual control task. We did, as I thought I indicated, determine the manual control transfer function experimentally. It did fit well with the cross-over model. The vehicle dynamics were compensated to be very close to a pure integrator. I can't remember what the roll-off frequency was. As far as the effects of motion cues on manual control, this is something that has been our bread and butter for about 12 years. There are a set of models in which we indicate the effects of semicircular canal and otolith stimulation on the cross-over model for manual control. In a nutshell, the vestibular cues result in a decreasing phase lag after crossover frequency and an increasing bandwidth. I would refer you to Shirley's and Curry's papers on this.* As far as the standard deviation across our subjects, that was indicated on the diagram. I can't give you from memory confidence intervals on the parameters of the transfer function.

OMAN: I'd just like to make an observation in response to Dr. Gualtierotti's question which may clarify this discussion with respect to the utility of the optimal estimator model. I think it's important that we all realize that this is an optimal estimator model and not an optimal control model. The internal model that you saw represented really contained only a model for the sensory dynamics and does not take into account the subject's knowledge of the effects of his own control. Now, when you get to consideration of the more complicated question which was presented in Larry's last slide, we're talking about a more complicated problem in which the system has to have internalized models of the body's own dynamics, and the vehicle's own dynamics in response to control inputs. Let's put it this way: the estimator model applies for perhaps the passive motion situation in particular. I think where we're at now is trying to extend the estimation concept to the control context, and when we do so I think we'll have perhaps more to say about closed-loop situations.

GUALTIEROTTI: I think that this is a fundamental problem because we are dealing here with a very important sector of research, namely, the significance of studying basic elements using integrated separation between the different systems. So, I think this discussion is quite relevant. I have been referring to the fact, for instance, that man, even in simple experiments like in respiration, will increase his oxygen consumption before starting to work. We are dealing here with trying to assess a system analysis, and we have to take into account all the elements. I remember, by the way, my fourth observation: what about vibration? This sled is going to be a nest of vibration. I know that. We have reason to think, for instance,

*Shirley, R.S. and Young, L.R. Motion cues in man-vehicle control. IEEE Trans. on Man-Machine Systems. *mm*-9:4, 1968 and Curry, R.E., Hoffman, W.C., and Young, L.R. Pilot Modelling for Manned Simulation. AFFDL-TR-76-124, 1976.

that some of the most important changes that we observed during the experiment of the frog in space, namely, the change of mode, might have been partly due to chance vibration that we had in the package. We are dealing here with an experiment that has additional unwanted elements, and we are dealing also with an integrated organism, the analysis of which is quite difficult to make.

YOUNG: (The question of the vibration on the space sled.) We are indeed very much concerned about it. There is, unfortunately, very little we can do about fully controlling the environment in Spacelab, however.

BÜTTNER: You pointed out that there is always a close similarity in duration and strength of nystagmus and circularvection, but that these two parameters can also be separated, the nystagmus and circularvection going in opposite directions. And this even becomes obvious during poor vestibular stimulation. Nystagmus velocity and motion sensation can also dissociate during vestibular stimulation. Groen [Laryngoscope 67:894, 1957] showed that time constants for nystagmus are usually longer as compared with those for motion sensation. These findings become important when one tries to compare motion sensation, nystagmus, and behavior of single neurons recorded in the alert monkey. Time constants of neurons in the vestibular nuclei in response to angular acceleration are as long as time constants of simultaneously recorded nystagmus. So, one always should keep the differences between psychophysics (motion sensation, circularvection) and nystagmus in mind, when one compares these data with single neuron recordings in alert animals.

YOUNG: I'm in complete agreement with you.

HONRUBIA: One question concerns the use of random signals to study the visual-vestibular interaction. Your view has enlightened me on the possibility that when you are doing experiments, you are testing the ability of this smooth pursuit system to maintain fixation in space. And, you are using a random signal which is not predictable. So that some of the failure of a linear system model to predict interaction of the visual and the vestibular system may be due to the fact that you are not using the adequate test for the system. We found that using random signals for smooth pursuit gives very poor results. Then, another question has to do with some experiments that Cliff Lau has done recently. When you are testing vestibular interaction you really cannot use linear systems analysis for the predictions of the test. When you are moving your head in the environment, the visual stimulus is actually the difference between the angle of the head to the space and the angle of the eye's compensatory movement to the space. The important issue is that the gain of these systems is not linear. For example, the gain of the optokinetic system is much greater for a small stimulus than for a large stimulus. During visual-vestibular interaction, you are minimizing the magnitude of the visual stimulus, in effect making this reflex operate in a region of greater gain. Under these conditions the combined input of the vestibuloocular reflex and visual system results in perfect matching or compensation. Cliff Lau's model prediction in this symposium will not be adequate if the gains are those of the larger stimulus measured independently. It was a very interesting situation that he observed, this strategy of the CNS in designing reflexes with nonlinear characteristics to result in a linearized performance.

YOUNG: Concerning the matter of the input to the eye movement system during head motions, it is clear that one has both retinal slip and presumably reconstructed eye velocity. The latter is provided, we believe, through the corollary discharge or efferent copy. We do have to take into account the movement of the target with respect to the head. As to the first questions of predictability, I think it's an error to think that a single sinusoidal stimulation is the simplest test to apply to the system. In fact, it's a very complicated one because it is predictable. It calls upon a higher level function to predict what the future target motion will be. It is more meaningful physiologically to test with random input than it is with predictable input.

FREE PAPERS

25

Vestibular Neuron Response Alteration with Repeated Angular Acceleration

P. KILENY, B.F. McCABE,
J.H. RYU, AND P.J. ABBAS

The purpose of this study was to investigate the effects of repeated and consecutive accelerations and decelerations on vestibular nuclei neurons in anesthetized and unanesthetized intact cats.

A progressive decrement of behavioral or neural response occurring with repeated presentations of an effective stimulus in the presence of an unaltered capacity to respond is defined as the habituation of the response. Under controlled conditions (allowing no interference with the transmission of the stimulus), the reoccurrence of the stimulus appears to be the sole extrinsic factor responsible for the diminution of the response. As in other plastic behavioral phenomena (i.e., classic conditioning), this process establishes temporarily a new stimulus-response relationship. Habituation is considered to be an elementary form of behavioral plasticity (33), a fundamental adaptive mechanism (27), or the simplest form of learning (18). Responses subject to habituation range from behavioral to single unit activity. All are bound together by a series of common characteristics which set habituation apart from other decremental phenomena (e.g., adaptation or end organ fatigue) such as specificity to modality and to spatial and temporal configuration of the input (33).

With repeated vestibular end organ stimulation there occurs a modification of the vestibuloocular reflex, the most common effect being a progressive decline in the magnitude of the various nystagmus parameters (6,8,12,14,20,23). This phenomenon occurs with both angular acceleration and caloric stimuli, and the final outcome is affected by extravestibular factors such as visual input and degree of arousal (6,8).

In spite of growing evidence of habituation of neuronal responses in sensory and motor systems (5,17,21,26,30), there have been few attempts to investigate this phenomenon in the vestibuloocular pathway (9,16).

PROCEDURES

Experimental Animals and Their Preparation

A total of 38 cats weighing between 2.0 and 4.0 kg were used in the present study. Preoperative and operative procedures were identical for all experimental animals, and were described elsewhere (29). In 19 cats, unit recording was carried out under general sodium-pentobarbital anesthesia. In a second group of 19, unit recording was carried out unanesthetized under pancuronium bromide (Pavulon Organon)—induced paralysis or as *encéphale isolé* preparations. Both preparations necessitated artificial ventilation. Recording was performed with tungsten or glass pipette electrodes driven through an opening in the dura mater through the intact cerebellum to the approximate region of the vestibular nuclei following stereotactic coordinates (3). After completion of the craniotomy in the unanesthetized group an L-shaped brass plate was mounted on the frontal bone and the scalp sutured back. The cats were allowed to recover from the operation for 2 to 7 days. On the day of the experiment a tracheotomy was performed under ether anesthesia and the cats were mounted in the stereotactic frame. A bar mounted on the stereotactic frame was fastened to the frontal bone-mounted L-shaped plate after the cat was already in the stereotactic position. This allowed the removal of the eye bars' tooth and sometimes the ear bars without altering the cat's position. The wounds were periodically infiltrated with 1% lidocaine hydrochloride.

The stimuli consisted of consecutive constant horizontal clockwise angular accelerations followed by equal clockwise decelerations (4–8°/s^2). Initial velocity was always zero. The peak velocity at the end of an acceleration was 90 or 100°/s.

Extracellular unit activity and induced nystagmus were evaluated for progressive response alterations occurring with stimulus repetition. In order to determine the stability of neurons, spontaneous resting discharge was recorded for 1024 s. This was followed by the experimental paradigm consisting of up to 70 repeated and consecutive accelerations and decelerations. If a decline in neuronal response was detected, dishabituation was usually attempted by dropping out one or two consecutive stimuli from the sequence or by substituting the standard acceleration-deceleration stimulus by a novel one, such as calorics. Unit response to a given stimulus was evaluated by integrating total discharge above resting during the stimulation period of inter-

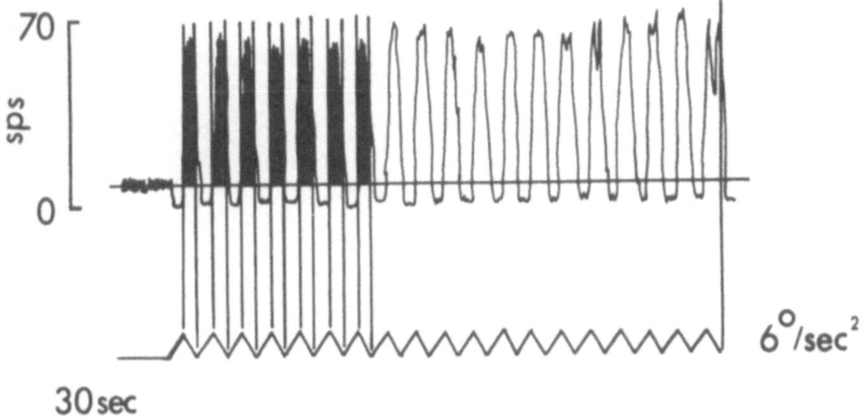

Fig. 25-1. Neuronal response measurement.

est. The perimeter of the area circumscribed by the discharge rate histogram time locked to stimulus period (shaded areas in Fig. 25-1) was traced with the probe of a sonic digitizer and an area corresponding to response magnitude was obtained. Nystagmus was retrieved as the electrooculographic recording of eye displacement, and total slow phase displacement was measured manually from these records for every acceleration and deceleration, as shown in Fig. 25-2.

RESULTS

Units in Anesthetized Cats

Thirty-five vestibular nuclei units recorded stereotactically from 19 anesthetized cats were examined for response modification. An examination of the series of discharge rate histograms elicited by the repeated and consecutive acceleration-deceleration stimuli revealed 17 units with no detectable response modification. The rest were divided into two apparently distinct patterns of response modification: either the response declined progressively with stimulus repetition as shown in Fig. 25-3 (14 units), or there was an apparent initial increase in response magnitude followed by a progressive decline, as shown in Fig. 25-4 (four units). In these and some of the following figures the upper trace is the frequency discharge histogram elicited by repeated and consecutive accelerations and decelerations. Stimulus velocity waveforms are illustrated in the middle trace. The bottom graph plots percent response (with respect to the second response) vs. stimulus number.

Dishabituation of the habituated neuronal response was attempted by a break in stimulation equivalent in duration to one to three stimuli in 14 neurons. In seven of these, dishabituation was attempted a second time with a second break. In four neurons dishabituation was attempted by interposing

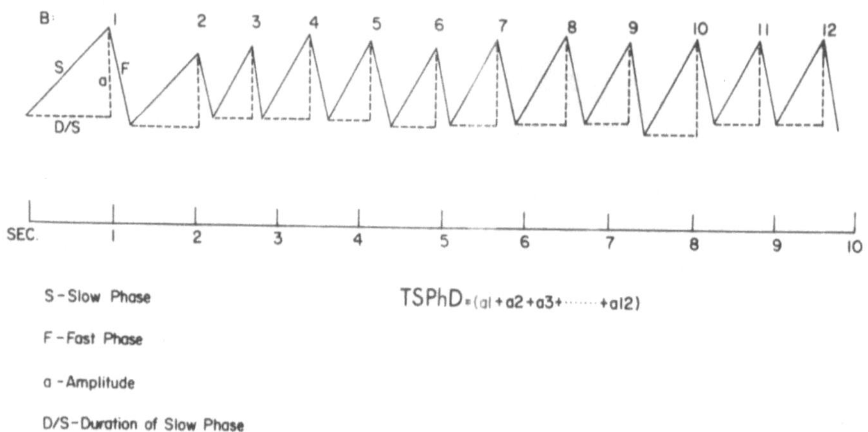

Fig. 25-2. Nystagmus parameters, and total slow phase displacement (TSPhD) calculation.

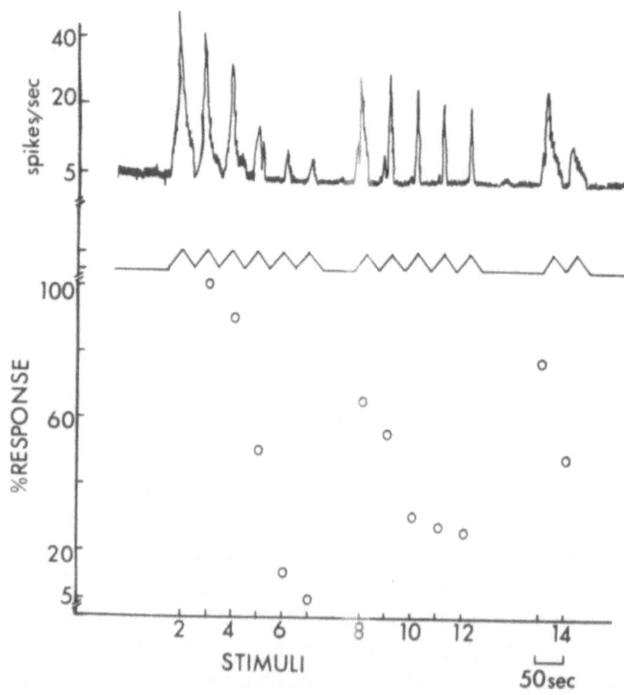

Fig. 25-3. Habituation, dishabituation, and rehabituation of an anesthetized cat unit. Successive discharge rate histograms (**upper trace**), angular velocity waveforms (**second trace**), and percent response change (**lower graph**).

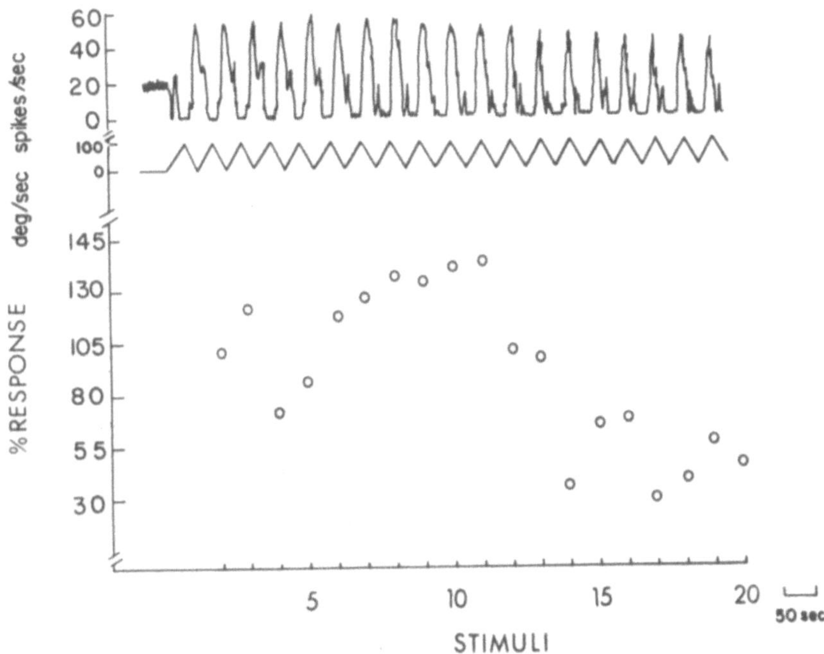

Fig. 25-4. Anesthetized cat unit manifesting an apparent initial enhancement of response followed by a progressive decline. Successive discharge rate histograms, stimulus waveform, and percent response change.

a novel stimulus, calorics. The effect of interrupting the stimulus train is illustrated in Fig. 25-3. A short break equivalent in duration to one standard acceleration-deceleration stimulus was effective twice in partially restoring the response. After six repetitions of the acceleration-deceleration stimulus, the response declined to about 10% of the control. Following this, one stimulus was dropped out of the sequence. The second stimulus following the short break elicited a response that was 50% of the control. With further stimulation the response dropped down to 25%. Another break equivalent in duration to two stimuli again restored the response partially.

The effect of ice water calorics on a habituated neuronal response is illustrated in Fig. 25-5. Each discharge rate histogram shown in this figure represents the sum of eight responses of a unit to consecutive $8°/s^2$ acceleration-deceleration stimuli. The angular velocity waveform of one acceleration-deceleration is shown in Fig. 25-5E. Figure 25-5A–C are the first, second, and fifth groups of eight responses, respectively. There is an obvious decline from Fig. 25-5A to Fig. 25-5C in both the area circumscribed by the discharge rate histogram and its amplitude. The discharge rate histogram shown in Fig. 25-5D represents the sum of eight responses obtained following the delivery of 10 cc ice water in the left ear of the anesthetized cat: the recovery is evident.

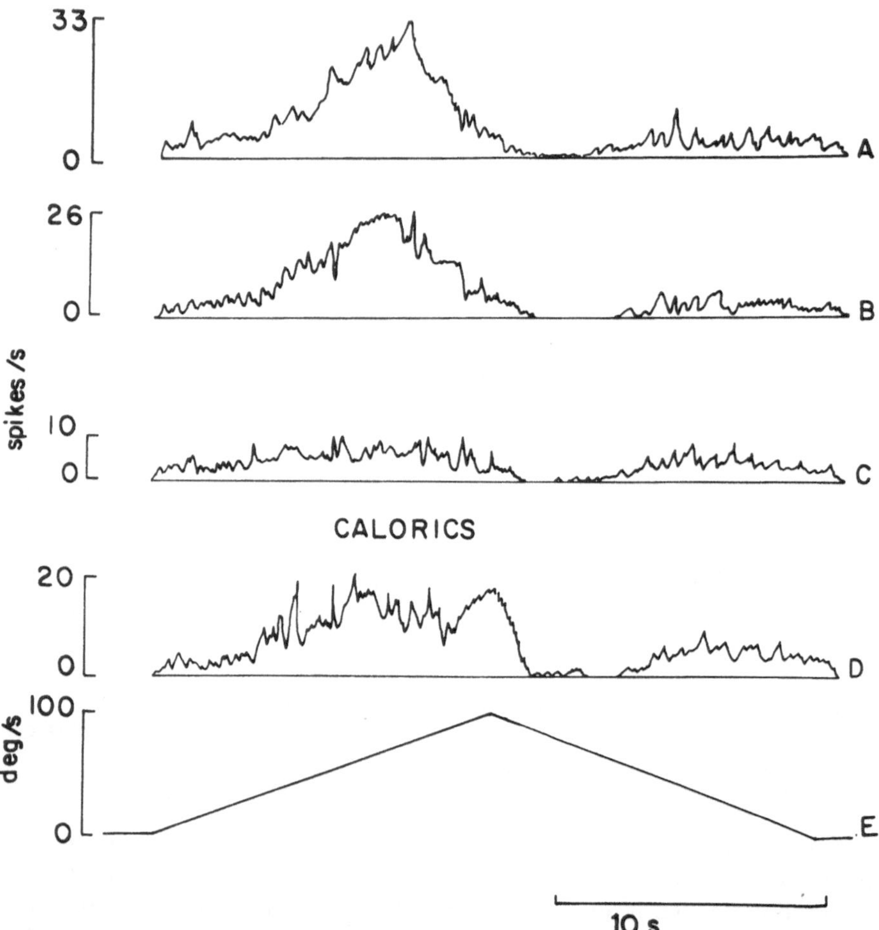

Fig. 25-5. The effect of ice water calorics on a depressed neuronal response following stimulus repetition. **A** Averaged discharge rate histogram elicited by stimuli 1–8. **B** Averaged discharge rate histogram elicited by stimuli 9–16. **C** Averaged discharge rate histogram elicited by stimuli 32–40. **D** Averaged discharge rate histogram elicited by stimuli 41–48 following ice water calorics. **E** Angular velocity pattern of stimulus.

Units in Conscious Cats

Forty-two units recorded from 19 unanesthetized cats were examined for response modification. Nystagmus was recorded simultaneously with eight units from five *encéphale isolé* preparations. In 20 units the paradigm was repeated twice.

Inspection of the data suggested three apparently distinct patterns of response modification. The unit illustrated in Fig. 25-6 shows a typical progressive and orderly response decline with stimulus repetition. This pattern was on a slower time scale than its equivalent found in anesthetized

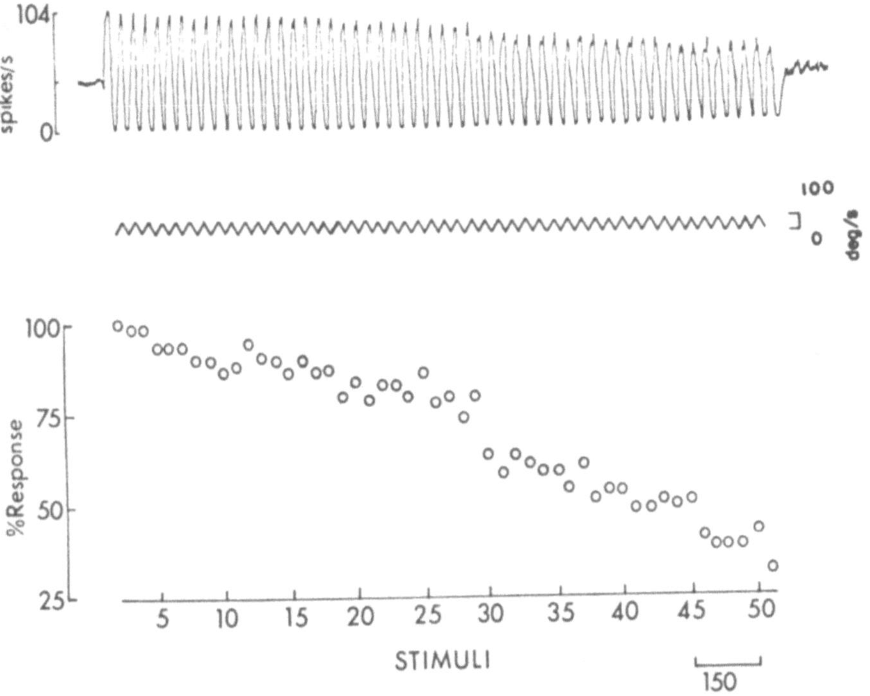

Fig. 25-6. Progressive response decline manifested by a conscious cat unit with stimulus repetition. **Upper trace** Successive discharge rate histograms. **Second trace** Angular velocity waveform. **Lower graph** Percent response change.

cats. Figure 25-7 illustrates responses from a typical unit showing an initial gradual increase in its evoked activity followed by a progressive decline. The third pattern is illustrated in Fig. 25-8. This unit did not alter its activity substantially with the first 30 stimuli. During the last 11 stimuli unit activity declined substantially in this case.

In eight units nystagmus was recorded simultaneously with unit activity for time segments of various lengths. Figure 25-9 illustrates a typical example of simultaneous changes of neuronal and nystagmus response expressed in percent response. The two curves appear similar, although the percentage change in activity was larger for the unit (upper portion of the graph) than for nystagmus. Both nystagmus and unit response increased initially, and then decreased back to about control level.

In order to determine the statistical significance of the observed response alterations and to provide an objective means for grouping the various response modification patterns and describe them mathematically, the data were submitted to a number of statistical analysis procedures.

Polynomial regression or trend analysis determined the general trend. Unit response modification patterns were divided by trend analysis into

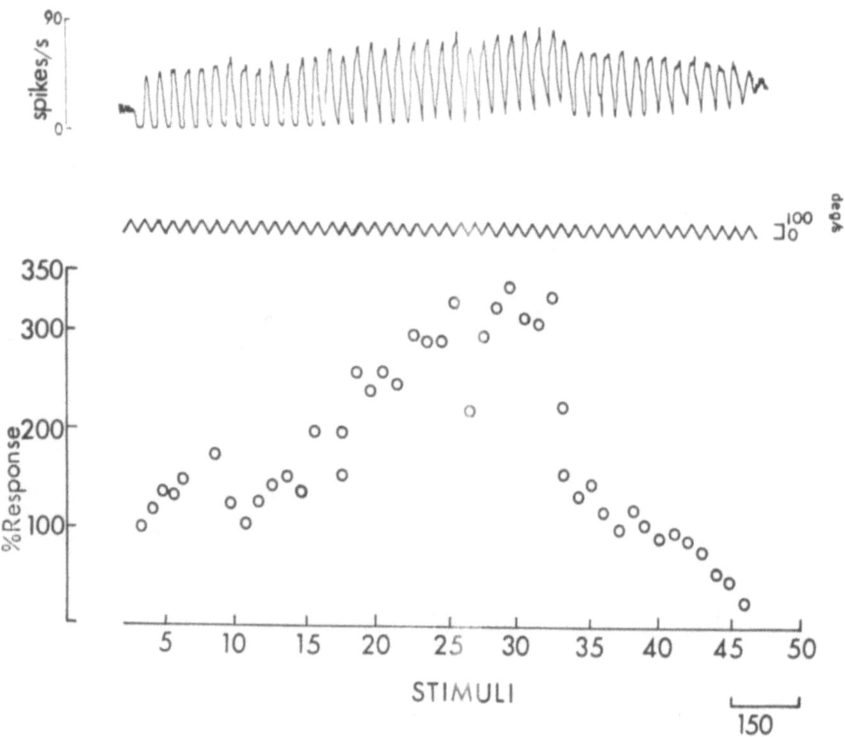

Fig. 25-7. An initial response enhancement followed by a progressive decline manifested by a conscious cat unit with stimulus repetition. **Upper trace** Successive discharge rate histograms. **Second trace** Angular velocity waveform. **Lower graph** Percent change in response.

linear and linear and quadratic (second order polynomial) trends. The majority of anesthetized units that did manifest response modification with stimulus repetition were classified as "linear." About two-thirds of the units from conscious cats were linear and quadratic (second order polynomial) and one-third linear. A chi-square significant at the 0.01 level confirmed the significant difference between the two groups of units. Results of polynomial regression are summarized in Table 25-1. For the "linear" units, the slope of the regression line was also calculated.

Table 25-1.
SUMMARY: TREND ANALYSIS.

	Number of Cases			
	Not Significant	**Linear**	**Quadratic**	**Quadratic and Linear**
UNITS: Anesthetized[a]	17	13	1	4
UNITS: Conscious[a]	16	7	3	17

[a] $\chi^2 = 9.42$ (significant at 0.01 level).

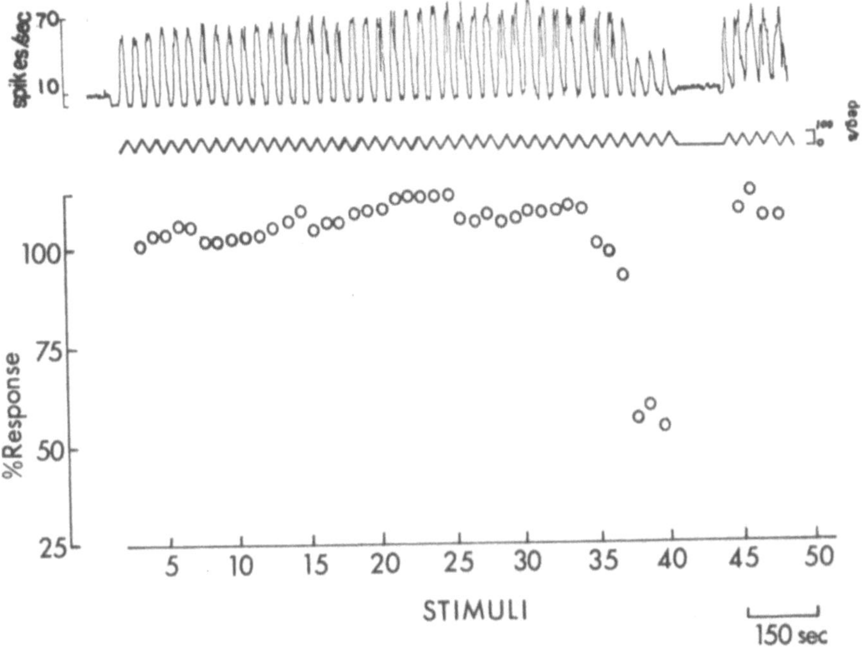

Fig. 25-8. Response decline with repeated stimulation and recovery following a short break equivalent in duration to three acceleration–deceleration stimuli in a conscious cat unit. Successive discharge rate histograms, stimulus waveform, and percent change in response with successive stimuli.

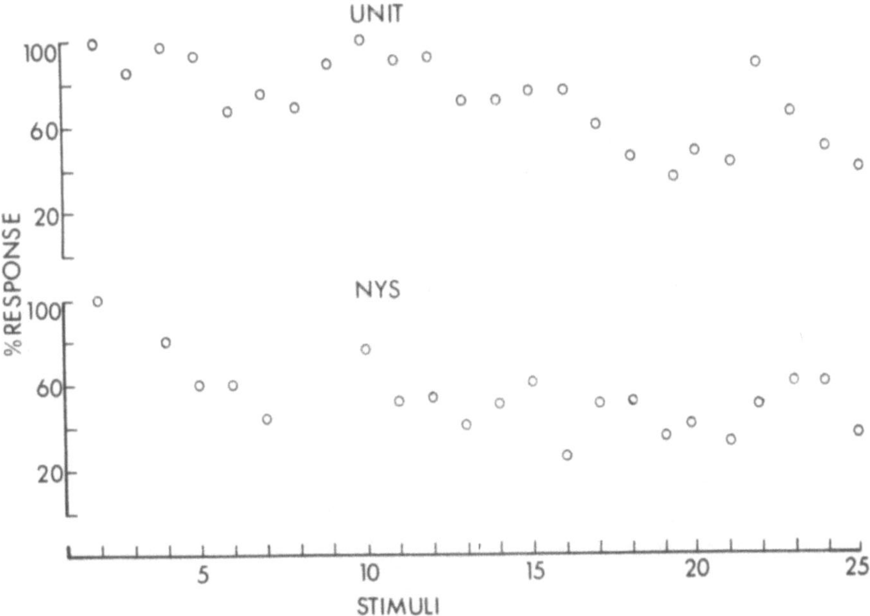

Fig. 25-9. Simultaneous changes in nystagmus and unit response with repeated stimulation. Both are expressed as percent response. $r = 0.69$; $p = 0.02$.

Table 25-2.
SUMMARY: REGRESSION ANALYSES.

	Linear Regression: Negative Slope		Nonlinear Regression 1[a]				Nonlinear Regression 2[b]		
			Positive Slope		Negative Slope		Positive Slope	Negative Slope	
	Mean	SD	Mean	SD	Mean	SD		Mean	SD
Anesthetized	−8.84	6.75	11.02	–	−16.60	–	NS	−8.27	1.01
Unit	$N = 14$		$N = 1$					$N = 3$	
Conscious	−2.72	1.95	8.52	10.39	− 6.38	2.53	NS	−5.90	.3.76
Unit	$N = 8$		$N = 8$					$N = 11$	

All slope scores in percent response, with respect to response No. 2. NS, not significant.
[a]Both positive and negative slopes significant.
[b]Only negative slope significant.

In order to determine whether the initial increase and the subsequent decrease observed in the units defined by a second degree polynomial were significant, data from those units were subjected also to a nonlinear broken line regression analysis which fit the data with two straight lines having different slopes and intercepts (2,15). The slope of each line and the break point were estimated. A 95% confidence interval was specified for each slope estimation. This served as a test of significance. The results of this procedure are summarized in Table 25-2. In Figs. 25-10–25-12, which illustrate the results of the broken line regression analysis, the abscissa is scaled by successive stimuli of equal durations. The ordinate is scaled by digitized area units proportional to the areas circumscribed by successive discharge rate histograms. Figure 25-10 illustrates a linear unit recorded from a conscious cat with the estimated regression line calculated by the analysis. Figure 25-11 illustrates the result of the broken line nonlinear regression analysis performed on a unit whose response modification was determined to be a second degree polynomial. It was determined that this unit manifested a significant increase in response followed by a significant decline. Figure 25-12 illustrates a unit with an initially unmodified response followed by a significant decline.

The relationship between nystagmus and unit response modification for the cases where both were simultaneously recorded was evaluated by calculating Pearson correlation coefficients for the pairs of simultaneously recorded response modalities. In three of the eight cases where this was performed, the correlation was not found to be significant. In the remaining five cases the correlation coefficients ranged between 0.44 and 0.69 and were found to be significant. In the case illustrated in Fig. 25-9 the correlation coefficient was 0.543 and was found to be significant. Table 25-3 summarizes the correlation between nystagmus and unit activity.

DISCUSSION

The vestibular nuclei act as a relay station between the vestibular end organs and the CNS. Therefore, it appeared appropriate to investigate neuronal

Table 25-3. SUMMARY OF NYSTAGMUS-UNIT CORRELATION.

Cat	Unit	*r*	Significance
49	#1	0.69	0.05
	#2	0.33	>0.05
51	#1	0.543	0.05
	#16	−0.10	0.05
46	#2	0.24	0.05
	#5	0.44	0.05
29	#1	0.56	0.05
32	#6	0.55	0.05

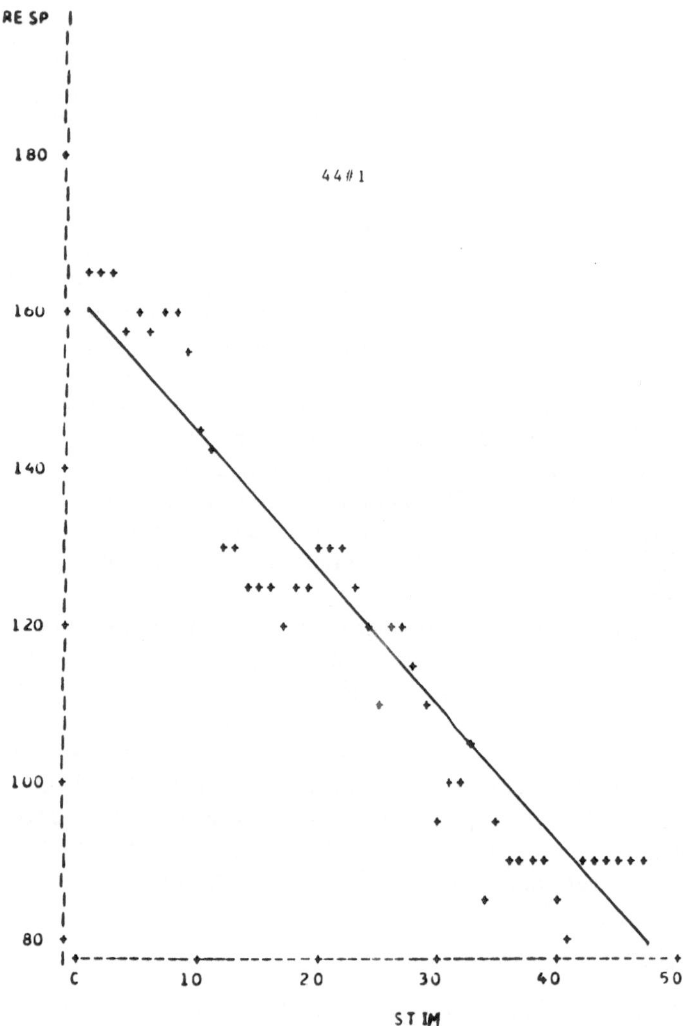

Fig. 25-10. Linear regression performed on the response alteration pattern with respect to successive stimuli of a conscious cat unit. *x-axis*; successive stimuli; *y-axis*; digitized area under discharge rate histogram in mm².

habituation at their level. Several studies have provided evidence that changes are likely to occur in the vestibular nuclei during the habituation of the vestibuloocular reflex. The generalization (transfer) of habituation from the trained to the untrained ear when eliciting nystagmus in the practiced direction (13,20) implicates the nuclear complex. Visual input has a marked effect upon habituation (7), and there is evidence for the modulating influence of visual input upon vestibular nuclei units (19). In addition, McCabe

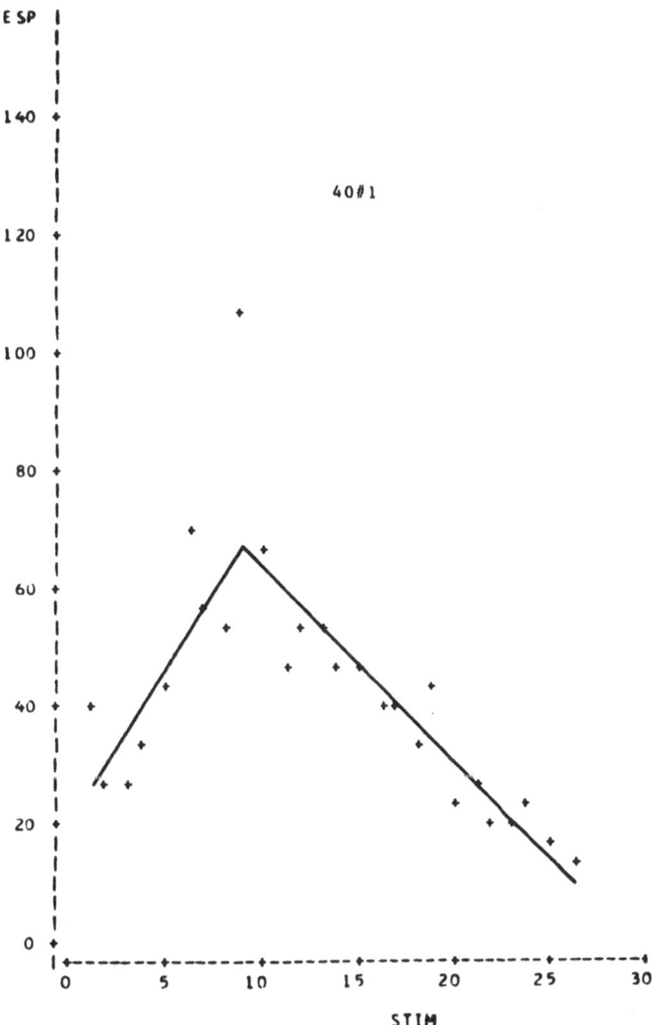

Fig. 25-11. Nonlinear broken line regression analysis performed on the response alteration pattern with respect to successive stimuli of a conscious cat unit. *x-axis*; successive stimuli; *y-axis*; digitized area under discharge rate histogram in mm^2. Positive slope = 12.65%; negative slope = −5.02%.

and Gillingham (25) showed dishabituation following lesions to the lateral vestibular nucleus.

Over 50% of neurons recorded from the anesthetized cats manifested significant changes with stimulus repetition. Dishabituation was also demonstrated in a number of cases. While most neurons recorded from the anesthetized cats showed relatively rapid and unimodal changes in response magni-

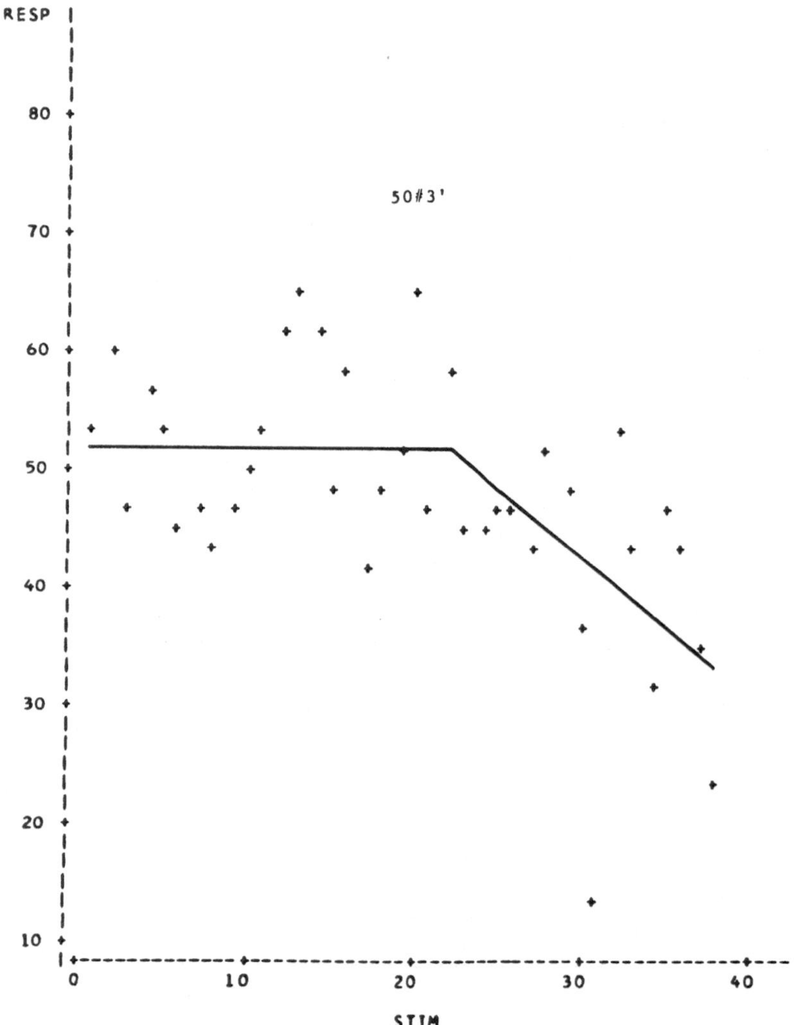

Fig. 25-12. Nonlinear broken line regression analysis performed on the response alteration pattern with respect to successive stimuli of a conscious cat unit. *x-axis*; successive stimuli; *y-axis*; digitized area under discharge rate histogram in mm^2.

tude with repeated stimulation, the changes seen in the conscious cat units were on a slower time scale. In addition, in a larger number of units those changes were bimodal. In approximately two-thirds of units recorded from the conscious cats, response decline was preceded by its enhancement or by a period of unaltered response.

An initial response increment followed by a gradual decline occurring with repeated stimulation as observed in certain conscious cat vestibular neurons has been reported to occur also in other systems with stimulus repetition.

Thus, during the habituation of the hindlimb flexor reflex in acute spinal cats, one group of spinal interneurons exhibited an initial response increase followed by a gradual decline (18), this being also the pattern of the reflex modification. The investigators attributed this pattern to a competitive interaction between incremental and decremental processes generated at the spinal interneuronal level. Some reticular neurons exhibited the same behavior with repeated stimulation of tectal, vestibular cortical, and cutaneous areas (26). The increment was attributed to a long-lasting facilitation through intrinsic reticuloreticular pathways. In the vestibular system some primary afferents in the frog manifested an exponential response decline with stimulus repetition; others maintained a constant response initially, which subsequently declined (16).

A simple model can be hypothesized. Initially, the repeated acceleration-deceleration stimuli elicited a general arousal reaction, which contributed toward tonic facilitation of certain vestibular nuclei neurons through the reticular system (4,10,11). Hence the initial increase in response, or the maintenance of a constant response in conscious cat units. Under barbiturate anesthesia, both ascending and descending impulses through the reticular system are blocked (1,24). Due to the relative inactivity of the reticular system in the anesthetized preparation, an initial significant response increment was not prevalent in our anesthetized cats.

After a number of stimulus repetitions, the nonspecific arousal reaction habituated (31) and the vestibular nuclei lost the excitatory input associated with the arousal reaction process. At this point, the inhibitory cerebellar input (22,28,32), which may have been overshadowed by the excitatory reticular input, may have become effective, and a progressive response decline may have taken over.

SUMMARY

This was a study of the effects of repeated angular accelerations on the activity of vestibular nuclei units in anesthetized and in conscious cats. The experimental animals were subjected to trains of repeated and consecutive constant acceleration-deceleration ramps (4–8°/s^2). In a few cases nystagmus was recorded along with extracellular unit activity.

The effect of stimulus repetition on vestibular neuronal activity consisted of one of the following:

1) No change in response.
2) Progressive response decline.
3) Initial maintenance of constant response level through the first part of the stimulus paradigm followed by a progressive decline.
4) An initial gradual response enhancement followed by a progressive decline.

This classification is based on results of polynomial and nonlinear broken-line regression analysis. Categories 3 and 4 were found predominantly in units recorded from conscious cats. The majority of neurons recorded from anesthetized cats that exhibited response changes with stimulus repetition manifested a progressive response decline. The response decline in anesthetized cat units was usually on a faster time scale than in those of conscious cats.

Correlations between unit response modification patterns and simultaneously recorded nystagmus were mostly of a moderate degree.

ACKNOWLEDGMENTS

This project was supported by the following: NINDS, Grant No. 06785; Iowa City VAH, Grant No. 103.87; and the Departments of Otolaryngology and Maxillofacial Surgery, and Speech Pathology and Audiology, of the University of Iowa.

References

1. Arduini, A. and Arduini, M.G.: Effects of drugs and metabolic alterations on brain stem arousal mechanisms. J. Pharmacol. (Kyoto) 110:76, 1954.
2. Barr, A.J., Goodnight, J.H., Sall, J.P., and Helwig, J.I.: A User's Guide to SAS-76. SAS Institute, Inc., Raleigh, N.C. 1976.
3. Berman, A.L.: The Brain Stem of the Cat. Madison, Wis., The University of Wisconsin Press, 1968.
4. Bizzi, E., Pompeiano, O., and Somogyi, I.: Spontaneous activity of single vestibular neurons of unrestrained cats during sleep and wakefulness. Arch. Ital. Biol. 102:308, 1964.
5. Buchwald, J.S. and Humphrey, G.L.: An analysis of habituation in the specific sensory systems. In Stelar, E. and Sprague, J.M. (eds.): Progress in Physiological Psychology. New York, Academic Press, 1973.
6. Collins, W.E.: Task-control of arousal and the effects of repeated unidirectional angular acceleration on human vestibular responses. Acta Otolaryngol. (Suppl.) 190:1, 1964.
7. Collins, W.E.: Effects on vestibular habituation of interrupting nystagmus responses with opposing stimuli. J. Comp. Physiol. Psychol. 64(2):308, 1967.
8. Collins, W.E.: Adaptation to Vestibular Disorientation. XII. Habituation of Vestibular Responses: An Overview. Report No. FAA-AM-74/3, Federal Aviation Administration, Office of Aviation Medicine, Civil Aeromedical Institute, Oklahoma City, Oklahoma, 1974.
9. Crampton, G.: Response of single cells in the cat brain stem to angular acceleration in the horizontal plane. In Graybiel, A. (ed.): First Symposium on the Role of the Vestibular Organs in the Exploration of Space. NASA SP-77. Washington, D.C., 1965.
10. Duensing, F. and Schaefer, K.P.: Die Neuronenaktivität in der Formatio Reticularis des Rhombencephalons beim Vestibularen Nystagmus. Arch. Psychiat. 196:402, 1957.
11. Dumont, S.: Effects extralabyrintiques sur l'activité des cellules des noyaux vestibularies. J. Physiol. (Paris) 52:87, 1960.

12. Fernandez, C. and Schmidt, R.S.: Studies in habituation of vestibular reflexes. III. A revision. Laryngoscope 72:939, 1962.
13. Fluur, E. and Mendel, L.: Habituation efference and vestibular interplay. III. Unidirectional rotatory habituation. Acta Otolaryngol. 57:81, 1964.
14. Forssman, B., Henriksson, N.G., and Dolowitz, G.A.: Studies on habituation of vestibular reflexes. VI. Habituation in darkness of calorically induced nystagmus, laterotorsion, and vertigo in man. Acta Otolaryngol. 56:663, 1963.
15. Gallant, A.R.: Nonlinear regression. Am. Statistician 29:73, 1975.
16. Goetemakers, R.: Vestibular adaptation in *Rana*. Adv. Otorhinolaryngol. 17:107, 1970.
17. Groves, P.M. and Thompson, R.F.: Habituation: A dual process theory. Psychol. Rev. 77:419, 1970.
18. Groves, P.M. and Thompson, R.F.: A dual-process theory of habituation: Neural mechanisms. In Peeke, H.V.S. and Herz, M.J. (eds.): Habituation, Vol. 2. Physiological Substrates. New York, Academic Press, 1973.
19. Henn, V., Young, L.R., and Finley, C.: Vestibular nucleus units in alert monkeys are also influenced by moving visual fields. Brain Res. 71:144, 1974.
20. Henriksson, N.G., Kohut, P., and Fernandez, C.: Studies on habituation of vestibular reflexes. I. Effect of repetitive caloric test. Acta Otolaryngol. 53:333, 1961.
21. Horn, G.: Changes in neuronal activity and their relationship to behavior. In Horn, G. and Hinde, R.A. (eds.): Short Term Changes in Neural Activity and Behavior. Cambridge, Mass., Cambridge University Press, 1970.
22. Ito, M., Highstein, S.M., and Fukuda, J.: Cerebellar inhibition of the vestibulo-ocular reflex in rabbit and cat and its blockage by picrotoxin. Brain Res. 17:524, 1970.
23. Lidvall, H.: Vertigo and nystagmus responses to caloric stimuli repeated at short and long intervals. Acta Otolaryngol. 53:507, 1961.
24. Magoun, H.W.: The Waking Brain. Springfield, Ill., Thomas, 1958.
25. McCabe, B.F. and Gillingham, K.K.: The mechanisms of vestibular suppression. Ann. Otorhinolaryngol. 73:816, 1964.
26. Peterson, B.W., Frank, J.I., Pitts, N.G., and Daunton, N.G.: Changes in responses of medial ponto-medullary reticular neurons during repetitive cutaneous, vestibular, cortical, and tectal stimulation. J. Neurophysiol. 39:564, 1976.
27. Pribram, K.H.: The limbic system's efferent control of neural inhibition and behavior. In Adey, W.R. and Tokizave, T.T. (eds.): Progress in Brain Research. Amsterdam, Elsevier, 1967.
28. Robinson, D.A.: Adaptive gain control of vestibuloocular reflex by the cerebellum. J. Neurophysiol. 39:954, 1976.
29. Ryu, J.H. and McCabe, B.F.: Neural activity in the vestibular nuclei of the cat. Ann. Otorhinolaryngol. (Suppl.) 9:1, 1973.
30. Segundo, J.P., Takenaka, T. and Encabo, H.: Somatic sensory properties of bulbar reticular neurons. J. Neurophysiol. 30:1221.
31. Sharpless, S. and Jasper, H.: Habituation of the arousal reaction. Brain 79:655, 1956.
32. Singleton, G.T.: Relationships of the cerebellar nodulus to vestibular function: A

study of the effects of nodulectomy on habituation. Laryngoscope 77:1579, 1967.

33. Thompson, R.F. and Spencer, W.A.: Habituation: A model phenomenon for the study of neuronal substrates of behavior. Psychol. Rev. 73:16, 1966.

Discussion

CRAMPTON: In the animals in which you recorded both nystagmus and neural discharge, did you use any special alerting techniques? There's recent reasonable evidence that, even when attempting to produce a maximum alerting using either amphetamine or methylphenidate, there is still a response decline. I would suggest, in continuing experiments, that maximum alertness be maintained.

KILENY: Yes, I agree. We did some preliminary recordings where we only recorded nystagmus, and due to lack of time I couldn't show you those recordings. But there's a problem with the *encéphale isolé* preparation, and certainly it is a good idea using some kind of alerting. However, we wanted to stay away from drugs as much as possible. Another aspect of this question that I would really like to point out is that in the future I would like to try to monitor at least general EEG activity in experiments with awake animals to estimate level of arousal.

26

Responses of Neurons in the Vestibular Nuclei of Awake Squirrel Monkeys During Linear Acceleration

ADRIAN A. PERACHIO

The role of the vestibular system in the control of postural reflexes has been investigated by the use of transient and sustained linear acceleration as forms of natural stimulation of the otolith receptors. The phasic motor response to linear acceleration is opposite to the tonic reflex induced by maintained tilting of the head (13). Moreover, translational motion may reveal different dynamic response characteristics of the reflex from that observed with time-varying tilts. A recent study has demonstrated the dynamic properties of the extensor reflex of decerebrate cats induced by linear acceleration along the horizontal and vertical axes. The reflex motor output exhibits decreasing gain and increasing phase lag with increased frequency of translational acceleration during side-to-side or fore-and-aft sinusoidal motion (2). Neither conduction velocity delays nor the dynamics of the macular afferents can account for the dynamic relationship between linear acceleration input and the motor reflex response output. This, along with additional considerations, led Anderson and his coworkers (2) to propose that neurons in the reticular formation may be involved in the process of integration.

A necessary step in the elucidation of a brain stem integrator mechanism is the development of information on the response dynamics both of the neurons of the vestibular nuclei that receive direct primary afferents from utricular and saccular receptors (6,12), as well as those with indirect macular input. Few data are available on the characteristics of the response of vestibular nuclei neurons to linear acceleration. Daunton and Melvill Jones (3) have reported on the behavior of neurons in the vicinity of the vestibular nuclei in decerebrate cats that were passively moved along their horizontal and vertical axes at a single stimulus frequency. The temporal relationship of the neural response to the acceleration input was variable across cells, so that both phase lags and phase advances were observed (7). In more recent studies, Daunton and colleagues (4) have replicated these results in intact paralyzed cats.

The purposes of the present study were (1) to record in an awake, restrained primate the responses of neurons in and around the vestibular nuclei during continuous sinusoidal fore-and-aft or side-to-side oscillatory acceleration; (2) to determine the directional sensitivity of responsive neurons; and (3) to describe, in a preliminary fashion, the dynamic response characteristics of a small number of cells tested over a limited stimulus frequency range.

Each of three squirrel monkeys used in these studies was prepared with an electrode guide platform stereotaxically implanted on the calvaria under sterile surgical procedures. This device was designed so that multiple penetrations of the target area with microelectrodes could be achieved. In addition, a feature for rigidly and painlessly restraining the head of the animal was incorporated in the frame of the implanted platform. Each recording session was limited to the investigation of neurons along a single electrode traverse. At the beginning of a recording session, the monkey was placed in a primate chair and its head was held at a fixed angle inclined 22° from the stereotaxic horizontal plane. The chair was positioned over a base that could be manually rotated so that the animal could be aligned along either of the two major axes of motion: fore-and-aft (x) and side-to-side (y). It was also possible to test informally the responses of neurons to angular acceleration around the yaw axis of the head, in the plane of the horizontal semicircular canals. The platform on which the chair was positioned was mounted on a cart that was driven on neoprene tires along a set of parallel rails. The cart was coupled to a DC torque motor controlled by a position servo through a cable/capstan arrangement.

The search for responsive neurons was carried out in the following typical fashion: As the microelectrode was advanced, the animal was continuously oscillated along a single axis in a sinusoidal fashion at a fixed fundamental frequency (e.g., 0.7 Hz $\pm$ 0.197 g peak acceleration). A distance of 20 cm was traversed during each half-cycle. Peak acceleration was $\pm$ 0.016 g at 0.2 Hz and $\pm$ 0.579 g at 1.2 Hz. Periodically, the chair was rotated 90°. All spontaneously active neurons were tested for responses to both x- and y-axis acceleration. When a responsive neuron was encountered, a marked modulation of its activity by the stimulus could be detected by visual and auditory monitoring of its discharge rate. All responsive units were tested at as many stimulus frequencies as possible for both axes of stimulation until the cell was lost or injured.

The following categories defined the relationship between the stimulus and the neural response: (1) x-Positive or x-negative responses were denoted by peak firing associated with forward and backward acceleration, respectively. Input to these neurons could have originated in either saccular or utricular maculae. (2) y-Positive or y-negative responses were characterized by peak firing associated with acceleration directed opposite to or toward the recording side; e.g., y-positive acceleration is directed toward the right for a neuron recorded on the left side of the brain. These responses could have derived from input originating in utricular receptors alone. (3) Biaxially sensitive neurons responded to acceleration along either axis; e.g., x-positive/y-positive, x-negative/y-negative, x-negative/y-positive, and so on.

The response pattern of most neurons was characterized by a regular modulation of the firing activity with a fairly constant phase relationship between maximum discharge rate and the peak of the sinusoidal acceleration input (Fig. 26-1). A number of cells exhibited a cessation of firing for approximately half of the stimulus cycle. A similar response "cutoff" has been

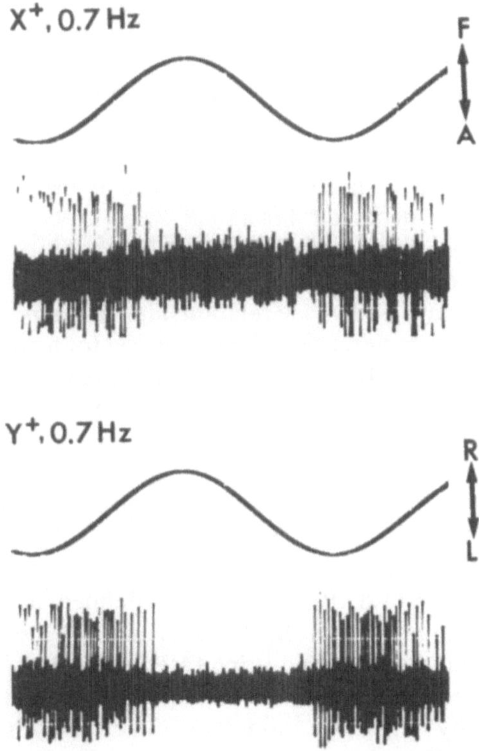

Fig. 26-1. Response of a neuron in the descending vestibular nucleus of a squirrel monkey to sinusoidal translational motion at a frequency of 0.7 Hz along the longitudinal (x) or transverse (y) axis of the animal. The upper trace in each half of the figure represents position; the lower trace is the extracellular recording of the discharge of the neuron. *F*/*A*; fore/aft; *R*/*L*; right/left.

reported for dynamic responses to sinusoidal roll in primary otolith afferents (1) and vestibular nuclei neurons of the cat (11), and for responses to oscillatory translational motion in the brain stem neurons in cat (3) and rat (9).

Most electrode traverses were directed toward the lateral and descending vestibular nuclei. Of the total of 49 neurons that responded to linear acceleration, 63% were located within or immediately ventral to these two nuclei. Approximately half of the responsive neurons in the vestibular nuclei were modulated by both fore-and-aft and side-to-side acceleration. These biaxially sensitive cells tended to fall into one of the following response categories as defined by the convention used in this study: (1) *x*-positive/*y*-positive or (2) *x*-negative/*y*-negative. Neurons associated with either category tended to be differentially represented in the lateral and descending nuclei. Thus, 80% of the neurons recorded in the descending nucleus that responded to linear acceleration were classified as *x*-positive/*y*-positive in terms of their peak firing with respect to peak acceleration. In the lateral vestibular nucleus, approximately 67% of the neurons affected by translational acceleration were classified as *x*-negative/*y*-negative. Thus, the majority of neurons in the lateral and descending vestibular nuclei exhibited responses to acceleration along both the horizontal and transverse body axes. Responsive neurons exhibited peak firing either during forward, contralaterally directed (*x*-positive/*y*-positive) acceleration or backward, ipsilaterally directed (*x*-negative/*y*-negative) acceleration. A small number of the remaining biaxially sensitive neurons were assigned to the *x*-negative/*y*-positive category, since firing occurred during backward or contralaterally pointing acceleration. These neurons were located in an area subjacent to the vestibular nuclei in the medullary reticular formation.

The sensitivity (gain) and phase relationships of the peak firing of the cells with respect to peak acceleration were computed for a number of cells tested over a range of stimulus frequencies. Fourier analyses were performed on data from multiple cycles of stimulation at each tested frequency. Gains were computed on the average maximum firing rate of the neuron during the excitatory phase of the cycle as represented by the amplitude of the fundamental. No corrections were made for neurons exhibiting silencing during a portion of the stimulus cycle. This may have resulted in an underestimation of the gain as compared to that computed for neurons that fired throughout the cycle (1,8,10). Phase was defined as the difference, in degrees, between the maxima of the fundamentals of the input (acceleration in m/s^2) and the output (averaged firing rate in impulses/s).

Neurons of all response types exhibited gain attenuations as a function of increasing frequency (Figs. 26-2 and 26-3). Since only the amplitude of displacement was fixed, peak acceleration increased with increasing stimulus frequency; thus, the linearity of the responses (superposition) was not determined. Phase relationships were not uniform across neurons. Both phase advances and lags were observed in different cells. Most neurons of each response category exhibited relatively constant phase lags in a range of 60° to 80° with respect to peak acceleration.

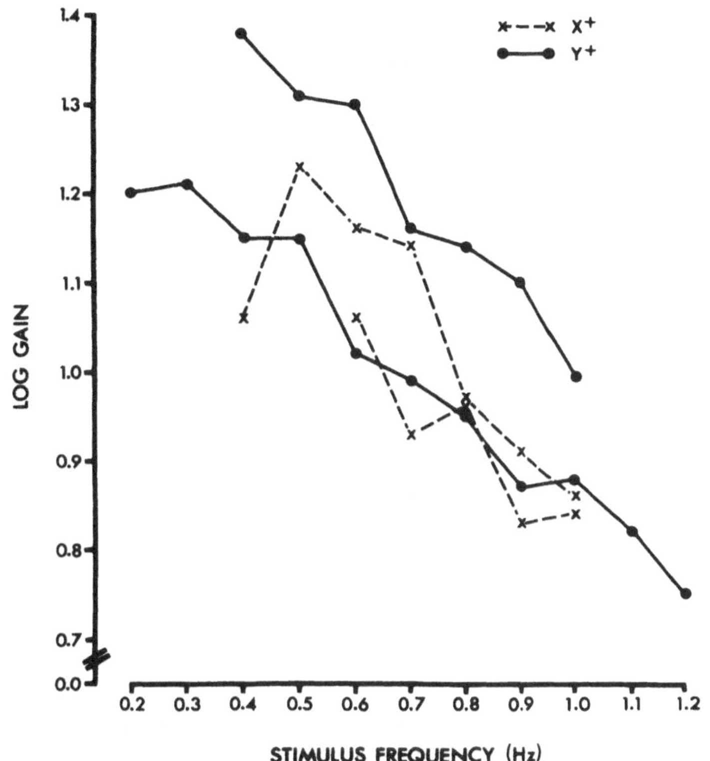

Fig. 26-2. Gain (log impulses/s per m/s²) of three vestibular nuclei neurons as a function of input frequency. X^+ designates peak firing associated with peak forward-directed acceleration along the animal's longitudinal axis. Y^+ indicates responses associated with acceleration along the animal's transverse axis directed contralateral to the side of the brain in which the neuron was located. Linear displacement for half of the cycle of sinusoidal movement is the same at all frequencies: 20 cm. One of the neurons responded to acceleration along both longitudinal and transverse axes (Fig. 26-1) over a frequency range of 0.4 to 1.0 Hz.

With regard to the relationship of brain stem vestibular neurons to the dynamic responses in postural reflexes in decerebrate cats (2), the neurons that were tested over a comparable frequency range exhibited similar gain characteristics to that found in the extensor muscle response to linear acceleration. This is particularly well illustrated for a neuron found in the descending vestibular nucleus that responded only during side-to-side translational motion (Fig. 26-4). The response, over a stimulus frequency range of 0.2 to 1.2 Hz, exhibited a decrement of approximately 9 dB, thus not fully accounting for the decrease in gain in the motor output, reported by Anderson et al. (2) to be as great as 14 to 20 dB. The phase lag of the vestibular neuron's response with increasing stimulus frequency was as great at 1.1 Hz as that observed in the cat extensor muscle response.

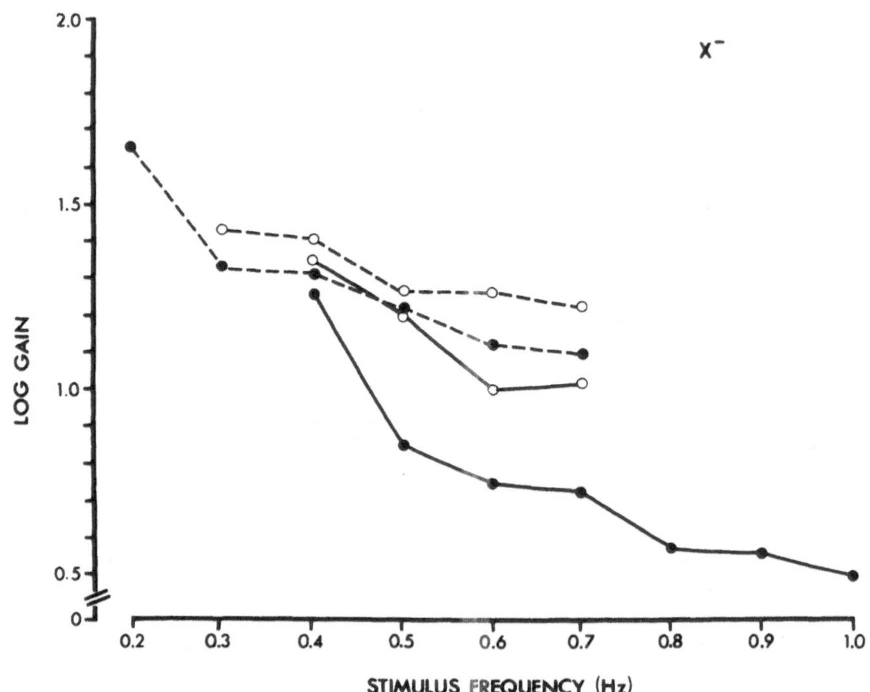

Fig. 26-3. Gain (as in Fig. 26-2) of four neurons, located in the region of the lateral vestibular nucleus, as a function of input frequency. X^- designates peak firing associated with backward-directed acceleration along the longitudinal axis of the head.

The directional sensitivity of the response of vestibular neurons to linear acceleration suggests a polarization vector that is best described as lying 45° between the longitudinal and transverse axes of the horizontal plane of the skull when the head is inclined so as to orient the major portion of the utricular maculae parallel to the plane in which the animal was moved (5). Most neurons in this sample were bidirectionally sensitive for acceleration along either horizontal axis. Different vectors for maximum response seemed to be represented disproportionately in the two nuclei best sampled in this study, i.e., the descending and lateral nuclei.

In those cells which exhibited increasing phase lag with increasing stimulus frequency, evidence can be found for the temporal integration observed in the extensor muscle reflex response to linear acceleration during translational motion. The gain attenuation found in all units tested over the limited frequency range used in the present study is also consistent with the decrement in peak response of the motor output of the cat extensor forelimb muscles with increasing frequency of translational acceleration. Thus, neurons in the vestibular nuclei may account for a portion of the dynamic response of postural muscles during time-varying linear acceleration. It

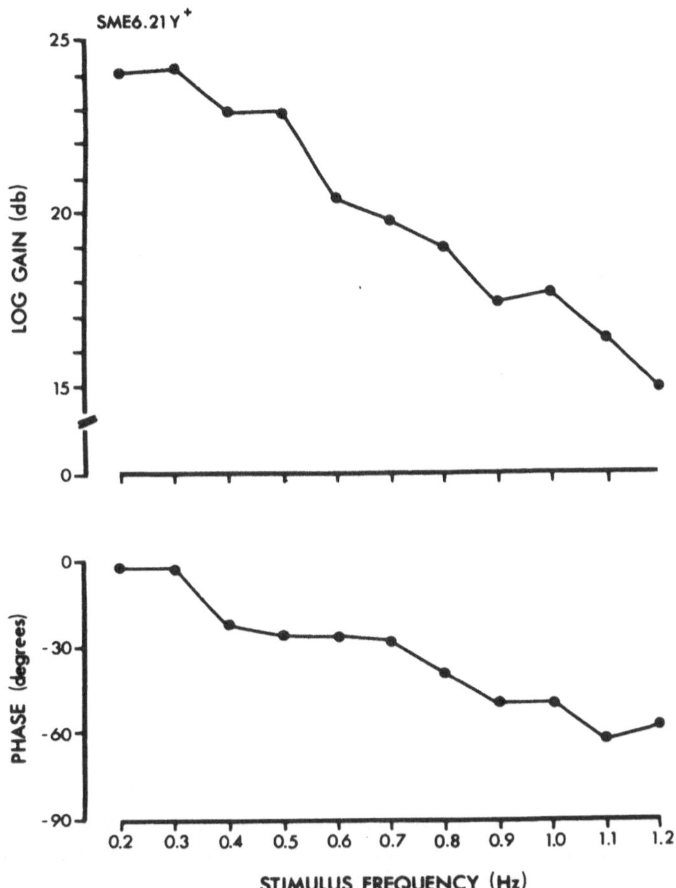

Fig. 26-4. Phase and gain relationships between input acceleration and the response of a neuron in the descending vestibular nucleus for linear acceleration along the transverse (y) axis, as a function of input frequency. Phase is the difference, in degrees, between peak positive linear acceleration and the maximum of the neuronal response in impulses/s.

remains to be demonstrated, however, that these neurons contribute to the vestibulospinal tracts and whether they are recipients of input from the primary afferents of the vestibular nerve.

Translational acceleration offers certain advantages as a natural vestibular stimulus. This may be particularly important for neurons that adapt rapidly to static or slowly varying tilts. More importantly, translational motion may reveal different aspects of the response dynamics of vestibular neurons. Otolith afferents in the superior division of the VIIIth nerve of the cat have been shown to have gain characteristics that are flat or increase with the frequency of oscillation about the roll axis of the head (1). Similarly, Schor (11) reported that the dynamic responses of vestibular nucleus neurons in the cat exhibit an increased sensitivity at higher frequencies of sinusoidal roll rotations. These effects are in contrast to the gain attenuation observed at higher frequencies of translational acceleration in the responses of vestibular nucleus neurons in the awake squirrel monkey.

ACKNOWLEDGMENTS

The author wishes to thank Dr. Steven Anschel for technical assistance and D. Rice and W. Bouris for their critical contributions in all phases of the study. This research was conducted at the Yerkes Regional Primate Research Center, Emory University, supported by NASA Grant NGR 11-001-045 and in part by N.I.H. Grant RR 00165.

References

1. Anderson, J.H., Blanks, R.H.I., and Precht, W.: Response characteristics of semicircular canal and otolith systems in cat. I. Dynamic responses of primary vestibular fibers. Exp. Brain Res. 32:491, 1978.
2. Anderson, J.H., Soechting, J.R., and Terzuolo, C.A.: Dynamic relations between natural vestibular inputs and activity of forelimb extensor muscles in the decerebrate cat. I. Motor output during sinusoidal linear accelerations. Brain Res. 120:1, 1977.
3. Daunton, N. and Melvill Jones, G.: Directional representation of horizontal and vertical acceleration in the neural activity of cat vestibular nuclei. Las Vegas, Aerospace Medical Association Meetings, 1973, p. 144.
4. Daunton, N.G., Thomsen, D.D., and Christensen, C.A.: Varying combinations of visual and vestibular inputs alter responses of otolith units in cat vestibular nuclei. Soc. Neurosci. 4:610, 1978.
5. Fernandez, C., Goldberg, J.M., and Abend, W.K.: Response to static tilts of peripheral neurons innervating otolith organs of the squirrel monkey. J. Neurophysiol. 35:978, 1972.
6. Gacek, R.: The course and central termination of first order neurons supplying vestibular end organs in the cat. Acta Otolaryngol. (Suppl.) 254:1, 1969.
7. Melvill Jones, G. and Daunton, N.: Comparison of brain stem neural responses to vertical and horizontal linear accelerations. Progress Report, NASA Ames Research Center, 1973.
8. Melvill Jones, G. and Milsum, J.H.: Frequency-response analysis of central vestibular unit activity resulting from rotational stimulation of the semicircular canals. J. Physiol. (Lond.) 219:191, 1971.
9. Perachio, A.A., Miller, D.S., Rice, D., and Bouris, W.: Effects of linear acceleration on neurons in rat vestibular nuclei. Soc. Neurosci. 3:545, 1977.
10. Schneider, L.W. and Anderson, D.J.: Transfer characteristics of first and second order lateral canal vestibular neurons in gerbil. Brain Res. 112:61, 1976.
11. Schor, R.H.: Responses of cat vestibular neurons to sinusoidal roll tilt. Exp. Brain Res. 20:347, 1974.

12. Stein, B.M. and Carpenter, M.B.: Central projections of portions of the vestibular ganglia innervating specific parts of the labyrinth in the rhesus monkey. Am. J. Anat. 120:281, 1967.

13. Wilson, V.J. and Peterson, B.W.: Peripheral and central substrates of vestibulospinal reflexes. Physiol. Rev. 58:80, 1978.

Discussion

PRECHT: There is a very interesting difference when we compare the results in the canal system with respect to the vestibulospinal reflexes and your data with the otolith; in the canal case, vestibulospinal tracts seem to contain velocity information. In the otolith system you already found information in the vestibular nuclei that could go without further processing to the muscle. So there is a very interesting difference between the canal and otolith data. In Shimazu's lab it has been found that cutting the descending m.l.f. had no effect on the response phase of motoneurons. This supports the idea that, in the canal system, the processing is probably happening in the reticular spinal system rather than by descending m.l.f.; but now in the otolith system it's apparently already there in the vestibular nuclei. A question that I would like to add is, did you find in other parts of the vestibular complex, like the rostral medial nucleus or superior nucleus, an otolith response? The reason why I ask is, we have seen in vestibular neurons recorded from these areas, and that project to the oculomotor system very often, gravity responses and canal responses in the same unit.

PERACHIO: Indeed, we have found neurons responsive to translational motions that also appear to have converging inputs from the horizontal canals. They tend to exhibit type I responses.

27

Visual Modulation of Otolith Responses: A Paradigm for the Study of Self-Motion Perception and Its Neural Substrate

N.G. DAUNTON, C.A. CHRISTENSEN, AND D.D. THOMSEN

Recent perceptual studies suggest that visual and vestibular inputs combine to give an organism information about its own self-motion (6,11). The importance of the visual input channel in self-motion perception is demonstrated by the occurrence of illusions of self-motion which are induced by visual stimulation alone. These illusions have been described for both rotational self-motion or circularvection (5) and translational self-motion or linearvection (1). Our laboratory is conducting neurophysiologic and behavioral studies of visual-vestibular interactions in animals, with emphasis on the interactions between visual and otolithic inputs which provide information about translational self-motion. These studies may contribute to our understanding of the role of visual-vestibular interactions in self-motion perception.

Our neurophysiologic work has been carried out in the chronically prepared cat, which is paralyzed with Flaxedil to eliminate artifacts produced by eye and body movements. Extracellular recordings are made in the area of the medial vestibular nuclei. Vestibular stimulation is provided by sinusoidal translational motion of a parallel swing. This swing, which is suspended from the ceiling by either wires or springs, provides accelerations of the same frequency and amplitude (0.59 Hz; 0.15 *g*) in three orthogonal planes: right-left, fore-aft, and up-down. The parallel swing is surrounded on three sides and bottom by a frame covered with a flower-and-dot pattern.

The frame, like the parallel swing, is suspended from either wires or springs, depending on the direction of motion required. This stimulus frame can be moved sinusoidally at the same amplitude and frequency, and in the same directions, as the parallel swing. The two stimulation devices can be moved together or separately, providing various combinations of visual and vestibular stimulation. Figure 27-1 shows an artist's representation of the apparatus.

VISUAL INFLUENCES ON OTOLITHIC UNITS

The following three conditions of stimulation were used in our initial investigations of otolith-visual interaction in the vestibular nuclei: (1) vestibular stimulation alone (parallel swing motion in the dark); (2) visual stimulation alone (visual frame motion in the light, parallel swing stationary); and (3) visual-vestibular stimulation (parallel swing motion in the light, visual frame stationary). The results of this study provide the first direct evidence that

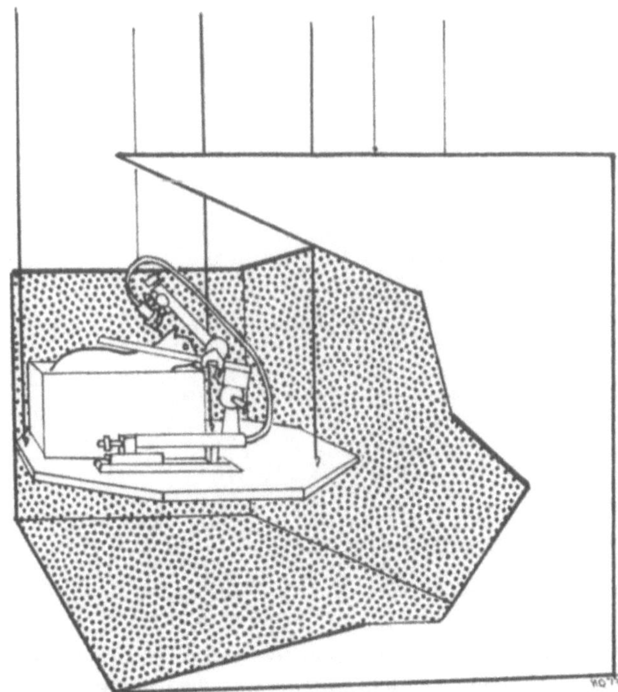

Fig. 27-1. Artist's representation of visual-vestibular stimulation apparatus. Cutaway view shows cat positioned on the parallel swing, which is surrounded by the visual stimulus frame. The parallel swing and frame are shown suspended from the ceiling by three wires for horizontal axis stimulation. For vertical axis motions specially tuned springs are substituted for the wires.

otolithic neurons in the vestibular nuclei are influenced by visual input (4). Seventy-six percent of the 45 units sampled in the area of the medial vestibular nuclei were modulated not only by motion of the parallel swing in one or more of the three axes of stimulation, but also by motion of the visual stimulus frame in the corresponding axis.

An example of a unit which responded to both visual and vestibular stimulation is shown in Fig. 27-2A. This neuron was modulated by vestibular stimulation alone, and also by movement of the visual stimulus past the stationary animal. It should be noted that the unit's responses to visual and vestibular stimuli were of opposite phase. The unit was maximally excited by movement of the visual stimulus in the direction opposite to the direction which provided maximal excitation when the platform was moved. Since the same displacement of the retinal image occurs when the animal is moved in one direction or the visual surround is moved in the opposite direction, the resulting visual and vestibular inputs provide the unit with congruent infor-

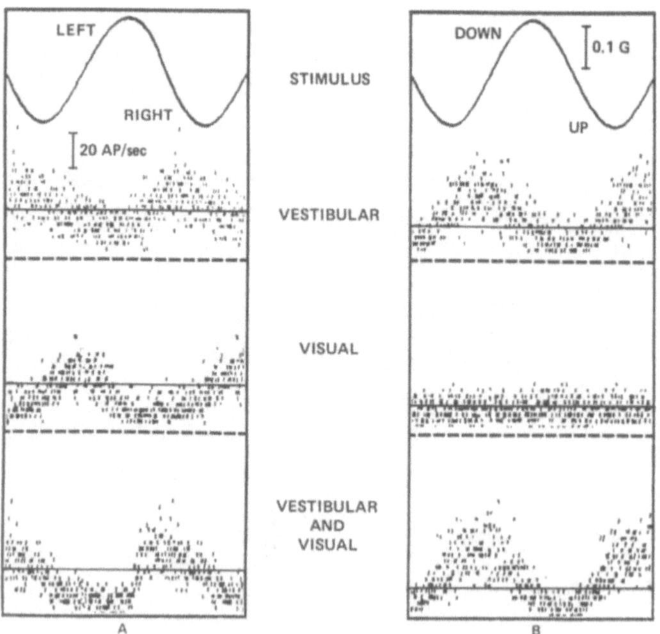

Fig. 27-2. **A** Response of a visually sensitive unit to motion in the right-left direction. **B** Response of a nonvisual unit to motion in the up-down direction. Both units were subjected to 60 cycles of stimulation under the following conditions: vestibular, visual, and combined visual-vestibular. Each point represents the average response rate in action potentials/s at that portion of the acceleration cycle. Solid horizontal line in each condition represents level of spontaneous activity. Peaks of the sine wave represent maximum stimulus acceleration points.

mation. In addition, the shape and phase of the unit's response to combined visual and vestibular stimulation were different from the shape and phase observed with either vestibular or visual stimulation alone, demonstrating that the two inputs have been integrated in the unit. These effects were displayed by all of the visually sensitive neurons in this study.

An example of a unit which was not modulated by visual stimulation is shown in Fig. 27-2B. This unit, like all the nonvisual units we studied, showed no response to visual stimulation and responded very similarly to the vestibular and combined visual-vestibular stimulus configurations.

Visual-vestibular interactions were common in each of the three axes of stimulation. However, units sensitive to vestibular stimulation in the fore-aft axis more often responded to visual input than did units sensitive to stimulation in either the right-left or up-down axes. Visual modulation was observed in 82% of the units responsive to fore-aft stimulation, but only 58% and 52% of units sensitive to right-left and up-down stimulation, respectively, showed visual modulation. The fact that the response was so prevalent in the fore-aft direction, in which no extraocular eye movements are induced, suggests that the visual modulation is not related to eye movements but to stimulation of the retina itself. This hypothesis is supported by evidence for the dissociation of eye movements and self-motion perception in studies with human subjects (2), and by demonstration of the independence of eye movement potentials and visual modulation of canal-related neurons in monkeys (12).

Twenty-four percent of the units tested in this study failed to respond to visual stimulation. This finding differs from those reported by Waespe and Henn (12) and Keller and Precht (8), who found that almost all of the canal-related units sampled in their study of the vestibular nuclei of monkeys and cats are modulated by visual inputs. There are two factors which may explain these differences. Perhaps fewer otolith-related than canal-related neurons receive visual inputs. However, the discrepant results are more likely explained by procedural differences among the studies. We have observed that visual modulation at the level of the vestibular nucleus depends on the attentive state of the animal. If the animal is not alert, the effect disappears. In the experiments cited above, amphetamine was used to maintain the animals in an attentive, alert state. However, we were reluctant to use amphetamine since it is known to have a direct effect on neurons in the vestibular system (9). Instead, we attempted to ensure that the subject was alert by arousing it with auditory and tactile stimulation before each 2-min period of data collection. This method of arousing the animal may produce a less consistent level of attention than the use of amphetamine; however, this method is free from possible confounding side effects. We are presently assessing the influence of attention on visual modulation by monitoring cortical potentials during unit recording. Control of the attentive state of the subject, whether by monitoring cortical potentials or by using alert, behaving animals, will make it possible to determine whether differences in the degree of visual interaction in canal- and otolith-related neurons actually exist.

Effects of Various Combinations of Visual and Vestibular Stimulation

To characterize the nature of the visual influence on otolith-related units, we examined the effects of several combinations of visual and vestibular inputs on the responses of a second sample of visually sensitive neurons in the vestibular nuclei ($N = 25$). We recorded the response of each unit to vestibular stimulation in the dark (V_D), and to visual-vestibular stimulation in the light (VV_1), as described in the previous experiment. In addition, two new stimulus conditions were presented. In one condition, responses were recorded during stimulation provided by movement of the parallel swing and the visual stimulus frame in a coupled configuration (VV_0). Retinal displacement of the visual image does not occur in this condition. In the other condition, visual and vestibular stimuli were applied in opposite directions (VV_2), resulting in twice the rate of image displacement that occurs in the VV_1 condition. Visual and vestibular inputs provided congruent information about the direction of the animal's motion in both the VV_1 and the VV_2 conditions; congruent information was not provided in the VV_0 condition. Stimulation in the VV_1 condition is similar to that experienced by an animal as it moves through a stationary visual environment. In the following discussion we have designated VV_1 as the normal, congruent condition, to distinguish it from the VV_2 condition.

Figure 27-3 shows the response of one unit to these four stimulus configurations. In this unit, the addition of the visual stimulus had the effect of increasing the gain of the response over that observed with vestibular stimulation alone. This effect was observed in most of the units tested. Units not displaying this effect tended to fall into a special class of neurons that showed responses of the same phase to visual and vestibular stimulation, unlike the majority of cells in which the visual and vestibular responses were of opposite phase. These neurons, which we have called "same phase" units, were excited by visual or vestibular stimulation in the same direction, which results in displacement of the retinal image in opposite directions.

Figure 27-4 summarizes the results of comparing the three VV conditions with the V_D condition for the frequently encountered opposite phase units ($N = 20$) and for the less frequently encountered same phase units ($N = 5$). Data presented in the upper portion of the figure represent the percentage of units in which the gain of the responses to the various VV conditions exceeded that observed in the V_D condition. The lower portion of the figure represents the percentage of units in which the sensitivity was greater with V_D stimulation alone. Comparisons of V_D with VV_1 and VV_2 show that the sensitivity of most opposite phase units was increased by the addition of visual input when that visual input resulted in displacement of the retinal image. The fact that stimulation in the VV_0 condition did not consistently result in an increase in sensitivity suggests that displacement of the retinal image, not just general illumination of the retina, is the factor producing the increase in gain. Although a significant difference in sensitivity was observed between VV_0 and VV_1, no such difference was found between the VV_1 and the VV_2

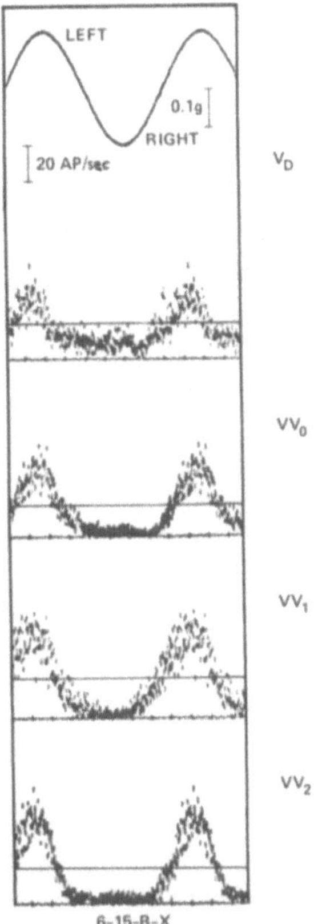

Fig. 27-3. Changes in firing rate of a visually sensitive otolith-related unit produced by sinusoidal vestibular stimulation in the dark (V_D) or combinations of visual-vestibular stimulation. VV_0, coupled movement of subject and visual stimulus (no retinal image displacement); VV_I, normal, congruent visual-vestibular stimulation produced by movement of the subject past a stationary visual stimulus; VV_2, movement of subject and visual stimulus in opposite directions. Each plot is the average of 60 cycles of stimulation. Solid horizontal lines represent level of spontaneous activity. Peaks of the sine wave represent maximum stimulus acceleration points.

stimulus configurations. This result suggests that the rate of the retinal displacement is not important, at least at this particular frequency of visual stimulation.

Exactly the opposite data profile is seen in the same phase units in Fig. 27-4. All of these units were more sensitive to vestibular stimulation alone (V_D) than to either of the conditions that produced retinal displacement of the visual image (VV_1 and VV_2). This finding is not unexpected. Sensitivity appears to increase with the addition of visual input if stimulation results in movement of the retinal image in the direction producing excitation of the unit at the same time that the animal is moved in the direction causing ves-

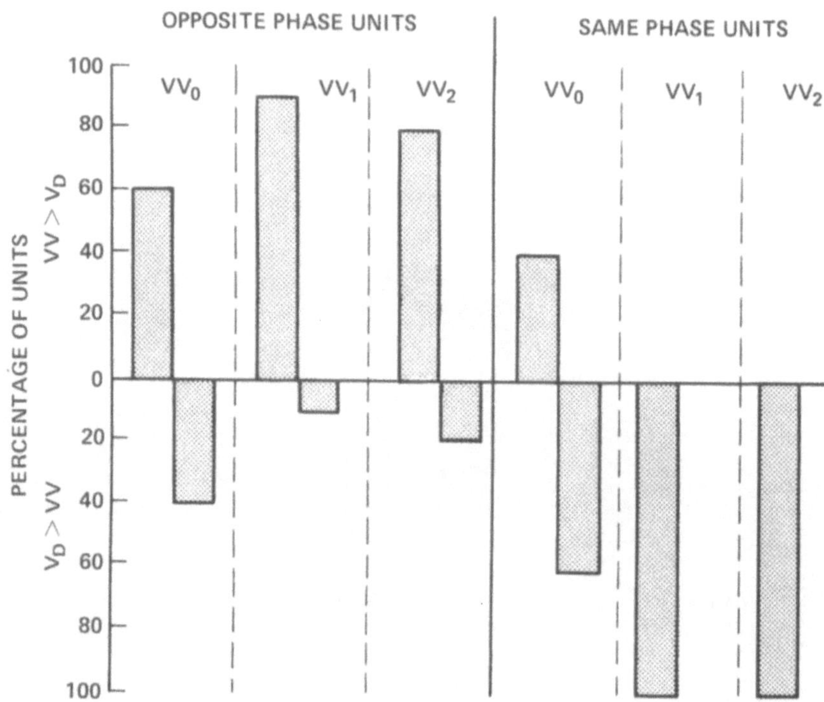

Fig. 27-4. A comparison of the gain of unit activity for vestibular stimulation in the dark (V_D) and combinations of visual-vestibular stimulation in opposite phase and same phase units. Percentages of units with greater gain for VV stimulation are represented in the upper portion of the figure; percentages of units with greater gain for V_D are represented in the lower portion of the figure.

tibular excitation. This stimulus configuration occurs with the VV_1 and VV_2 stimulus conditions for opposite phase units. However, these two stimulus configurations fail to produce simultaneous vestibular and visual excitation in same phase units. Instead, one input tends to increase the unit's activity, while the other input acts to decrease it. From these data it is not possible to determine whether the reduced sensitivity observed in the congruent VV conditions relative to V_D reflects an additive interaction of the visual and vestibular inputs, or whether the influence of the visual channel is diminished when conflicts occur, as the nonlinear models of Young (13) and Zacharias (14) predict. Either an additive or a nonlinear model would predict the results we obtained in same phase cells.

Since the number of same phase units examined in this experiment was small, these data can only be taken as suggestive. However, it should be noted that a small number of same phase units has also been recorded in canal-related neurons in the vestibular nuclei (12) and in the ventroposterior nucleus of the thalamus (3), suggesting that these cells comprise a distinct category of neurons in both the canal and otolith portions of the vestibular system. Such neurons should be especially responsive to combinations of visual and vestibular stimulation in which the subject and visual stimulus move in the same direction. This stimulus configuration occurs frequently in ordinary experience when, for example, the visual periphery of a driver is stimulated by an automobile passing in the same direction. Due to the small number of same phase units studied in this experiment, their data have been excluded from the analyses presented below.

Besides increasing neuronal sensitivity to translational motion, the addition of visual input appears to have a further effect. As demonstrated by the unit represented in Fig. 27-3, phase shifts occurred in the VV conditions relative to the phase recorded in the V_D condition. A significant number of units in all VV conditions showed shifts in phase from their responses in the dark, although the number of units showing these phase shifts was greatest in the VV_1 condition.

Data on phase shifts for individual neurons are represented in Fig. 27-5. The data show that, even at the single frequency of translational motion used in this experiment, the responses of otolith-related neurons show considerable variation in phase. However, given the phase of each unit's response to V_D stimulation, the addition of the normal, congruent visual input (VV_1) caused the phase of the response of most units to shift toward maximum velocity. Statistical analysis of the data shows that a significant number of units shifted toward maximum velocity with VV_1 stimulation, but for the VV_0 and VV_2 conditions, no consistency in the direction of phase shifts was observed.

In summary, data from opposite phase units show that the addition of visual stimulation increases the sensitivity of most otolith-related neurons to motion stimuli when the visual and vestibular inputs are congruent and result in displacement of the retinal image (VV_1 and VV_2). The addition of normal, congruent visual input also causes a shift in the phase of the response such that most units respond more closely in phase with maximum velocity. Thus, the addition of visual information may make the vestibular system a better sensor of the velocity of translational self-motion. This finding complements that reported by Dichgans et al. (7), who have shown that the addition of visual input allows canal-related units in the vestibular nuclei to respond not only to angular acceleration, but also to rotational velocity.

INFLUENCE OF VISUAL INPUT ON THE POSTURAL RESPONSES OF PIGEONS

The function of visual-vestibular interactions can best be assessed in combined neurophysiologic and behavioral investigations which examine the perceptual, postural, and oculomotor function of awake, behaving animals. Although we have not yet undertaken these combined experiments, we have examined the influence of visual-vestibular interactions on postural responses in pigeons. This work was conducted by Martha McCarty (10), a graduate student in our laboratory. In this experiment, pigeons sat on a

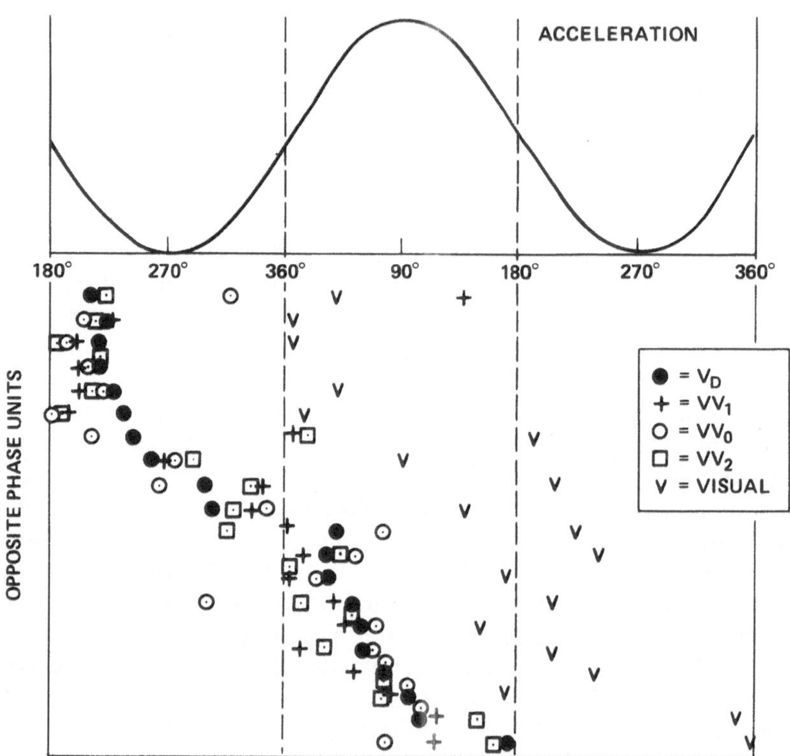

Fig. 27-5. Phase of the response of individual opposite phase units to visual, vestibular, and various combinations of visual and vestibular stimulation. Stimulus acceleration profile is represented by the sine wave at the top of the figure. Maximum stimulus velocity occurs at the points represented by the dashed lines.

perch instrumented with a strain gauge to measure the torque applied by the bird's feet as it maintained its balance. The cage and perch were placed on a parallel swing and were surrounded by a visual stimulus frame covered with a striped pattern. Both the parallel swing and visual stimulus frame could be moved sinusoidally in the fore-aft direction, either together or separately, providing translational motions of approximately 0.59 Hz and 0.15 *g*.

An example of the results of this study, shown in Fig. 27-6, demonstrates that visual stimulation was as effective as vestibular stimulation in evoking postural responses in pigeons. Similar balancing torques were applied whether the stimulus was actual fore-aft motion of the swing, or motion of the striped pattern alone. This investigation suggests that animals may experience an illusion of linear self-motion similar to that experienced by human subjects tested under similar conditions of visual and vestibular stimulation (1).

The next step in our research program is to combine neurophysiologic,

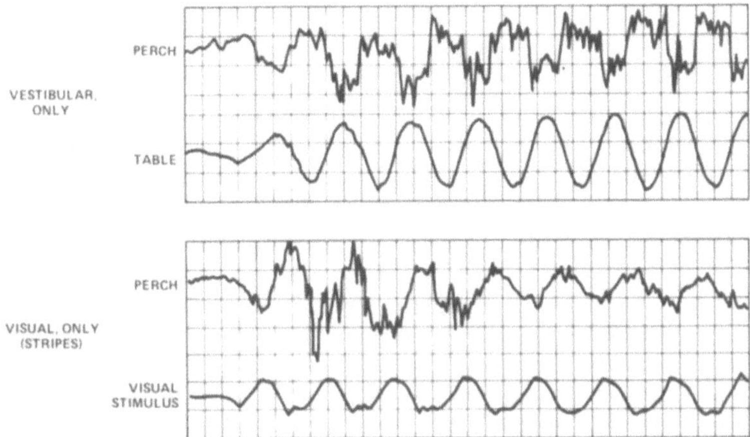

Fig. 27-6. Balancing response of a pigeon elicited by movement of the platform or the visual stimulus frame. Balancing responses were recorded by an instrumented perch; stimulation was monitored by accelerometers attached to the platform or the visual frame.

postural, and perceptual studies in a single animal. We are currently designing such studies using squirrel monkeys trained to detect their own self-motion. The basic paradigm described in this paper will be used to investigate visual and vestibular interactions in these animals. In addition, analyses of the harmonic content of single unit responses, and histologic verification of the location of the neurons studied, will be used to distinguish functional subgroups of neurons which receive both visual and otolithic inputs. These studies should begin to clarify the nature of visual-otolithic interactions at the level of the vestibular nucleus and delineate the oculomotor, postural, and perceptual consequences of these interactions.

References

1. Berthoz, A., Pavard, B., and Young, L.R.: Perception of linear horizontal self-motion induced by peripheral vision (linearvection): basic characteristics and visual-vestibular interaction. Exp. Brain Res. 23:471, 1975.
2. Brandt, T., Dichgans, J., and Koenig, E.: Differential effects of central vision versus peripheral vision on egocentric and exocentric motion perception. Exp. Brain Res. 16:476, 1973.
3. Büttner, E, and Henn, V. Thalamic unit activity in the alert monkey during natural vestibular stimulation. Brain Res. 103:127, 1976.
4. Daunton, N.G. and Thomsen, D.D.: Otolith-visual interactions in single units of cat vestibular nuclei. Neurosci. Abs. 2:1057, 1976.
5. Dichgans, J. and Brandt, T.: Visual-vestibular interaction and motion perception. Bibl. Ophthalmol. 82:327, 1972.

6. Dichgans, J. and Brandt, T.: The psychophysics of visually induced perception of self-motion and tilt. In Schmitt, F.O. and Worden, F.G. (eds.): The Neurosciences. Cambridge, Mass., MIT Press, 1974, pp. 123–129.
7. Dichgans, J., Schmidt, C., and Graf, W.: Visual input improves the speedometer function of the vestibular nuclei of the goldfish. Exp. Brain Res. 18:319, 1973.
8. Keller, E. and Precht, W.: Persistence of visual response in vestibular nucleus neurons in cerebellectomized cats. Exp. Brain Res. 32:591, 1978.
9. Kirsten, E., Schoener, E., and Wang, S.: Effects of D-amphetamine on single vestibular neurons. J. Pharmacol. Exp. Ther. 191:377, 1974.
10. McCarty, M. and Daunton, N.G.: Vestibular and visual effects on postural responses of pigeons. Paper presented at the Western Psychological Association Meeting, Seattle, April 21–24, 1977.
11. Pavard, B. and Berthoz, A.: Perception du mouvement et orientation spatiale. Travail Humain 39:207, 1976.
12. Waespe, W. and Henn, V.: Neuronal activity in the vestibular nuclei of the alert monkey during vestibular and optokinetic stimulation. Exp. Brain Res. 27:523, 1977.
13. Young, L.: On visual-vestibular interaction. In: Fifth Symposium on the Role of the Vestibular Organs in Space Exploration. NASA SP-314, 1970, pp. 205–210.
14. Zacharias, G.: Motion Sensation Dependence on Visual and Vestibular Cues. Ph.D. Thesis. M.I.T., 1977.

Discussion

CORREIA: Two questions. First, what is the magnitude of the angular acceleration you obtain from your angular accelerometers at the end-points of excursion of your parallel swing? Second, I wonder if you would comment on why we should expect any high frequency dynamics of the otolith organs to *y*-axis translations?

DAUNTON: I'm not sure I understand that.

CORREIA: Well, functionally we would expect high frequency maculocollic or maculoocular responses to movement of the linear acceleration vector due to gravity in the sagittal head plane, since this frequently occurs as we run and our head pitches back and forth. However, tilt of the head left and right is usually a low frequency maneuver. As I perceived your results, you showed otolithic responses to reasonably high *y*-axis translational frequencies. I wonder if these responses might not have been produced by stimulation of the semicircular canals and from subsequent convergence of semicircular canal afferents on otolithic neurons in the vestibular nuclei.

DAUNTON: I think that this is very unlikely. Although I do not have the figures for the end-point accelerations, if you are talking about stimulation in the right-left direction, we have seen some units which are very, very sensitive. You need to move the table only about an inch to get good modulation of the unit. Under those conditions I don't think you would have significant angular accelerations at the turnaround points.

28

Cerebellectomy in Goldfish Prevents Adaptive Gain Control of the VOR without Affecting the Optokinetic System

JOHN O. SCHAIRER AND MICHAEL V.L. BENNETT

The vestibuloocular reflex (VOR) stabilizes the visual image on the retina. Head rotations are detected by the vestibular apparatus and the eyes are moved in the opposite direction to compensate. For ideal compensation the eye velocity is equal and opposite to the head velocity. If compensation is not perfect the visual image slips on the retina. Retinal slip activates the optokinetic reflex which counteracts the slip, providing further stabilization of the visual image. The optokinetic reflex, however, is slow compared to the VOR and is of primary value at low velocities of rotation.

In addition to causing optokinetic eye movements, retinal slip occurring during head movements slowly modifies the VOR in the appropriate manner to reduce slip (15,18,20,21). The gain of the VOR can be defined as the ratio of eye angular velocity to head angular velocity in the opposite direction measured in the dark. If the amount of retinal slip during head rotations in the light is altered by having the subject wear glasses or by moving the visual field as well as the head, the gain of the VOR (measured in the dark) then changes in the direction that would reduce slip if the visual field were illuminated. This process has been termed adaptive gain control of the VOR (20).

Several experimental procedures have been used to alter the relationship between the rotation of the head and the concurrent apparent motion of the visual surround. Telescopic spectacles which magnify the image 2 × require eye movements twice as large as a given head movement in order to have a perfectly stabilized visual image on the retina. Since wearing 2 × spectacles increases the VOR toward 2 (20), we will term this (and other situations requiring a VOR gain of 2 for retinal stabilization) a paradigm that trains the gain toward 2. By the same reasoning, an animal wearing 0.5 × spectacles is being trained toward 0.5 and an animal wearing no spectacles is being trained toward 1. Reversing prisms require eye movements exactly opposite to those of an animal wearing no spectacles, hence they train an animal toward −1. It is important to keep in mind that the VOR is defined by the response to vestibular stimulation alone and hence is always measured in the dark.

Each of these training situations can be simulated by moving the visual surround and the animal in a coordinated fashion. Training the gain toward 2 consists of rotating the visual surround equal and opposite to rotation of the animal. Training toward 1 consists of rotating the animal with the visual surround stationary. Not mentioned above, training the gain toward zero is performed by rotating the animal and the visual surround together. In each of these paradigms there is immediate, direct influence on eye movements when the visual field is illuminated (lights on), which is ascribable to optokinetic reflexes.

Chronic combined visual and vestibular stimulation in one of these paradigms produces adaptive gain changes in the response to purely vestibular stimulation. Adaptive gain control of the VOR has been demonstrated in humans (7–10,18), monkeys (20), cats (19,21), rabbits (15), chicks (11), and goldfish (22).

As the earliest of these reports of VOR gain changes were appearing, Ito (12,13) argued that unlike closed loop control systems such as the optokinetic system, the open loop VOR would need careful gain control to maintain reasonably accurate responses. Closed loop systems are inherently insensitive to internal gain variations, whereas open loop systems are directly affected by gain variations. Ito proposed that the cerebellum was responsible for gain control in the VOR. For this function the cerebellum would need a visual input which would convey the information of errors in the action of the VOR. As Ito (12,13) surmised, visual inputs to the cerebellum do occur (16,17).

Subsequently, Ito studied eye movements in the rabbit, which was rotated while viewing a vertical slit lamp. The lamp was either held stationary or rotated equal and opposite to the animal. With the stationary lamp retinal stabilization would occur with a VOR gain of + 1. With the rotating lamp the appropriate gain would be − 1 (14). Modest changes from the VOR measured in the dark were seen when the slit lamp was on, there being some tendency to follow the slit. These changes were eliminated by removal of the vestibulocerebellum. When the rabbit was subjected to several hours of training toward −1 or +1 with the vertical slit lamp (15), there were modest changes in VOR in the appropriate direction. This modification of the VOR was also eliminated by removal of the vestibulocerebellum.

Robinson demonstrated that removal of the vestibulocerebellum abolished VOR gain control in the cat (21). His animals were fitted with reversing prisms and allowed to wander about the lab for many days. Large reductions in gain were observed over the first few days. Vestibulocerebellectomy in trained animals abolished established gain reductions, and there was no return of the gain to low levels as the animals continued to wear the prisms.

We report here that cerebellar lesions can prevent adaptive gain changes of the VOR in goldfish. The same lesions do not affect responses to sinusoidal optokinetic stimulation or the immediate visual influence on the response to rotation in the light during training toward zero.

METHODS

Goldfish were restrained in a clear plastic, cylindric tank surrounded by a 20-in. diameter, black-and-white striped drum illuminated from above (Fig. 28-1). The tank sat on a platform which was sinusoidally rotated with a variable speed motor. It produced a constant 20° amplitude sine wave with a

Fig. 28-1. The experimental setup. As seen on the right the goldfish is restrained in the clear plastic tank which rests on a platform suspended from above. A black field coil for the search coil system can be seen in front of the fish. A section of the striped drum has been removed. A variable speed motor (left) drives the platform and the drum through eccentrics and connecting rods. Water is perfused over the fish's gills and through the tank. A reservoir and some of the plumbing can be seen in the background.

variable period from 64 s (0.0156 Hz) to 2 s (0.5 Hz). The drum was held stationary or rotated with the platform, thus providing training toward 1 or 0, respectively.

Position of the left eye was monitored by a magnetic search coil. The prewound and precalibrated search coil was tied to pins inserted in the cornea on either side of the pupil. Goldfish lack an eyelid and appear to tolerate the procedure very well.

A typical eye position trace produced by rotation of the animal is shown in Fig. 28-2A. It is easy to see the slow phase response to the sinusoidal rotation between the superimposed saccades. Electronically differentiating the eye position signal gives the eye velocity trace (Fig. 28-2B). Saccades now become interruptions in the trace and the sinusoidal nature of the response is fully revealed. Gains were measured directly from velocity traces by measuring the peak-to-peak amplitude of the trace and dividing by the known peak-to-peak amplitude of the stimulus, which in this case was

$$2\pi fA = 2\pi \times 0.125 \text{ Hz} \times 20° = 31.4°/\text{s}.$$

Training lasted 4 to 7 h and was interrupted by short, about 1 min, dark periods to determine the gain of the VOR. Test periods were separated by at least one-half hour. At the beginning and end of training the gain of the VOR was measured in the dark at a range of frequencies from 0.0156 to 0.5 Hz.

Eight of the ten operations were performed under Tricaine methane-sulfonate anesthesia; two fish were operated on without anesthesia. Goldfish tolerate such operations quite well; they generally only react to removal of the skin over their skull.

Operations were performed by first removing a piece of the skull over the cerebellum and optic tectum. The fat which covers the brain was aspirated away until the dorsal surface of the brain could be seen clearly. The brain lesions were made by aspiration under visual control through a dissecting microscope. After lesioning, the space over the brain was filled with stopcock grease and the skull resealed either with a dental cement cap or by gluing the flap of bone back on with Permabond, a brand of cyanoacrylate glue. Four of the eight anesthetized fish were allowed a day to recover from the operation before assessment of the VOR gain control system began. Eye coils were placed on all the other fish during the operation. Their VOR gain control systems were assessed as soon as recovery from the anesthesia was indicated by a normal pattern of saccadic eye movements (5).

RESULTS

VOR Gain Control and Optokinetic Reflexes in Unoperated Animals

An example of the adaptive gain changes of the VOR that exist in the goldfish is illustrated in Fig. 28-2C. Each of 12 fish was rotated for 6 h together with the striped drum. This procedure drives the gain toward zero. Before training was started the gain was measured in the dark (open circles). When

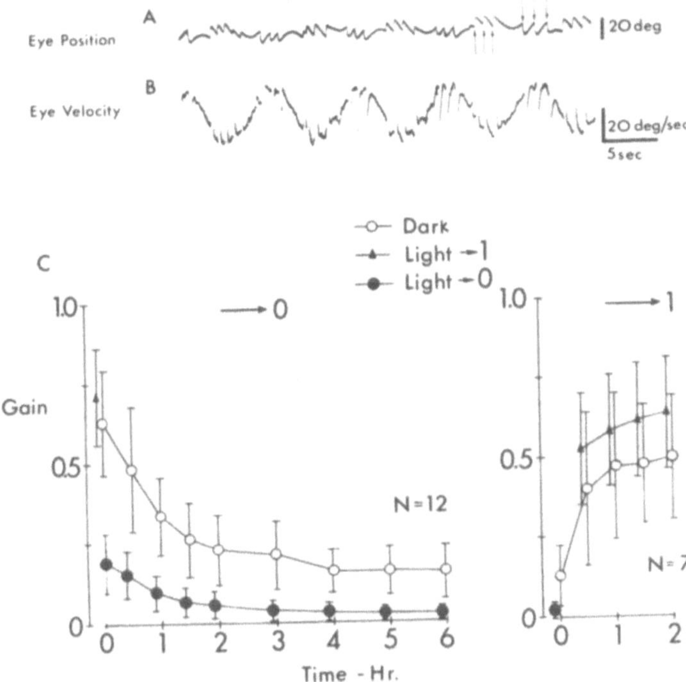

Fig. 28-2. Eye movements and VOR gain changes in unoperated fish. **A** Sample eye position record. These eye movements were observed during sinusoidal rotation with a frequency of 0.125 Hz and an amplitude of 20° (40° peak to peak). The arrows at the right point to saccades superimposed on the slow movements. **B** Eye velocity given by electronic differentiation of the position trace shown in **A**. Saccades made interruptions in the trace as the velocities went off scale. **C** Gain changes during and after training toward zero of normal unoperated fish. Gains were measured during rotation of the fish in the light with (1) a stationary striped drum (*filled triangles*), i.e., training toward 1; and (2) the striped drum moving with the fish (*filled circles*), i.e., training toward 0. Gains were also measured during rotation of the fish in the dark (*open circles*), i.e., assessment of the VOR. Means and standard deviations for 12 fish during 6 h of training toward zero are shown on the left. Means and standard deviations for seven of them during 2 h of training toward 1 are shown on the right.

the lights came on again the striped drum was moving with the fish (filled circles) and training toward zero was started. The gain measured within two cycles of rotation after the lights were turned on was reduced to about 26% of the VOR gain measured just before the lights were turned on. This decrease presumably occurs because the VOR-generated eye movements produce retinal slip in this situation and the retinal slip counteracts the eye movement through the optokinetic system.

During the first 2 h of rotation with the surround, the VOR gain, which was always measured in the dark, decreased rapidly. The subsequent 4 h of

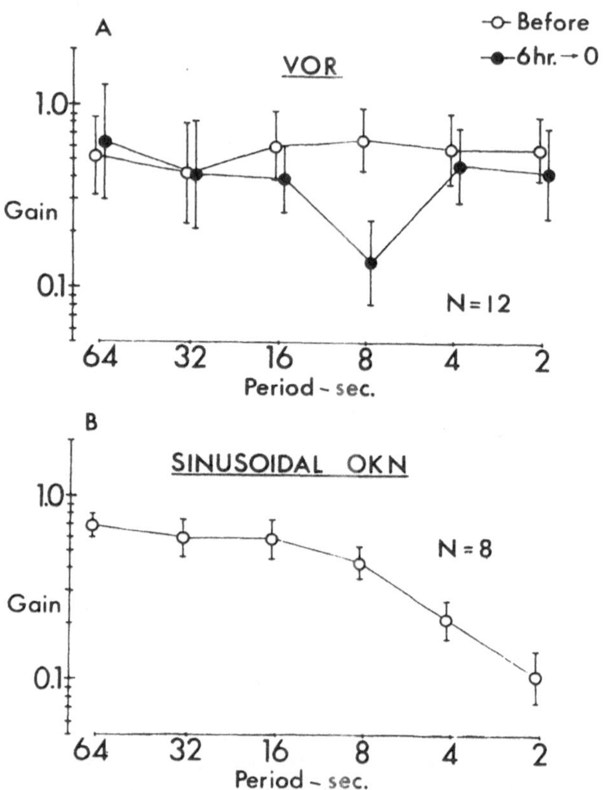

Fig. 28-3. **A** Frequency responses of the VOR of normal unoperated fish measured in the dark before (*open circles*) and after (*filled circles*) 6 h of training toward 0. The gain reduction is predominantly at the training frequency. Means and standard deviations are plotted for eight fish. **B** The frequency response to sinusoidal optokinetic stimulation of 20° amplitude (40° peak to peak) in normal unoperated animals before training. Means and standard deviations are plotted for eight fish.

training produced more modest changes. Gains measured in the light during training toward zero decreased in a similar manner.

At the end of 6 h of training the gain toward zero, seven of the fish were rotated inside a stationary illuminated drum. The VOR gain rapidly increased during this 2-h training toward 1.0.

At the beginning of the experiment and after 6 h of training toward zero, the frequency response of the VOR gain was assessed (Fig. 28-3A). The frequency response measured before training shows little or no effect of frequency over the range of measurement. However, after 6 h of training the VOR gain had decreased primarily at the training frequency (0.125 Hz). Analysis of variance with repeated measures also showed significant decreases at 0.063 and 0.5 Hz ($p < 0.01$ and $p < 0.05$, respectively).

The direct optokinetic response was measured during sinusoidal rotation of the striped drum around stationary fish at various frequencies (Fig. 28-3B). Gains decreased with increasing frequency which at constant amplitude is equivalent to increasing maximum velocity.

The gains of the sinusoidal optokinetic responses at the training frequency in the goldfish are at least an order of magnitude higher than gains in responses to corresponding optokinetic stimuli in the Dutch Belted rabbit (1–3).

VOR Gain Control in Operated Animals

The cerebellum in goldfish has a corpus which can be seen from the dorsal surface (Fig. 28-4D) and a valvula cerebellum which extends forward under the optic tecta and over the tegmentum. Removal of a large part of the corpus cerebellum abolishes adaptive gain changes of the VOR.

Three classes of lesions were carried out: (1) near total removal of the corpus cerebellum; (2) partial lesions of the corpus cerebellum; and (3) sham operations where a portion of the tectum was removed.

In three fish the corpus cerebellum was totally aspirated or nearly so (Fig. 28-4A). These fish showed no significant gain reduction during 5 to 7 h of training toward zero (Fig. 28-5, asterisks; because of variability between fish, each fish's gain is normalized with respect to its initial value). A linear regression line was fit to the time course of the VOR gain from each fish. The slope of that line was tested for significant deviation from zero. The data from two of these three fish had slopes which were not significantly different from zero. The third fish had a small but significant increase in gain over 5 h of training toward zero ($p < 0.01$).

In three fish partial cerebellar lesions were made (Fig. 28-4B and C; Fig. 28-5B and C). These lesions were on the dorsal and dorsocaudal surface of the corpus. One of these fish showed no significant gain change during training and one showed a small gain increase (Figs. 28-4B and 28-5B). The third fish with a partial lesion (Figs. 28-4C and 28-5C) did show a gain reduction after 2 h of training. This change was in the range of those observed in unoperated animals. To ensure that the reduction in gain was not due to deterioration of the animal, it was transferred to the condition of training toward 1. After 2 h, two-thirds of its original gain had returned. Although this increase in gain was slower than those of unoperated fish (Fig. 28-2C), the gain increase clearly indicates that the reduction with training toward zero was a real change in the VOR gain and not caused by deterioration of the fish.

In the sham operations portions of the brain which do not participate in either the VOR or the optokinetic system were lesioned (Figs. 28-4D and 28-5D). Usually a piece of the optic tectum was removed which has been shown to have no effect on optokinetic nystagmus (6). In four sham-

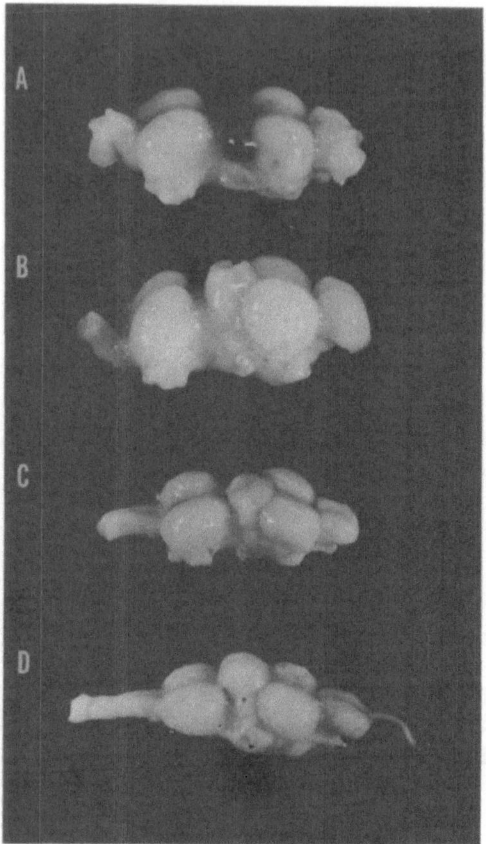

Fig. 28-4. Views of the right sides of four lesioned brains. Four major lobes can be seen of which only the cerebellum is unpaired. From anterior to posterior (right to left) they are the forebrain, the optic tectum, the corpus cerebellum, and the vagal lobes. The training data from each example are shown in Fig. 28-5. **A** A nearly complete removal of the corpus cerebellum. **B** A partial lesion of the dorsal corpus cerebellum which prevented VOR gain reduction during training. **C** A partial lesion of the caudodorsal corpus cerebellum which did not prevent VOR gain reduction. **D** A small lesion of the dorsal optic tectum which did not prevent VOR gain reduction.

operated controls, the gain was decreased by the training procedure in a completely normal manner (Fig. 28-5, filled circles).

These ten operated goldfish formed two definite groups. Five of them showed clear, normal reductions in gain with training toward zero. The other five showed no significant reduction in gain; two of them actually showed small increases.

The range of initial postoperative gains was quite large; from 0.45 to 1.45.

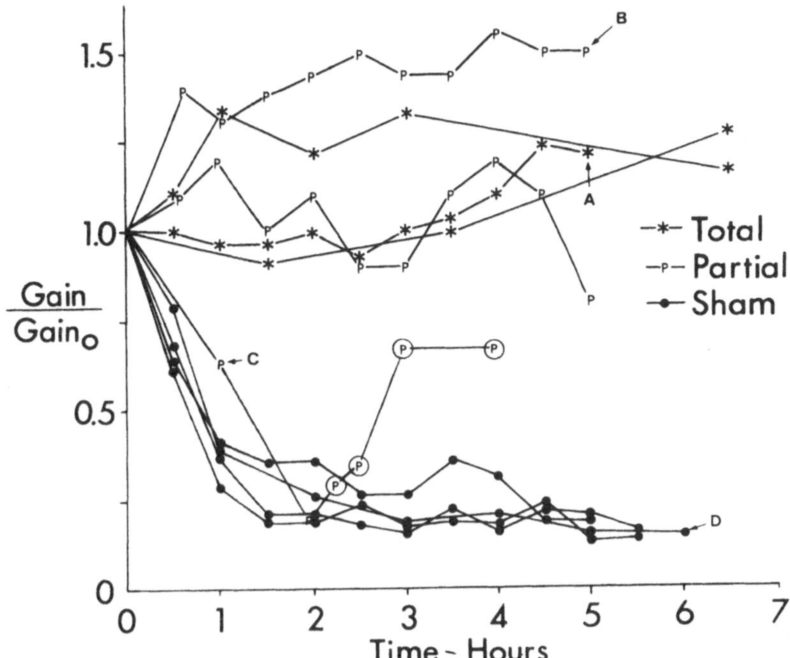

Fig. 28-5. Gain changes at the training frequency for the ten operated fish. For each fish gain is normalized with respect to its initial value. Data from the three classes of lesion are shown: "total" cerebellectomies (*asterisks*), where the corpus cerebellum was totally removed or nearly so; partial lesions of the corpus cerebellum (*P*); and sham controls which had portions of the optic tectum lesioned (*filled circles*). Data from the four fish whose brains are shown in Fig. 28-4 are labeled correspondingly *A*, *B*, *C*, and *D*. In one fish with a partial lesion training toward 0 produced a marked gain reduction in 2 h. It was then trained toward 1 for 2 h (*circled P*). The gain largely recovered indicating that the gain reduction was not due to morbidity.

The initial gains measured at the training frequency of the five fish which showed no gain reduction had a significantly larger variance between them than did the initial gains of the five fish which showed gain reductions ($F = 9.28$; $p < 0.05$). This result may indicate that previous gain adjustments made while the fish were swimming in a normal environment were eliminated by the lesions.

The mean initial gain for the five operated fish which showed normal gain changes was 0.96 ± 0.13, which is significantly higher than the initial gain for 12 normal fish (0.63 ± 0.16; $p < 0.05$; t test). This difference may reflect an arousal effect of the operation. The mean initial gain of the operated fish not showing gain reductions was greater than that of normals but the difference was not significant. A trend would have been obscured by the increased variance.

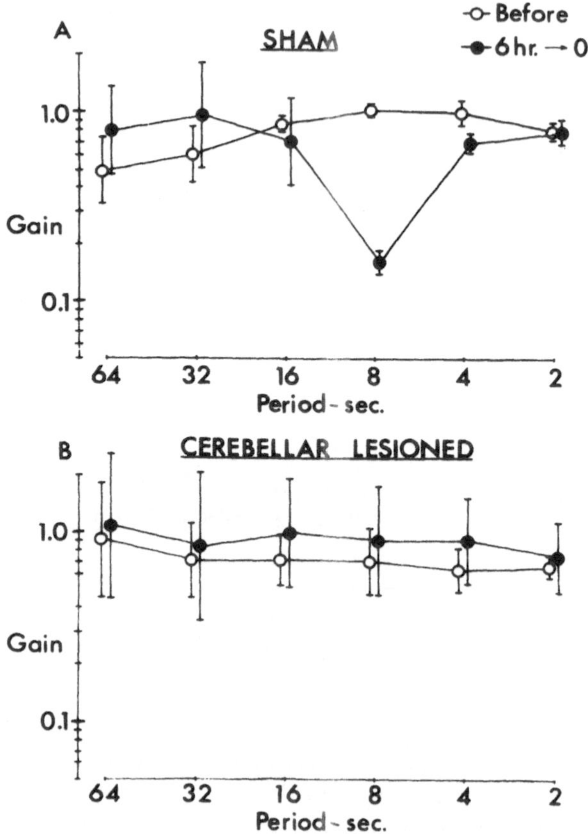

Fig. 28-6. Frequency response of the VOR taken before (*open circles*) and after 6 h of training toward 0 (*filled circles*). **A** Data for four of the sham-operated controls. The gain reduction was predominantly at the training frequency as in unoperated animals. **B** Data for four of the cerebellar-lesioned animals which did not exhibit gain changes at the training frequency. There was no significant gain reduction at the other frequencies. Means and standard deviations are plotted for each group.

Frequency responses of the VOR were determined before and after training the operated fish. The sham-operated fish showed the same predominance of gain reduction at the training frequency as seen in unoperated fish (Fig. 28-6A). Four fish that showed no gain reduction at the training frequency also showed no gain reduction at other frequencies (Fig. 28-6B). No significant differences were seen between the gains measured before and after training (analysis of variance with repeated measures). There may be an increase in gain but there are too few fish to achieve statistical significance.

In order to assess the effect of the cerebellectomies on visual input to the oculomotor system, the optokinetic response to sinusoidal rotation of the

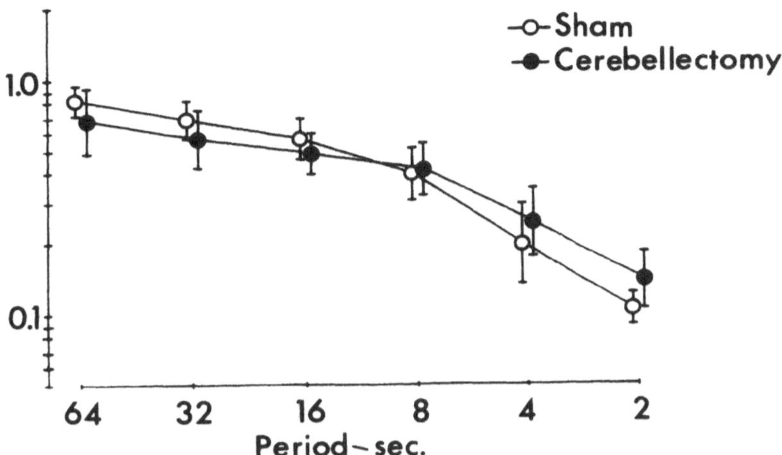

Fig. 28-7. The frequency response to sinusoidal optokinetic (OKN) stimulation: similarity of sham-operated animals and operated animals not showing VOR gain changes. The graph shows data from four sham-operated controls (*open circles*) and data from four animals with cerebellar lesions which blocked reduction in VOR gain (*filled circles*). Means and standard deviations are plotted for each group.

striped drum around stationary fish was measured (Fig. 28-7). No significant difference was found between operated and unoperated fish whether or not the latter showed gain reductions.

Another measure of optokinetic inputs is the immediate decrease in gain seen when the striped drum is illuminated while moving with the fish (Fig. 28-8). The degree of this suppression in operated fish not showing gain reductions was within the range of that observed in normal unoperated fish measured within the first two cycles of training toward zero. (The mean and standard deviation for normals are shown on the left in Fig. 28-8.) In fish not showing gain reductions, the response when the lights were on maintained a fairly constant level for each fish over 5 to 7 h of training toward zero. This result is in contrast to the continuous reduction in the gain measured with the lights on in normal unoperated fish (Fig. 28-2C). It seems reasonable to conclude that the slow progressive reduction in gain measured with the lights on in normal fish reflects the reduction in vestibular influence seen when the lights were off. Similarly, the lack of reduction in gain measured when the lights were on in the cerebellectomized fish presumably reflects the lack of reduction in gain measured when the lights were off. Thus, the immediate optokinetic influence on eye movements appears to be intact in both of two different stimulus paradigms: rotation of the drum around the stationary fish and rotation of the fish and drum together.

A further indication of the specificity of the lesions causing loss of gain control is that saccadic eye movements were apparently unaffected.

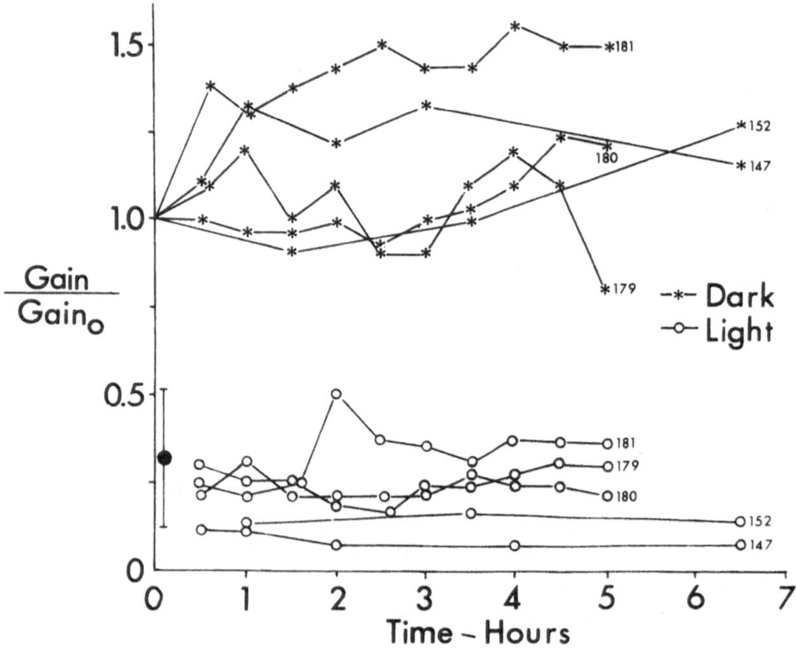

Fig. 28-8. Optokinetic effect during training toward 0 of operated animals not showing VOR gain reduction. The gains of the VOR (measured in the dark) are replotted from Fig. 28-5 (*asterisks*). The gains when the striped drum was illuminated (*open circles*) were much smaller and remained quite constant during training. These gains were in the range of those measured from the first two cycles of rotation during training toward 0 of normal unoperated animals (mean, *filled circle,* and standard deviation are shown at the lower left). In unoperated animals, however, the gain when the striped drum was illuminated decreased during training as the VOR gain decreased (Fig. 28-2**C**).

DISCUSSION

We have shown that in the goldfish, VOR gain reductions due to chronic rotation together with the visual surround can be completely prevented by partial cerebellar removal. Indeed, the VOR gain actually increased in two fish.

These lesions have no detectable effect on the sinusoidal optokinetic response or on the immediate reduction in gain measured in the light when the visual field moves with the fish. We cannot exclude that the optokinetic system requires a different portion of the cerebellum than that removed in these experiments, although Easter also found that very extensive lesions of the cerebellum did not prevent optokinetic responses in goldfish (6).

The goldfish is similar to the rabbit and cat in that adaptive gain changes can be abolished by cerebellar lesions. However, unlike the rabbit, lesions which blocked adaptive gain changes did not affect the immediate visual influence on eye movements. This discrepancy may be due to species differences or differences in experimental design. A reduction in optokinetic response has been seen in monkeys with flocculus lesions (25). Moreover, after floccular lesions a reduction of the immediate visual influence on calorically induced vestibular nystagmus was found by Takemori and Cohen (23,24). However, the monkey is a foveate animal which may have different visual influences over eye movements than exist in nonfoveate animals such as rabbits and goldfish.

The optokinetic response of the goldfish is much stronger than that of the rabbit. This difference may explain why the magnitude of the gain change during training was greater and faster in goldfish than in the albino rabbit (15). An additional factor may be the reversed optokinetic response from the anterior visual field that is found in the albino rabbit (4).

The goldfish is an especially promising animal for studying adaptive gain control because of the rapid, strong modification of the VOR and ease of handling. The cerebellar cortex is smooth and unconvoluted and lesions affecting adaptive gain control should be easily localized. Also, as we have shown here, vision in the goldfish affects responses to rotation through two separable circuits; the long-term visual influence of adaptive gain control requires a portion of the cerebellum that is not necessary for immediate optokinetic responses.

ACKNOWLEDGMENTS

The data in this paper are from a thesis submitted in partial fulfillment of the requirements for the degree of Doctor of Philosophy in the Sue Golding Graduate Division of Medical Sciences of the Albert Einstein College of Medicine. J.O.S. was partially supported by Training Grant 5T32GM7288 from the National Institutes of Health. The research was supported in part by grants HD-04248, NS-12627, and NS-07512 from the NIH and BNS 77-24420 from the National Science Foundation.

References

1. Baarsma, E.A. and Collewijn, H.: Vestibulo-ocular and optokinetic reactions to rotation and their interaction in the rabbit. J. Physiol. 238:603, 1974.
2. Collewijn, H.: Optokinetic eye movements in the rabbit: input-output relations. Vision Res. 9:117, 1969.
3. Collewijn, H.: Latency and gain of the rabbit's optokinetic reactions to small movements. Brain Res. 36:59, 1972.
4. Collewijn, H., Winterson, B.J., and Dubois, M.F.W.: Optokinetic eye movements in albino rabbits: inversion in anterior visual fields. Science 199:1351, 1978.
5. Easter, S.S.: Spontaneous eye movements in restrained goldfish. Vision Res. 11:333, 1971.
6. Easter, S.S.: Horizontal eye movements in fish. In Ali, M.A. (ed.): Vision in Fishes. New York, Plenum, 1975.
7. Gonshor, A. and Melvill Jones, G.: Plasticity in the adult human vestibulo-ocular reflex arc. Proc. Can. Fed. Biol. Soc. 14:11, 1971.

8. Gonshor, A. and Melvill Jones, G.: Changes of human vestibulo-ocular responses induced by vision reversal during head rotation. J. Physiol. 234:102P, 1973.

9. Gonshor, A. and Melvill Jones, G.: Short term adaptive changes in the human vestibulo-ocular reflex arc. J. Physiol. 256:361, 1976.

10. Gonshor, A. and Melvill Jones, G.: Extreme vestibulo-ocular adaptation induced by prolonged optical reversal of vision. J. Physiol. 256:381, 1976.

11. Green, A.E. and Wallman, J.: Rapid change in gain of vestibulo-ocular reflex in chickens. Soc. Neurosci. Abstr. 4:492, 1978.

12. Ito, M.: Neurophysiological aspects of the cerebellar motor control system. Int. J. Neurology 7:162, 1970.

13. Ito, M.: Neural design of the cerebellar motor control system. Brain Res. 40:81, 1972.

14. Ito, M., Shiida, T., Yagi, N., and Yamamoto, M.: Visual influence on rabbit horizontal vestibulo-ocular reflex presumably affected via cerebellar flocculus. Brain Res. 65:170, 1974.

15. Ito, M., Shiida, T., Yagi, N., and Yamamoto, M.: The cerebellar modification of rabbits' horizontal vestibulo-ocular reflex induced by sustained head rotation combined with visual stimulation. Proc. Jap. Acad. 50:85, 1974.

16. Maekawa, K. and Simpson, J.I.: Climbing fiber activation of Purkinje cells in the flocculus to impulses transferred through the visual pathway. Brain Res. 39:245, 1972.

17. Maekawa, K. and Simpson, J.I.: Climbing fiber responses evoked in the flocculus by visual pathway stimulation in the rabbit. In: Proceedings of the XXVth International Congress of Physiological Sciences, Munich, 1971, p. 1060.

18. Melvill Jones, G.: Plasticity in the adult vestibulo-ocular reflex arc. Philos. Trans. R. Soc. Lond. [Biol.] 278:319, 1977.

19. Melvill Jones, G. and Davies, P.: Adaptation of cat vestibulo-ocular reflex to 200 days of optically reversed vision. Brain Res. 103:551, 1976.

20. Miles, F.A. and Fuller, J.H.: Adaptive plasticity in the vestibulo-ocular responses of the rhesus monkey. Brain Res. 80:512, 1974.

21. Robinson, D.A.: Adaptive gain control of vestibulo-ocular reflex by the cerebellum. J. Neurophysiol. 39:954, 1976.

22. Schairer, J.O. and Bennett, M.V.L.: Adaptive gain control in vestibulo-ocular reflex of goldfish. Soc. Neurosci. Abstr. 3:485, 1977.

23. Takemori, S. and Cohen, B.: Visual suppression of vestibular nystagmus in rhesus monkeys. Brain Res. 72:203, 1974.

24. Takemori, S. and Cohen, B.: Loss of visual suppression of vestibular nystagmus after flocculus lesions. Brain Res. 72:213, 1974.

25. Zee, D.S., Yamazaki, A., and Gucer, G.: Ocular motor abnormalities in trained monkeys with floccular lesions, Soc. Neurosci. Abstr. 4:168, 1978.

Discussion

YOUNG: Despite the lack of statistical significance, there does appear to be a consistent effect of cerebellectomy on the overall gain level.

SCHAIRER: I think maybe I could say something about the variance here. There are two parts to it. One is that when you do the cerebellectomy the gains tend to have a rather wide variation. The initial gain after cerebellectomy ran from 0.45 up to 1.45. So, removing the cerebellum may release the gains from the previously adapted level. The other thing that happens is that some of these gains get very big at low frequencies. This is a very strange effect that I have seen more consistently when I have rotated animals toward zero for 5 h a day for 5 days in a row. And I also see this every once in a while with just 6 h of training. The gains at these low frequencies go up to three or four. It's as if they've gotten so sensitive that they just sense change in direction and they produce a giant response.

OMAN: When you were stimulating these animals at a given frequency and then stopped the stimulation, did you ever notice pendulous eye movements continuing for several cycles? In other words, is the frequency selectivity, at least in the reduced gain case, related to learning the stimulus pattern?

SCHAIRER: That's a very interesting question. Because as I said this morning, if you take the fish and simply rotate the stripes for a number of hours about him, or in the reversal condition, where you rotate the fish at half the speed of the stripes, you do see continued movements after stopping rotation. In the dark the animal is not moving, but he moves his eyes back and forth in slow phases as if he was being sinusoidally rotated. I never see this effect in fish after training the gain toward zero or two.

29

Vestibular Nuclei Activity and Nystagmus in the Alert Monkey and Their Relation to Optokinetic and Vestibular Stimulation

U. BÜTTNER, U.W. BUETTNER, AND V. HENN

The concept of the function of vestibular nuclei in vestibuloocular reflex, posture control, and motion sensation has expanded considerably over the last few years. Earlier experiments on anesthetized or decerebrate animals concentrated on determining the transfer characteristics of vestibular nuclei neurons in response to angular acceleration (19,29,30,31) and time constants of the vestibular nuclei were found to be similar to those of the vestibular nerve (3,13,29). This suggested a relatively simple relay function of the vestibular nuclei in transmitting nerve activity.

Experiments performed in alert animals, however, revealed that the vestibular nuclei are a highly complex system integrating information from different sensory sources and interfacing them with the motor output. Besides being activated by the vestibular end organ, vestibular nuclei neurons are strongly and reliably modulated by proprioceptive stimuli to limb and neck afferents (14,28) and to moving visual fields which cause optokinetic nystagmus (2,11,18,22,34). These experiments demonstrate that activity changes in the vestibular nuclei can occur independently of those in the vestibular nerve.

Vestibular and optokinetic stimulation lead to eye movements in the form of nystagmus. This motor response is thought to be mainly caused by activity changes in the vestibular nuclei (37). In alert animals it is possible to exactly control the sensory stimuli (vestibular, optokinetic), to monitor the neuronal and motor (nystagmus) response, and then to correlate these different parameters.

Experiments will be described in which neuronal activity was measured in the vestibular nuclei during vestibular and optokinetic stimulation. In order to test whether activity changes only reflect the motor response, some animals were subjected to anesthesia, while others were trained to suppress their nystagmus. This permits the manipulation of the motor response, while the same sensory stimuli are applied. It also allows us to determine whether vestibular nuclei activity is only related to sensory input or is also dependent on the oculomotor (nystagmus) output.

METHODS

Experiments were performed on chronically prepared alert rhesus monkeys (*Macaca mulatta*). Details have been described previously (7). Monkeys were seated with their head fixed in a primate chair on a servo-controlled turntable. Single neurons were recorded with varnished tungsten microelectrodes. Horizontal and vertical eye positions were monitored by chronically implanted DC silver-silver chloride electrodes. A strain gauge for measuring head torque was mounted on the head holder.

For vestibular stimulation the monkey was rotated about a vertical axis in complete darkness, using a velocity trapezoid: $5-10°/s^2$ acceleration; 40–100°/s constant velocity for at least 50 s; and $5-10°/s^2$ deceleration. The turntable was enclosed by an optokinetic cylinder covered with vertical black and white stripes. During optokinetic stimulation this cylinder rotated with velocities between 3 and 100°/s around the stationary monkey.

In some experiments the same monkeys were anesthetized with pentobarbital (30 mg/kg i.p.) and exposed to identical vestibular stimulation conditions.

For nystagmus suppression monkeys were trained to fixate a small spot of light using the paradigm of Wurtz (38). The fixating spot was attached to the turntable. It was stationary during optokinetic stimulation and rotated with the monkey during vestibular stimulation. To test neuronal behavior during nystagmus, the fixating light was removed.

Neuronal activity, horizontal and vertical eye position, head torque, turntable and cylinder velocity, and training parameters were stored on magnetic tape. Analysis was performed off-line. Instantaneous and average frequency (running average over 250 to 1000 ms) of the neuronal activity together with horizontal eye position and velocity (first derivative of eye position) and turntable and cylinder velocity were written out on a six-channel rectilinear oscillograph. With respect to vestibular stimulation, time constants of decay for neuronal activity and for slow phase nystagmus velocity were determined as the time elapsed between the maximal activity at the end of the acceleration period and the point at which neuronal activity or slow phase nystagmus velocity had returned to $1/e$ above resting discharge. During optokinetic stimulation, frequency increase was determined over a period of 20 to 30 s after a steady activity level was reached.

RESULTS

Neuronal Activity in Response to Vestibular Stimulation

Recordings Under General Anesthesia

Quick phases of nystagmus cease during light anesthesia, and only a tonic deviation of the eyes can be observed which is opposite to the direction of angular acceleration. With deeper levels of anesthesia no eye movements can be elicited. Neurons recorded in the vestibular nuclei receiving an input from the horizontal semicircular canals can be classified as type I or type II neurons (12). During acceleration to the ipsilateral side type I neurons increase their activity, which returns to resting discharge level with an exponential decay during rotation at constant velocity. During deceleration neuronal activity decreases, often to zero, and returns to the resting discharge level, again with an exponential time course, when the monkey is at rest (Fig. 29-1). Type II neurons exhibit a mirror-like behavior. This is the typical pattern of response to a step in velocity found in all neurons of the anesthetized monkey, and it is similar to that described for the vestibular

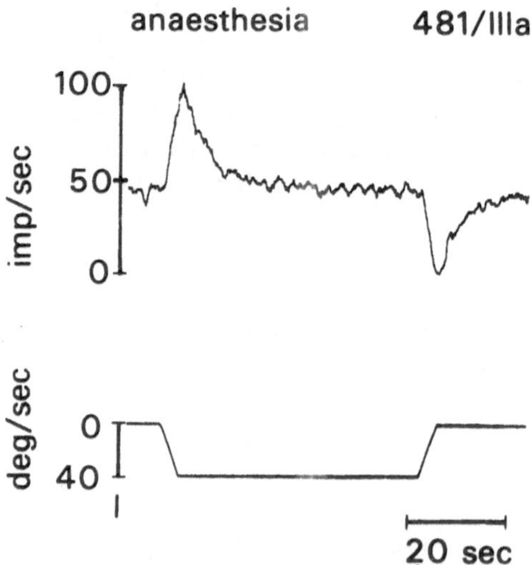

Fig. 29-1. Response of a type I neuron from the left vestibular nuclei to angular acceleration in the anesthetized monkey. **Upper trace** Neuronal activity (running average). **Lower trace** Turntable velocity. Resting discharge, 46 impulses/s; acceleration, $10°/s^2$; constant velocity rotation, 40°/s; deceleration, $10°/s^2$. Neuronal activity decays after termination of acceleration with a time constant of 6 s. From Buettner, V.W., et al.: J. Neurophysiol. 41:1614, 1978.

nerve of the squirrel monkey (16) and the vestibular nuclei of the decerebrate cat (30). Time constants for the decay of neuronal activity, after termination of acceleration, are 5 to 6 s for vestibular nuclei neurons in anesthetized rhesus monkeys (7); these are the values also found for the vestibular nerve of the squirrel monkey (16).

Recordings from Alert Monkey Exhibiting Vestibular Nystagmus

During spontaneous eye movements about 50% of vestibular nuclei neurons of the alert monkey show an additional modulation with fast eye movements and/or eye position (7,15,26). These eye movement–related activity changes occur in addition to the responses caused by vestibular stimulation.

An alert monkey responds to a step in velocity with nystagmus beating toward the direction of rotation. During rotation with constant velocity in the dark nystagmus slow phase velocity decreases exponentially until all nystagmus ceases (Fig. 29-2). A reverse pattern of nystagmus is induced during and after deceleration. Time constants for the decay of slow phase nystagmus velocity range between 10 and 35 s, depending on the state of

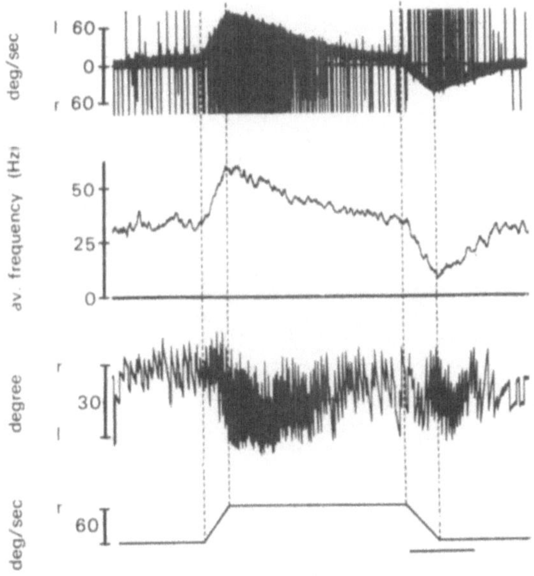

Fig. 29-2. Type I vestibular nuclei neuron recorded in the alert monkey during rotation about a vertical axis in complete darkness. **First trace** Eye velocity (first derivative of the horizontal eye position, shown in the third trace); fast phase velocity is clipped off at an arbitrary level. **Second trace** Neuronal activity as running average. **Third trace** Horizontal eye position. **Fourth trace** Turntable velocity: acceleration 7.5°/s², constant velocity rotation 60°/s. At the end of acceleration both neuronal activity and slow phase nystagmus velocity return to baseline levels with a time constant of about 30 s. Time scale is 20 s. From Büttner, V., et al.: In Hauske, G. and Butenandt, E. (eds.): Kybernetik 1977. München, Oldenbourg Verlag, 1978.

habituation. Vestibular nuclei neurons recorded in the alert monkey during vestibular nystagmus show similar long time constants. The time courses of nystagmus slow phase velocity and vestibular nuclei activity are always closely correlated (Fig. 29-2; ref. 7). These long time constants were found for all neurons, independent of whether they were type I or type II, or whether the neurons were modulated with eye movements (7).

Recordings from Alert Monkeys During Suppression of Vestibular Nystagmus

Trained monkeys are able to suppress vestibular nystagmus in the dark, when they are exposed to velocity trapezoids as described above. Activity of vestibular nuclei neurons was investigated for the same acceleration stimulus with and without suppression of nystagmus. With nystagmus the time courses of slow phase velocity and neuronal activity were similar. Nystagmus suppression did not alter neuronal activity during the acceleration or deceleration phase. However, during the constant velocity period neuronal activity returned much faster to levels of spontaneous activity (Fig. 29-3). Time constants during nystagmus suppression were 5–6 s as compared to 10–35 s during nystagmus. Thus, time constants of vestibular nuclei neurons

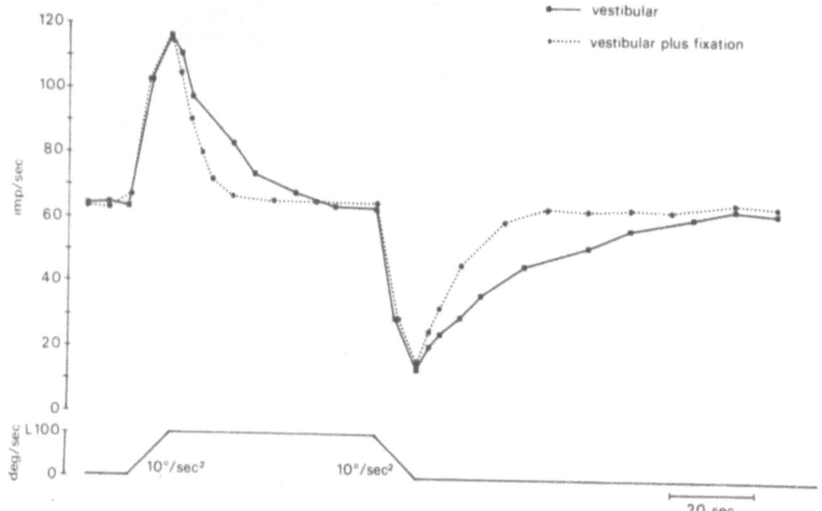

Fig. 29-3. Influence of nystagmus suppression on activity of a type I vestibular nuclei neuron during rotation about a vertical axis in complete darkness. Values for acceleration, constant velocity rotation, and deceleration are indicated at the bottom. Neuronal responses of the same neuron (average of 3 trials) are shown during vestibular nystagmus (*solid line*) and nystagmus suppression (*dotted line*). Maximum level of activation or inhibition is similar, but neuronal activity returns to baseline levels with a shorter time constant (5–6 s) when nystagmus is suppressed. When nystagmus was allowed, time constants were 15 s (after acceleration) and 25 s (after deceleration). Nystagmus slow phase velocity showed the same asymmetry and similar time constants.

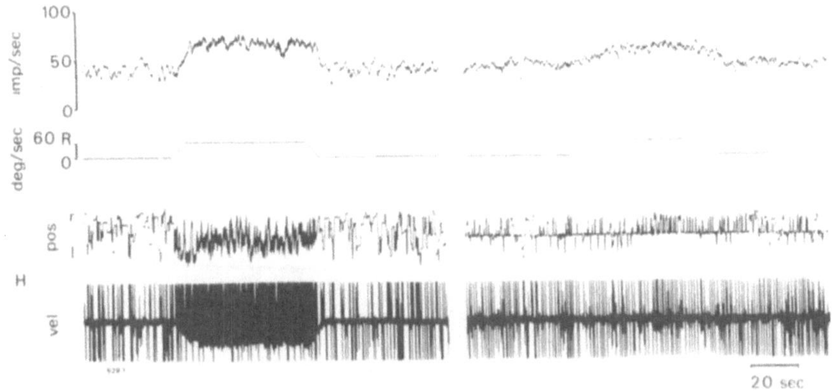

Fig. 29-4. Response of a vestibular nuclei neuron (type I) to optokinetic stimulation leading to nystagmus (*left side*) and during nystagmus suppression (*right side*). **First trace** Running average of neuronal activity. **Second trace** Velocity of optokinetic stimulus. **Third trace** Horizontal eye position, the first derivative (eye velocity) of which is shown in the fourth trace. Velocity of fast phases is clipped off at an arbitrary level. During suppression of optokinetic nystagmus neuronal activity increases less than during optokinetic nystagmus. From Buettner, V.W., and Büttner, V.: Exp. Brain Res. 37:581 1979.

do change with different behavioral tasks, although the vestibular stimulus remains the same. During nystagmus suppression time constants are similar to those obtained under anesthesia, when no eye movements are possible (6).

Vestibular Nuclei Neurons in Response to Optokinetic Stimulation

Recordings During Optokinetic Nystagmus

All vestibular nuclei neurons can be modulated by adequate optokinetic stimuli (34). Optokinetic and vestibular stimuli have to move in opposite directions in order to elicit the same activation or inhibition in a neuron, and then the vestibular or optokinetic nystagmus beats into the same direction. The neuronal modulation during optokinetic stimulation lasts as long as the stimulus is given (Fig. 29-4). A linear relationship between stimulus velocity and increase of neuronal activity can be shown for velocities up to 60°/s; above this, neuronal activity saturates (Fig. 29-5; ref. 34).

Recordings During Suppression of Optokinetic Nystagmus

During suppression of optokinetic nystagmus neurons in the vestibular nuclei still respond with an increase or decrease of neuronal activity depending on the direction of stimulus movement (6). The activity changes are, however, less than the response during optokinetic nystagmus under otherwise identical conditions (Fig. 29-4). Neurons were usually tested at

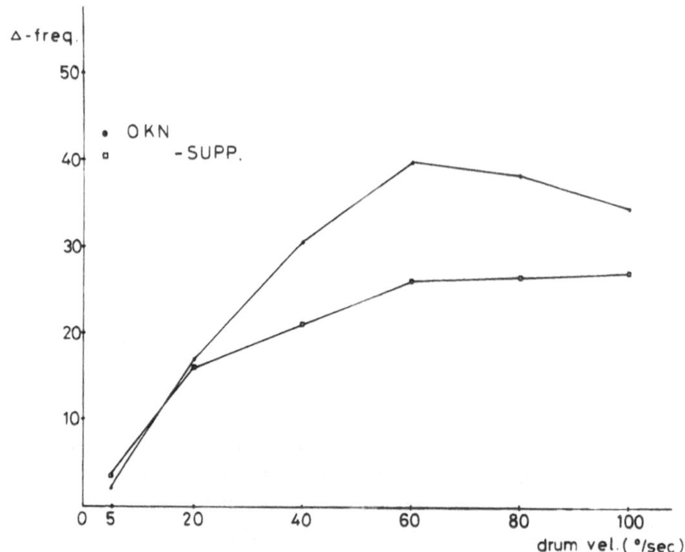

Fig. 29-5. Averaged responses of nine vestibular nuclei neurons during optokinetic nystagmus (*filled circles*) and their responses during nystagmus suppression (*open squares*). *Abscissa*, stimulus velocity; *ordinate*, neuronal activity increase above resting discharge. Low stimulus velocities cause small activity changes under both conditions. At higher velocities the activity increase during nystagmus is considerably higher than during nystagmus suppression. Both responses tend to saturate at stimulus velocities above 60°/s.

stimulus velocities of 40 and 60°/s. If the activity increase at these velocities is set at 100% during nystagmus, then during suppression of nystagmus, the activity increase on average is reduced to about 60% (Figs. 29-4 and 29-5). The variation in this suppression effect for different neurons is wide.

Retinal slip velocity is only a few degrees per second during nystagmus, but approaches stimulus velocity during nystagmus suppression. Neurons were therefore also tested at low stimulus velocities (3 and 5°/s). At these velocities activity during nystagmus and nystagmus suppression is small under both conditions (Fig. 29-5).

DISCUSSION

Results show that vestibular nuclei activity is related not only to sensory input (vestibular or optokinetic), but also to the oculomotor output (nystagmus). In all situations investigated vestibular nuclei activity was different depending on whether nystagmus was allowed or suppressed.

Vestibular Stimulation

When vestibular nystagmus is abolished, either by anesthesia or fixation suppression, the vestibular nuclei carry a signal similar to that of the vestibular nerve. Time constants for the decay of neuronal activity are 5 to 6 s, values also found in the vestibular nerve of the squirrel monkey (16). If vestibular nystagmus is allowed, time constants are always longer, ranging from 10 to 35 s. Similar values are obtained when sinusoidal stimuli are used and the transfer functions calculated (7).

Skavenski and Robinson (32) pointed out the differences in the time course of vestibular nerve activity and nystagmus response in alert animals and postulated a "low frequency compensation" in the brain stem to overcome these differences. Our single neuron recordings in the alert monkey demonstrate that this "low frequency compensation" is already present in the vestibular nuclei.

The exact mechanism of how this is achieved is not known. All evidence suggests that vestibular nerve activity in the monkey is similar in the alert and anesthetized state (20,24). The cerebellum is probably not involved since its ablation does not alter time constants of nystagmus (27). It seems likely that an interaction between the vestibular nuclei and the surrounding reticular formation, which is known to provide ample reciprocal connections (5), plays a decisive role. All procedures which suppress nystagmus (anesthesia, fixation) also abolish the effects of the "low frequency compensation" seen in the vestibular nuclei. It was suggested that nystagmus activity itself is responsible for the longer time constants via a direct feedback mechanism (1,33). However, there are other possibilities just as likely. No experiment has been devised so far which could unequivocally decide in favor of one of the possibilities.

It also should be emphasized that suppression of vestibular nystagmus leads to no or only minor activity changes in the vestibular nuclei during acceleration (Fig. 29-3; ref. 21). This requires additional information processing between the vestibular and oculomotor nuclei in order to cancel vestibular activity during acceleration, if nystagmus is suppressed. A neuronal signal, which could provide this cancellation, was recorded in the flocculus (23). However, the exact site of interaction in the brain stem has yet to be determined. A more pronounced suppression of vestibular activity can be observed during conflicting visual-vestibular stimulation (36).

Optokinetic Stimulation

Vestibular nuclei neurons are still modulated in their activity to optokinetic stimulation when nystagmus is suppressed. This shows that activity is not just a reflection of the occurrence of nystagmus.

When stimulus velocities above 20°/s are used and nystagmus is suppressed, neuronal activity is substantially attenuated (Figs. 29-4 and 29-5). During nystagmus suppression retinal slip velocity equals stimulus velocity. During optokinetic nystagmus retinal slip (stimulus minus eye velocity) is much less, probably not exceeding 10% of the stimulus velocity up to 60°/s (10). Figure 29-5 shows that neuronal activity cannot directly be related to retinal slip velocity, since neuronal response during nystagmus suppression decreases also with lower stimulus velocities. Therefore, response differences at higher velocities cannot be due to stimulation above the optimal response range for retinal slip velocities. This shows that vestibular nuclei neurons encode a signal, which is partially determined by the velocity of the stimulus motion and partially by the oculomotor response. The mode of interaction of these two parameters is not known and nonlinear interactions seem possible. With a linear mode of interaction the neuronal response would be mainly determined by the nystagmus parameter (Fig. 29-5). This is also the conclusion, if one compares neuronal responses during optokinetic nystagmus and after-nystagmus (35). The transition from optokinetic stimulation (visual input plus nystagmus) to after-nystagmus in the dark does not lead to a sudden change of neuronal activity; rather, it decreases slowly together with the nystagmus. In conclusion, under certain stimulus conditions a fixed relationship between neuronal activity in the vestibular nuclei and nystagmus velocity can be established. However, these two parameters can also be dissociated. This shows that their causal connection is a complex one and that activity in the vestibular nuclei cannot alone explain all the phenomena of nystagmus or its suppression.

Nystagmus and Psychophysics

In humans, large field optokinetic or vestibular stimulation leads to the sensation of self-motion (25,39). Strength and direction of sensation often match nystagmus velocity, but these two parameters can also be dissociated (4). This raises the question, what do the vestibular neurons encode? Groen (17) showed that time constants for motion sensation after angular acceleration are considerably shorter than those for nystagmus; whereas the time constants for nystagmus and vestibular nuclei neurons are similar, as described earlier for the monkey. In humans, the strength of the sensation of self-motion remains essentially unchanged if nystagmus is suppressed by fixation (4). In the monkey, neuronal responses in the vestibular nuclei were always smaller during suppression. If the monkey data are valid for humans also, additional information processing between the vestibular nuclei and their cortical projection area is required. A precise vestibular signal has been recorded in the parietal cortex of the monkey (8). Quantitative investigations to provide answers to these intricate questions still have to be done.

What Parameters Are Encoded in the Vestibular Nuclei?

Experiments in alert monkeys demonstrate complex interactions at the level of the vestibular nuclei, even if only three parameters are considered; vestibular input, visual input, and eye movement output. Other parameters were kept constant, i.e. afferents from other sensory systems, especially limb and neck. It could be shown that vestibular neurons in the monkey respond in a similar manner to vestibular or visual stimuli, whenever these stimuli are likely to lead to the sensation of self-motion in humans. However, activity is also modulated by the presence or absence of nystagmus in the monkeys. This underlines the important role of the vestibular nuclei in relation to the motor system. It also questions the century-old partition of the nervous system into sensory, central, and motor divisions. It rather stresses the complex multilevel interaction, which can only be observed if experiments are performed in alert behaving animals. It is to be expected that with more parameters controlled, the picture described above will look oversimplified. It is a long way from the old portrait of the vestibular nuclei as a relay station for certain reflexes to the realization of what complex information processing actually takes place there.

ACKNOWLEDGMENT

This investigation was supported by Swiss National Foundation for Scientific Research SNF 3.233.77 and 3.343.78.

References

1. Allum, J.H.J. and Graf, W.: Time constants of vestibular nuclei neurons in the goldfish: a model with ocular proprioception. Biol. Cybern. 28:95, 1977.
2. Allum, J.H.J., Graf, W., Dichgans, J., and Schmidt, C.L.: Visual-vestibular interactions in the vestibular nuclei of the goldfish. Exp. Brain Res. 26:463, 1976.
3. Blanks, R.H.L., Estes, M.S., and Markham, C.H.: Physiological characteristics of vestibular first-order canal neurons in the cat. II. Response to constant angular acceleration. J. Neurophysiol. 38:1250, 1975.
4. Brandt, T., Dichgans, J., and Koenig, E.: Differential effects of central versus peripheral vision on egocentric and exocentric motion perception. Exp. Brain Res. 16:476 1973.
5. Brodal, A.: Anatomy of the vestibular nuclei and their connections. In Kornhuber, H.H. (ed.): Handbook of Sensory Physiology, Vol. VI/1. Berlin, Springer-Verlag, 1974, pp. 239–352.
6. Buettner, U.W. and Büttner, U.: Vestibular nuclei activity in the alert monkey during supression of vestibular and optokinetic nystagmus. Exp. Brain Res. 37:581, 1979.
7. Buettner, U.W., Büttner, U., and Henn, V.: Transfer characteristics of neurons in the vestibular nuclei of the alert monkey. J. Neurophysiol. 41:1614, 1978.
8. Büttner, U. and Buettner, U.W.: Parietal cortex (2v) neuronal activity in the alert monkey during natural vestibular and optokinetic stimulation. Brain Res. 153:392, 1978.
9. Büttner, U., Waespe, W., and Miles, T.S.: Transfer characteristics of the vestibular system determined from nystagmus and neuronal activity in the alert monkey. In Hauske, G., and Butenandt, E (eds.): Kybernetik 1977. München, Oldenbourg Verlag, 1978, pp. 126–136.
10. Cohen, B., Matsuo, V., and Raphan, T.: Quantitative analysis of the velocity

characteristics of optokinetic nystagmus and optokinetic after-nystagmus. J. Physiol. 270:321, 1977.

11. Dichgans, J. and Brandt, T.: Visual vestibular interaction and motion perception. In Dichgans, J. and Bizzi, E. (eds.): Cerebral control of eye movements and motion perception. Bibl. Ophthalmol. 82:327, 1972.
12. Duensing, F. and Schaefer, K.P.: Die Aktivität einzelner Neurone im Bereich der Vestibulariskerne bei Horizontalbeschleunigungen unter besonderer Berücksichtigung des vestibulären Nystagmus. Arch. Psychiat. Nervenkr. 198:225, 1958.
13. Fernandez, C. and Goldberg, J.M.: Physiology of peripheral neurons innervating semicircular canals of the squirrel monkey. II. Response to sinusoidal stimulation and dynamics of peripheral vestibular system. J. Neurophysiol. 34:661, 1971.
14. Fredrickson, J.M., Schwarz, D.W.F., and Kornhuber, H.H.: Convergence and interaction of vestibular and deep somatic afferents upon neurons in the vestibular nuclei of the cat. Acta Otolaryngol. (Stockh.) 61:168, 1966.
15. Fuchs, A.F. and Kimm, J.: Unit activity in vestibular nucleus of the alert monkey during horizontal angular acceleration and eye movement. J. Neurophysiol. 38:1140, 1975.
16. Goldberg, J.M. and Fernandez, C.: Physiology of peripheral neurons innervating semicircular canals of the squirrel monkey. I. Resting discharge and response to constant angular accelerations. J. Neurophysiol. 34:635, 1971.
17. Groen, J.J.: Cupulometry. Laryngoscope 67:894, 1957.
18. Henn, V., Young, L.R., and Finley, C.: Vestibular nucleus units in alert monkeys are also influenced by moving visual fields. Brain Res. 71:144, 1974.
19. Jones, G.M. and Milsum, J.H.: Frequency-response analysis of central vestibular unit activity resulting from rotational stimulation of the semicircular canals. J. Physiol. 219:191, 1971.
20. Keller, E.L.: Behavior of horizontal semicircular canal afferents in alert monkey during vestibular and optokinetic stimulation. Exp. Brain Res. 24:459, 1976.
21. Keller, E.L. and Daniels, P.D.: Oculomotor related interaction of vestibular and visual stimulation in vestibular nucleus cells in alert monkey. Exp. Neurol. 46:187, 1975.
22. Keller, E.L. and Precht, W.: Persistence of visual response in vestibular nucleus neurons in cerebellectomized cat. Exp. Brain Res. 32:591, 1978.
23. Lisberger, S.G. and Fuchs, A.F.: Role of primate flocculus during rapid behavioral modification of vestibuloocular reflex. I. Purkinje cell activity during visually guided horizontal smooth-pursuit eye movements and passive head rotation. J. Neurophysiol. 41:733, 1978.
24. Louie, A.W. and Kimm, J.: The response of 8th nerve fibers to horizontal sinusoidal oscillation in the alert monkey. Exp. Brain Res. 24:447, 1976.
25. Mach, E.: Grundlinien der Lehre von den Bewegungsempfindungen. Leipzig, Engelmann, 1875 (reprint: Amsterdam, Bonset, 1967).
26. Miles, F.: Single unit firing patterns in the vestibular nuclei related to voluntary eye movements and passive head movements in conscious monkeys. Brain Res. 71:221, 1974.

27. Robinson, D.A.: Adaptive gain control of vestibuloocular reflex by the cerebellum. J. Neurophysiol. 39:954, 1976.

28. Rubin, A.M., Liedgren, S.R.C., Milne, A.C., Young, J.A., and Fredrickson, J.M.: Vestibular and somatosensory interaction in the cat vestibular nuclei. Pfluegers Arch. 371:155, 1977.

29. Schneider, L.W. and Anderson, D.J.: Transfer characteristics of first and second order lateral vestibular neurons in gerbil. Brain Res. 112:61, 1976.

30. Shimazu, K. and Precht, W.: Tonic and kinetic responses of cat's vestibular neurons to horizontal angular acceleration. J. Neurophysiol. 28:1014, 1965.

31. Shinoda, Y. and Yoshida, K.: Dynamic characteristics of responses to horizontal head angular acceleration in vestibuloocular pathways in the cat. J. Neurophysiol. 37:653, 1974.

32. Skavenski, A.A. and Robinson, D.A.: Role of abducens neurons in vestibuloocular reflex. J. Neurophysiol. 36:724, 1973.

33. Sugie, N. and Jones, M.G.: A model of eye movements induced by head rotation. I.E.E.E. Trans. Systems Man Cybern. 1:251, 1971.

34. Waespe, W. and Henn, V.: Neuronal activity in the vestibular nuclei of the alert monkey during vestibular and optokinetic stimulation. Exp. Brain Res. 27:523, 1977.

35. Waespe, W. and Henn, V.: Vestibular nuclei activity during optokinetic afternystagmus (OKAN) in the alert monkey. Exp. Brain Res. 30:323, 1977.

36. Waespe, W. and Henn, V.: Conflicting visual-vestibular stimulation and vestibular nucleus activity in alert monkeys. Exp. Brain Res. 33:203, 1978.

37. Waespe, W. and Henn, V.: The velocity response of vestibular nucleus neurons during vestibular, visual, and combined angular acceleration. Exp. Brain Res. 37:337, 1979.

38. Wurtz, R.H.: Visual receptive fields of striate cortex neurons in awake monkeys. J. Neurophysiol. 32:727, 1969.

39. Young, L.R., Dichgans, J., Murphy, R., and Brandt, T.: Interaction of optokinetic and vestibular stimuli in motion perception. Acta Otolaryngol. (Stockh.) 76:24, 1973.

Discussion

CORREIA: I notice that your phase data based on a nystagmic response indicate that in the mid-range of frequencies, there is zero phase between head velocity and slow phase eye velocity. If I remember correctly, Skavenski and Robinson [J. Neurophysiol. 36:724, 1973] found a constant phase advance of 3° or so over the frequency range 0.04 to 1.5 Hz. My question is: generally, did you find a zero phase difference between head velocity and slow phase eye velocity for the mid to high test frequencies you used?

BÜTTNER: The highest frequency investigated was 0.5 Hz. At that frequency neuronal activity in the vestibular nuclei showed a phase advance of 10 to 20° relative to turntable velocity. In our recordings we always found that the phase of the vestibular nuclei neurons, similar to that of the vestibular nerve fibers [Fernandez and Goldberg: J. Neurophysiol. 34:661, 1971], seldom came down to zero, resulting in a phase difference of about 10° compared with the nystagmus velocity. For the nystagmus velocity we found a phase advance of 0 to 3° at 0.1 Hz and higher frequencies, allowing an error of 1 to 2° with the method of our measurements. We feel that our data are compatible with those of Skavenski and Robinson [J. Neurophysiol. 36:724, 1973].

30

The Influences of Head Orientation and Bilateral Semicircular Canal Plugs upon the Vertical Vestibuloocular Reflex of the Rabbit

N. H. BARMACK AND V. E. PETTOROSSI

Vestibuloocular reflexes have been studied in a variety of invertebrates and vertebrates. Because of the accuracy with which these reflexes can be measured and the precision with which vestibular stimuli can be controlled, study of these reflexes has yielded useful descriptions of how specific populations of afferents encode vestibular stimulation as well as how this information is processed by the central nervous system.

The horizontal vestibuloocular reflex (HVOR) has been studied most extensively. This reflex can be attributed almost exclusively to the activation of primary afferents originating from the horizontal semicircular canals (3,7,11,14,15).

The input to the vertical vestibuloocular reflex has multiple origins. For example, rotation of a human subject about a longitudinal axis (roll) would be expected to stimulate activity in the anterior and posterior semicircular canals. A roll would also stimulate the otoliths by modulating the orientation of otolithic hair cells with respect to the linear acceleration of gravity. Such a stimulus evokes torsional movements of the frontally located human eyes, but evokes vertical movements of the laterally placed eyes of the rabbit. It might be expected that the VVOR of rabbits would reflect its multiple vestibular origins. The present experiment examines how signals from the semicircular canals and the otoliths combine to produce the VVOR during sinusoidal roll tilt. The contribution of the semicircular canals to this reflex was studied under conditions in which the otolithic information was eliminated by changing the longitudinal axis of the rabbit with respect to gravity. For example, by vertically aligning the longitudinal axis of the rabbit so that the rabbit's nose points up, the orientations of the planes of polarization of the hair cells of both the utricular and saccular maculae are changed, so that the linear acceleration of gravity now acts orthogonally to these planes (10,12). The contribution of the otoliths to the VVOR in isolation of the semicircular canals was studied by surgically plugging the semicircular canals, thus rendering them dysfunctional (6,16).

METHODS

Both albino and Dutch Belted rabbits, weighing 1–2 kg, were used in these experiments. In a preparatory operation, each rabbit was anesthetized with ketamine hydrochloride (50 mg/kg IM) and halothane. The rabbit's head was aligned in a stereotactic apparatus so that the lambda suture was 1.5 mm below the bregma suture. Two inverted stainless steel 10-32 screws were fixed to the closed calvarium with dental cement. The dental cement also encapsulated four 2-56 anchor screws. The two larger screws mated with holes in a steel rod which was used subsequently to immobilize the head at the center of rotation of a biaxial rate table. The rod was tilted in the posterior-anterior direction at an angle of 12° in order to align the plane of the horizontal semicircular canals with the horizontal plane of the apparatus.

The horizontal and vertical vestibuloocular reflexes (HVOR and VVOR) were tested in unanesthetized rabbits. The head was fixed in the center of rotation of the rate table. The body was firmly encased in "egg carton" foam rubber. The rate table was servo-controlled in the horizontal and vertical planes. Vestibular stimulation was always performed with both eyes of the rabbit completely occluded by three layers of black cloth.

Eye position was measured with an infrared light projection technique. One or both eyes were anesthetized topically with proparacaine hydrochloride. Small suction contact lenses, on which light-emitting diodes (LED) were mounted, were attached to the eyes. The LEDs projected narrow beams of infrared light onto photosensitive *x-y* position detectors which were mounted 3 to 5 mm from the tip of the LEDs in situ. The output of the position detector was proportional to the position of the incident centroid of infrared light. Eye velocity was measured by electronic differentiation of the eye position signal.

The gain of the vestibuloocular reflexes was determined from measurements of the peak eye velocities attained during each half-cycle of sinusoidal stimulation. For the HVOR, $G = (V_R + V_L)/2V_{T\max}$, where V_R = peak velocity of compensatory eye movement to the right; V_L = peak velocity of compensatory eye movement to the left; and V_T = table velocity. The phase of the HVOR (eye position + 180° re head position) was measured at each

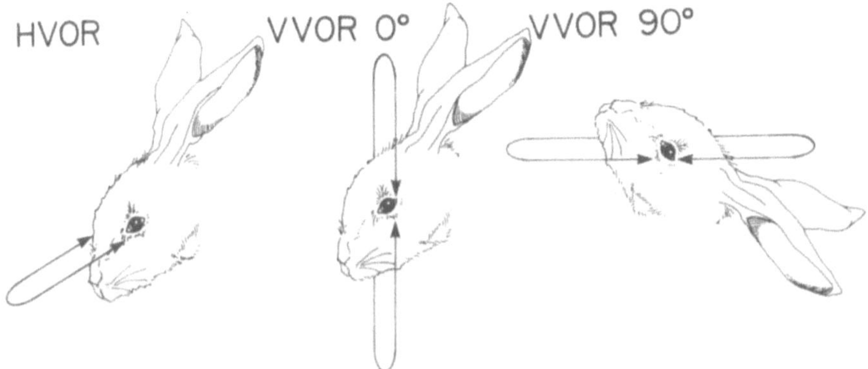

Fig. 30-1. Positions of the rabbit during vestibuloocular reflex testing. The vertical vestibuloocular reflex (VVOR) was tested when the animal was prone (VVOR 0°) and when the animal was positioned with its nose pointing up (VVOR 90°).

half-cycle where $V_T = 0$. Gain and phase were measured at frequencies of sinusoidal stimulation of 0.005–0.8 Hz and at a fixed amplitude of ± 10°.

In some experiments, the vertical vestibuloocular reflex was measured with the rabbit positioned "nose up" (VVOR 90°), so that the longitudinal axis was parallel with rather than orthogonal to the linear acceleration of gravity. These different modes of vestibular stimulation are illustrated in Fig. 30-1.

The semicircular canals were plugged bilaterally in different combinations. The orientation of the semicircular canals is illustrated in Fig. 30-2. Different pairs of these canals were plugged under microscopic control by drilling a small opening in the bony canal and inserting a 0.15- to 0.25-mm thick, 0.5- to 0.75-mm long, spindle-shaped silver wire into the bony lumen, thereby compressing the membranous duct. The opening was gently packed with bone wax to prevent loss of perilymph. In some rabbits, operations were performed sequentially on the anterior semicircular canals (Fig. 30-2, *1* and *2*). When this dual operative procedure was employed, the second plugs were made closer to the ampullae, after the effects of the initial plugs had been evaluated. In other rabbits, serial operations were performed on the posterior and anterior canals (Fig. 30-2, operations *4* and *1*).

RESULTS

A comparison of the eye movements evoked by vestibular stimulation in the horizontal and vertical planes reveals obvious differences in the characteristics of the HVOR and VVOR (Fig. 30-3). At low frequencies (0.01–0.02 Hz) of constant amplitude (± 10°) sinusoidal stimulation, the HVOR has a low gain and a large phase lead. As the frequency is increased the gain is increased and the phase lead is reduced. At frequencies of stimulation above

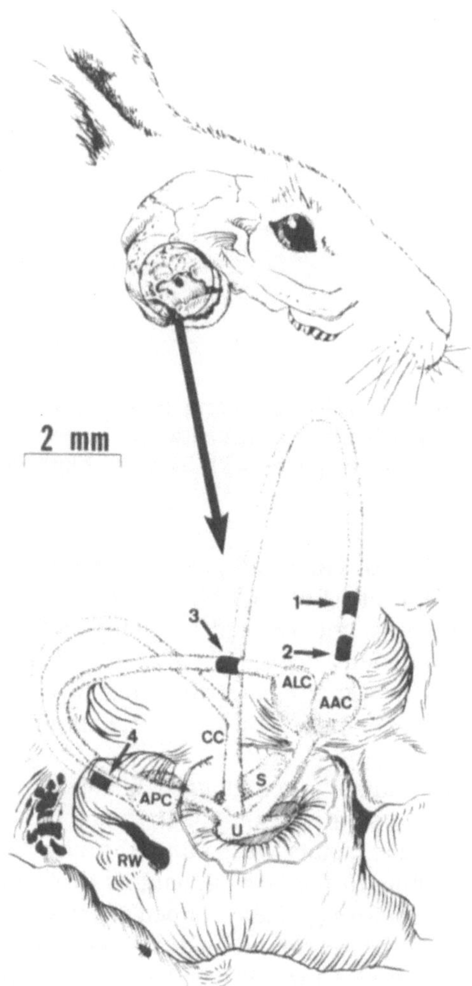

Fig. 30-2. Illustration of the right labyrinth of the rabbit. Note that the ampullae of the anterior (*AAC*) and lateral (*ALC*) semicircular canals are illustrated as sharing a common duct to the utricle (*U*). This common duct is based on the anatomy of the bony canals and may be two separate membranous passages. The numbers (1–4) refer to locations of plugs in different experiments. *S*, saccule; *CC*, common crus; *APC*, ampulla of posterior canal; *RW*, round window.

0.02 Hz the compensatory eye movements of the HVOR are repeatedly interrupted by anticompensatory resetting eye movements. By contrast, the VVOR has a higher gain and only a small phase lead at low frequencies of stimulation (0.01–0.02 Hz). This gain is increased somewhat at higher frequencies without any significant change in phase. The compensatory eye movements of the VVOR are never interrupted by anticompensatory resetting eye movements. The differences in gain and phase between the HVOR

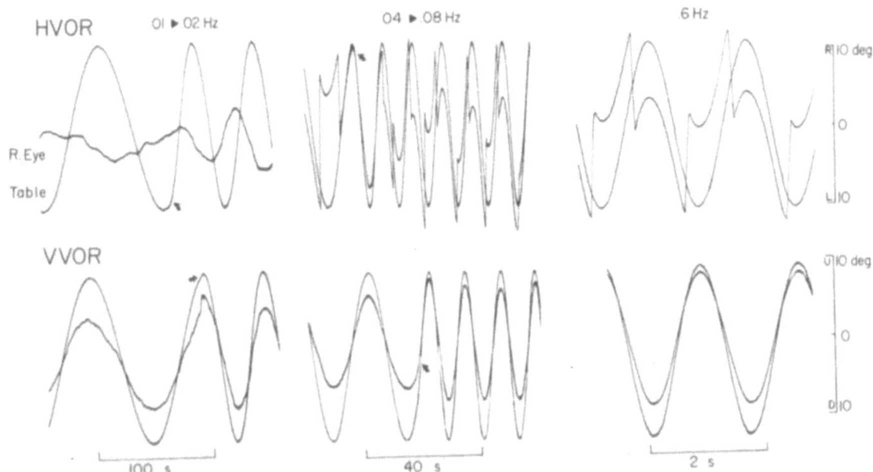

Fig. 30-3. Eye movements evoked by sinusoidal vestibular stimulation about the vertical and longitudinal axes. The horizontal vestibuloocular reflex (*HVOR*) has a low gain and a large phase lead to low frequencies of stimulation. The vertical vestibuloocular reflex (*VVOR*) has a larger gain and smaller phase lead at low frequencies (0.01–0.04 Hz). The arrows indicate transitions between different frequencies of stimulation. The calibrations indicate the position of the right eye. The table position signal has been inverted.

and VVOR confirm the expectation that the VVOR receives a contribution from the otoliths as well as the vertical semicircular canals (Fig. 30-4). This contribution appears to enhance the low frequency gain and to interact with information which originates from the semicircular canals in a way which prevents the occurrence of anticompensatory resetting eye movements.

The hypothesized otolithic origin of the low frequency component of the VVOR can be evaluated by orienting the rabbit "nose up." In this condition the gravity vector is orthogonal to the planes of polarization of most of the hair cells of both the utricle and saccule (Fig. 30-1, *VVOR 90°*). Therefore, oscillation of the rabbit about its longitudinal axis would no longer modulate the gravitational acceleration acting upon these hair cells. This procedure renders the phase and gain of the VVOR 90° equivalent to the HVOR measured in the same animals (Figs. 30-5 and 30-6), and also causes the appearance of anticompensatory resetting eye movements.

If it is assumed that information which originates from the semicircular canals and the otoliths combines vectorially to produce the VVOR, then it should be possible to obtain an estimate of the otolithic component of this reflex by subtracting vectorially the VVOR 90° (which contains only semicircular canal information) from the VVOR 0° (which contains information originating from both the otoliths and semicircular canals). Such a polar subtraction is illustrated in Fig. 30-7. These data suggest a relatively high gain (0.4–0.5) for the otolithic component of the VVOR at low frequencies

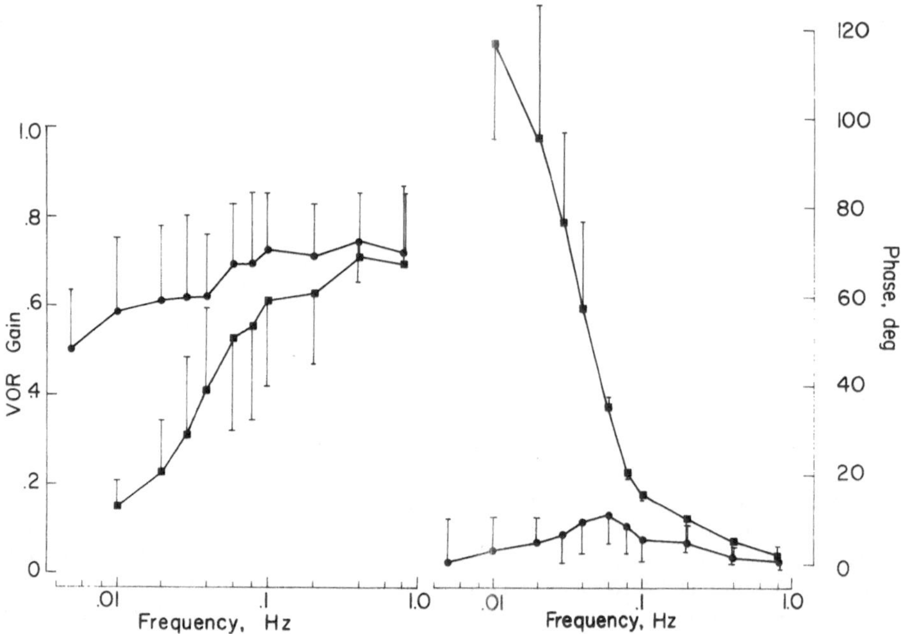

Fig. 30-4. Comparison of the HVOR and VVOR. The polar average phase and gain for the HVOR (*squares*) and VVOR (*circles*) were computed for 21 rabbits. The larger of the standard deviations (+ or −) is indicated for each data point.

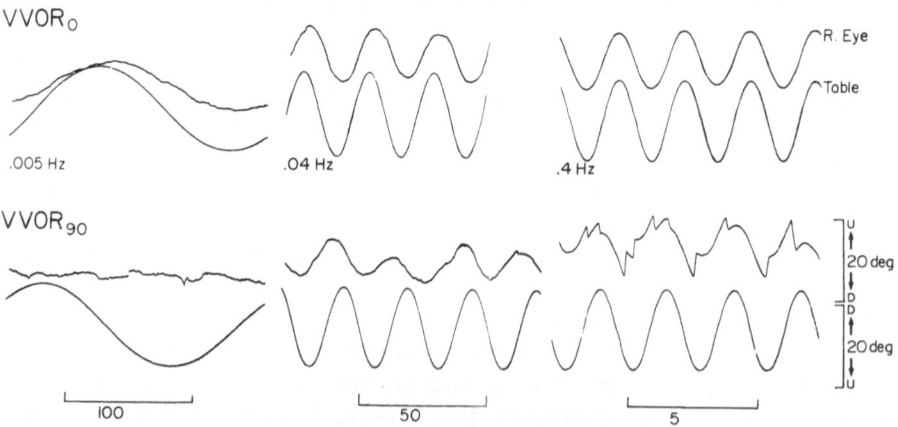

Fig. 30-5. Vertical eye movements evoked by sinusoidal roll stimulation with the longitudinal axis in two different orientations: VVOR O° (prone) and VVOR 90° ("nose up"). In contrast to the VVOR O°, the gain of the VVOR 90° is reduced at 0.005 Hz and anticompensatory resetting eye movements are evoked at 0.4 Hz.

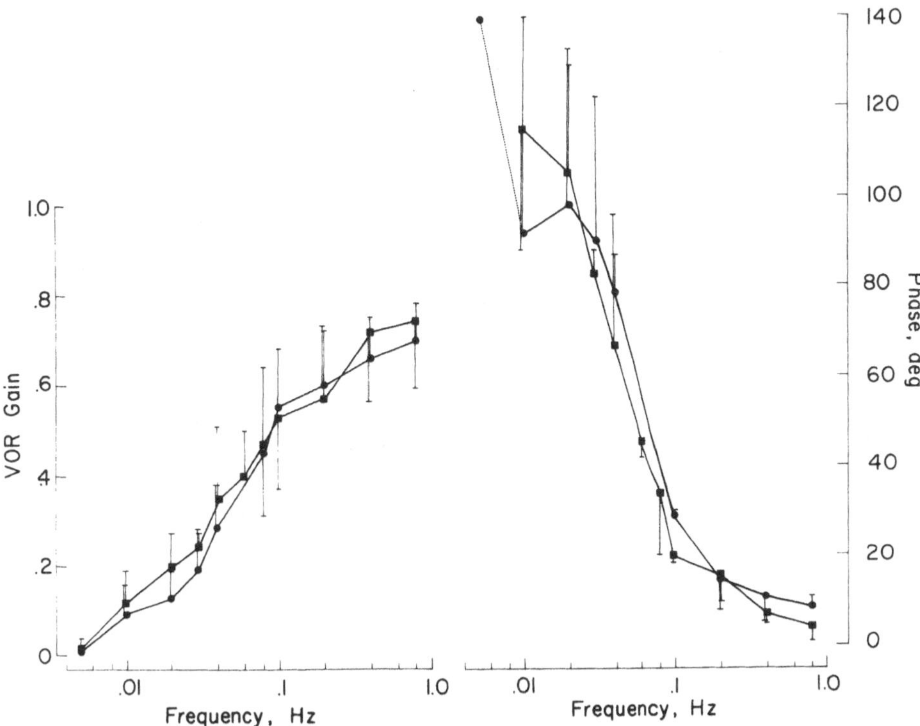

Fig. 30-6. Comparison of the HVOR and VVOR 90°. The average polar phase and gain were computed for nine rabbits. Only one rabbit had any measurable gain of the VVOR 90° at a frequency of 0.005 Hz. This data point is connected to other data points by a dotted line. The larger of the standard deviations (+ or −) is indicated for each data point.

of stimulation (0.005–0.01 Hz) and a gain of less than 0.2 at stimulus frequencies above 0.1 Hz. The predicted phase shows an increasing lag as the stimulus frequency is increased. The large standard deviations for the phase predictions at higher frequencies are probably due to the small otolithic contribution to the VVOR at these frequencies. Hence, small errors in the measurement of gain at these frequencies could cause large errors in the predicted phase of the otolithic component of the VVOR. The predicted gain of the otolithic component of the VVOR derived from vectorial subtraction (VVOR-VVOR 90°, or VVOR-HVOR) is in reasonable agreement with the measured VOR of rabbits to sinusoidally varied linear accelerations (4). However, due to technical limitations, measurements of the VOR at low frequencies of linear acceleration, where the present results indicate the highest otolith gain, have not been made (Fig. 30-7, *filled diamonds*). The estimate of the otolith contribution to the VVOR obtained in the present experiment is at variance with a description of otolith function

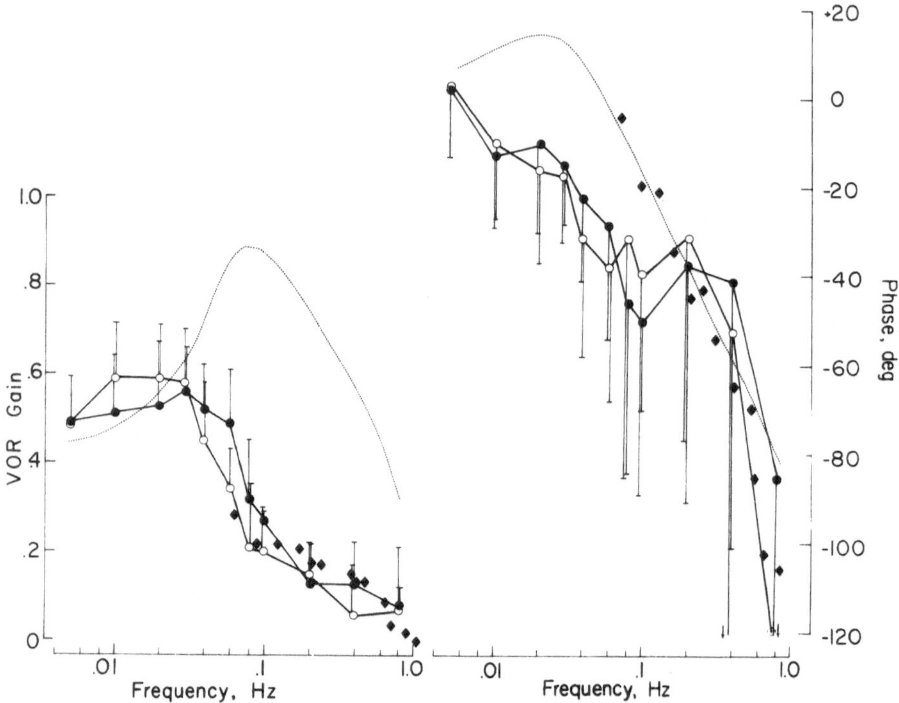

Fig. 30-7. Derived prediction of the otolithic contribution to the VVOR. The filled circles indicate the predicted frequency response of the otolith, obtained by polar subtraction of the VVOR 90° from the VVOR. The open circles indicate the prediction of the otolith frequency response obtained by polar subtraction of the HVOR from the VVOR. The dotted line indicates the frequency response of the otoliths obtained from a descriptive model of Young and Meiry (18). The solid diamonds indicate the frequency response obtained from stimulation of the VOR with linear accelerations (4). $N = 9$ rabbits. The larger of the standard deviations (+ or −) is indicated for each data point.

(Fig. 30-7, *dotted line*), which suggests a higher gain and higher frequency response of the otoliths (18).

Both the anterior and posterior semicircular canals would be expected to contribute to the VVOR. A dissection of temporal bones of three rabbits indicated that the plane of the anterior canal of the rabbit forms an angle of approximately 50° with the sagittal plane of the skull, while the plane of the posterior canal forms an angle of about 40° with the sagittal plane. The circumferences of the bony passages of all three semicircular canals were measured in three rabbits by threading an 80-μm wire through each of the canals. The mean circumferences for each semicircular canal were as follows: horizontal, 14.3 mm; posterior, 14.7 mm; and anterior, 19.2 mm. Therefore, because of its greater angle with respect to the sagittal plane and its greater circumference, the anterior canal would be expected to be activated more by a roll stimulus than the posterior canal. This relative difference in the activation of the anterior and posterior canals should be manifest in measurements of the VVOR.

Both the anterior and posterior semicircular canals were plugged separately and in combination to determine their relative contributions to the VVOR. There were no obvious postural disturbances following bilateral canal plugs when the rabbit was left undisturbed. However, if the head of the rabbit was forcefully displaced in the plane of the plugged canals, a transient oscillation of the head continued in that plane for a few seconds. For plugs of the horizontal semicircular canal, this oscillation was from side to side. Following bilateral plugs of the anterior canals, an external perturbation caused the head to rotate. Following bilateral plugs of the posterior canals, an external perturbation caused the head to move in the sagittal plane, almost as if the rabbit was acknowledging the perturbation by nodding. Bilateral plugs of either anterior semicircular canals or the posterior semicircular canals caused a reduction in gain and an increased phase lag of the VVOR which was particularly evident at intermediate stimulus frequencies (Fig. 30-8). As the frequency was increased to the 0.1- to 0.8-Hz range, the phase lag developed into a phase lead without any further reduction in gain. This phase lead exceeded the normal values at frequencies of 0.4–0.8 Hz. The reduction in

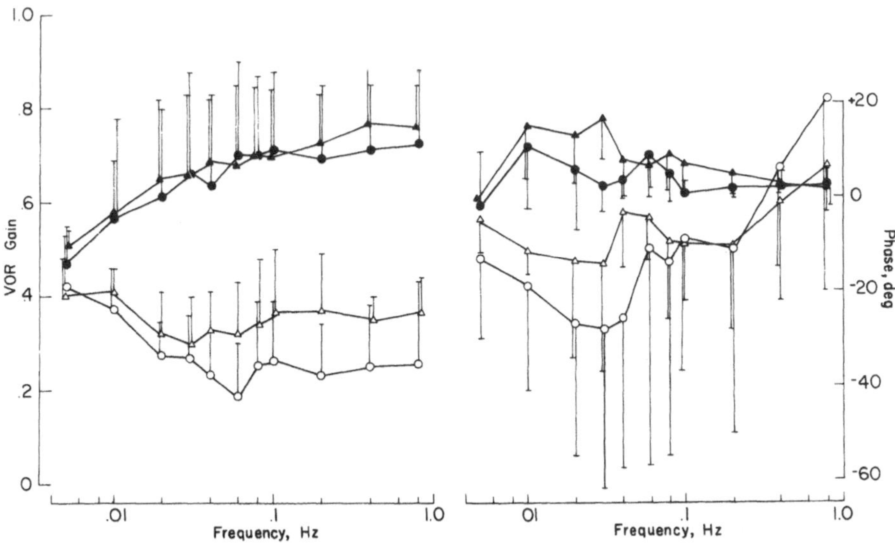

Fig. 30-8. Comparison of the effect of bilateral plugs of the posterior or anterior canals on the VVOR. Open circles illustrate the mean phase and gain of the VVOR of nine rabbits which received bilateral plugs (Fig. 30-2, operation *1*) of the anterior canals. The open triangles illustrate the mean phase and gain of four rabbits which received bilateral plugs of the posterior canals (Fig. 30-2, operation *4*). The filled symbols indicate the normal control VVOR for each of the two groups. The larger of the standard deviations (+ or −) is indicated for each data point.

gain and the increase in phase lag over the intermediate frequency range (0.02–0.1 Hz) were relatively greater for the anterior semicircular canal–plugged rabbits than for rabbits with posterior canal plugs (Fig. 30-8).

In five rabbits, two-stage bilateral plugs of both the anterior and posterior semicircular canals were made. These rabbits evinced a reduced VVOR gain and a phase lag which increased as the frequency of stimulation was increased (Fig. 30-9). At stimulation frequencies above 0.2 Hz, the gain was less than 0.05 and the phase lag exceeded 100°. In three of these rabbits the anterior canals were replugged, moving the plugs closer to but not encroaching upon the ampullae of the anterior semicircular canals (Fig. 30-2, operation *2*). In these rabbits the phase lag was further decreased at all frequencies tested, suggesting that there might have been some residual

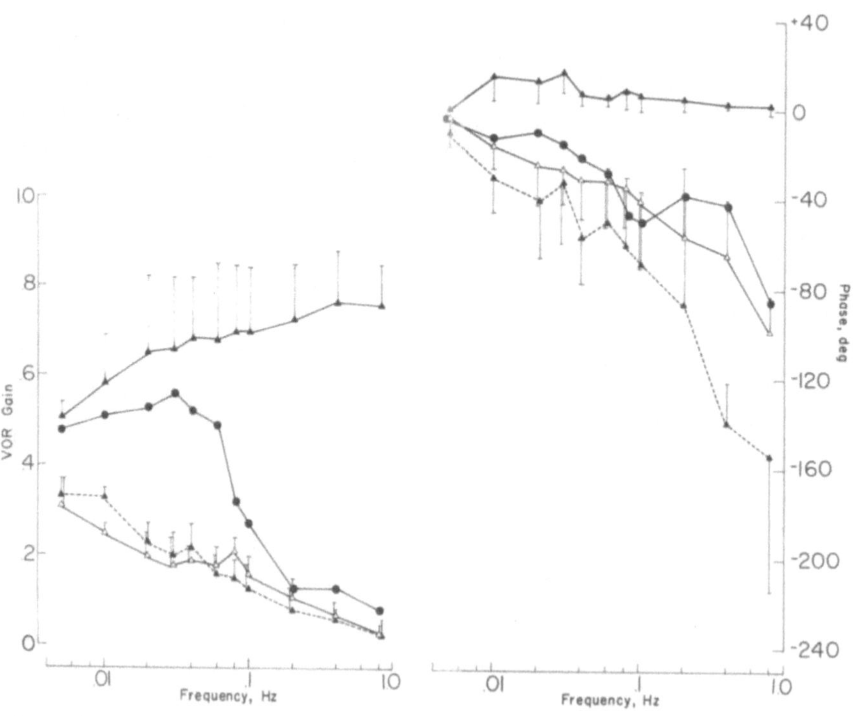

Fig. 30-9. The influence of bilateral plugs of the posterior and anterior canals on the VVOR. Filled triangles connected by solid lines indicate the normal VVOR of five rabbits. The open triangles indicate the VVOR obtained after bilateral plugs of the anterior and posterior canals (operations 1 and 4). The filled triangles connected by dotted lines indicate the VVOR of three of the rabbits following an additional operation in which the anterior canals were plugged closer to the ampullae (Fig. 30-2, operation *2*). The solid circles indicate the response of the otolithic component of the VVOR predicted assuming linear summation of canal and otolith information (Fig. 30-7).

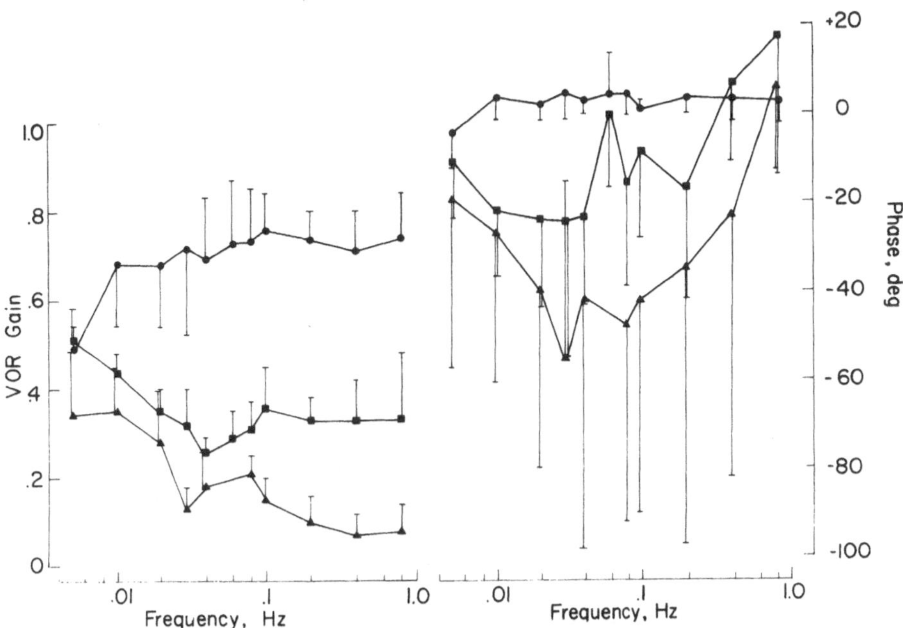

Fig. 30-10. The influence of bilateral anterior canal plugs on the VVOR as a function of the distance of the plugs from the anterior canal ampulla. Filled circles illustrate the control VVOR of four rabbits. Filled squares illustrate the VVOR of these rabbits after bilateral plugs of the anterior canals (Fig. 30-2, operation *1*). Filled triangles illustrate the VVOR of the same rabbits following an additional operation in which the anterior canal plugs were made closer to the ampullae (Fig. 30-2, operation *2*). The larger of the standard deviations (+ or −) is indicated for each data point.

function of the anterior semicircular canals when they were plugged more than 1 mm from the ampullae.

The efficacy of the anterior semicircular canal plugs as a function of their distance from the ampullae was examined in four rabbits. First the influence on the VVOR of bilateral plugs remote from the ampullae was examined (Fig. 30-2, operation *1*). Then the anterior semicircular canals were plugged again closer to the ampullae (Fig. 30-2, operation *2*). The second operation caused a further reduction in gain and a greater phase lag at all frequencies tested (Fig. 30-10). It was of interest to note that at stimulation frequencies of 0.4–0.8 Hz, the VVORs of rabbits with remote anterior canal plugs (operation 1) evinced phase leads which were larger than the phase leads recorded from intact rabbits.

DISCUSSION

The present experimental results have confirmed the hypothesized multiple contributions of the vestibular apparatus to the VVOR. We have demonstrated that there is a significant otolithic contribution to the VVOR at frequencies of stimulation below 0.2 Hz. This otolithic contribution can be removed by orienting the rabbit "nose up." In this orientation the linear acceleration of gravity would be acting orthogonally to the major planes of polarization of hair cells of both the utricular and saccular maculae. The utricular maculae lie within a few degrees of the planes of the horizontal semicircular canals. Utricular hair cells have polarization vectors which are horizontally directed, with a preponderance of excitatory polarizations directed mediolaterally (12). These morphologic data have been corroborated by an electrophysiologic investigation, which in addition suggested an even greater preponderance of mediolaterally directed excitatory polarization vectors (8). The saccular maculae are oriented almost orthogonally to the utricular maculae. The plane of the saccular maculae is inclined dorsoventrally by an angle of about 10° with respect to the sagittal plane (12). Hair cells of the saccular maculae are polarized vertically, with a preponderance of excitatory polarizations directed ventrodorsally. It is probable that only utricular hair cells contribute to the otolithic component of the VVOR when the rabbit is oriented in the normal position (Fig. 30-1, *VVOR 0°*). Symmetric roll deviations of the head about this position cause a maximal change in the linear gravitational acceleration acting in the plane of polarization of utricular hair cells. To achieve an equivalent modulation of linear gravitational acceleration in the plane of polarization of the saccular hair cells, it would be necessary to establish an orientation in which the intraaural axis would be parallel with rather than orthogonal to the gravity vector (i.e., left side or right side down). We have made measurements of the VVOR of rabbits rolled symmetrically about this orientation. The resulting function is neither descriptive of a purely semicircular canal input (HVOR or VVOR 90°), nor is it similar to the normal VVOR (VVOR 0°). These differences might be due to the nonreciprocal stimulation of the saccular hair cells by a roll stimulus. Because the utricular maculae are oppositely polarized (mediolateral), the activity originating from one utricle is increased by changing the orientation of the head about the longitudinal axis, and the activity originating from the opposite utricle is decreased. However, due to the nearly parallel polarization of the saccular maculae (ventrodorsal), a roll stimulus simultaneously increases the activity originating from both saccules as the rabbit is rolled from a position at which the intraaural axis is parallel with the linear acceleration of gravity ("side up") to a more supine position (inverted), and decreases the activity during a roll from a "side up" orientation toward a more prone orientation.

The gain and phase of the otolithic component of the VVOR were obtained by subtracting vectorially the VVOR 90° (canal input) from the VVOR 0° (canal input + otolith input) (Fig. 30-7). The validity of this procedure depends upon the degree to which these inputs combine linearly. The gain and phase of the otolithic component of the VVOR derived from this procedure are in reasonable agreement with the measurements by Baarsma and Collewijn (4) of the VOR evoked by sinusoidal linear accelerations of rabbits. However, because of technical limitations these authors did not measure the VOR at frequencies below 0.06 Hz. The present data and the electrophysiologic data of others suggest that the otoliths are most sensitive at these lower frequencies (9,13).

There was a discrepancy between the estimate of the low frequency VVOR gain obtained by the vectorial subtraction and the low frequency VVOR gain measured from rabbits which had received bilateral plugs of the anterior and posterior semicircular canals (Fig. 30-9). It is possible that the semicircular canals contribute to the low frequency gain of the VVOR 0° combination with an otolithic input. However, this semicircular canal signal may be below "threshold" for many central neurons without a concurrently modulated otolithic input. In other words, a nonlinearity due to a "threshold" might cause an overestimate by vectorial subtraction of the VVOR gain at lower frequencies of stimulation. This possible facilitative interaction between semicircular canal and otolithic inputs would be eliminated by plugs of the semicircular canals.

The efficacy of the canal-plugging procedure in preventing modulation of the output of the plugged anterior semicircular canal appears to be dependent upon the proximity of the plug to the ampulla (Figs. 30-9 and 30-10). The residual function of plugged canals was usually small and may have escaped detection by less sensitive behavioral techniques employed by previous investigators (6,16). Our data suggest that the plugging technique might not be entirely effective in preventing primary afferent modulation of semicircular canal origin, particularly at higher frequencies of stimulation. Therefore, electrophysiologic experiments which use the plugging technique to prevent semicircular canal afferent modulation should be interpreted with caution (1,2,17).

Finally, it appears that there are interesting interactions between the otoliths and the semicircular canals in which the otoliths suppress anticompensatory resetting eye movements in the vertical plane. These anticompensatory resetting eye movements are characteristic of the HVOR and can also be evoked by electric stimulation of the horizontal ampullary nerve (5). The absence of such anticompensatory resetting eye movements during stimulation of both the otoliths and semicircular canals (VVOR 0°), and the appearance of these eye movements when only the vertical semicircular canals are stimulated (VVOR 90°), suggest that both classes of afferents have access to a central subset of neurons contributing to the generation of anticompensatory resetting eye movements.

SUMMARY

The phase and gain of the horizontal (HVOR) and vertical (VVOR) vestibuloocular reflexes of rabbits were compared. In contrast to the HVOR, the VVOR had a higher gain (eye velocity/head velocity) and a smaller phase lead (eye position + 180° re head position) at low frequencies of sinusoidal angular acceleration (± 10°; 0.005–0.05 Hz). At higher frequencies, (0.05–0.8 Hz), the HVOR and VVOR were equivalent in both gain and phase. It was demonstrated that the low frequency gain of the VVOR could be attributed to an otolithic input by measuring the VVOR in rabbits oriented "nose up," so that the linear acceleration of gravity acted orthogonally to the planes of polarization of the otolithic hair cells. The VVOR measured under this condition was identical to the HVOR in both gain and phase. The assumption that the signals originating from the semicircular canals and otoliths combine linearly to produce the VVOR was examined by testing the VVOR of rabbits which had received bilateral plugs of the anterior and posterior semicircular canals. Such plugs caused a greater reduction in gain and a greater increase in phase lag than would be predicted by a simple vectorial sum of semicircular canal and otolith inputs. Finally, it was demonstrated that the efficacy of a plug of the anterior semicircular canal was, in part, dependent upon the distance of the plug from the ampulla.

ACKNOWLEDGMENTS

This research was supported by N.I.H. Grant EY-00848 and by the Oregon Lions Sight and Hearing Foundation. V.E.P. is the recipient of C.N.R.-N.A.T.O. Fellowship and Oregon Lions Sight and Hearing Foundation Fellowship.

References

1. Abend, W.K.: Functional organization of the superior vestibular nucleus of the squirrel monkey. Brain Res. 132:65, 1977.
2. Abend, W.K.: Response to constant angular accelerations of neurons in the monkey superior vestibular nucleus. Exp. Brain Res. 31:459, 1978.
3. Baarsma, E.A. and Collewijn, H.: Vestibulo-ocular and optokinetic reactions to rotation and their interaction in the rabbit. J. Physiol. 238:603, 1974.
4. Baarsma, E.A. and Collewijn, H.: Eye movements due to linear accelerations in the rabbit. J. Physiol. 245:227, 1975.
5. Cohen, B., Goto, K., and Tokumasu, K.: Return eye movements, an ocular compensatory reflex in the alert cat and monkey. Exp. Neurol. 17:172, 1967.
6. Correia, M.J. and Money, K.E.: The effect of blockage of all six semicircular canal ducts on nystagmus produced by dynamic linear acceleration in the cat. Acta Otolaryngol. 69:7, 1970.
7. Fernandez, C. and Goldberg, J.M.: Physiology of peripheral neurons innervating semicircular canals of the squirrel monkey. II. Response to sinusoidal stimulation and dynamics of peripheral vestibular system. J. Neurophysiol. 34:661, 1971.
8. Fernandez, C. and Goldberg, J.M.: Physiology of peripheral neurons innervating otolith organs of the squirrel monkey. I. Response to static tilts and to long-duration centrifugal force. J. Neurophysiol. 39:970, 1976.
9. Fernandez, C. and Goldberg, J.M.: Physiology of peripheral neurons innervating

otolith organs of the squirrel monkey. III. Response dynamics. J. Neurophysiol. 39:996, 1976.

10. Flock, A.: Structure of the macula utriculi with special reference to directional interplay of sensory responses as revealed by morphological polarization. J. Cell Biol. 22:413, 1964.
11. Goldberg, J.M. and Fernandez, C.: Physiology of peripheral neurons innervating semicircular canals of the squirrel monkey. I. Resting discharge and response to constant angular accelerations. J. Neurophysiol. 34:635, 1971.
12. Lindeman, H.H.: Studies on the morphology of the sensory regions of the vestibular apparatus. Ergeb. Anat. Entwgesch. 42:1, 1969.
13. Loe, P.R., Tomko, D.L., and Werner, G.: The neural signal of angular head position in primary afferent vestibular nerve axons. J. Physiol. 230:29, 1973.
14. Mayne, R.: A systems concept of the vestibular organs. In Kornhuber, H.H. (ed.): Handbook of Sensory Physiology, Vol. 6. Vestibular System. Berlin, Springer-Verlag, 1974, pp. 493–580.
15. Melvill Jones, G. and Milsum, J.H.: Frequency-response analysis of central vestibular unit activity resulting from rotational stimulation of the semicircular canals. J. Physiol. 219:191, 1971.
16. Money, K.E. and Scott, J.W.: Functions of separate sensory receptors of nonauditory labyrinth of the cat. Am. J. Physiol. 202:1211, 1962.
17. Schor, R.H.: Responses of cat vestibular neurons to sinusoidal roll tilt. Exp. Brain Res. 20:347, 1974.
18. Young, L.R. and Meiry, J.L.: A revised dynamic otolith model. Aerospace Med. 39:606, 1968.

Discussion

GOLDBERG: Wouldn't your rotations in the roll plane also influence the posterior semicircular canal?

BARMACK: Yes. We worried about that question. The posterior canal, as I indicated, contributes to the vertical vestibuloocular reflex. However, if plugs are made in the anterior canals remote from the ampullae, then destruction of the posterior canal ampullae does not eliminate a residual canal gain of 0.1 to 0.2.

GOLDBERG: Did you ever plug the horizontal canal, in addition to the vertical canals?

BARMACK: We plugged both the posterior and the horizontal.

GOLDBERG: Was histology done to confirm that the membranous canal was occluded?

BARMACK: We saved the skulls of the rabbits and confirmed that the plugs fully occluded the intended canals—in some cases we made plugs that were over 1.5 mm long, because we worried about the plugs being incomplete. However, we did not perform histologic analysis to determine if our plugging procedures caused any degeneration of the vestibular nerve.

GOLDBERG: Was there actually a growth of bone across the canal?

BARMACK: We have examined the bony labyrinths of rabbits in which bone wax fully occluded the intended canal. No. In fairness to Money, his technique of measuring labyrinthine reflexes might not have been sufficiently sensitive to reveal residual canal function.

SCHOR: I have previously studied otolith responses in canal-plugged cats. The plugging was performed using Ken Money's technique, which involves opening the bony and membranous labyrinth at a point remote from the ampulla and filling the ends with bone dust. Such a procedure leads to new bone formation which very effectively seals the canals. What is the nature of the bony labyrinth in the rabbit—is it very hard, as in the cat, or is it spongy? Also, have you ever plugged all six canals? Now, in the cat, this is an extremely dense bone. It's the hardest bone in the entire body. I know in some other animals, there are great advantages in working in something other than the cat because the bone of rabbit labyrinth is much softer. I'm not that familiar with the rabbit. My questions are, what is the nature of the bone in the rabbit and how long do you wait after plugging the canals before you measure the vestibuloocular reflex?

BARMACK: Your first question: "Is the temporal bone of the rabbit hard?" The answer is yes. The second question: "Have we ever plugged all six canals?" The answer is no. We have only plugged four canals at a time, never six. The plugging has been done with bone wax, with a combination of bone wax and silver wires, or with a small spindle of silver formed by melting some silver wire. We worried about damaging the ampullae by tamping too much bone wax into the canal, or by drilling through the canals and causing leakage of both endolymph and perilymph. Each of our plugging techniques yielded equivalent results when tested within a 48-h postoperative period. However, it is our impression that the plugs made with silver spindles caused less trauma to the system and resulted in less leakage of perilymph.

SCHOR: How do you know that you have, in fact, occluded flow through the canals? Have you histologically examined any of your preparations?

BARMACK: We did not examine the temporal bones specifically with the idea of looking for bone growth. It did not occur to me to look for that.

LOWENSTEIN: What was the reason for not plugging the posterior canal?

BARMACK: Well, we did.

LOWENSTEIN: The posterior canal?

BARMACK: Yes.

LOWENSTEIN: I thought you destroyed the posterior ampulla?

BARMACK: We did both. Yes. We wanted a stronger argument than that. We wanted to eliminate the possibility of attributing the residual function to the posterior canal. If there was residual function following destruction of the ampullae of the posterior canal, then this residual function could only be attributable to the anterior canals.

OMAN: The data are interesting. It just strikes me, though, that it's a long way, even in the rabbit, from the peripheral vestibular system to the output of the vestibular ocular reflex. And I wonder...aren't there other possible hypotheses that one could invoke to explain these data other than the mechanical events in the periphery? For example, suppose that there was a differential effect on the canal. Canal plugging is a very gross technique. If one does it close to the ampulla perhaps there could be a disturbance in the spontaneous activity in that canal, whereas if one was more gentlemanly and operated a little bit farther away, one could simply knock down the response and maybe you could account for the kinds of gain shift that you see as an adaptive response of the animal to a visual-vestibular conflict or something like that. You're recording the response after surgery and the animal has had a chance to learn to reinterpret perhaps its own vestibular output. The second question I wanted to ask is, you conclude that the blockage of the free flow of endolymph in one of the canals impedes the flow of endolymph in all canals. And you mentioned that you think that flow of fluids in one canal will induce flow in another? Am I correct?

BARMACK: Yes.

OMAN: I would agree with you in a case where you, say, cannulated one canal, which is not an inertial stimulus, and looked for flow coupling in an orthogonal canal. But I don't believe that's the case if one uses an inertial stimulus and the canal is truly orthogonal. In that case the flow in the second canal, which is driven by an inertial pressure gradient around the canal, is zero, because the inertial pressure gradients are balanced symmetrically. Is it because the canals are not orthogonal?

BARMACK: Yes. In the rabbit the horizontal and anterior canals are not orthogonal. There is an angle of about 95° between them. The horizontal and anterior canals may be more nearly orthogonal in other species. In answer to your first question, the effects of the canal plugs are immediately observable after the animal recovers from anesthesia. So, it is not a long-term adaptive effect. That is not an explanation that I consider reasonable.

GOLDBERG: Two experimental observations have a bearing here. First, the angular acceleration response of canal afferents is determined by the cosine of the angle between the effective canal plane and the plane of motion [Estes, M.S., Blanks, R.H., Markham, C.H.: J. Neurophysiol. 38:1232, 1975; Abend, W.K.: Brain Res. 132:65, 1977]. No response is seen when the two planes are orthogonal to one another. The cosine relation is founded on well-established physical principles and may almost be considered dogma. Second, Goldberg and Fernandez [Acta Otolaryngol. (Stockh.) 80:101, 1975] found, as had been assumed by Money and Scott [Am. J. Physiol. 202:1211, 1962], that canal plugging does abolish the angular acceleration response of canal afferents without greatly affecting their resting discharge. Dr. Barmack sees a residual reflex after plugging the appropriate canals. He concludes that the plugging does not render the canals entirely insensitive to angular accelerations. Another possibility is that the residual reflex is due to the activation of some other receptor. The simplest way to investigate the alternatives is to record from the afferents innervating the plugged canal.

BARMACK: I disagree with you. I don't think that there is any dogma that we can't open up to question. Furthermore, the residual contribution of the plugged anterior canal is small and would be difficult to observe electrophysiologically. Perhaps if people examined their electrophysiologic records they might see such an effect. Finally, the fact that I was referring to is not an electrophysiologic fact. I was referring to an anatomic fact.

LOWENSTEIN: I hate to add to the confusion. He was working on rabbits. You were working on squirrel monkeys. Now I have plenty of experience with working on two different elasmobranchs. In the dogfish, we found that in fact if we aligned the horizontal canal and positioned it in the horizontal plane we got responses from the vertical canal. But not in the ray. So the question is, are these canals really orthogonal?

GOLDBERG: Professor Lowenstein is correct in stating that the vertical canals are, in some animals, not orthogonal to the horizontal canal. Dr. Barmack raised another question, viz., do canals respond when they are orthogonal to the plane of angular acceleration? I claim that they do not.

LOWENSTEIN: I see.

BARMACK: I did not make that claim.

GOLDBERG: In all of your experiments, if a canal is orthogonal to the plane of motion you will get fluid pull in some other canal and within the plane of motion and that does not contribute to any fluid motion in the canal that is orthogonal. It's well based in physical theory and it's also established in physiologic observation.

BARMACK: But the horizontal and anterior canals are not orthogonal in the rabbit.

31

Ewald's Second Law of Labyrinthine Function and the Vestibuloocular Reflex

VICENTE HONRUBIA, YOUNG S. KIM,
HERMAN A. JENKINS, CLIFFORD G.Y. LAU,
AND ROBERT W. BALOH

There is general agreement that the functioning of the vestibuloocular reflex (VOR) arc is primarily determined by the characteristics of the nerve responses to labyrinthine organ stimulation. Throughout the years, physiologists and clinicians have endeavored to abstract significant features from the VOR response which would indicate whether or not the inner ear is functioning normally. One such characteristic of labyrinthine function, known for almost 100 years, which applies when there is only one normal labyrinth, is embodied in Ewald's second law. This law predicts that when the normal horizontal semicircular canal is stimulated by angular acceleration which causes endolymph displacement toward the ampulla [ampullopetal (AP) displacement], greater VOR responses will be produced than by acceleration which causes ampullofugal (AF) displacement. Changes in first order neuron firing rates in primates in response to constant angular acceleration which produces AP fluid displacement were recently shown to be approximately 1.5 times greater than those induced by accelerations in the opposite direction, thus providing physiologic support for Ewald's law (3,4). Furthermore, whereas firing rates could be considerably increased by large AP stimuli, AF stimuli soon silenced the spontaneous firing, thus creating a significant directional difference in the vestibular neuron response.

Measurement of VOR responses in human subjects in our laboratory, with the use of precisely controlled rotatory stimuli, confirmed Ewald's second law (1,8) when large magnitudes of head accelerations were used. However, small stimuli, contrary to the predictions from vestibular nerve studies, failed to demonstrate statistically significant asymmetry in VOR response. This may be due to the large variability in VOR responses masking the expected asymmetry at lower magnitudes of stimulation (2). These studies were performed with sinusoidal rotations of 0.05 Hz.

However, Mathog (7) and Wolfe et al. (12), using different frequencies, detected greater asymmetries in VOR responses at lower frequencies of rotation. In view of this situation, we conducted a theoretical study to obtain a better understanding of the physiopathologic principles underlying the production of VOR responses in unilaterally labyrinth-defective (UL) subjects. For this we performed a theoretical linear-system analysis of VOR function based on the results of Fernandez' and Goldberg's data for primary semicircular canal neurons (3,4). According to this analysis (summarized in the Appendix), not only should the gain of the VOR be greater during AP than AF rotation, with gain being defined as the ratio of peak slow component eye velocity to peak head velocity, but the VOR response in UL patients and animals should have the following significant features: (1) the ratio of AP to AF gains should change with the frequency of rotation, significantly being the greatest at lower frequencies; (2) the VOR gain in normal subjects should be equal to the addition of the AP and AF gains in UL subjects while the peak-to-peak eye velocity in normal subjects should be twice that of UL subjects; and (3) the time constants of the VOR responses in UL subjects should be different from those of normal subjects. Consequently, because of the differences in time constants the phase angle relationship between the head rotation and the VOR response should be different in normal and UL subjects. Phase angle measurements indicate the difference between the angular position of the head, or the head velocity, and the position or velocity of the eye.

This report summarizes the results of the comparative study in animals and human subjects to test the above predictions.

METHODS

Data for the analysis of VOR responses in normal and UL subjects were obtained from humans and rabbits. The human subject group consisted of ten persons with normal bilateral vestibular function and seven patients having a complete loss of function in one labyrinth: six as a result of vestibular nerve section during excision of a tumor and one patient as a consequence of ischemic vestibulopathy, leaving the affected side unresponsive to caloric irrigation. The subjects were tested in the dark, seated on a platform which was rotated sinusoidally at frequencies of 0.0125, 0.025, 0.05, 0.1, and 0.2 Hz, 60°/s peak head velocity. Additional tests at 30° and 120°/s peak head velocity were conducted at 0.05 Hz. The VOR responses were recorded using electrooculography. During the tests, the subjects were instructed to keep their eyes open with their gaze fixed straight ahead. Mental arithmetic tasks were assigned to maintain alertness. Electrooculographic calibrations were made before and after each rotatory test.

The VOR responses in five healthy rabbits were obtained using the scleral search coil technique of Robinson (9) to monitor eye position. Each animal was rotated in the dark at frequencies of 0.01, 0.02, 0.04, 0.08, 0.1, 0.2, and 0.4 Hz and peak head velocity of 60°/s with additional tests conducted at 30° and 120°/s peak head velocity at 0.1 Hz.

Subsequent to the completion of these tests, the right horizontal semicircular canal of each rabbit was opened under general anesthesia and plugged with guttapercha following the technique of Lorente de Nó (6). The test battery was repeated 1 week after surgery at which time only one animal exhibited spontaneous nystagmus. With this exception, none of the animals developed the typical signs of labyrinthectomy, i.e., changes in posture, torsion of the head, and head nystagmus. The tests were repeated again 1 year after surgery. The asymmetric VOR responses of the UL subjects were evaluated with the aid of a heuristic model (see Appendix) developed on the basis of the differential directional asymmetry observed in vestibular nerve fiber recordings. The model response describing the slow component eye velocity, $\dot{\theta}_e(t)$, as a function of time (t) over the period (T) of the sinusoidal rotation in an ideal case of left labyrinthectomy, is shown in Fig. 31-1. The amplitude of the eye velocity is seen to attain different maximum values

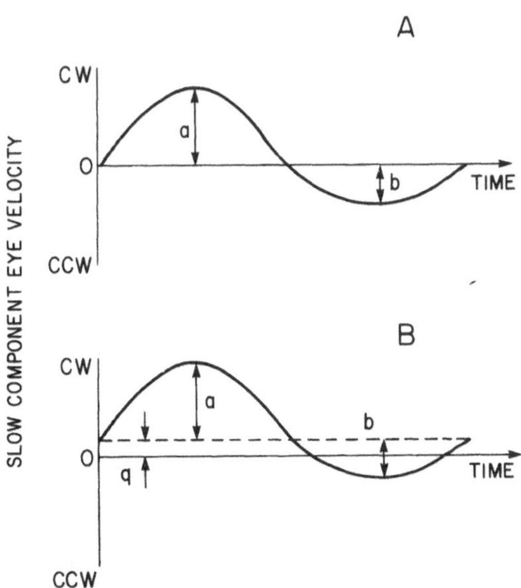

Fig. 31-1. The VOR response of the UL subject without dc bias (**A**) and with dc bias (**B**).

depending on the direction of rotation. The general equation describing $\dot{\theta}_e(t)$ is given by:

$$\dot{\theta}_e(t) = \begin{cases} q + a \sin \omega t, \ 0 \leq t < T/2 \\ q + b \sin \omega t, \ T/2 \leq t < T \end{cases}$$

where a and b represent the peak eye velocity due to stimulation in the AP (CW $\dot{\theta}_e$) and AF (CCW $\dot{\theta}_e$) directions, respectively, and q represents a constant velocity offset, e.g., as a result of spontaneous nystagmus. In the absence of spontaneous nystagmus, $q = 0$ and the values of the peak eye velocity simply equal a and b for AP and AF stimulation, respectively (Fig. 31-1A). However, in the presence of a constant velocity offset, the value of the observed peak eye velocity equals $a + q$ during AP stimulation and $b - q$ during AF stimulation.

The collection and analysis of the eye movement data were conducted with the aid of a PDP-11 minicomputer system. The human electrooculographic and rabbit search coil signals were digitized and stored in computer files. Subsequent data processing was performed off-line.

The digitized eye position data were differentiated by a computer program to obtain the instantaneous eye velocity, $\dot{\theta}_e$. Those data associated with the fast component of the nystagmus were removed and a Fourier analysis was performed on the remaining slow component eye velocity data. The fundamental and second harmonic Fourier coefficients were fitted to the data using a least-squares method. The higher order terms, which accounted for less than 5% of the total waveform energy, were neglected; the resulting error was deemed insignificant. The Fourier coefficients were then used to compute the values of a, b, and q. The AP and AF gains of the eye velocity ($\dot{\theta}_e$) with respect to the head velocity ($\dot{\theta}_h$) were obtained from the values of a, b, and q. The computer program also provided information about the phase angle relationship between the fundamental component of eye velocity and the head velocity.

RESULTS

Comparison of the characteristics of VOR responses between normal and UL subjects can be made by inspection of Figs. 31-2 and 31-3, showing animal data, and Fig. 31-4, showing human data.

Figure 31-2 illustrates 1-week-postoperative measurements, in four canal-plugged rabbits, of AP and AF gain, calculated with and without bias (see Methods). The mean gain and standard deviation in five normal animals are included for comparison. During AF stimulation, gain measurements of the canal-plugged animals are more than two standard deviations below the normal mean, whereas during AP stimulation, although more variable, most are more than one standard deviation below the normal mean. The normal range was smaller at mid-frequencies (0.04 to 0.2 Hz), making these stimuli the most successful in distinguishing normal from abnormal responses. Since

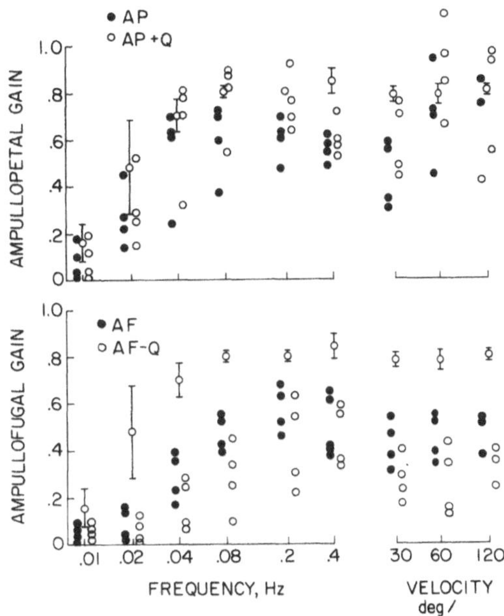

Fig. 31-2. Scatter plots of AP (*top*) and AF (*bottom*) gains of 1-week-postoperative rabbits. Open and filled circles represent the gains with and without bias, respectively. Each point represents a mean value over several half-cycles. Open circles with bars represent the mean (± S.D.) gain in normal animals. For normals, clockwise and counterclockwise rotations showed no significant statistical differences and were combined.

none of the animals had spontaneous nystagmus, even after repeated tests, differences in the VOR gain are interpreted to reflect differences in the excitability of the reflex during AP and AF displacement of the fluids in the normally functioning semicircular canal. Furthermore, after the VOR gain of each rabbit's response was calculated with the bias q, the differences between AP and AF responses were even greater. The q value, determined from the Fourier coefficients of the least-squares fit, is therefore not simply a measure of spontaneous nystagmus. Table 31-1 summarizes the results obtained in UL animals 1 week after operation. The AP/AF ratio, computed without the bias q, is greatest at the lowest frequencies as predicted by the model.

Figure 31-3 shows results for the three animals available for testing 1 year postoperatively. Both the AP gain and the q bias are diminished compared to animals 1 week after surgery, resulting in a diminution of differences between AP and AF gains with the overall VOR gain being approximately half that of normal animals.

Similar data for human subjects are presented in Fig. 31-4. As with rabbits, gain measurements in both normal and UL subjects increased with in-

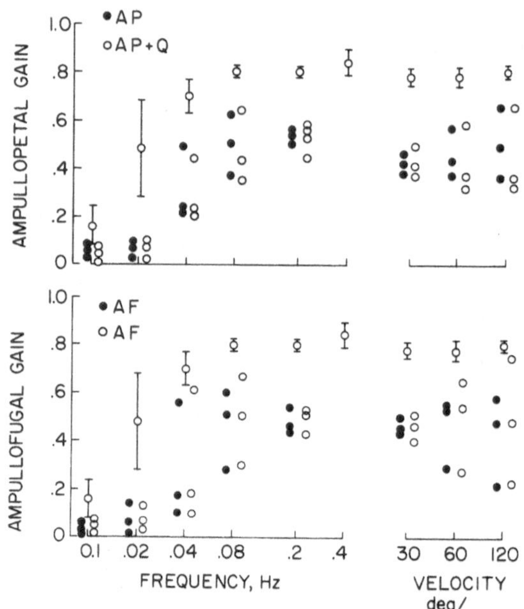

Fig. 31-3. Scatter plots of AP (*top*) and AF (*bottom*) gains of 1-year-postoperative rabbits. Open and filled circles represent the gains with and without bias, respectively. Each point represents a mean value over several half-cycles. Open circles with bars represent the mean (± S.D.) gain in normal animals. For normals, clockwise and counterclockwise rotations showed no significant statistical differences and were combined.

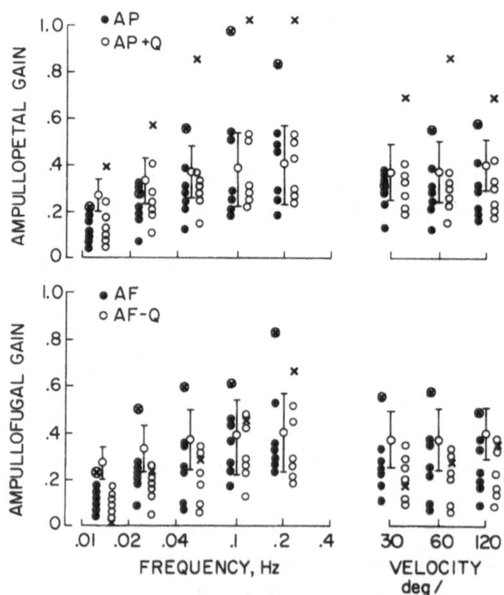

Fig. 31-4. Scatter plots of AP (*top*) and AF (*bottom*) gains of seven UL patients. Open and filled circles represent the gains with and without bias, respectively. Each point represents one test in each subject. Open circles with bars represent mean (±S.D.) VOR gains for normal subjects. All patients are more than 2 years postoperative except one (X's) who is 1 week postoperative.

Table 31-1.
VOR GAINS AND AP/AF RATIOS FOR 1-WEEK-POSTOPERATIVE RABBITS.

	Mean Gain			AP/AF	
Frequency	**Normal**	**AP**	**AF**	**Experimental**	**Model**
0.01	0.16	0.16	0.06	2.63	1.41
0.02	0.47	0.34	0.13	2.63	1.33
0.04	0.70	0.57	0.30	1.89	1.22
0.08	0.80	0.64	0.46	1.39	1.16
0.20	0.80	0.63	0.54	1.16	1.13
0.40	0.84	0.60	0.50	1.20	1.13

Summary of VOR gains and AP/AF ratios for animals 1 week postoperatively. Columns 1–5 are mean gain for normal animals, mean AP and AF gains for UL animals, experimentally determined AP/AF ratios, and AP/AF ratios predicted by the model, respectively.

creasing frequency of rotation. Surprisingly, in some cases, both AP and AF gains in patients were greater than in normals. The mean AF gain was significantly ($p < <0.05$) below normal at 0.0125 Hz only, whereas the mean AP gain was significantly lower at 0.0125, 0.025, and 0.05 Hz. Table 31-2 summarizes AP and AF gains recorded in patients and VOR gains in normal subjects. Mean gains, at lower frequencies, were smaller in patients than in normals. The ratio of AP to AF gains was close to one at all frequencies, which was at variance with the prediction of the model (*column* 5). Six of the patients, tested more than 2 years after surgery, had minimal spontaneous nystagmus and small q values which had little effect on the differences between AP and AF gains. One patient, tested 1 week after surgery, had a very large q value (16.0°/S at 0.05 Hz). When his gain measurements were computed with q, the difference between AP and AF gains was much larger than when q was omitted.

A common method of evaluating clinical vestibular tests is estimation of directional preponderance (DP). This measure is defined as the ratio of the difference in the response in each direction to the sum of the responses:

Table 31-2.
VOR GAINS AND AP/AF RATIOS FOR HUMAN SUBJECTS.

	Mean Gain			AP/AF	
Frequency	**Normal**	**AP**	**AF**	**Experimental**	**Model**
0.0125	0.27	0.13	0.13	1.00	1.39
0.025	0.33	0.22	0.25	0.88	1.28
0.05	0.37	0.30	0.27	1.11	1.19
0.10	0.42	0.42	0.35	1.20	1.15
0.20	0.40	0.43	0.40	1.08	1.13

Summary of VOR gains and AP/AF ratios for human subjects. Columns 1–5 are mean gain for normal subjects, mean AP and AF gains for UL patients, experimentally determined AP/AF ratios, and AP/AF ratios predicted by the model, respectively.

DP = (AP − AF) / (AP + AF). For this purpose, AP and AF gains were computed as $(a + q) / |\dot{\theta}_h|$ and $(b - q) / |\dot{\theta}_h|$, which are equivalent to the measurements clinicians could make from electronystagmographic recordings. With this criterion, patient groups were statistically different from the normal population at all test frequencies except 0.2 Hz. Individually, however, there were significant exceptions. One patient was consistently within the normal range and the number of patients whose DP measurements were outside two standard deviations from the mean of normals varied with frequency of rotation: five patients at 0.0125, and four at 0.025 and 0.05 Hz (60°/s). Although it did not identify all patients as abnormal, the DP measurement was most effective of all measurements evaluated in identifying abnormal reflex responses in UL subjects.

A summary of phase measurements in normal and UL subjects is shown in Fig. 31-5. In the normal rabbits, phase angle differences between eye velocity and head velocity increased, approaching 180°, at the higher frequencies of 0.2 to 0.4 Hz, indicating that the eye movement was compensatory at these frequencies. In humans, gaze compensation was not achieved even at 0.2 Hz, the highest frequency used.

Due to a defective labyrinth, there was always a decreased lag in phase angle in animals and, with some exceptions, in human subjects. The variance associated with mean phase values in normal subjects was relatively small.

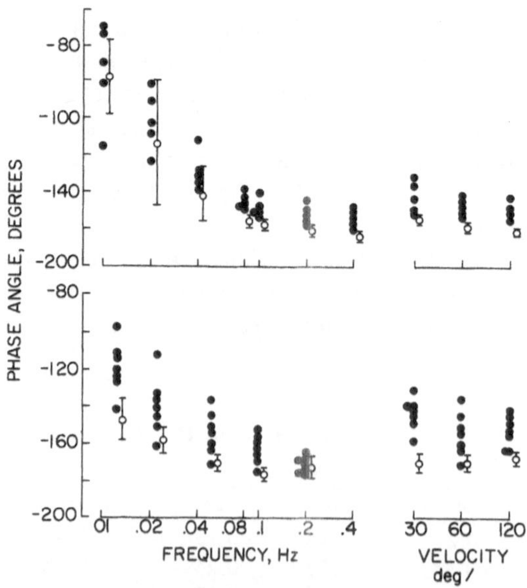

Fig. 31-5. Scatter plots of the phase relationships between VOR responses and head stimulation for 1-week-postoperative rabbits (*top*) and UL patients (*bottom*). Open circles with bars represent mean (± S.D.) phase measurements for normal subjects.

and the differences between the normal and UL subjects were more obvious than were the differences in gain. Mean phase angle values in patients were statistically significantly different from those in normals at three lower frequencies.

Time constants for rabbits and humans were obtained from the Bode plots. In normal rabbits, the time constant obtained with the use of asymptotic approximation to fit the data was 6.6 s from the gain plot and 7.6 s from the phase plot. In the canal-plugged animals, the time constant was 5.9 s from the AP gain plot and 3.9 s from the AF plot. In the normal humans, our data did not cover frequencies low enough to obtain an estimation of time constants. In patients, however, the frequency response characteristics underwent such a change that time constants could be computed. The 3 dB points of the gain plots of AP and AF responses occurred near 0.020 Hz ($T = 7.9$ s) and 0.024 Hz ($T = 6.63$ s), respectively. From the phase angle plot measurements, the estimated time constant was 7.9 s, corresponding to the frequency at which the eye velocity lagged behind the head velocity by 135°.

DISCUSSION

The model discussed in the Appendix predicts that VOR gains will be higher for AP than AF stimulation in UL patients and animals, with the greatest differences at the lowest frequencies of rotation. The average AP and AF gains in UL animals 1 week after surgery (Fig. 31-6) followed this prediction in all cases with the greatest differences occurring between 0.02 and 0.08 Hz and the highest AP/AF ratio at the lower frequencies (Table 31-1). In contrast, the difference between AP and AF responses in humans was not as large as expected (Fig. 31-7).

A possible explanation of the different results in humans is suggested by comparing the animal data 1 week and 1 year after surgery. Figure 31-6 shows that the differences between AP and AF responses were much greater 1 week (*top*) than 1 year (*bottom*) after surgery, suggesting that the process of CNS compensation consists of decreasing the AP gain to equalize the VOR response to each direction of rotation. This interpretation is consistent with recent findings showing the ability of the CNS to change the VOR gain to adapt the reflex response to new environmental conditions such as wearing reversing prisms (5,10).

The dashed and solid lines in Figs. 31-6 and 31-7 represent model predictions for AP and AF gains, respectively, in UL subjects based on the assumptions that (1) each labyrinth is responsible for the normal VOR response in the same proportion as the nerve response from the end organ; and (2) the normal VOR is the addition of AP and AF signals from each ear. In UL animals 1 week after the operation and in UL humans, the observed AP and AF responses in general were higher than those predicted by the model. If, in the normal state, each vestibular nerve were to make the same con-

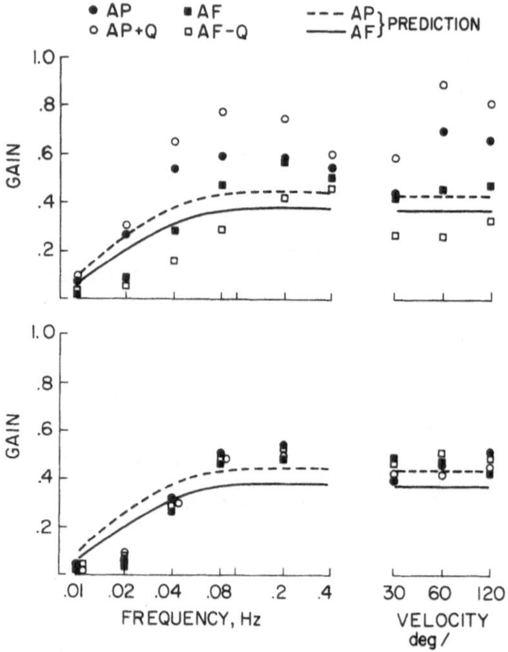

Fig. 31-6. Scatter plot of average AP (*circles*) and AF (*squares*) gains calculated with (*open symbols*) and without (*filled symbols*) bias for 1-week- (*top*) and 1-year- (*bottom*) postoperative rabbits. Solid and dashed lines represent model predictions.

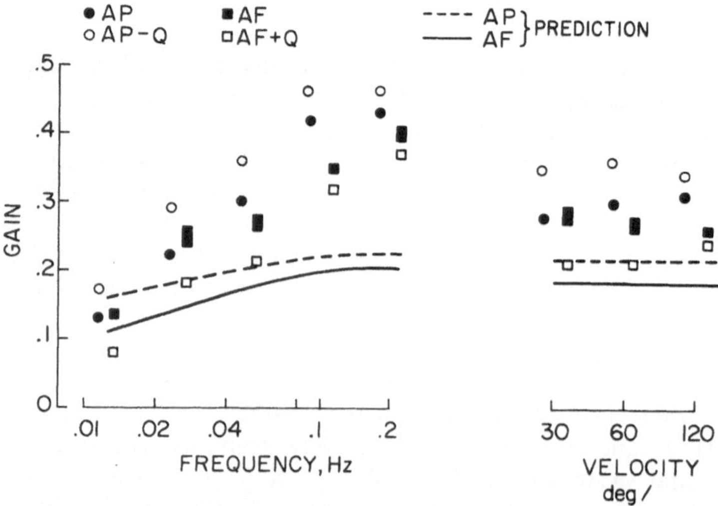

Fig. 31-7. Scatter plot of average AP (*circles*) and AF (*squares*) gains calculated with (*open symbols*) and without (*filled symbols*) bias for UL patients. Solid and dashed lines represent model predictions.

tribution to the VOR response that it does in the UL state, the normal VOR gain would be greater than unity in many cases—a physiologic anomaly. The data suggest that the assumption of algebraic combination of the nerve outputs to produce the normal reflex response is incorrect in accounting for the observed VOR responses.

The phase angle measurements followed the predictions of the model outlined in the Appendix. Phase lead in normal subjects was greatest at the lowest frequencies of stimulation and there was a greater phase advance in the UL animals and patients. In both animals and patients, the normal range of phase angle measurements narrowed at higher frequencies and the mid-frequency range was most effective in distinguishing normal from abnormal responses. Unfortunately, however, phase measurements do not provide crucial clinical information concerning the side on which the lesion is located.

From a practical point of view, the results indicate that the effectiveness of rotatory tests in the diagnosis of peripheral labyrinthine disorders depends on the time course of the disease process. The complete absence of one labyrinth will not necessarily result in statistically abnormal responses if the lesion is longstanding. It is reasonable to expect that less than complete dysfunction would result in less difference between AP and AF responses as suggested by our preliminary studies on patients with partial unilateral labyrinthine dysfunction (2).

The statistical differentiation of the patients' responses from the normal subjects' responses is enhanced by computing the gains with the bias q using the DP formula. Although this may appear a convenient method of distinguishing normal from abnormal responses, there could be a number of reasons for the existence of bias. For example, the single patient who was tested immediately after surgery had a strong spontaneous nystagmus and the largest q values of all the patients. However, the precise relationship between spontaneous nystagmus and bias has not been established. If the relationship is strong, DP measurement may provide the same information that is available from recording spontaneous nystagmus.

Spontaneous nystagmus of other than UL causes (e.g., congenital nystagmus) would result in a large q value and significant DP. Alternatively, the large q values in rabbits, tested 1 week after surgery, which had no spontaneous nystagmus, illustrate that bias can be produced by causes other than spontaneous nystagmus, perhaps as an expression of earlier CNS compensatory mechanisms. In both animals and humans with chronic UL lesions, the bias values are small and the differences in gain between AP and AF stimulation of the intact labyrinth are too small to be consistently distinguished from the responses of normal subjects.

This study exemplifies the dilemma faced by the neurologist in evaluating patients with UL lesions of long standing. An understanding, aided by models, of normal labyrinthine function is helpful in explaining the physiopathologic changes that are observed in the VOR responses. We suspect, however, that many changes are due to compensation by the CNS, and

Since the ratio of m_e/m_i is known (Table 31-1), Eq. 31-4 can be solved to obtain the predicted values of m_e and m_i in terms of VOR response for different frequencies of rotation after one of the labyrinths has been made inoperative. The model predictions thus obtained are shown in Figs. 31-6 and 31-7. The predictions of this model for the case in which one horizontal semicircular canal is blocked or one labyrinth is destroyed are depicted in Fig. 31-8, diagrams *II* and *III*, respectively. In the canal-blocked animal, we assume that the spontaneous firing of the vestibular nerve is unaffected, and therefore in the absence of any input due to rotation, the CNS is in a balanced condition with equal input from each vestibular nerve. During physiologic stimulation, however, the CNS input should reflect the difference in the values of m_e and m_i.

When the vestibular nerve is sectioned or the labyrinth destroyed, the CNS labyrinthine input is solely that from the remaining ear. At rest, there is an imbalance due to the spontaneous firing from the normal labyrinth. During rotation, the imbalance is modulated by the response to the AP or AF stimulation.

There are two basic asymmetries that can be expected in the VOR response because of the absence of one labyrinth, as illustrated in Figs. 31-1 and 31-8. One asymmetry is caused by the directional sensitivity to clockwise and counterclockwise rotations embodied in Ewald's second law with the eye velocity reaching different peak amplitude values during each half-cycle of rotation (Fig. 31-1A). This would be the asymmetry due to semicircular canal blockage (Fig. 31-8, *II*). The other results from the presence of a constant velocity bias reflecting the portion of the normal vestibular nerve resting activity not compensated by the CNS (Fig. 31-1B). This would be the asymmetry due to sectioning of the vestibular nerve (Fig. 31-8, *III*). With the head at rest, this asymmetry results in the clinical picture of labyrinthectomy, including the presence of spontaneous nystagmus. During rotatory physiologic stimulation of the functioning labyrinth, the induced VOR responses modulate the spontaneous nystagmus; this needs to be taken into consideration for the estimation of the VOR gain (Fig. 31-8, *III*). According to our model, immediately after sectioning, the VIIIth nerve should not be able to reverse the spontaneous nystagmus. The most that stimulation of the remaining canal could produce is complete inhibition of the spontaneous firing and thereby suppression of the nystagmus. In the case of the squirrel monkey (11), it can be estimated that during rotation at 0.4 Hz it would take a stimulus of approximately 300°/s peak velocity to suppress the spontaneous nystagmus. Assuming a normal VOR gain of 0.8, the eye velocity due to the AF response would be approximately 55°/s.

Because of the difference in time constants of the AP and AF cupular responses, there should be a different phase angle relationship between the head and the nerve response. In the CNS, when the output from the two labyrinths combines to excite, for example, the left abducens motoneurons, the stimulating effect is less than it would have been if the signal from the two

The schematic representation of these postulates is shown by diagram *I* in Fig. 31-8. Note the different sign of input lines at Σ representing the excitatory action of the contralateral labyrinth and the inhibitory action of the ipsilateral labyrinth on a given ocular motoneuron. Even though the input from each labyrinth is asymmetric, because of the synergistic mode of operation of the system, the input to the ocular motoneurons during each half-cycle of rotation is symmetric. The output signal V_o of the equivalent transfer function of the motoneuron represents the symmetric rate of modulation of the resting potential of the right abducens neuron or the rate of change of action potential firing frequency. The CNS processing includes an integration of vestibular signals not shown in the diagrams. The steady-state gain of the VOR (VOR_g) can be determined experimentally from measurement of the eye velocity during the slow component of the vestibular-induced nystagmus. Under our assumption, this value is approximately that of the slope ($m_e + m_i$) of the final transfer curve relating the eye velocity to the head velocity, such that:

$$VOR_g = m_e + m_i \qquad [31\text{-}4]$$

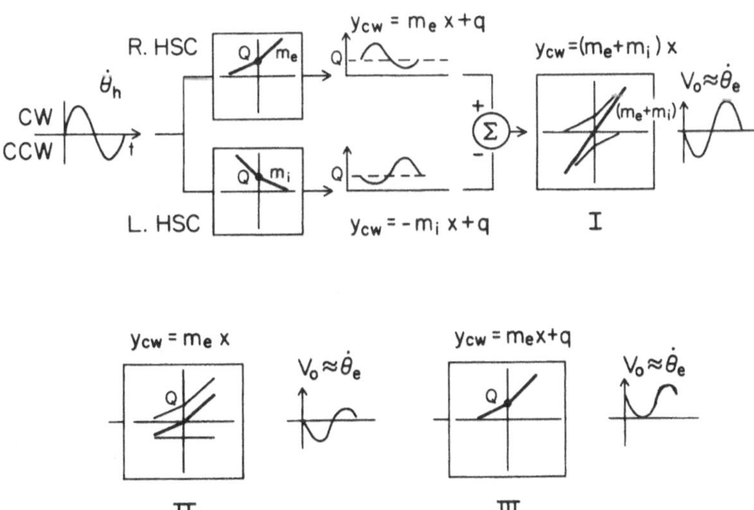

Fig. 31-8. Diagrams illustrating the main neuroanatomical elements in the production of VOR responses. A description of the signal transformation taking place during sinusoidal head rotation is at top. A sinusoidal signal ($\dot\theta_h$) stimulates the semicircular canals (*HSC*) to produce a nerve response in each labyrinth during AP and AF cupular deflections, shown in middle. The combined output of the inner ear results in a reflex response in terms of eye velocity ($\dot\theta_e$) which is compensatory to the head rotation in normal subjects (*I*). Changes in the VOR response resulting from blocking one semicircular canal or sectioning the vestibular nerve are illustrated in diagrams *II* and *III*, respectively.

Discussion

YOUNG: I'd like to ask your reaction to the point about the spontaneous nystagmus and the assumed difference taken between the resting discharge of the two complementary canals. It seems to me not likely that the average resting discharge is equal on the two sides. Consequently, it is not likely that, in the absence of any stimulus, the output of that summation would be exactly zero. And yet, most human subjects, and most of your animals, do not show a spontaneous nystagmus in the dark. I would propose that the nystagmus measure is a crude one. First of all, there is normally a compensation for this directional imbalance. Only under situations in which that compensation is effectively removed does one see the underlying imbalance. Finally, nystagmus is an awfully gross measure of imbalance, which requires so much compensatory eye deviation that repeated fast phases are required. If one uses more sensitive measures, for example . . . subjective sensation.

HONRUBIA: Your comments deal with an important point. I think it is the experience of most clinicians, and I would like to defer the final answers to my colleagues or some of the other clinicians here, that most normal persons have a small spontaneous nystagmus. In patients, spontaneous nystagmus is in the opposite direction. But then I don't know what to tell you about the nystagmus as a gross measurement. I think everybody agrees to that. And that's why we are putting so much effort into quantifying and controlling the stimulus and performing statistical analysis with nystagmus data. Now, the point that you brought out about the subjective sensation is one that I would like to know more about, because I had the same feeling as you, that a measure of the subjective sensation may have less variance, and be more reliable than nystagmus measurements to quantify labyrinth imbalance. But I think I am not familiar enough with the literature in this issue to be in complete agreement with you.

YOUNG: In attempting to look at the differences in the set points from each labyrinth, for rotation, we found it was only necessary to allow the subject to try to maintain himself stationary in space on a chair while putting in some random zero mean disturbance, and watch the way he drifts. Every subject drifts. And the direction of drift is consistent. The question is, of course, does drift have to do with vestibular function? When we compared it to clinical directional preponderance measured calorically, there is a statistical correlation, with a huge variance of low r value. I believe that the low correlation is inherent in the grossness of the caloric test and not in the drift measurements.

GOLDBERG: I wrote down that in your canal-plugged animals the gain as measured for excitatory ampullopetal is 0.7 and for ampullofugal is 0.4. And it adds up to much more that the normal gain. Could you tell us what the normal gain was in your situation?

HONRUBIA: Normal gain was 0.8.

labyrinths had been exactly in phase. The estimated error from the assumption of an algebraic instead of vectorial addition is 2% at 0.01 Hz and less at higher frequencies.

Time constants of the system can be obtained from experimental data. During rotation with frequencies $\omega >> 1/T_1$ the gain should reach a constant asymptotic value when the eye movement is perfectly compensatory; the gain will be 30% less than the maximum at $\omega = 1/T_1$ at which time the velocity of the compensatory eye movement will lag behind the head velocity by 135°. Time constants in UL subjects will tend to be smaller, reflecting the time constant of one cupular direction of rotation.

References

1. Baloh, R.W., Honrubia, V., and Konrad, H.R.: Ewald's second re-evaluated. Acta Otolaryngol. 83:475, 1977.
2. Baloh, R.W., Sills, A.W., and Honrubia, V.: Impulsive and sinusoidal rotatory testing. A comparison with results of caloric testing. Laryngoscope 89:646, 1979.
3. Fernández, C., and Goldberg, J.M.: Physiology of peripheral neurons innervating semicircular canals of the squirrel monkey. II. Response to sinusoidal stimulation and dynamics of peripheral vestibular system. J. Neurophysiol. 34:661, 1971.
4. Goldberg, J.M. and Fernández, C.: Physiology of peripheral neurons innervating semicircular canals of the squirrel monkey. I. Resting discharge and response to constant angular accelerations. J. Neurophysiol. 34:635, 1971.
5. Gonshor, A. and Melvill Jones, G.: Extreme vestibulo-ocular adaptation induced by prolonged optical reversal of vision. J. Physiol. 256:381, 1977.
6. Lorente de Nó, R.: Über die Drehreflexe auf die Augenmuskeln. Acta Otolaryngol. 12:238, 1927.
7. Mathog, R.H.: Testing of the vestibular system by sinusoidal angular acceleration. Acta Otolaryngol. 74:96, 1972.
8. Owens, D.E., Jenkins, H., and Honrubia, V.: Differences between ampullofugal and ampullopetal stimulation of the horizontal semicircular canal. Confirmation of Ewald's second law. Trans. Am. Acad. Ophthalmol. Otolaryngol. 82:ORL 197, 1976.
9. Robinson, D.A.: A method of measuring eye movement using a scleral search coil in a magnetic field. IEEE Trans. Biomed. Electron. BME 10:137, 1963.
10. Robinson, D.A.: Adaptive gain control of vestibulo-ocular reflex by the cerebellum. J. Neurophysiol. 39:954, 1976.
11. Skavenski, A.A. and Robinson, D.A.: Role of abducens neurons in vestibulo-ocular reflex. J. Neurophysiol. 36:724, 1973.
12. Wolfe, J.W., Engelken, E.J., Olson, J.W., et al.: Vestibular responses to bithermal caloric and harmonic acceleration. Ann. Otol. Rhinol. Laryngol. 87:861, 1978.

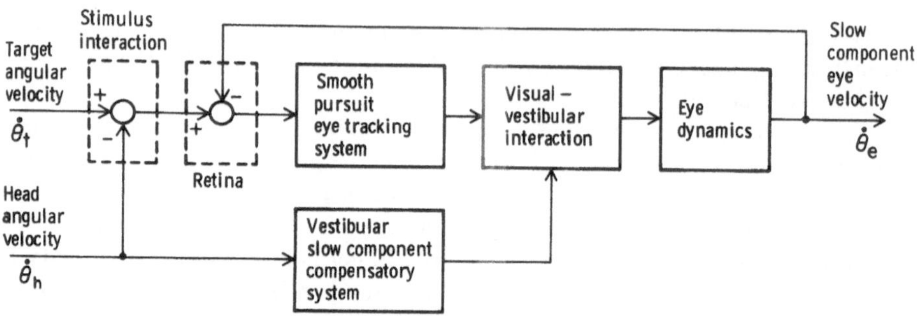

Fig. 32-1. Functional block diagram of the model for visual-vestibular interaction.

resents the VOR which includes the semicircular canal dynamics and central nervous system processing. The anatomic equivalent of the box labeled *Visual-vestibular interaction* probably takes place at the vestibular nuclei (5), as mentioned in several papers in this symposium.

In the present study, the model shown in Fig. 32-1 is validated in the rabbit for various combinations of optokinetic and vestibular stimuli.

METHODS

For these experiments, six normal Dutch rabbits were subjected to the following stimuli:

1) Sinusoidal head rotation in the dark at a peak velocity of 15°/s and at various frequencies from 0.005 Hz to 8 Hz.
2) Sinusoidal rotation of the optokinetic drum with the head fixed in space at a peak velocity of 15°/s and at various frequencies from 0.005 Hz to 0.04 Hz. This test evaluates the SPS of rabbits.
3) Sinusoidal head rotation in the light with a stationary optokinetic drum as the background at a peak velocity of 15°/s and at various frequencies from 0.005 Hz to 8 Hz.
4) Sinusoidal optokinetic stimuli at 0.025 Hz and at peak velocities from 3.75 to 30°/s; sinusoidal head rotation in the dark and in the light at 0.025 Hz and at peak velocities from 7.5 to 60°/s.

The eye position was monitored by the magnetic search coil technique (3) and analyzed with the aid of a PDP-11 computer system. Figure 32-2 illustrates the method of data analysis. The nystagmus recording (*trace A*) was digitally differentiated to obtain the instantaneous eye velocity (*trace B*). With the help of a computer program (4), the fast components of nystagmus were identified and then removed. The remaining slow component eye velocity (*trace C*) was then sinusoidally curve-fitted to obtain an estimate of the gain and phase. The head velocity is shown at the bottom (*trace D*).

32

A Model For Visual-Vestibular Interaction

CLIFFORD G.Y. LAU, VICENTE HONRUBIA, AND ROBERT W. BALOH

During head movement in the light, clear vision is maintained by the combined responses of several systems including the fixation smooth pursuit system, the optokinetic system, and the vestibuloocular reflex (VOR). The pursuit tracking system has been known to have low-pass characteristics. On the other hand, the vestibuloocular reflex has been known to have a decreased response at low frequencies. During simultaneous stimulation of the visual and vestibular systems, the gain is almost 1 for a wide range of frequencies.

A model was previously proposed for the interaction of the visual system and the vestibuloocular reflex (1,2), as shown in Fig. 32-1. In this model, the target angular velocity and the head angular velocity relative to space are the input and the slow component eye velocity relative to the head is the output. During head movement in the light, there is an equivalent movement of the target in the opposite direction. The geometry of the target movement is represented by the box labeled *Stimulus interaction*.

The box labeled *Retina* represents the visual system which is assumed to detect the retinal velocity error. The box labeled *Smooth pursuit eye tracking system* (SPS) represents the processing of the retinal velocity error by the central nervous system to arrive at an appropriate eye movement control command. The *Vestibular slow component compensatory system* rep-

only by following their time course can the compensatory mechanisms be elucidated. This information could then be incorporated into better models of vestibular function. Until this is done, clinicians must be cautious in applying principles derived from normal studies to the pathologic state.

APPENDIX

The analysis of the VOR responses in UL subjects was performed through the use of a heuristic model that incorporated the information obtained by Fernández and Goldberg (3,4) describing the directionally sensitive responses of squirrel monkey vestibular nerve fibers during rotational stimulation. The transfer function describing the nerve response is given by:

$$H(s) = \frac{T_A s}{1 + T_A s} \cdot \frac{(1 + T_L s)}{(1 + T_1 s)(1 + T_2 s)} \qquad [31\text{-}1]$$

where $T_A = 80$ s, $T_1 = 5.7$ s, $T_L = 0.05$ s, and $T_2 = 0.003$ s are the time constants necessary to fit the data. For the range of frequencies used in the present study, between 0.01 and 1.0 Hz, Eq. 31-1 reduces to:

$$H(s) = \frac{1}{(1 + T_1 s)} \qquad [31\text{-}2]$$

Accordingly, for sinusoidal accelerations $\alpha(t) = \alpha_{max} \sin(\omega t + \dot{\theta})$, the slope ($m$) of the transfer function relating the firing rate, in spikes/s, of a "typical" slow-adapting vestibular neuron to the angular head velocity τ_n, in °/s, is given by:

$$m = S\omega/(1 + (\omega T_1)^2)^{-1/2} \qquad [31\text{-}3]$$

where S is the steady state sensitivity coefficient in (spikes/s) (°/s²) and ω is the angular frequency in radian/s. Significantly, Fernández and Goldberg found that S assumed different values during AP stimulation, in which $S_e = 2.44$ (spikes/s)/(°/s²), and AF stimulation, in which $S_i = 1.6$ (spikes/s)/(°/s²). Likewise, the value of T_{1e} was direction-dependent with $T_{1e} = 6.73$ s and $T_{1i} = 5.24$ s for AP and AF stimulation, respectively. As shown in Fig. 31-8, the transfer curve describing the steady-state output (y) from the vestibular nerve has two regimes: during AP stimulation, the output is given by $y_e = m_e\dot{\theta}_h + q$, whereas during AF stimulation, the neural output is given by $y_i = m_i\dot{\theta}_h + q$, where m_e and m_i are the slopes of the curves and q is the spontaneous firing rate of the vestibular neuron. Evidently, from Eq. 31-3, the values of m_e and m_i are frequency-dependent. The ratio m_e/m_i is close to unity at high frequencies but greater at the lower frequencies (Table 31-1, *column 6*). During head rotation, the CNS input from the horizontal semicircular canals can be thought of as consisting of the algebraic summation of the input from each of the two canals. Upstream in the CNS, if we assume a synergistic mode of operation of the two canals to control the motion of the eyes, represented by the summing point Σ, the final transfer curve is linear with the slope equal to the sum of $m_e + m_i$; thus $y = (m_e + m_i)|\dot{\theta}_h|$.

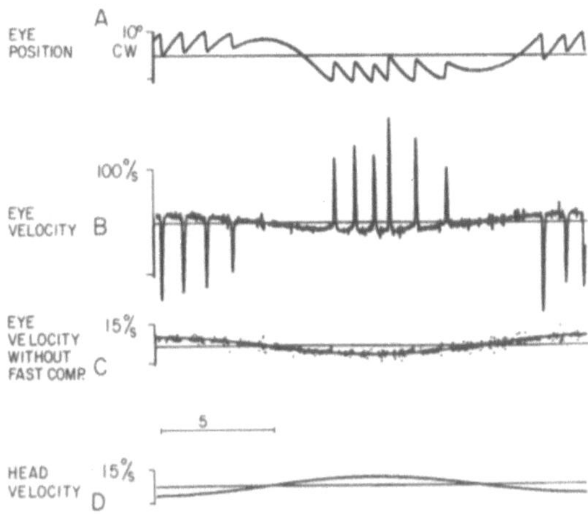

Fig. 32-2. Method of data analysis. Instantaneous eye velocity is obtained by differentiating the eye position (nystagmus) record.

RESULTS

Measurements of the gain and phase for each of six rabbits at different frequencies during the sinusoidal head rotation in the dark are shown in Fig. 32-3. The gain is defined as the ratio of peak slow component eye velocity to peak head velocity. The solid line represents the third order linear time-invariant transfer function fitted to the data with a long time constant (T_1) of 3.2 s and a short time constant (T_2) of 0.032 s. A third time constant (adaptation) (T_a) was also obtainable because of the low frequency of stimulus used (0.005 Hz) and the stability of the magnetic search coil technique. The adaptation time constant for the rabbit is about 10.6 s. In the midrange of frequencies, the gain is about 0.82 and the phase is about −180°.

For the optokinetic nystagmus (OKN), the gain is defined as the ratio of peak slow component eye velocity to peak optokinetic drum velocity. Figure 32-4 shows the gain and phase for the rabbit's SPS for different frequencies. The first order transfer function, fitted to the data, has a time constant (T_4) of 10 s. At low frequencies, the gain is about 0.84, approaching the gain for constant velocity optokinetic nystagmus, approximately 0.9 for the rabbit at these velocities.

During sinusoidal head rotation in the light, both the visual and vestibular systems are stimulated. The eye movement response under this condition has been called the visual-vestibuloocular reflex (VVOR). Figure 32-5 shows the frequency response characteristics defined in terms of slow component eye velocity to head velocity for sinusoidal head rotation in the light. The gain is almost 1 and the phase is −180° for a wide range of frequencies.

These linear time-invariant system transfer functions can now be used to formulate a control system model for visual-vestibular interaction, as shown

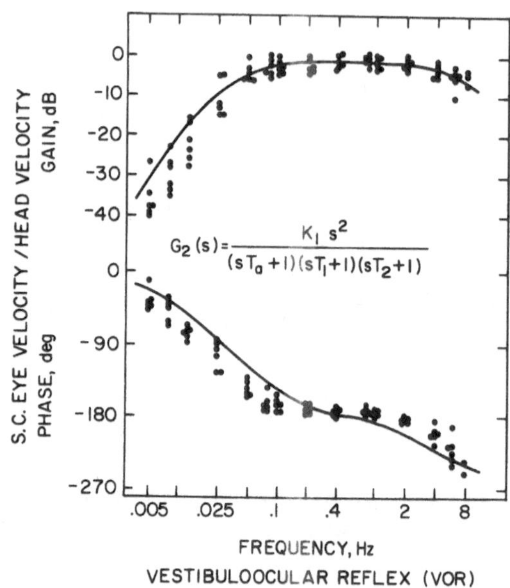

Fig. 32-3. Gain and phase characteristics for the rabbit's vestibuloocular reflex at various frequencies. Solid line is the least-squares fit with linear time-invariant model.

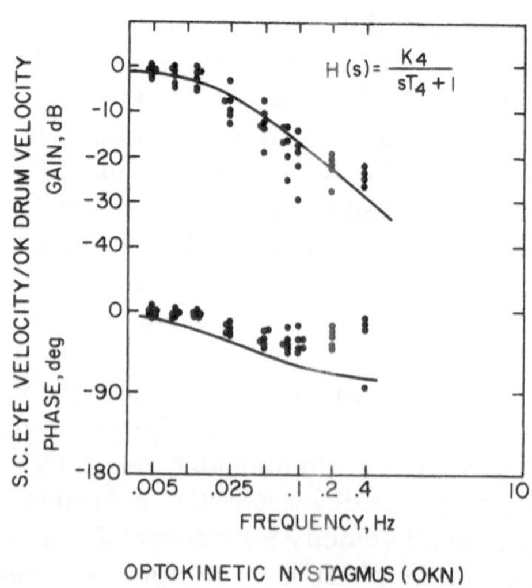

Fig. 32-4. Gain and phase characteristics for the rabbit's optokinetic nystagmus.

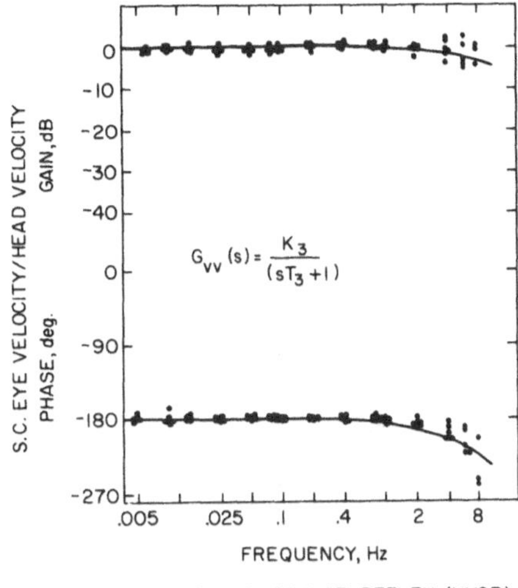

Fig. 32-5. Gain and phase characteristics for the rabbit during sinusoidal head rotation in the light, or the visual-vestibuloocular reflex.

in Fig. 32-6. In this model, $G_1(s)$ represents the open-loop transfer function of the visual pursuit system, and $G_2(s)$ represents the transfer function of the VOR. As a first order approximation, the visual-vestibular interaction is assumed to be a linear summation of the visual signal and the vestibular signal at the interaction center. The output of the system, slow component eye velocity, is given by the equation

$$\dot{\theta}_e = \frac{G_1(s)}{1 + G_1(s)}\dot{\theta}_t + \frac{G_2(s) - G_1(s)}{1 + G_1(s)}\dot{\theta}_h \qquad [32\text{-}1]$$

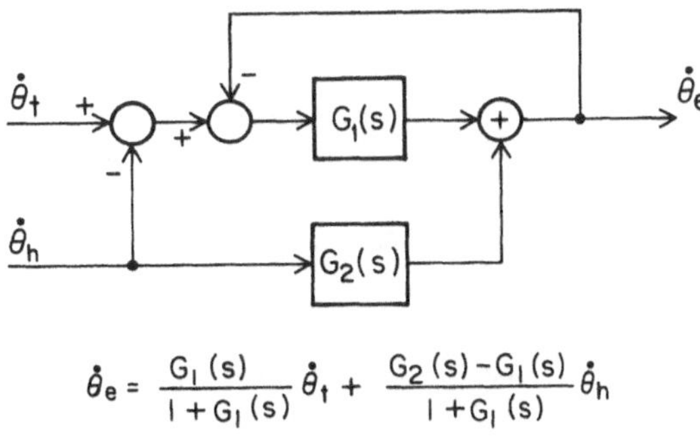

Fig. 32-6. Linear control system model for visual-vestibular interaction.

The first term represents the smooth pursuit eye movement response to a smoothly moving target in the absence of head movements. The second term represents the eye movement response to head rotation. For sinusoidal head rotation in the light, the transfer function is given by

$$G_h(s) = \frac{G_2(s) - G_1(s)}{1 + G_1(s)} \qquad [32\text{-}2]$$

In other words, the transfer function is no longer that of the VOR alone, $G_2(s)$, but is modified by the open-loop transfer function of the smooth pursuit system, $G_1(s)$.

The transfer function $H(s)$ obtained in Fig. 32-4 is the closed-loop transfer function of the smooth pursuit system. The open-loop transfer function, $G_1(s)$, can be obtained analytically by

$$G_1(s) = \frac{H(s)}{1 - H(s)} \qquad [32\text{-}3]$$

if $1 - H(s)$ is invertible.

Based on the transfer functions $G_1(s)$ and $G_2(s)$ obtained here, it is possible to compare the model prediction with the VVOR data, as shown in Fig. 32-7. The dashed lines show the frequency response of the VOR, OKN, and VVOR. At low frequencies, the model-predicted VVOR response is almost the same as the response of the OKN, and at high frequencies the model-predicted VVOR response is almost the same as the response of the VOR. This is consistent with the observation that the visual system dominates at low frequencies, whereas the vestibular system dominates at high frequencies. However, in the region of the crossover, the model prediction is not as good. In order to find an explanation for this discrepancy, another series of

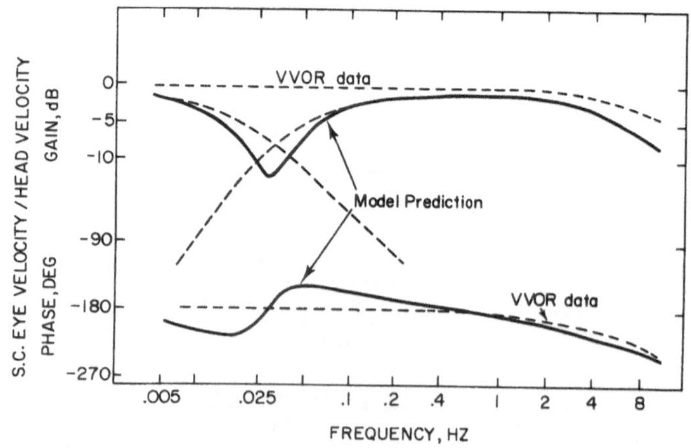

Fig. 32-7. Comparison of model predictions with the data from the visual-vestibuloocular reflex.

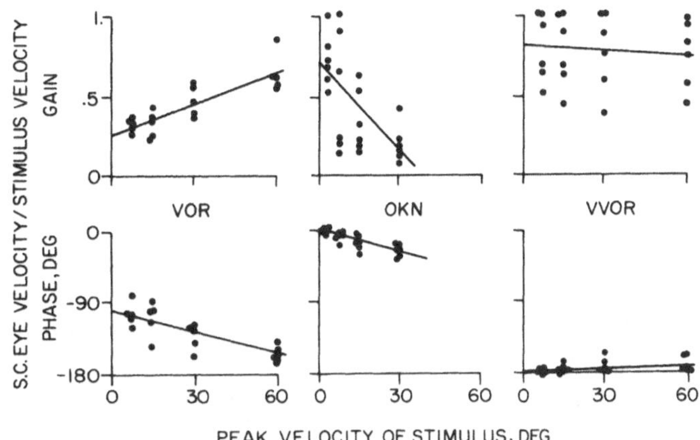

Fig. 32-8. Gain and phase nonlinearities for the rabbit's VOR, OKN, and VVOR at 0.025 Hz for different peak velocities of stimuli.

experiments was conducted to investigate more closely the eye movement response at the crossover frequency 0.025 Hz.

In the first set of experiments, the peak stimulus velocity during the VOR, OKN, and VVOR tests was held at 15°/s for different frequencies. In the second set of experiments the stimulus frequency was kept at 0.025 Hz, but the peak velocity varied from 3.75 to 30°/s for the OKN tests and from 7.5 to 60°/s for the VOR and VVOR tests. Figure 32-8 shows the gain and phase of the VOR, OKN, and VVOR as a function of the peak velocity of the stimulus. The plot on the left shows that the slow component system gain of the VOR increases as the magnitude of the stimulus increases, and also the phase is approximately −180° or very close to being compensatory. On the other hand, the plot in the middle shows that the OKN gain decreases for larger magnitude stimuli. In other words, the optokinetic system does not respond to fast moving stimuli for the rabbit. In contrast to these findings the two plots on the right show that for the VVOR, the gain is almost the same for different magnitudes of stimuli.

The observed nonlinearity for the VOR and OKN accounts for the discrepancy between the model predictions and the data. The model predictions were based on a linear system response for the VOR and the optokinetic system. Without corrections, the model predicted a VVOR gain of 0.35 at 0.025 Hz. However, with corrections for the nonlinear response, the model-predicted VVOR gain is increased to 0.72, which is much closer to the experimental VVOR gain of 0.81.

DISCUSSION

The linear model for visual-vestibular interaction (2) was based on data from normal human subjects and patients with cerebellar degeneration. The same model has been shown here to be valid for the rabbit. Within the limitations of the linear system model, the model predictions are surprisingly close to the observed data. At low frequencies, the visual system dominates so that the eye movement response to head rotation in the light is almost the same as the response to optokinetic stimulus. However, at high frequencies, the vestibular system dominates so that the eye movement response to head rotation in the light is almost the same as the response to head rotation in the dark.

In the rabbit, the responses of the VOR and OKN have been found to be nonlinear. For the VOR, the gain is higher for larger magnitudes of peak head velocity. On the other hand, for the OKN, the gain is decreased for larger magnitudes of peak optokinetic drum velocity. During head movements, the line of gaze is partially compensated by the eye movement response from the VOR. The effective optokinetic stimulus is reduced by these vestibular eye movements. When the peak velocity of the head rotation is low, the equivalent optokinetic stimulus is also low. Subsequently the OKN gain is high. When the peak velocity of the head rotation is high, the VOR gain is high so that the effective optokinetic stimulus is again low. Even though the responses of both the VOR and the OKN are nonlinear, the result of the interaction is such that the gain for the VVOR is almost constant for different magnitudes of stimuli. The nonlinear response seems to be preprogramed to maintain good gaze stability.

So far, validation of the visual-vestibular interaction model has been limited to measuring the eye movement responses to vestibular and optokinetic stimulation.

To explain visual-vestibular interaction in animals with a fovea the model must be expanded to include the interaction between the fixation smooth pursuit system, the optokinetic system, and the vestibuloocular reflex, as shown in Fig. 32-9. The hypothesis is that there are two separate visually controlled pursuit systems. The fovea provides the main input to the fixation smooth pursuit system, whereas the optokinetic system receives its input primarily from the peripheral retina. The fixation pursuit system responds rapidly whereas the optokinetic system responds slowly, as evidenced by the long time constant of optokinetic after-nystagmus.

The equivalent control system model is shown in Fig. 32-10. In this generalized model, the inputs are the angular velocity of the fixation point relative to space ($\dot{\theta}_{fp}$), the optokinetic drum velocity relative to space ($\dot{\theta}_{ok}$), and the head velocity relative to space ($\dot{\theta}_h$). The output is the slow component eye velocity ($\dot{\theta}_e$) relative to the head, and is given by

$$\dot{\theta}_e = \frac{G_3}{1 + G_1 + G_3}\dot{\theta}_{fp} + \frac{G_1}{1 + G_1 + G_3}\dot{\theta}_{ok} + \frac{G_2 - (G_1 + G_3)}{1 + G_1 + G_3}\dot{\theta}_h \quad [32\text{-}4]$$

The first term represents the response to movements of a fixation point, the second term represents the response to the optokinetic stimulus, and the third term represents the response to head movements.

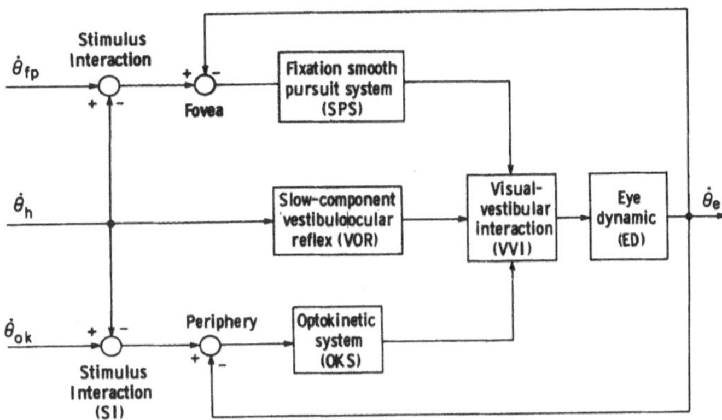

Fig. 32-9. Functional block diagram of a model for the interaction of fixation smooth pursuit, optokinetic system, and the vestibuloocular reflex.

At low frequencies (< 0.1 Hz) the gain of fixation smooth pursuit is almost 1, so that the open-loop gain G_3 is very large. For example, if one assumes a smooth pursuit gain of 0.98, the open-loop gain G_3 is 49. Similarly, assuming a closed-loop optokinetic gain of 0.8, the open-loop gain of the optokinetic system G_1 is 4. By comparison, the VOR gain G_2 at low frequencies is less than 0.5.

Thus, in general, at low frequencies the fixation smooth pursuit system dominates the optokinetic system which in turn dominates the vestibular eye movements. At high frequencies where the smooth pursuit and optokinetic

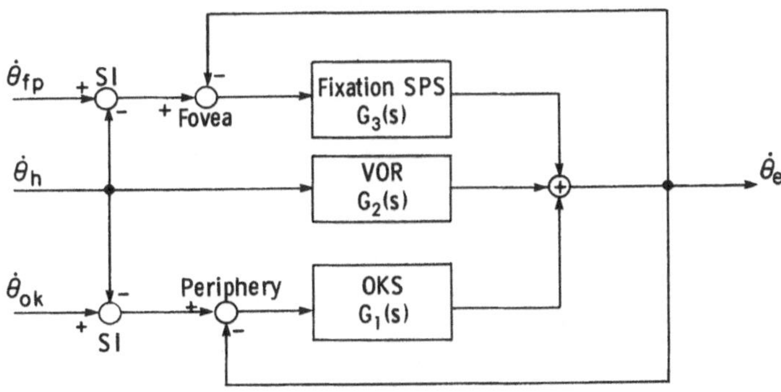

$$\dot{\theta}_e = \frac{G_3}{1 + G_1 + G_3}\dot{\theta}_{fp} + \frac{G_1}{1 + G_1 + G_3}\dot{\theta}_{ok} + \frac{G_2 - (G_1 + G_3)}{1 + G_1 + G_3}\dot{\theta}_h$$

Fig. 32-10. Control system model for the interaction of smooth pursuit, optokinetic system, and vestibuloocular reflex.

gains are low, the vestibuloocular reflex dominates. This model accounts for a large number of observations concerning the effect of vision on vestibular eye movements, and provides a good theoretical basis for the design of experiments involving the interaction of fixation smooth pursuit, optokinetic system, and the vestibuloocular reflex.

ACKNOWLEDGMENT

We sincerely thank Ms. Marilyn Oreck for her editorial aid. This study was supported by NIH Grant NS 08335..

References

1. Lau, C.G.Y.: Modeling of visual-vestibular interaction and the fast components of nystagmus. Ph.D. Dissertation. Santa Barbara, University of California, 1978
2. Lau, C.G.Y., Honrubia, V., Jenkins, H.A., Baloh, R.W., and Yee, R.D.: Linear model for visual-vestibular interaction. Aviat. Space Environ. Med. 49(7):880, 1978.
3. Robinson, D.A.: A method of measuring eye movement using a scleral search coil in a magnetic field. IEEE Trans. Biomed. Electr. 10:137, 1963.
4. Sills, A.W., Honrubia, V., and Kumley, W.D.: Algorithm for the multiparameter analysis of nystagmus using a digital computer. Aviat. Space Environ. Med. 46:934, 1975.
5. Waespe, W. and Henn, V.: Neuronal activity in the vestibular nuclei of the alert monkey during vestibular and optokinetic stimulation. Exp. Brain Res. 27:523, 1977.

Discussion

BÜTTNER: Did you determine the VOR of the rabbit in the dark?

LAU: The VOR experiment is done completely in the dark.

BÜTTNER: Yes. And what time constant did you obtain?

LAU: The time constant of the VOR, the long time constant, is about 3.2 seconds.

BÜTTNER: As far as I remember H. Collewijn also measured the VOR in rabbits and found different time constants depending on the type of experiment. When steps of angular velocity were applied, the time constant of decay for nystagmus velocity was 14 sec. When animals were sinusoidally rotated at different frequencies, the time constant was as short as 3.3 sec (Baarsma and Collewijn, J. Physiol [Lond.] 238:603, 1974). Could you comment on this?

LAU: It's hard to say whether a linear time invariant model can completely explain the data as you describe the difference between a step response and a sinusoidal response. So, maybe there's a problem with modeling the VOR as a linear time invariant system.

Index

A

D

E

F

G

H

I

K

L

P

R

S

T

U

V

W

Z